OBJECTS TO COMPONENTS

With the Java™ Platform

Art Gittleman

California State University Long Beach

D1613891

Scott/Jones Inc.
P.O. Box 696,
El Granada, California 94018
Voice: **650-726-2436**
Facsimile: **650-726-4693**
E-mail: **scotjones2@aol.com**
Web page: **//www.scottjonespub.com**

ISBN 1-57676-035-9

ISBN: 1-57676-035-9

1 2 3   Y

Text Design and Composition: Cecelia G. Morales
Cover Design: Heather Bennett
Copyediting: Cathy Baehler
Book Manufacturing: Malloy Lithographing, Inc.

Scott/Jones Publishing Company
Publisher: Richard Jones
Sponsoring Editorial Group: Richard Jones and Cathy Glenn
Marketing & Sales: Barbara Masek, Lynne McCormick, Hazel Dunlap, Donna Cross, Michelle Windell
Business Operations: Chuck Paetzke, Michelle Robelet, Cathy Glenn, Michelle Windell

Additional Titles of Interest from Scott/Jones

Computing with Java
   by Art Gittleman

Starting Out With C++, Second Edition
Starting Out With C++, Brief Edition
Starting Out With C++, Alternate Edition
   by Tony Gaddis

Modern Fortran 77/90/2000
Introduction to Programming Using Visual Basic, Second Edition
   by Gary Bronson

C by Discovery, Second Edition
   by L. S. Foster

Assembly Language for the IBM PC Family, Second Edition
   by William B. Jones

The Visual Basic 6 Coursebook, Fourth Edition
QuickStart to VBScript
ShortCourse in HTML
QuickStart to DOS for Windows 9X
   by Forest Lin

Advanced Visual Basic 6, Second Edition
   by Kip Irvine

HTML for WWW Developers
Server-Side Programming for WWW Developers
Client-Side Programming for WWW Developers
   by John Avila

Lab Activities for the WWW: Annual Editions
   by Paula Ladd and Ralph Ruby

The Access 2000 Guidebook, Third Edition
   by Maggie Trigg and Phyllis Dobson

Building Applications with Microsoft Office and Visual Basic
   by Ron Gilster and Karen Braunstein-Post

The A+ Self Study Guide
The Complete Computer Repair Textbook, Second Edition
   by Cheryl Schmidt

The Windows 98 Textbook Series: Short Course, Standard, and Extended Editions
   by Stewart Venit

The Windows NT Server Lab Manual
   by Gerard Morris

Contents

CHAPTER 3 *Object-Oriented Concepts 85*

CHAPTER 6   *Graphical User Interfaces   211*

CHAPTER 11 *Java Database Connectivity (JDBC)*   445

Preface

The nature of computing is changing. Enterprise systems serve thousands of users via world-wide networks. Audio and video liven communication. Automobiles, telephones, and televisions add computing devices to enhance their capabilities. Java's phenomenal growth reflects its suitability for the broader notion of computing.

The title of this book signifies a path from objects to components covering many of the concepts important in the new world of computing. After an introductory chapter showing what Java can do and including a review of Java's C-like features, Chapter 2 introduces object-oriented programming while Chapter 3 delves deeper into object-oriented concepts. Chapter 4 treats arrays and vectors, ending with an example on callbacks that motivates the next three chapters on event-driven programming.

Chapters 5, 6, and 7 present the concepts of event-driven programming, arranged by difficulty for ease of learning. The graphics in Chapter 5 only requires overriding the paint method. Graphical user interfaces in Chapter 6 use the Java event model with high-level events and their one-method interfaces, while Chapter 7 covers low-level mouse, key, and window events with their more complex multi-method interfaces. These chapters introduce HTML and applets, and show how to do standalone graphics.

Chapter 8 combines topics involving data, prefaced by the necessary introduction to exception handling. After covering text, binary, and object I/O, it shows how to pass data to programs and applets, concluding with recursion, and the linked list and stack data structures.

One approaches advanced topics with joy in Java because the language and its libraries make multimedia, network, and database programming much easier. Chapter 9 starts with threads and concurrent programming, and continues with animation, images, and sound.

Chapter 10 develops networking from the simplest case in which Java handles the protocol details to the use of sockets, the development of simple browsers and web servers, and distributed computing with RMI.

The Java database libraries insulate programmers from the details of specific database systems. Before using Java, Chapter 11 introduces relational databases and SQL statements, then uses Java to connect to a database and retrieve information from it. Complete programs illustrate the use of metadata, prepared statements, and transactions. This chapter concludes with a GUI for users to query a database.

Having covered object-oriented programming, event-driven programming, applets, graphical user interfaces, exception handling, I/O, concurrent programming, multimedia, networking, and database access, the book concludes with Chapter 12 on the JavaBeans™ component technology which allows developers to create applications from configurable, prebuilt components. Writing every program from scratch, even using object-oriented techniques, will never meet the increasing demand for software. Component technologies hold the promise of better software faster. Moving from objects to components, this text provides the concepts aspiring computer professionals will need to keep pace with computing's remarkable growth.

Background

This text assumes a prior programming course, preferably using C or C++. Section 1.6 covers the Java basics in a way that should be familiar to those with a course in C or C++. Appendix A adds further details and may enable well-prepared readers without prior C or C++ to use this text. For those with little or no prior programming background, the author's text *Computing with Java: Programs, Objects, Graphics*, Scott/Jones, 1998, may be more appropriate.

Features

Annotated Example Programs

Examples facilitate learning. This text contains over 100 complete, annotated programs including sample output. I could use comments to annotate the code, but the detailed comments necessary for teaching purposes would obscure the code. For that I reason I much prefer to label significant lines and then present the extended comment in a Note immediately following the code. These notes also serve to separate discussions of code from the explanations of concepts in the textual material, making the text easier to follow. Some of the Notes repeat the code they annotate, to minimize students flipping pages back-and-forth as they study. For ease of use, the programs are on a disk included with this text.

Exercises

The many varied exercises allow the learner to assimilate concepts and techniques actively.

Test Your Understanding exercises provide an immediate opportunity to use the ideas presented. By putting exercises in each section I target each new idea as it occurs. Answers to many of these appear at the end of the text and the answers to most of the rest can be found on the disk included with the text. Some, labeled **Try It Yourself** encourage the learner to experiment with the examples, modifying them to better understand the principles involved.

Skill Builder Exercises at the end of each chapter provide various matching, fill-in the blanks, and code-tracing exercises to build a working knowledge of the material covered. Answers to all of these appear at the end of the text.

Critical Thinking Exercises at the end of each chapter provide multiple choice questions that force students to think carefully about the concepts. Answers to all appear at the end of the text.

A **Debugging Exercise** in each chapter contains a program with errors which are to be found and corrected if possible. I include the code for these challenging exercises on the disk included with this text to encourage students to attempt them.

Program Modification Exercises occur as soon we develop examples complex enough to modify fruitfully. Much of a professional programmer's work involves modifying existing code, so I believe in giving students early practice at it. Marked **Putting It All Together Exercises** revisit examples from previous chapters in the light of the new ideas in the current chapter.

Program Design Exercises appear in quantity to provide a variety of assignments at all levels of difficulty to enable students to develop design and programming ability.

Programming Projects appear in the last three chapter to give more challenging assignments on advanced topics.

Tips and Style

The tips help the student avoid common mistakes and misconceptions while the style notes give students guidance on appropriate program formatting. I use these judiciously, hoping that students will remember and use them, rather than be overwhelmed by an overly long list.

The Big Picture

A beginner at chess knows the moves but a master sees the big picture to use those moves effectively. I introduce Big Picture boxes to help students make the transition from beginners to masters, integrating details into concepts they can use effectively in solving problems.

Unified Modeling Language (UML)

I use the UML notation, introducing students to this standard object-oriented modeling language, and to the techniques of object-oriented software engineering.

Build Your Own Glossary

The glossary terms listed at the end of each chapter are provided on a file on the disk included with this text. Students learn the terminology by adding definitions of these terms to that file.

Using This Text

At the start of each chapter I suggest some sections that might be omitted if desired and list the sections, if any, that depend on them later in the text. The chapter dependencies are:

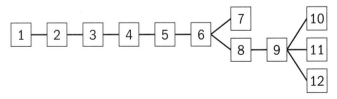

Versions of Java

Since Java was first released in 1995 there have been three major versions, 1.0, 1.1, and 1.2 (now called the Java 2 Platform), using Sun's numbering. The first, 1.0, used an older event model that I do not cover in this text. Version 1.1 made a number of changes, including a new event model. The Java 2 platform made fewer changes but added many new facilities. All programs in this text work with either version 1.1 or Java 2 (formerly 1.2).

Some Java packages are not part of the core, but are standard extensions. I chose not to include any of these. The Swing user interface components fit into this category. To include them in a book of this length would require omitting one of the networking, database, or JavaBeans chapters which I feel are more important, especially given that I introduce and use the AWT components which provide a basis for later study of Swing.

Acknowledgments

Thanks go especially to my wife, Charlotte, and daughter, Amanda, for providing the joy that sustains me during the long hours this project has taken. My Java students supply great feedback on my presentations and found many typographical errors. Cathy Glenn, Cecelia Morales, Cathy Baehler, Hazel Dunlap, and the entire Scott/Jones team make what might otherwise be an onerous task into a satisfying experience while turning the raw materials I present into a finished product. I admire and thank them for their superb support. Richard Jones sees the big picture and helps me to see it, too. He has been an indispensable collaborator.

Many fine reviewers made insightful comments and suggestions that helped me immensely. They include:

Russell Abbott
California State Los Angeles

Dwight Barnette
Virginia Tech

Joseph Bergin
Pace University

Stuart Brian
Holy Names College

Ann Burroughs
California State University Humboldt

Peter Casey
Central Oregon CC

Qiyang Chen
Montclair State University

Chakib Chraibi
Barry University

Frank Coyle
SMU

Kathy Cupp
Oklahoma County CC

Andrew Downs
Tulane University

Rod Farkas
CC of Allegheny County

Eric Gossett
Bethel College

Simon Gray
Ashland University

Andrew Harris
IUPUI

Mark Harris
Appalacian State University

Mark Lattanzi
James Madison University

Sandeep Mitra
SUNY Brockport

John Neitske
Truman State University

Robert Plantz
Sonoma State University

Timothy Price
IUPUI

James Shaw, Jr.
Texas State Tech College

Ray Springston
University of Texas Arlington

Jack Van Luik
Mt Hood CC

Class Testing

We never know if a book "works" until we've taught out of it. Jack Van Luik and his students at Mt. Hood Community College class tested this book in manuscript form. Their feedback helped bring this book to the quality of a Second Edition.

Web Site

My web site for this book is at

`http://www.engr.csulb.edu/~artg/java`

where an errata list will be posted.

Typos and Errors

Despite class testing this material repeatedly, undoubtedly some typos have escaped detection. Please send any discovered to me via the mail link on my web site, or in care of thepublisher at scotjones2@aol.com.

1

What Java Can Do

A s we study computing, we are challenged to keep up with rapid technological change. With Java we can use the basic techniques of programming that have brought us to this point and go forward with object-oriented, interactive, graphical, event-driven programming, using networking, the World Wide Web, databases and component technology that take us to the future.

After an overview of Java, this chapter shows how to run Java applications and applets, then contrasts program-driven with event-driven applications, and procedural with object-oriented programming. Java supports all these programming paradigms. Our AtmScreen, a simple example of a Java event-driven, object-oriented program, can be improved using networking, databases, and Java Beans to become a full-fledged Internet application. The AtmScreen motivates our study in the remaining chapters of the concepts needed for programming in an Internet world.

OBJECTIVES

◆ Survey Java's history, its features, and how it works.
◆ Edit, compile, and run a Java standalone application.
◆ Edit, compile, and run a Java applet.
◆ Be aware of the difference between program-driven and event-driven programming, both of which Java supports.

◆ Be aware of the difference between character mode and graphics mode programs, both of which Java supports.

◆ Be aware of the difference between procedural and object-oriented programming, both of which Java supports.

◆ Be aware of Java's support for enterprise applications using networking, databases, and component technology.

◆ Understand the differences in basic constructs between Java and C or C++.

◆ Be able to write procedural programs in Java.

1.1 ▪ Overview of Java

History

FORTRAN and **COBOL** were among the first high-level languages, introduced in the late 1950s. Both are still used today, FORTRAN for scientific applications and COBOL for business. **Pascal**, designed for teaching, introduced in the early 1970s, and its successors **Modula-2** and **Modula-3** influenced the design of Java. Much of Java's basic syntax is modeled after C, developed in the early 1970s, and its successor, C++, developed beginning in the 1980s.

The **C** language became very popular due in part to efficiency; developers can write C programs that perform well. The **C++** language, as its name suggests, adds features to C to enable the writing of larger programs using object-oriented programming. While C is a relatively small language, C++ is large and complex. The team at Sun Microsystems who were developing what was to become Java tried at first to use C++, but found it did not meet their needs for a language to develop software for devices such as cellular phones and digital assistants. They began the development of a new language, originally called Oak, in the early 1990s coinciding with the growth of the Internet and the World Wide Web. Seeing greater opportunities for their new language in network programming they changed their focus and created Java.

Java's Features

Java's designers list its important characteristics as:

• simple, object-oriented, and familiar

• robust and secure

• architecture neutral and portable

• high performance

• interpreted, threaded, and dynamic

Simplicity and Robustness

Java gets its simplicity by omission of some of the complexities such as pointers and memory management that plague C and C++ programmers. However C and C++ programmers will find Java familiar as it borrows much of its syntax from them. We will consider Java's object-oriented features in detail in this text.

A **robust** program is reliable, running without crashing due to programming errors, erroneous input, or failures of external devices. By remaining simple, Java eliminates the source of many errors. It checks that data is being properly accessed, and provides exception handling to manage unexpected events. Security, especially when communicating across a network with unknown and potentially untrustworthy sources, requires careful measures which Java implements.

Architecture and Portability

Being **architecture neutral** means a compiled Java program will run on a variety of processors using various operating systems. A programmer can create a program, compile it, and make it available over the Internet to users all over the world who download that program and run it on many different processors and operating systems. We will explore this further when we discuss how Java works. Besides architecture neutrality, an important component of portability, Java specifies the language so as to reduce the implementation dependencies prevalent in other languages.

As Java develops it attains higher levels of performance. Its designers hope in time to match the performance of the most efficient current languages while still maintaining portability. In network applications, communications delays lessen the need for speed of code execution.

Multitasking

Multitasking operating systems allow users to switch between several applications running at the same time; for example, editing one program while downloading another from the Internet. Java **threads** allow a Java program to perform multiple tasks simultaneously. An animation can run while the user is entering data in a form, for example. Java can link dynamically, while the program is running, to library code it needs.

In sum, Java is a general purpose programming language well-suited to interactive and network programming. Since its introduction in 1995 it has been enormously successful.

How Java Works

Java uses a **compiler**, but (typically) does not translate a high-level program to the machine instructions of each specific processor. The Java creators designed a machine, called the **Java Virtual Machine (JVM)** because it is not usually implemented in hardware. The Java compiler takes as input the high-level Java program and produces as output an equivalent program written using the instruction set of the JVM. If our processor is a JVM, we can run this program very efficiently; if not we use an interpreter to run the JVM code on our processor. Figure 1.1 shows the compilation of a Java program to a JVM program followed by the interpretation of the JVM program to produce the results on two different processors.

The Java developers call the code for the JVM **byte code**. The advantage of producing byte code instead of code for each different processor is that the same byte code will run anywhere that has a Java byte code interpreter. We can download a Java program which has been compiled to byte code for the JVM on one type of machine and run it on a much different type of machine. For networking, this feature of Java is invaluable. The disadvantage of producing byte code is slower execution, because running byte code using a software interpreter is a slower process than running code for a specific processor directly on that hardware. New advances in compilation technology are reducing execution time.

FIGURE 1.1 Compiling and executing a Java program

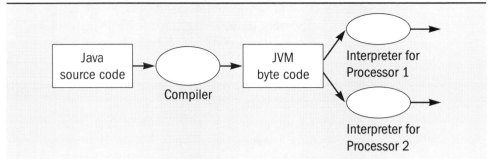

The Big Picture

Java was created in the mid-1990s, basing much syntax on C and C++ and using concepts from several other languages. Its creators characterize it as simple, object-oriented, familiar, robust, secure, architecture neutral, portable, high performance, interpreted, threaded, and dynamic.

Java source compiles to byte code interpreted by a Java Virtual Machine. The byte code runs anywhere which makes Java very popular for Internet applications in which diverse machines communicate.

TEST YOUR UNDERSTANDING

1. From what language did Java derive much of its basic syntax?

2. List Java's features as portrayed by the developers of Java.

3. The Java compiler (usually) translates Java programs to equivalent programs using the instruction set of which machine?

4. What is the advantage of translating a Java program to byte code?

1.2 ▪ Running Java Programs and Applets

Java programs fall into two main categories: applications and applets. An application, often called a **standalone application**, runs on its own, whereas an **applet** is a small application embedded in a web page (see Section 5.1 for an introduction to the writing of web pages). A Java **interpreter** in the browser runs the applet when the browser loads the web page containing that applet.

Various vendors provide tools with which to develop Java programs. Each has its own steps used to create, compile, and run Java programs and applets.[1]

---

[1] See Appendix E for instructions on using some integrated development environments. This note describes the use of the Java Development Kit (JDK) provided at no charge by Sun Microsystems, Inc. at its web site,

`http://java.sun.com/`

continued on page 5

Editing, Compiling, and Running a Java Standalone Application

To create our first Java program, we can use an integrated development environment or a simple editor available on our machine.[2] Example 1.1 shows a standalone application which just displays the sentence "Hello World!" when we run it.

EXAMPLE 1.1

Hello.java

```
/* Displays "Hello World!"
 */

public class Hello {
    public static void main(String [] args) {
        System.out.println("Hello World!");
    }
}
```

Output
```
Hello World!
```

The file name for this program must be `Hello.java`, where case is important. Java considers the name `hello.java` as different from the name `Hello.java`. We will explain the elements of this program in Section 1.6, and just use it here as an example of how to compile and run a Java application.

To compile this program using the JDK[3] use the command

```
javac Hello.java
```

On Windows systems click on the Start button, Programs, and Command Prompt (MS-DOS Prompt in Windows 95) to get the console window from which to enter the compilation command. The name `javac` represents the Java compiler which translates the Java program

---

Sun provides implementations for the Windows 95, Windows 98, Windows NT, and Sun Solaris operating systems. They list implementations by others for additional platforms. The documentation requires a separate download.

To make it easier to compile and run Java programs, users of the JDK should add the directory containing the Java compiler and interpreter to the path environment variable which lists all the places the system should look for command programs. In Windows 95 and Windows 98 we can edit the `autoexec.bat` file, adding the following line at the end

```
set path=c:\jdk###\bin;%path%
```

where `jdk###` is the directory in which the JDK is installed. (Replace ### by the number of the Java version used, `jdk1.2` for version 1.2, for example.) In Windows NT, click on the Start button, Settings, Control Panel, and the System icon. Click on the Environment tab and the Path variable. Add the directory

```
c:\jdk###\bin
```

to the list in the value box, where `jdk###` is the directory in which the JDK is installed, and the Java version number replaces the ###. Then click the Set button and the OK button.

[2] The Notepad editor on Windows machines can be used to create and edit programs for use with the JDK. Save the program using the `.java` extension.

[3] See Appendix E for instructions on compiling using an integrated development environment.

`Hello.java` to an equivalent byte code program, `Hello.class`, for the JVM. Files with the `.class` extension are byte code files.

Once we have compiled Example 1.1 we can execute the code. To run this program using the JDK,[4] enter the command

```
java Hello
```

in the console window. The name `java` designates the Java interpreter that executes the byte code file `Hello.class`.

☛ **TIP**   When compiling, we include the extension in the file name, as in:

```
javac Hello.java
```

but when executing the byte code, we omit the extension, as in:

```
java Hello
```

Editing, Compiling, and Running a Java Applet

An applet is simply a type of Java program we can embed in a web page. We edit and compile it just as we did the Java application in Example 1.1. For our first applet we display the message "Hello World!" Example 1.2 shows the code for this applet. Do not try to understand this code now; we will explain it in Chapter 5.

EXAMPLE 1.2   **HelloApplet.java**

```
/* Displays the message "Hello World!" in
 * a browser or applet viewer window.
 */

import java.applet.Applet;
import java.awt.Graphics;
public class HelloApplet extends Applet {
    public void paint(Graphics g) {
        g.drawString("Hello World!",30,30);
    }
}
```

The World Wide Web (WWW) uses the Hypertext Markup Language (HTML) (introduced in Section 5.1) to create web pages. HTML uses tags enclosed in angle brackets, <>, to format a web page.

To run an applet, we create an HTML file which can be as simple as that of Figure 1.2.

FIGURE 1.2   An HTML file for the applet of Example 1.2

```
<applet code=HelloApplet.class width=300 height=200>
</applet>
```

---

[4] See Appendix E for instructions on running Java programs using an integrated development environment.

Once we have compiled the applet code we can display the applet using a browser or applet viewer. To execute the applet using the JDK[5] we load the HTML file of Figure 1.3 into a browser or the applet viewer[6] which is a program provided by Sun to test applets. The

```
appletviewer HelloApplet.html
```

command, where HelloApplet.html contains the file shown in Figure 1.3, will run the applet of Example 1.2. Figure 1.3 shows the applet window, 300x200 pixels, in which we draw "Hello World!"

FIGURE 1.3 The applet of Example 1.2

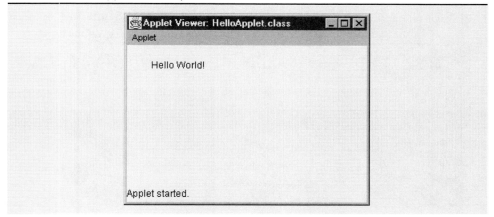

A Graphical Applet

Applets, usually embedded in web pages, typically use graphics and have a user interface. Our first applet in Example 1.2 was very simple. Example 1.3 demonstrates a graphical applet which we use in the next section to illustrate event-driven programming.

EXAMPLE 1.3	**Sort.java**

```
/* Generates 10 random numbers between 0 and 99,
 * displaying them in a bar chart. Inserts the next number
 * in numerical order with respect to the numbers to
 * the left of it, until all 10 numbers are ordered
 * from smallest to largest. Simplifies Example 6.20.
 * The code is on the disk included with this text.
 */
```

After we compile the applet using the command `javac Sort.java`, we can display it using the command `appletviewer Sort.html`,[7] or open it in a browser. Initially the display

---

[5] See Appendix E for instructions on executing Java applets using an integrated development environment.

[6] Using an applet viewer will ensure that the applet can be viewed. Some browsers are not configured with the latest versions of Java. A Java 1.1 (or higher) enabled browser should be used to view the applets in this text.

[7] Integrated development environments have their own applet viewers or call a browser to display applets.

will contain a pink square with a Sort button in the top center. Pressing the Sort button displays 10 rectangles of randomly chosen heights shown in Figure 1.4.

After displaying the initial data the applet changes the label on the button to Next. When the user presses the Next button the next value, starting with the second, is inserted in numerical order with respect to the numbers to the left of it. The first time the user presses the Next button, the applet will move the second bar in front of the first if it is smaller than the first. The second time the user presses the Next button, the applet will insert the third bar from the left in its proper place with respect to the first two, and so on. Figure 1.5 shows the applet after the user pressed the Next button four times.

FIGURE 1.4 The ten numbers to be sorted in Example 1.3

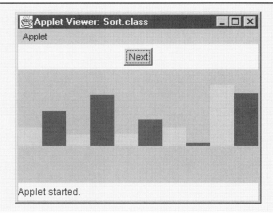

FIGURE 1.5 The Sort applet after inserting four numbers

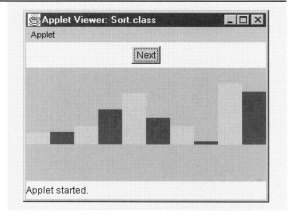

☛ **TIP**

Java has undergone several revisions in its short life, starting out as version 1.0, progressing to versions 1.1 and 2 (formerly known as 1.2). Within these major releases were several minor releases, so version 1.1 has had releases 1.1.1, 1.1.2, . . . and so on. Version 1.1 made many changes to version 1.0; version 2 added much to 1.1 but made fewer changes. Programs written using earlier versions of Java will compile and run using later versions. All programs in this text are compatible with versions 1.1 and 2 of Java. Use version 1.1 or 2 or later while using this text.

The Big Picture

....

Standalone Applications
Run using the Java interpreter to call the main method.

Applets
Run in an HTML file displayed by a browser or tested with an applet viewer.

TEST YOUR UNDERSTANDING

TRY IT YOURSELF   5.   Compile and run the Hello.java program of Example 1.1.

TRY IT YOURSELF   6.   Compile and run the HelloApplet.java applet of Example 1.2.

TRY IT YOURSELF   7.   Compile and run the Sort.java applet of Example 1.3 using the code found on the disk included with this text.

1.3 ▪ Program-Driven vs. Event-Driven Applications

Older programming languages support the program-driven style of programming in which our program executes from beginning to end following the steps of our code much as a cook follows a recipe. Java permits program-driven code, and further supports event-driven programming in which our program, like the firemen in the station, wait for an event to spur them to action.

The Character Mode

For many years computers had no graphical interfaces. Characters were the means of inputting data to the computer and getting results from it. Computer operators could enter commands from a typewriter-like console, but the most common means of input was punched cards. A programmer typed at a keypunch machine which would punch holes in the card representing the characters typed. Initially, alphabets only included uppercase letters because of the limited number of punch codes.

During these years, computers executed programs in batch mode. The computer operator put together a batch of programs, using the card reader to feed them to the computer which executed the programs one after another, printing a listing of each program and the results. When the operating system started a program, that program had control of the processor until it finished executing. We call such execution **program-driven** because the program continues executing the program, one statement after the other until the program ends or a fatal error occurs.

Executing such a program is like preparing a dish from a recipe. The cook is in charge, executing the steps of the recipe until the dish is complete. Figure 1.6 shows the "program" for the preparation of a simple omelet.

FIGURE 1.6 The "program" for preparing an omelet

```
Get the ingredients.
Get the cookware needed.
Crack the eggs into a bowl.
Add salt.
Beat the eggs.
Melt some butter in the frying pan.
Pour the eggs into the frying pan.
Turn the eggs over.
Add the cheese, folding the eggs over.
Serve the omelet.
Clean up.
```

Program-driven applications are still a very important part of computing. Preparing the payroll does not require user intervention. The payroll program takes the input regarding hours worked and rate of pay and prepares the paychecks and other payroll records. As another example, the Java compiler takes the source program as its input and produces the byte code for the JVM with no need for a graphical user interface, operating in **character mode** in which input and output consists of characters.

The Graphical User Interface (GUI)

With monitors and operating systems capable of displaying graphics, we can develop programs which interact with the user via a **graphical user interface (GUI)**. The operating system responds to the user's input, conveying it to the program which passively waits for user input to request it to provide some service. We represent these user requests as events. An **event-driven** program includes code to respond to messages informing it about events involving its window, such as button presses or data entry.

Program-Driven Applications	**Event-Driven Applications**
Execute code in a step-by-step fashion like a recipe.	Wait for the user to generate an event. Include code to respond to that event.
Transform input into output, without user intervention.	Respond to user actions.
Typically use character mode.	Typically have a graphical user interface.

Example 1.3, the Sort program, illustrates event-driven programming. This program initially displays a window with a Sort button and a pink canvas. The program waits passively until the user presses the Sort button, at which time the operating system relays this event to the program which displays a bar chart showing ten numbers from 0 to 99 as shown in Figure 1.4. Again the Sort program waits passively until the user presses the Next button, at which time the operating system passes this event to the program which inserts the next bar into numerical order with respect to the bars to its left. After the user inserts the remaining nine numbers, the program disables the Next button, waiting passively until the user terminates the program.

To contrast an event-driven with a program-driven application, we rewrite Example 1.3 to create 10 numbers at random, display a list of the numbers, sort the numbers in numerical order from smallest to largest and display the resulting sorted list. The SortCommand program of Example 1.4 does not require a GUI. It generates its own data, performing the computation without interference, like a cook preparing an omelet, finally listing the results in the console window.

EXAMPLE 1.4   **SortCommand.java**

```
/* Generates 10 integers from 0 to 99 at random.
 * Sorts these numbers in order from smallest to
 * largest, displaying the sorted numbers.  Simplifies
 * Example 4.6 to illustrate a program-driven application.
 * Complete code is on the disk included with this text.
 */
```

Compiling and running Example 1.4 follows the steps outlined for Example 1.1. The output will look like

```
The data to sort is {8,80,21,36,68,80,65,7,3,43}
The sorted data is {3,7,8,21,36,43,65,68,80,80}
```

although the numbers will change each time we run the program, because we generate them randomly. The console window in which the output appears may look different on different operating systems and development environments. In the Windows operating system, on an x86 type processor, the output may appear in a console window as shown in Figure 1.7.[8]

FIGURE 1.7 The console window for entering commands

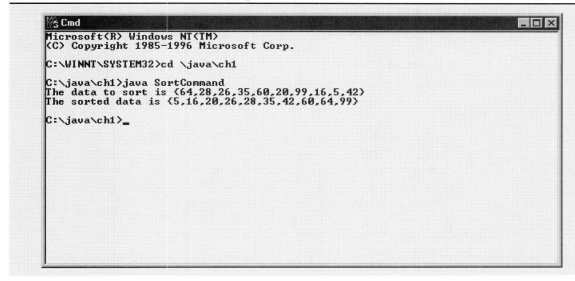

```
Cmd                                                            _ □ ×
Microsoft(R) Windows NT(TM)
(C) Copyright 1985-1996 Microsoft Corp.

C:\WINNT\SYSTEM32>cd \java\ch1

C:\java\ch1>java SortCommand
The data to sort is (64,28,26,35,60,20,99,16,5,42)
The sorted data is (5,16,20,26,28,35,42,60,64,99)

C:\java\ch1>_
```

☛ TIP

The command prompt window of Figure 1.7 is a character mode window. Only characters, not graphics, appear in a limited choice of fonts. By contrast, the graphics windows of Figures 1.4 and 1.5 allow us to draw shapes in various sizes and colors, text in various fonts, sizes, and colors, and to add user interface components such as buttons.

The Big Picture

The program-driven approach suits applications such as compilers which require no user interaction. Event-driven programs suit user interfaces whose components respond to user interactions such as pressing a button or making a selection. With the advent of more powerful computers with graphics capabilities, event-driven applications have taken a more important place in computing.

TEST YOUR UNDERSTANDING

8. Illustrate the idea of a program-driven application by writing a "program" to wash a car, in the spirit of the "program" for preparing an omelet in Figure 1.6.

---

[8] On x86 based processors running Windows 95, Windows 98 or Window NT, the user can change back and forth between a full screen command prompt and a command prompt window, as shown in Figure 1.7, by hitting the [ENTER] key while pressing the [ALT] key.

9. Illustrate the idea of an event-driven application by writing a "program" for the operation of a fire station, including responses to several emergencies.

TRY IT YOURSELF 10. Compile and run the `SortCommand.java` program of Example 1.4, using the code on the disk included with this text.

1.4 ■ Procedural vs. Object-Oriented Programming

We contrast two styles of programming, procedural and object-oriented. Java supports them both. The newer object-oriented style models our everyday intuition and, if used well, can make software more reusable and easier to maintain.

Procedural Programming

Procedural programming emphasizes the procedures used to manipulate the data. Think of a bunch of rocks that cannot do anything by themselves. We can use a throw procedure, which takes a rock and tosses it, a juggle procedure which takes three rocks (small ones) and juggles them, or a sculpt procedure that carves a rock into a sculpture.

These procedures operate on the rocks. If we call our rocks `rock1`, `rock2`, and `rock3`, informally we might say, `"throw rock1"` or `"juggle rock1, rock2, and rock3"` or `"sculpt rock2"`. In a programming language notation, these commands might look like

```
throw(rock1);
juggle(rock1, rock2, rock3);
sculpt(rock2);
```

As programmers, we would spend our time writing code to implement these operations, giving much less regard to the data.

We might describe the throw procedure as

```
Pick up the rock;
if (right-handed) {
    Raise the right arm overhead;
    Pull the right arm back then step forward with the
        left foot, shifting weight to the right foot;
    Swing the right arm forward while shifting weight to the left foot,
        releasing the rock at its highest point;
    Complete the motion of the right arm;
    Return left foot to its original position;
    if (recycling) fetch rock;
}
else {
    // same process with left and right reversed
}
```

When we test this procedure on some sample rocks, we find it works for rocks that fit conveniently in one hand, but requires modification for larger rocks. We may need to use two hands to lift and toss the rock. With a very heavy rock, we need to bend our knees and use other muscles to lift the rock. In addition, optionally, we may want to generate more force by throwing from a running start, but this modification only works for rocks light enough to carry.

Writing a throw procedure that covers all the variations can become a logical nightmare of choices and choices within choices. Each new modification becomes more difficult to make because the structure of the procedure gets more and more complex with each improvement.

We could name separate methods for each kind of rock, so our code would look like

```
if (the rock is a pebble)
    use the throwPebble procedure;
else if (the rock is a chunk)
    use the throwChunk procedure;
else if (the rock is a blob)
    use the throwBlob procedure;
```

Using this approach requires us to modify the code if we add a new type of rock, perhaps an artificial, low-density toy that requires yet another throwing method. We would then add

```
else if (the rock is a modelToy)
    use the modelToy procedure;
```

to our code. Similar complexities arise when we try to describe the juggle or sculpt procedures. By focusing on the data more, object-oriented programming alleviates many of these problems.

Object-Oriented Programming

Object-oriented programming combines data and procedures into an object that has its own attributes and behavior. To transform our rock example, we define a Rock class with operations throw, juggle, and sculpt. We are defining a type of object, whose operations tell us what we can ask it to do. If we have three Rock objects, rock1, rock2, and rock3, informally we can send the throw message to rock1, a sculpt message to rock2, or a juggle message to rock1, with the names, rock2 and rock3, of the other rocks juggled with rock1. In a programming language, these invocations might look like

```
rock1.throw();
rock2.sculpt();
rock1.juggle(rock2, rock3);
```

The object invokes one of its behaviors, in contrast to the procedural approach in which a procedure operates on data.

The power of object-oriented programming shows up when we address the complexities of the operations. We can specialize the rock type to have several subtypes, say pebble, chunk, and blob. Because each is a rock it has all of the rock operations, but may define them differently. A pebble will implement its throw behavior as described above, while a chunk, being larger, will require using two-hands. The heavier blob will require bending down to lift it, and a two-handed toss without raising it over the head.

As programmers we need not worry about such distinctions, leaving each type of object to implement its own throw behavior. We can invoke

```
rock1.throw();
```

to throw our rock and it will use the correct version of the throw operation. For example, if rock1 is a chunk, we will use two hands to throw it. The same invocation works if we add a ModelToy subtype. As programmers, we just ask the rock to throw and it will use the correct procedure depending on what type of rock it is.

Procedural Programs	Object-Oriented Programs
Focus on the procedures needed to solve the problem.	Focus on the objects needed to solve the problem.
Separate functions and data.	Combine data and operations in objects.
Must check many cases to deal with different types of data.	Leave it to objects to provide the correct operations for their type.

The Big Picture

In a procedural program, expect to see a lot of logic used to test the data and do different operations depending on the type of the data.

In an object-oriented program, expect to see a type definition which defines permissible operations for objects of that type. Much of the program will consist of objects invoking one another's behaviors.

TEST YOUR UNDERSTANDING

11. A puppet show and a stage play are two types of performances. Explain which type is more analogous to procedural programming, and which to object-oriented programming.

1.5 ▪ Networking, Databases, and Component Technology

To program in the age of the Internet, we need the object-oriented and event-driven programming concepts covered in Chapters 2 through 7. In this section, we first present an AtmScreen example which uses these techniques to simulate an interactive banking system. We then introduce the networking, database, and component technology concepts, to be covered in detail in Chapters 8-12, which we could use to extend this example to an Internet application.

Using Object-Oriented and Event-Driven Programming

We present our AtmScreen example as an applet to serve as a preliminary version of a simple home banking program. A customer would download this applet from the bank's web site and perform some transactions. We design a program-driven version of the Atm system in Section 3.5. In the event-driven version in Example 1.5, the user presses the Start button, and then follows the instructions to continue the transaction.

Using the applet viewer to test the program results in just the screen of Figure 1.8, Figure 1.9 shows the initial screen embedded in a web page with instructions above it. An applet is part of a web page; Figure 1.10 shows the HTML file (see Section 5.1 for an explanation of the HTML constructs) used to run Example 1.5. We include the code for Example 1.5 in Appendix C and on the disk included with this text. Running it provides another demonstration of event-driven programming.

EXAMPLE 1.5 **AtmScreen.java**

```
/* Illustrates event-driven, object-oriented programming, and can be extended
 * using networking, databases, and Java Beans, to provide remote banking.
 */
```

FIGURE 1.8 AtmScreen using the applet viewer

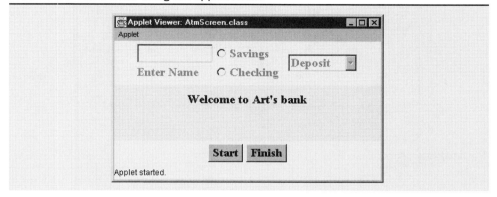

FIGURE 1.9 The AtmScreen in a web page using a browser

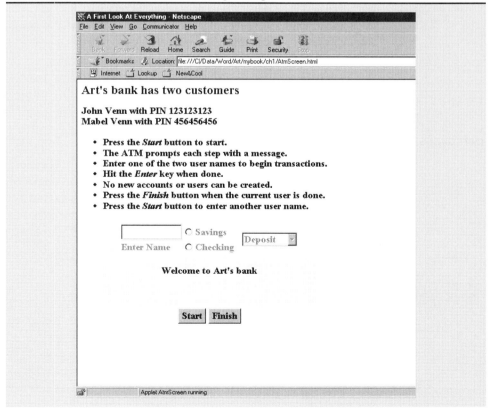

FIGURE 1.10 The HTML file used to run Example 1.5

```
<title>  A First Look At Everything  </title>
<h2> Art's bank has two customers <br>  </h2>
<h3>    John Venn with PIN 123123123  <br>
          Mabel Venn with PIN 456456456    </h3><p>
<ul>
     <li> Press the <em>Start</em> button to start.
     <li> The ATM prompts each step with a message.
     <li> Enter one of the two user names to begin transactions.
     <li> Hit the <em>Enter</em> key when done.
     <li> No new accounts or users can be created.
     <li> Press the <em>Finish</em> button when the current user is done.
     <li> Press the <em>Start</em> button to enter another user name.
<p>
<applet code=AtmScreen.class width=400 height=200>
</applet>
```

Networking

Whether using email or e-commerce, computing is intimately connected with **networks**. We use computers not in isolation, but as a means of global communication.

We can test an applet on our local machine, so Example 1.5 runs fine without any network connection, and using only data set in the program. If Example 1.5 was extended, making it into a remote banking program, we could use a browser to download a web page, from a bank's site, which contains the applet. We would run the applet on our local machine but access data on the web site which was the source of the applet.

To develop a variety of applications that communicate with remote sites, we need to use Java's input and output libraries covered in Chapter 8, its threads for concurrent processing which we present in Chapter 9, and the Java networking methods we explain in Chapter 10.

Databases

In Example 1.5 we included the data we used. In banking applications, **databases** store the data used. These databases may reside on separate machines dedicated to handling database transactions. Business applications typically require access to data stored in a database.

Java provides the Java Database Connectivity (JDBC) methods to allow programmers to use databases on the local machine or at remote sites. In Chapter 11, we show how to connect to databases using Java. We could then improve Example 1.5 to use a database to store the bank account and customer data.

Component Technology

Hardware designers have catalogs of parts they can use to design systems. Unfortunately, software designers have almost always built new systems from scratch, using few pre-built components. In fairness, software systems can be tremendously complex, and previously developed components were hard to find, and difficult or impossible to adapt to the proposed design.

However, skyscrapers cannot be built without powerful tools, and neither can large software systems at the rate needed in the modern world. New approaches to **component technology** address the difficulties of the past and suggest a future in which building software systems from components will be the rule and not the exception.

The simple parts of our AtmScreen example could be components. A button, for example, could be a component that we could customize to meet our needs, and use in our program. We might include a calculator in our banking applet as an aid to the customer. Rather than designing a calculator ourselves, we could buy a calculator component, customize it to meet our needs, and add it to our applet.

To communicate with a remote database, we might buy a component rather than doing that programming ourselves. In the sorting applet of Example 1.3, we might have used a pre-built chart component rather than designing our own. With enough components available we might not have to code at all, but could design the entire system using pre-built components. In the future there may be two types of programmers, those who build components using languages like Java, and those who design applications using pre-built components.

The Java Beans technology, covered in Chapter 12, lets us build components designers can use either with a visual tool for combining components into applets, or as part of Java programs. After the survey of Java basics in the next section, we present the Java features needed to develop powerful Internet applications, starting with objects in Chapter 2 and continuing to the Java Beans component technology in Chapter 12.

The Big Picture

In this text, we start with objects and then cover event-driven programming. The AtmScreen applet shows an object-oriented event-driven application, simulating an automated banking system. To fully implement such a system, we need the networking, database, and component technology concepts covered in the remaining chapters. With Java we can build the interactive Internet applications essential now, and crucial in our computing future.

TEST YOUR UNDERSTANDING

TRY IT YOURSELF   12. Run Example 1.5 to complete a deposit, a withdrawal, and a get balance transaction.

1.6 ▪ Comparing Procedural Programming in Java and C++

Our Java journey has a long way to go to reach object-oriented programming, event-driven programming, networking, databases, and component technology. Starting with procedural programming allows us to transfer programming experience from C, C++, or other languages to Java. The constructs used in procedural programming are basic to all Java programming. We present several examples of Java programs, comparing and contrasting features with those found in C and C++, and discuss those constructs which occur in the examples. By studying the examples in this section, reading Appendix A, and studying the extended chapter summary of Java basics, a C or C++ programmer should be able to write basic Java programs such as those needed to solve the programming exercises at the end of this chapter.

The Structure of a Procedural Program

We start our comparison with the code from Example 1.1

```java
public class Hello {
    public static void main(String[] args) {
        System.out.println("Hello World!");
    }
}
```

which uses a main method, like the main function of C++, but there are a number of differences from C++ in the form of even this simple program. In Java, we put the main method inside a class. The primary purpose of a Java class is to define a type of object, but we also use it for procedural programming by declaring our methods as static.

For now, we declare all our classes as public. The name of a public class must be the same (with the same case) as the file name. Thus the file name for this program must be Hello.java. Unlike C++, Java does not use include files; it uses the class name to find the code for the class.

☞ **TIP**   Because Java requires the external file name be the same as the class name for a public class, there can be at most one public class in a file.

In C++ we can declare the main function to return **void** or **int** and can omit its formal parameter, but in Java we always declare the main method as

```java
public static void main(String[] args)
```

The only options are where to put the array brackets, or which name to use for the formal parameter. Thus

```java
public static void main(String args[])
```

would be correct, as would

```java
public static void main(String[] arguments)
```

Java runs a standalone program by executing the code inside the main method, so, as in C++, a standalone program must include a main method.

In Java, out names the standard output file, usually the screen, and

```java
System.out.println("Hello World");
```

displays Hello World and moves the cursor to the beginning of the next line. Had we wanted to continue outputting text on the same line we could have used

```java
System.out.print("Hello World");
```

instead. We will explain these output statements more fully later in the text.

☞ **TIP**   You may need to break a long string into parts to display it. For example, the statement

```java
System.out.println("This line is so very long we in no
    way at all can get Java to print it as one string");
```

will give an error, but we can replace it by

```java
System.out.println("This line is very long, but I'm going"
    + "to break it up so I can display it without error");
```

which will work fine.

The Big Picture

....

The overall structure of a simple procedural program is

```
public class ClassName {
    public static void main(String[] args) {
        // put code here
    }
}
```

Our next program illustrates the use of variable declarations, Java selection and loop statements, console input, and packages. Let us look at Example 1.6 and then we will discuss the constructs it uses.

EXAMPLE 1.6   **Max.java**

....

```
/*  Finds the maximum of nonnegative invoice items
 *  the user enters.  The user enters a negative item to
 *  indicate that no more data is available. The program
 *  displays the maximum value.
 */

import iopack.Io;
public class Max {
    public static void main(String [] args) {
        double item;              // the next item
        double maxSoFar;          // the max so far

        System.out.println("Enter nonnegative invoice items");
        System.out.println
            ("Enter a negative value to terminate the input");
        item = Io.readDouble("Enter the first item");
        if (item >= 0) {                                            // Note 1
            maxSoFar = item;                                        // Note 2
            while (item >= 0) {                                     // Note 3
                item = Io.readDouble("Enter the next item");
                if (item > maxSoFar)                                // Note 4
                    maxSoFar = item;
            }
            System.out.println("The maximum is " + maxSoFar);      // Note 5
        }
        else
            System.out.println("No input provided");               // Note 6
    }
}
```

Output—First Run

```
Enter nonnegative invoice items
Enter a negative value to terminate the input
Enter the first item:  49.23
Enter the next item:   16.78
Enter the next item:   92.14
Enter the next item:   32.75
```
continued

```
Enter the next item:   -1.00
The maximum is 92.14
```

Output—Second Run
```
Enter nonnegative invoice items
Enter a negative value to terminate the input
Enter the first item:   -1.00
No input provided
```

Note 1: The user indicates the end of the input by entering a negative item. We continue finding the maximum only if the first item is nonnegative.

Note 2: We save the first item as the current maximum. It is certainly the largest of those items we have seen so far.

Note 3: In this loop, we read an additional item as long as the current item is non-negative.

Note 4: If the new item is greater than the largest item we have seen so far, we update maxSoFar to save this item.

Note 5: When one operand is a string, the + operator is string concatenation. Java converts the value of the variable maxSoFar to a string and appends it to the end of "The maximum is ".

Note 6: In this case, the first item the user enters is negative.

☞ STYLE   Capitalize class names such as Max, but start method names, such as main, and variables names, such as number1, with lower-case letters.

Familiar Features

Some parts of Example 1.6 look very familiar to C and C++ programmers. We look at these first.

Types

Java uses type names that occur in C and C++. The declaration

```
double item;
```

looks exactly like a C or C++ declaration, but there are some differences. The identifier follows the rules for Java identifiers listed in Appendix A. Java requires type **double** to use 64 bits, while in C++ the size of **double** values is implementation dependent.

Besides type **double**, Java has the **int** (32-bit), **char** (16-bit), **short** (16-bit), **byte** (8-bit), **long** (64-bit), **boolean**, and **float** (32-bit) primitive types. We list some properties of these types in Appendix A.

Operators

Java uses operator symbols that occur in C and C++. In the expressions

```
item >= 0
item > maxSoFar
maxSoFar = item
```

the operators >=, >, and = are the familiar greater-than-or-equal, greater-than, and assignment operators of C and C++. Most of Java's operators will be familiar to C and C++ programmers. Appendix F shows the Java operators, with their precedence relations. The value of relational expressions such as

```
item >= 0
```

always has type **boolean**, and is **true** or **false**.

Control Structures

Java uses control structures that occur in C and C++. The **if-else** statement

```
if (item >= 0) {
    // find max
}
else
    // print something
```

and the **while** loop

```
while (item >= 0) {
    // continue finding max
}
```

are the same as their C and C++ counterparts, except the condition, item >= 0 in this example, must be of type **boolean**, and cannot be an integer.

Comments

Java supports both styles of C++ comment.

```
/* this is a
   multi-line comment.
*/

// this is a single line comment
```

The Big Picture

....

Some Java Features Similar to C++

◆ Type declarations (but each primitive type has a fixed size).
◆ Operators (same symbols, +, - , *, /, %, ++, −, <, >, ==, !=, $$, | |)
◆ Control Structures, if-else, switch, while, for, do-while, (but test expressions must have type boolean).
◆ Comments (// and /* */)

Unfamiliar Features

Some features of Example 1.6 do not look as familiar to C and C++ programmers.

String Concatenation

We can **concatenate** two Java strings using the + sign. Concatenating two strings appends the second to the end of the first. Here are some examples:

Statement	**Output**
`System.out.println("House"+"boat");`	`Houseboat`
`System.out.println(76+" trombones");`	`76 trombones`
`System.out.println("number2 is now "+number2);`	`number2 is now 40`

In the expression `76+" trombones"`, the first argument, `76`, is an integer which Java converts to a string before concatenating `" trombones"` to it. In the expression `"number2 is now "+number2`, the second argument, `number2`, is an integer variable whose value Java converts to a string `40`.

Character Mode Input[9]

Java does not make it easy to read keyboard input values. One needs to use concepts that will be covered later in this text. For use now, we provide methods to input **int**, **double**, **char**, **long**, and **float** values. The `readInt` method, used to input integers, takes one argument, a prompt string, and returns an integer value. We use the prompt to tell the user what to input. The line

```
x = Io.readInt("Enter an integer");
```

prints the prompt and waits for the user to type an integer on the same line after the prompt.

```
Enter an integer:  25
```

After typing the number 25, the user hits the Enter key to transmit the value and store it in the variable x.

We do not declare the `readInt` method inside a class in our program, so we need to prefix the method name, `readInt`, with an identifier to show whose method it is. For the `readInt` method we add the `Io` prefix, showing that `readInt` belongs to the `Io` class.[10] Use the dot to separate the class name, `Io`, from the method name, `readInt`. If we had defined the `readInt` method in our own class we could have called it using `readInt("Enter an integer")`, but because the `Io` class defines it we call it with the `Io` prefix, as in `Io.readInt("Enter an integer")`.

The `readInt` method allows the user to enter one integer value. We call `readInt` for each value we wish the user to enter. If the user enters characters that do not represent an integer, the `readInt` method will indicate an error and prompt the user to re-enter the input. Thus the program will not fail if the user enters an incorrect type of data.

---

[9] We use character-mode input and output in Chapters 1-4, leaving graphical input until we cover the necessary event-handling techniques. We prefer to treat event-handling carefully rather than to hide it using the opaque machinery necessary at this point.

[10] The name Io derives from the initials i for input and o for output. We use an uppercase I because we begin all class names with an uppercase letter. The Io class was written by the author and is not part of the Java library.

This is what will happen if the user enters `cat` instead of 25.

```
Enter an integer:  cat
Error -- input an integer -- Try again
Enter an integer:
```

Now the user has a chance to correct the error and enter an integer value.

The `readDouble` method, used in Example 1.6, works similarly.

A First Look at Packages[11]

Java organizes code into **packages** which correspond to directories in the computer's file system. Each Java class is contained in a package. The Java library includes many packages. The `System` class, used in Examples 1.1 and 1.6, is in the `java.lang` package. Java automatically searches this package to find classes. We have not explicitly put our programs into a named package, therefore Java includes them in the default (unnamed) package consisting of all classes in the current directory in which we are working.

In Example 1.2, we used the `Applet` class which is in the `java.applet` package. When Java sees the name Applet, it will look in the default package, the current directory, and in the `java/lang` directory, checking for classes in the `java.lang` package, but it will not find the `Applet.class` file. We must include the `import` statement

```
import java.applet.Applet;
```

to tell Java where to find the `Applet` class code.

In Example 1.6, we use our `Io` class we wrote to make console input and output easier, putting it in a package named `iopack`. To use the `Io` class in our programs we add an `import` statement

```
import iopack.Io;
```

to tell the Java compiler the directory name in which to find it. Any `import` statements come before class definitions in our program. Our `import` statement tells the compiler to look for the `Io` class in the `iopack` package. Each Java development environment has its own way to tell the compiler where to find the packages referred to in `import` statements.[12]

We use the **Unified Modeling Language (UML)**[13] to represent object-oriented constructs. Figure 1.11 shows the UML notation for package diagrams. The rectangle with a tab denotes a package. We show the package name in the tab and the classes of the package inside the larger rectangle. The `Max` class of Example 1.6 is in a default package consisting of all classes in the same directory in which it is contained.[14] The dashed arrow shows a dependency in which the default package uses a class from the `iopack` package. We designate this dependency as `<<imports>>` showing the default package imports `iopack`.

---

[11] We consider packages further in Section 3.4.

[12] Using the JDK one can copy the `iopack` directory from the disk that comes with the text to the directory containing your program. See Appendix E for instructions on using `iopack` with other development environments and for an alternative way to use packages with the JDK.

[13] See `http://www.rational.com/` for information about the Unified Modeling Language.

[14] In Section 3.4 we will see how to put our classes into a named package.

FIGURE 1.11 Package diagram

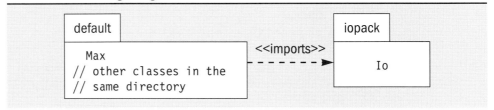

Our next example illustrates the use of Java methods as used in procedural programming.

EXAMPLE 1.7   **Growth.java**

```
/* Input:   Rate, a double value giving the yearly percent interest rate.
 *          Balance, a double value giving the amount deposited.
 *          Years, an int value giving the time period for the deposit.
 *          Repeat, 'Y' to repeat, 'N' to quit.
 *
 * Output:  The total amount in the account at the end of the time period.
 */

import iopack.Io;
public class Growth {
    public static double amount(double rate, double balance, int years) {
        for(int i = 1; i <= years; i++)
            balance *= (1+ rate/100);                                   // Note 1
        return balance;
    }
    public static void main(String [] args) {
        double rate;           // yearly percent interest rate
        double balance;        // amount on deposit
        int years;             // time period of the deposit
        char repeat;           // 'Y' to repeat, 'N' to quit

        do {                                                           // Note 2
            rate = Io.readDouble("Enter the percent interest rate");
            balance = Io.readDouble("Enter the initial balance");
            years = Io.readInt("Enter the time period for the deposit");
            System.out.print ("The balance after " + years + " years is ");
            Io.println$(amount(rate,balance,years));                   // Note 3
            repeat = Io.readChar("Enter 'Y' to repeat, 'N' to quit");
        } while (repeat == 'Y' || repeat == 'y');
    }
}
```

Output

```
Enter the percent interest rate:  4.5
Enter the initial balance:  1000
Enter the time period for the deposit:  7
The balance after 7 years is $1,360.86
Enter 'Y' to repeat, 'N' to quit:  n
```

Note 1: We update the balance adding in the interest earned each year. We use the mixed-type expression `rate/100` to convert from a percent to a decimal.

Note 2: The **do-while** loop is familiar from C and C++.

Note 3: The `Io.println$` from the package `iopack`, found on the disk included with this text, displays the decimal value in a dollars and cents format, with a leading dollar sign and two places after the decimal point. We print the final balance returned by the `amount` method.

Other Features

We discuss the features illustrated in Example 1.7 and some additional features we use later in the text.

Methods

A **method** contains code to perform an operation. The corresponding C and C++ term is **function**; Java makes the distinction because methods always occur inside of a class. Unlike C and C++, Java has no global functions.

Like the `main` method, methods used in procedural programming are declared `static`. They operate on data passed to them as arguments, as we see in the `amount` method of Example 1.7.

☞ **TIP**

When calling methods which use the `static` modifier[15] from a method of the same class, as in Example 1.7, where we call `amount` from `main`, we do not need to use the class prefix. Optionally, we could use the class name prefix, calling the `amount` method as `Growth.amount`. When calling static methods from a method of a different class, for example, calling `Io.readInt` in Example 1.7, we must use the class prefix to tell Java which class contains the method.

Constants

We declare constants as `public`, `static`, and `final`, as in

```
public static final int SIZE = 5;
```

using the C style of naming constants using uppercase identifiers. The `static` modifier indicates that the constant is defined for the entire class and is not part of any object. The `final` modifier means that we cannot change its value, effectively making it a constant. If `SIZE` is declared in a class named `Stuff`, to use `SIZE` outside of the class we refer to it as `Stuff.SIZE`.

Easy Output

The `Io` class has methods,

```
print(double d, int n)
```

and

```
println(double d, int n)
```

---

[15] In Section 2.3 we will see that methods using the static modifier are called class methods and will contrast them with instance methods used in object-oriented programming.

which will display the number d with n places after the decimal point, and methods

```
print$(double d)
```

and

```
println$(double d)
```

which will display the **double** d in dollars and cents, including the dollar sign, $. Use `print` or `print$` to remain on the same line for further output, and the `println` or `println$` methods to go to the next line for further output.

Mixed-Type Expressions

Usually a numeric expression uses all variables and literals of type **int**, or all type **double**. When we mix **int** and **double** in an expression, as in 2.54 + 361, Java converts the **int** to a **double** by changing 361 to 361.0. Although we do not lose any information by converting 361 to 361.0, it does require a change inside memory because 361 has an internal representation which is a lot different than that for 361.0.

Java will also convert from **int** to **double** in an assignment statement. For example, in

```
double d = 4;
```

Java will assign the **double** value 4.0 to the variable d. However, in contrast to C and C++, Java will not automatically convert a **double** to an **int** for the purpose of assigning it to an integer variable, because a **double** value may be out of the range of values an **int** variable can hold.

The general rule Java follows for these implicit conversions of one primitive type to another is that any numeric value can be assigned to any numeric variable whose type supports a larger range of values. To force a conversion we can use an explicit type cast, as for example

```
double d = 3.14159;
int i = (int) d;
```

Random Numbers and Other Math Methods[16]

The `Math` class of the `java.lang` package has methods for familiar operations such as raising a number to a power, computing its square root, and performing trigonometric calculations. For example,

```
Math.sqrt(2.0);
```

returns the square root of 2.0, while

```
Math.pow(2.0,3.0);
```

returns two raised to the third power.

Sometimes we cannot predict the outcome of an event precisely, but can specify a probability it will occur. For example, in tossing a fair coin, we cannot predict the outcome of any one toss, but we can say there is a 50 percent chance it will be heads and a 50 percent chance it will be tails, meaning in a long series of coin tosses we expect about half of each outcome.

---

[16] See Appendix A for a list of selected methods from the `Math` class.

We do not usually use computers to toss coins, but we can use computer-generated **random numbers**, a sequence of numbers which appear to have no pattern, to simulate coin tossing, and many other events such as the arrival of traffic at an intersection or customers at a bank.

The method, Math.random(), returns a double value between 0 and 1. Each time we call this method we get a different number between 0 and 1, and the numbers appear to be randomly scattered in that interval. In Example 1.8, we get five random numbers, using them to simulate coin tossing. We represent a **random number** between 0 and .5 as heads, and from .5 to 1.0 as tails, assuming the coin is fair with each side equally likely of appearing.

EXAMPLE 1.8   **Coin.java**

```java
/* Uses random numbers to simulate coin tossing.
 */

public class Coin {
    public static final int TRIALS = 5;                              // Note 1

    public static void tossCoin() {
        double toss = Math.random();
        System.out.print("The random number is " + toss + " which represents ");
        if (toss < .5)
          System.out.println("HEADS");
        else
          System.out.println("TAILS");
    }
    public static void main(String [] args) {
        for (int i = 0; i < TRIALS; i++)
            tossCoin();
    }
}
```

Output—First Run
```
The random number is 0.0011150808194488881 which represents HEADS
The random number is 0.589115261279429 which represents TAILS
The random number is 0.3299002133848955 which represents HEADS
The random number is 0.1676852828205404 which represents TAILS
The random number is 0.9738772851058562 which represents HEADS
```

Output—Second Run
```
The random number is 0.14875143393514823 which represents TAILS
The random number is 0.5910311808541433 which represents HEADS
The random number is 0.968743132262657 which represents HEADS
The random number is 0.6498168301891447 which represents HEADS
The random number is 0.694254199519966 which represents HEADS
```

Note 1:   Using a named constant allows us to easily find the uses of that constant in the code. Using a numerical value such as 5 in the **for** loop would not express the meaning that it represents the number of coin tosses.

The Big Picture

Procedural programming in Java has many similarities to procedural programming in C and C++. The examples in this section show these similarities and some differences which we noted in the text. Some useful observations for getting started writing Java programs:

◆ Java uses the + operator for string concatenation.
◆ The `System.out.println` statement can output any primitive type.
◆ The `Io` class (added by the author) makes it easier to do console I/O.
◆ The `import` statement tells Java where to find the byte code for a class.
◆ A `static` method is like a C or C++ function.
◆ The `Math` class contains Math library methods such as `sqrt`.

Appendix A summarizes basic Java constructs.

TEST YOUR UNDERSTANDING

13. How must we declare the `main` method in Java? Which parts of the declaration can we change?

14. Describe an important difference between type **double** in Java and type **double** in C++.

15. What is the purpose of a Java `import` statement?

16. Describe an important difference between Java and C++ in the type of expression allowed in an **if** test.

17. What is the value of `Math.pow(4.0,2.0)`?

18. Give an example of an implicit type conversion which would be legal in C++, but not in Java.

Chapter 1 Summary

Java maintains many features of earlier languages, modeling its basic syntax on the C language syntax, but both simplifies and extends earlier languages in useful ways. The Java developers describe Java as simple, object-oriented, familiar, robust, secure, architecture neutral, portable, high performance, interpreted, threaded, and dynamic. Its development coincided with a rapid growth of the Internet. Java is designed to be especially useful for network programming.

Sun implements Java using both a compiler and an interpreter. The compiler translates the Java source code to byte code, which uses an instruction set for the Java Virtual Machine. The "virtual" in the name means this machine is usually not implemented in hardware. Rather, an interpreter executes the byte code directly. The advantage of such an approach is the byte code can be transported across networks to machines with different types of processors and it will run on any of these machines as long as there is a Java interpreter for the JVM. The disadvantage is a somewhat slower execution of code.

Java programs fall into two main categories, applications and applets. Applications stand on their own, while applets, embedded in web pages, are viewed in a browser or applet viewer.

We edit and compile applications and applets similarly, but we call the Java interpreter to execute the byte code for an application, while the browser uses its own JVM interpreter to execute the byte code for an applet.

Program-driven applications are like the cook making an omelet; they perform one step after another maintaining control until finished. Such programs use the character mode for input and output. By contrast, event-driven programs are more like firemen waiting for a call at which point they respond as appropriate. They often use graphical user interfaces to allow users to generate events, such as button presses, to which they respond.

Procedural programming develops procedures which act on separate data. This style of programming requires a procedure to test what type of data it is using and to modify its algorithm to support each type. By contrast, in object-oriented programming, we place the operations together with the data in an object. When we ask an object to perform an operation it uses the method defined for its particular type. Object-oriented programs require less modification to work with new types, and more closely model our understanding of the problem.

Java suits an Internet world. It is portable and has facilities for communicating across networks, accessing databases, and programming using pre-built components. Our AtmScreen example suggests an online banking application which would take advantage of Java's many capabilities we will explore in this text.

Extended Summary of Java Basics

To begin writing Java programs, we need to know the basic elements of the Java language. We name our data and other items using identifiers which must start with a letter (including underscore, _, and dollar sign, $), followed by letters or digits or both, and can be of any length. Java is case-sensitive, distinguishing between identifiers fruit and Fruit, for example. Keywords, such as **int**, are reserved for special uses and cannot be used as identifiers. Java uses the Unicode character set which contains 38,885 characters, including all the commonly used ASCII characters.

A variable holds data the program uses. Every Java variable has a type and a name which the programmer must declare. We can initialize a variable in its declaration which will give that variable an initial or starting value. We use the assignment statement to change a variable's value during the execution of the program.

To perform computations, Java provides the binary arithmetic operators +, -, *, /, and %, and the unary arithmetic operators + and -. Java uses precedence rules to evaluate arithmetic expressions without having to clutter them with too many parentheses. Multiplication, division, and remainder have higher precedence than addition and subtraction. Java has operators, -=, *=, /=, and %=, that combine assignment with the other arithmetic operators, and increment and decrement operators, ++ and --, which come in either prefix or postfix forms. We can use the print and println statements to display our results.

The simplest form for a program puts the code in a main method which is itself enclosed in curly braces in the definition of a class. The main method is static, meaning it is part of the class, not part of any object. The Java interpreter executes the code in our main method.

In addition to the main method we can use other static methods in our programs. A method contains the code for an operation. We use parameters to communicate with methods and to make them more flexible. A method can return a value, the result of the operation. We use the `return` statement to specify the result.

The `Io` class, found on the disk included with this book, adds some methods to make Java input easier. We use the `readInt` method to read an integer value from the keyboard. The `readInt` method is part of the `Io` class, so we need to refer to it as `Io.readInt`, prefixing the method name with the class name in which it is found.

Java uses the keyword **int** for the integer data type. Integer variables can hold values of up to 10 decimal digits. Java provides the **boolean** type which has values **true** and **false**. One can use scientific notation to express decimals, so we can write `.000000645` as `6.45E-7`. The type **double** represents decimal numbers with 16-digit accuracy. We can declare variables of type **double**, and perform the usual arithmetic operations of +, -, *, and /.

Java displays numbers of type **double** without using exponents for numbers between .001 and 10,000,000 or between −.001 and −10,000,000, and uses scientific notation otherwise, making very small or very large numbers easier to read. We added methods to our `Io` class to input **double** values, to display them with a specified number of places after the decimal point, and to display dollars and cents.

If an operator, say +, has one **int** operand and one **double** operand, then Java will convert the **int** operand to a **double**, and add the two doubles, producing a result of type **double**. Java will not automatically convert a **double** to an **int**, but we can use an explicit type cast should we need this conversion.

The character type, **char**, internally uses Unicode to represent the many characters used in different locales. In this text we use the ASCII character set which includes lower- and uppercase letters, numerals, punctuation, various operators, and special characters some of which are non-printing control characters. Java uses single quotes to represent characters such as 'A'. The backslash, '\', is an escape character signaling to Java the next character is a special character such as '\n', the `newline` character. We can declare variables of type **char**, and input character data from the keyboard.

The **long** type allows us to use integers of up to 19 digits, roughly twice as many as the 10 digit maximum for type **int**. The **long** type has the arithmetic operators +, -,*, /, and %; we represent literals of type **long** using the suffix L or l to distinguish these values from those of type **int**. For decimal numbers, the type **float** uses less precision, six to seven digits, than the type **double**, and also uses less space internally. Each **float** literal is suffixed with an F or an f to distinguish it from a value of type **double**.

The `Math` class contains methods to compute a number of mathematical functions including powers, square roots, absolute value, maxima, minima, floor, and ceiling. There are methods for evaluating the natural logarithm, exponential, and trigonometric functions. A random number generator enables us to do simulations of events that occur with certain probabilities. In using any of the methods from the `Math` class we prefix the method name with the class name, `Math`, as for example in `Math.sqrt(2.0)` which computes the square root of two.

To make our programs more flexible, we need to make decisions based on the value of a test condition which can be either **true** or **false**. We write our test conditions as relational or equality expressions. The relational expressions use the operators <, >, <=, and >= and produce **boolean** values. The equality expressions use the operators == and != and also produce **boolean** values. We can include arithmetic operators in our test conditions. Java will use precedence rules to help evaluate such expressions, with arithmetic expressions having higher precedence than relational expressions.

The `conditional AND,` `conditional OR` and `NOT` operators take **boolean** operands and produce a **boolean** value. An `AND` expression is **true** only when both its operands are **true**, an `OR` expression is **true** when either or both of its operands are **true**, and a `NOT` expression is **true** when its operand is **false**. We use these operators to write more complex conditions, useful as tests in `if-else` and `while` statements.

Once we know how to write a test condition, we use the `if` and `if-else` statements to make choices based upon the result of a test condition. In the `if` statement, we execute the next statement if the condition is **true**, and skip it if the condition is **false**. The `if-else` statement gives us two alternative statements, one to execute if the test condition is **true**, and the other to execute if the test condition is **false**. Each statement can be a simple statement or a code block enclosed by curly braces.

To choose among multiple alternatives we can use nested `if-else` statements. A final `else` without any following conditional test can be used to handle the case when all the previous conditions are false. Ambiguity can arise when nesting `if` and `if-else` statements. Java uses a rule that pairs each `else` with the preceding `if`, but programmers can override this rule by enclosing a nested `if` statement in curly braces.

A `switch` statement is a better choice than nested `if-else` statements to handle more than a few alternatives. We mark each alternative in the code with a `case` label. When Java executes a `switch` statement, it jumps to the code at the `case` label specified by the value of the `switch` variable, and continues executing code from that point on. `Break` statements separate one alternative from the other. A `break` statement causes Java to jump to the end of the `switch` statement, bypassing the code associated with any cases that follow that break. The `default` label will handle any alternatives not covered by other `case` labels. We often use the `switch` statement to provide a menu of choices for the user.

To complete our set of basic control structures we have the `while` loop to handle repetition. While the test condition is **true**, Java repeats the body of the `while` loop, which can be either a simple Java statement or a block enclosed by curly braces.

The `for` statement and the `do-while` statement make it easier to write programs requiring repetition. The `for` statement has four parts: `initialize`, `test`, `update`, and `body`. The `initialize` part initializes, and may declare, an index variable that identifies each repetition. The `test` condition evaluates to **true** or **false**; Java repeats the execution of the body, a simple statement or a block enclosed in curly braces, as long as the condition is **true**. After each repetition, Java evaluates the `update` expression, which often increments the value of the index variable. The flexible `for` statement allows a number of variations in its use. It is ideal when we have a fixed number of repetitions.

The `do-while` statement is like the `while` statement, but it checks the test condition after executing the loop body instead of before; the loop body will be executed at least once. Java executes the body of a `do-while` statement, then terminates the loop if the test condition is **false**, but evaluates the body again if the test is **true**. The repetition will only terminate when the test condition becomes **false**. In writing a `do-while` statement, as well as in writing a `for` statement or a `while` statement, we must be very careful to make sure the test condition eventually fails or the loop will never terminate.

Build Your Own Glossary

Find the uses, in this chapter, of the terms below. Enter a definition of each in the `glossary.txt` file on the disk included with this text.

applet	database	network
architecture neutral	event-driven	object-oriented programming
byte code	FORTRAN	package
C	graphics user interface	Pascal
C++	(GUI)	procedural programming
character mode	interpreter	program-driven
COBOL	Java Virtual Machine (JVM)	random number
compiler	method	robust
component technology	Modula-2	standalone application
concatenation	Modula-3	Unified Modeling Language
console	multitasking	(UML)

Exercises

The exercises in this chapter are intended to assist in the transition from other languages to Java. They focus on basic Java constructs and procedural programming. Refer to Appendix A for a summary of basic Java constructs.

Skill Builder Exercises

1. What will be the output when the following code fragment is run?

```
int x = 12, y = 14, z;
z = y / x +7;
x = z * z;
System.out.println(x);
```

2. What will the following program fragment output?

```
int x=10, y=12, r;
if (y > x) {
    int t = y;
    y = x;
    x = t;
}
while (y != 0) {
    r = x % y;
    x = y;
    y = r;
}
System.out.println(x);
```

3. Rewrite the following `switch` statement using `if-else` statements.

```
switch(i) {
    case 1:
            j += 2;
            break;
```

```
      case 3:
            j -= 5;
            break;
      case7:
      case10:
            j *= 17;
            break;
      default:
            j = 0;
}
```

Critical Thinking Exercises

4. Fill in the blanks in the following:
   a. If x++ evaluates to 3, the value of x before the evaluation was _____, and its value after the evaluation will be _____.
   b. If ++x evaluates to 3, the value of x before the evaluation was _____, and its value after the evaluation will be _____.

5. How many times does the body of the while statement below get executed?

```
int x = 3;
while (x < 9)
   x += 2;
x++;
```
   a. 6
   b. 3
   c. 4
   d. 9
   e. none of the above

6. Which of the following for statements computes the same value for sum as shown below?

```
for (int x = 0; x < 15; x+=2) sum += x + 5;
```
   a. for (int x = 5; x < 20; x+=2) sum += x;
   b. for (int x = 5; x < 20; sum += x-2) x += 2;
   c. for (int x = 0; x < 15; sum += x+3) x += 2;
   d. all of the above
   e. none of the above

7. Which of the following do statements is equivalent to

```
y = x + 7;
x++;
while (x < 9) {
    y = x + 7;
    x++;
}
```
   a.
```
y = x + 7;
x++;
do {
    y = x + 7;
    x++;
} while (x < 9);
```
   b.
```
do {
    y = x + 7;
    x++;
} while (x < 9);
```
   c.
```
do {
    y = x + 7;
    x++;
} while ( x < = 9);
```
   d. none of the above

Debugging Exercise

8. The program below attempts to calculate the total commission received by a salesperson who earns 7 percent on sales of product A which total less than $40,000 and 10 percent of the amount above $40,000. For example, a sale of $50,000 would earn a commission of $3800. The salesperson receives a commission of 5 percent on sales of product B under $20,000, 6.5 percent on the amount of sales over $20,000 but under $50,000, and 7.5 percent on the amount over $50,000. Find and correct any errors in this program.

```java
public class Commission {
    public static void main(String [] args) {
        double salesOfA = Io.readDouble("Enter the amount of Product A sales");
        double salesOfB = Io.readDouble("Enter the amount of Product B sales");
        double amount = 0;
        if (salesOfA < 40000.00)
            amount += .07 * salesOfA;
        else
            amount = .1 * (salesOfA - 40000.0);
        if (salesOfB < 20000.00)
            amount += .05 * salesOfB;
        else if (salesOfB > 20000.00 || salesOfB < 50000.00)
            amount += 1000  + .065 * (salesOfB - 50000.00);
        else
            amount = .075 * (salesOfB - 50000.00);
        System.out.print("The commission is ");
            Io.println$(amount);
    }
}
```

Program Design Exercises

9. Write a Java program which computes the weekly pay for an employee. Input the item of hours worked. The employee receives $7.50 per hour for the first forty hours and $11.25 per hour for each additional hour.

10. Write a Java program to make change. Enter the cost of an item which is less than one dollar. Output the coins given as change, using quarters, dimes, nickels, and pennies. Use the fewest coins possible, For example, if the item cost 17 cents, the change would be three quarters, one nickel, and three pennies.

11. Write a Java program which inputs the prices of a box of cereal and a quart of milk at store A and the prices of the same items at store B. The program should output the total cost of three boxes of cereal and two quarts of milk at the store with the lower cost. Either store is acceptable if the two prices are equal.

12. Write a Java program to find the minimum of a sequence of nonnegative numbers entered by the user, where the user enters a negative number to terminate the input.

13. Write a Java program to convert kilograms to pounds or ounces. There are .45359237 kilograms in one pound, and 16 ounces in one pound. If the weight is less than one

pound, just report the number of ounces. Thus 3.4 kilograms converts to 7.4957181 pounds, while .4 kilograms converts to 14.109587 ounces.

14. Write a Java program to find both the maximum and the minimum of a sequence of nonnegative numbers entered by the user. In this solution, before reading the numbers, ask the user to input how many numbers will be input.

15. Write a Java program to find the maximum of a sequence of nonnegative numbers entered by the user. In this solution, before reading the numbers, ask the user to input how many numbers will be input.

16. Write a Java program to find the average of a sequence of nonnegative numbers entered by the user, where the user enters a negative number to terminate the input.

17. Write a Java program to enter the interest rate, the initial balance, and the final balance of an account, and calculate the number of years it will take before the account balance exceeds the specified final balance. Assume the interest is credited to the account at the end of each year.

18. Write a Java program to find the sum of the squares of a sequence of numbers entered by the user.

19. Suppose an annual inflation rate of 4 percent. Because of inflation, an item that costs $1.00 today will cost $1.04 one year from now. (We assume, for simplicity, the item we consider will rise in price exactly at the rate of inflation.) Write a Java program which inputs the cost of an item and outputs its cost three years from now.

20. Suppose you borrow $1000 at 12 percent annual interest, and make monthly payments of $100. Write a Java program to calculate how many months it will take to pay off this loan. (Each month you pay interest on the remaining balance. The interest rate is 1 percent per month so the first month you pay $10 interest and $90 goes to reduce the balance to $910. The next month's interest is $9.10, and $90.90 is applied to reduce the balance, and so on. The last month's payment may be less than $100.)

21. Write a Java program that inputs the radius of a circle and outputs its area. Allow the user to repeat the calculation as often as desired. Use Math.PI for the value of pi. The area of a circle equals pi times the square of the radius.

22. Write a Java program that inputs the radius of a circle and outputs its circumference. Allow the user to repeat the calculation as often as desired. Use Math.PI for the value of pi. The circumference of a circle equals pi times the diameter. The diameter of a circle equals twice the radius.

23. Write a Java program that inputs the radius of the base of a circular cylinder and its height and outputs its volume. Allow the user to repeat the calculation as often as desired. Use Math.PI for the value of pi. The volume of a cylinder equals the height times the area of the base.

24. The ancient Babylonians used a divide and average method for computing the square root of a positive number x. First estimate the square root by some value r; any positive estimate will do. Then compute the quotient, x/r. Averaging r and x/r gives a better estimate, so continue the process, dividing and averaging until the estimates agree to the desired number of places.

For example, to compute the square root of two

estimate 1	divide 2/1 = 2	average (1+2)/2 = 1.5
estimate 1.5	divide 2/1.5 = 1.33	average (1.5+1.33)/2 = 1.415
estimate 1.415 ...		

and so on. Write a Java program to compute the square root of a number input by the user. Use the divide and average method and stop after ten repetitions of the divide and average steps. Compare your result with the value produced by the sqrt method of the Math class.

25. Write a Java program to compute square roots, as described in Exercise 24, but stop the repetitions when two successive estimates differ by less than 1.0E-6. Also output the number of repetitions of the divide and average process.

26. The greatest common divisor (gcd) of two integers is the largest positive number that divides evenly into both numbers. For example, gcd(6,9) = 3, gcd(4,14)=2, and gcd(5,8) = 1. The Euclidean algorithm computes the gcd by a repetitive process. Find the remainder resulting from dividing the smaller number into the larger. Repeat this process with the smaller number and the remainder until the remainder is zero. The last non-zero remainder is the greatest common divisor. For example, to find the gcd of 54 and 16, the steps are

$$54 \% 16 = 6$$
$$16 \% 6 = 4$$
$$6 \% 4 = 2$$
$$4 \% 2 = 0$$

so gcd(54,16) = 2. Write a Java program to compute the greatest common divisor of two integers.

27. When we convert a fraction of the form 1/n, where neither two nor five divide n, to a decimal, we find the digits repeat a pattern over and over again. For example,

1/3 = .333333333333...	repeat pattern 3
1/7 = .142857142857142857...	repeat pattern 142657
1/37 = .027027027...	repeat pattern 027

The number of digits in the pattern for 1/n, called the period, is equal to the number of zeros in the smallest power of 10 that has a remainder of one when divided by n. For example, to find the number of digits on the pattern for 1/37, we calculate

$$10 \% 37 = 10$$
$$100 \% 37 = 26$$
$$1000 \% 37 = 1$$

Write a Java program to find the length of the repeating pattern for fractions 1/n, where neither two nor five divide n.

28. Write a Java program that converts currencies between British currency of pounds and pence, in which one pound contains 100 pence, and U.S. currency in dollars and

cents. Assume an exchange rate of 1.6595 U.S. dollars per British pound. Give the user a menu to choose the type of conversion. Allow the user to repeat as often as desired.

29. Write a Java program to perform geometric calculations. Let the user choose whether to find the area of a circle (see Exercise 21), the circumference of a circle (see Exercise 22), or the volume of a cylinder (see Exercise 23). Allow the user to repeat as often as desired.

30. Suppose you are able to pay $400 per month to buy a car. Write a Java program to determine if you can afford to buy a car which costs $15,000 if the interest rate is 6 percent and you make payments for 48 months. (*Hint:* Each month, determine how much of the $400 payment will be used to pay interest, then deduct the remaining payment from the principal.)

31. Generalize Exercise 30 to have the user input the size of the payment, the price of the car, the interest rate, and the number of monthly payments.

32. The Sturdy company invests $100,000 in a project that earns 10 percent compounded annually. Assuming the interest is allowed to compound, what is the value of the investment after seven years? Use the formula

    `V = P(1 + r/100)`^{N}

    where r is the interest rate, N is the number of years, P is the initial investment, and V is the value after N years.

33. The Sturdy company is evaluating an investment that will return $400,000 at the end of five years. The company wants to earn an interest rate of 20 percent compounded annually. How much should they pay for this investment? Use the formula

    `P = V / (1 + r/100)`^{N}

    where V is the investment's value after N years and r is the interest rate.

34. Use random numbers to simulate the toss of a fair coin. Letting the user input the number of tosses, report the percentage of outcomes that are heads. Allow the user to repeat the calculation as often as desired.

35. Calculate pi by throwing darts. In the figure below, the area of the circle divided by the area of the square is equal to `pi / 4`. Throw darts randomly at the square, counting the total number of darts thrown, and the number of darts that land inside the circle. The ratio of the latter to the former is an estimate of `pi /4`. Multiplying that ratio by four gives an estimate for pi. Use random numbers to simulate dart throwing. Get two random numbers, for the x and y coordinates. If $(x-0.5)^2 + (y-0.5)^2 < 1$, then the dart landed inside the circle. Let the user input the number of dart throws.

36. Calculate pi using the series `pi / 4` = 1 - 1/3 + 1/5 - 1/7 + 1/9 - ... and so on. Output the estimate of pi after computing 100, 1000, 10000, and 100000 terms of the series.

37. Compute e^x where

    $e^x = 1 + x/1! + x^2/2! + x^3/3! + \ldots \ldots + x^n/n! + \ldots$

    and $n! = n(n-1)(n-2) \ldots \ldots 1$, the product of the integers from one to n. (We pronounce the expression $n!$ as n `factorial`.) Let the user input the value of x, of type **double**, to use. Continue adding terms until the difference of successive terms is less than 1.0 `E-6`. Compare your answer with the Java method `Math.exp(x)`. (*Hint:* Compute each term from the previous one. For example, the fourth term is $x/3$ times the third term.)

2

Getting Started with Object-Oriented Programming

In this chapter we begin object-oriented programming by giving an intuitive sense of the object concept, then showing how to create and use objects, illustrating with `String` methods. Using two simple examples, a restaurant and an automated teller, we introduce object-oriented design with use cases and scenarios, which let us identify the objects in the system. An object-oriented program shows that control and interactions are found in each object rather than in a master controller.

Looking at these design examples shows us the need for class definitions to define the state and responsibilities of our objects. Writing a simple `BankAccount` class introduces the basic concepts of instance variables and methods, constructors, and overloading. We then complete the restaurant code to show a complete object-oriented program while introducing some UML diagrams. We conclude by discussing object composition, designing classes that include other classes as fields. In the next chapter, we continue with object-oriented concepts, including inheritance, polymorphism, abstract classes, interfaces, and design with use cases and scenarios to complete the ATM example.

OBJECTIVES

◆ Understand the concept of an object.

◆ Create and use objects.

◆ Learn selected `String` methods.

◆ Understand Java primitive type variables hold values, while object variables hold references.
◆ Begin object-oriented design with use cases, scenarios, and UML diagrams.
◆ Understand the difference between a concept and its instances.
◆ Understand that a class defines a type of object.
◆ Write a class, including instance variables and methods, and constructors.
◆ Use object composition.

2.1 ■ Using Objects

We introduce the object concept with an intuitive analogy to a vending machine, and then illustrate with string objects.

Defining an Object

An **object** has state, behavior, and identity. We sometimes speak of our state of mind, happy, sad, angry, meaning a particular condition we are in. Humans are rather complex objects. For simplicity let us deal with inanimate objects such as vending machines, or even conceptual objects such as bank accounts or character strings.

A vending machine has a state. It may be operational or broken. It may have plenty of candy or be out of candy. It may be able or unable to make change. Some information about its state may be public, for example a big sign may state it is out of order, or we may be able to see the number of candy bars available. Other information may be private, such as the amount of coffee available, or the amount of change.

FIGURE 2.1 A vending machine object

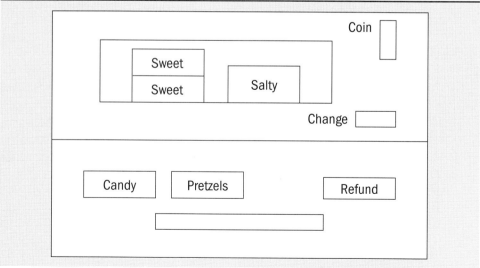

A vending machine has behavior. It does not initiate behavior, but it does provide services which we think of as its behavior. For example, we may select coffee, or candy, or ask for a refund.

Most of the vending machine's behavior is public. After all, the purpose of a vending machine is to provide services. However, there might be some private behavior, used only internally, such as filling water from a pipe when needed.

A vending machine has an identity. Another vending machine may have exactly the same state and provide the same services, but it is a different machine.

Although we do not (yet) have vending machines for programs, they do at least provide an instructive model for object-oriented programming. When we select coffee, we do not need to know the details of how the vending machine provides the coffee, whether the coffee is brewed in a big urn, whether instant coffee is mixed with hot water, or even as young children may think, whether little people in the machine make it.

As customers, we use the vending machine's services without needing to know the details of the machine's construction. The buttons or levers of the vending machine provide an interface to its behavior. Vending machine users do not have to manage the details of how to make the coffee or of how to dispense the candy.

Vending machines are handy, especially when we are hungry or thirsty, but we can use Java objects to illustrate the same principles. Java libraries provide many types of objects. For example, the String class, in the java.lang package, has many operations for manipulating strings of characters.

Strings

Before getting to the Java details, we look at a string informally as we did the vending machine. Inside a String object is a sequence of characters, such as "Java is fun". The state of a String object is private. Perhaps the letters of the string are wooden blocks numbered to indicate their position in the string, so block 'J' would have number 1, block 'a' would have number 2, and so on. Users of String objects do not need to know how the characters are represented.

Strings have many public operations to provide their services. We can ask the length of a string, or ask for the first position which contains the letter 'a'. The drawing of the String object, "Java is fun" in Figure 2.2 does not have a window to see inside, because strings do not show any of their state; it is all private. Strings provide many operations; we show only a few.

FIGURE 2.2 A String object (for "Java is fun")

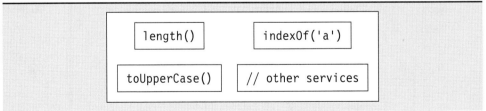

When we press the length() button in Figure 2.2 we get the result of 11 because "Java is fun" has 11 characters. The result of an operation depends on the state of the object. For this object the indexOf('a') operation returns 1. The first 'a' occurs as the second character, but for technical reasons we start the numbering with 0, so 'J' is at index 0, and 'a' is at index 1.

Strings in Java never change. The String object in Figure 2.2 will always represent "Java is fun". When we execute the toUpperCase() operation, we do not change the object, but rather we get a new String object representing "JAVA IS FUN" shown in Figure 2.3.

FIGURE 2.3 A String object (for "JAVA IS FUN")

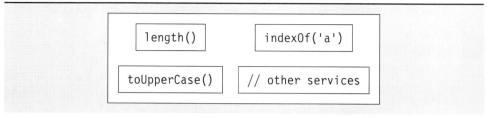

Because the state is hidden, Figure 2.3 looks just like Figure 2.2, but it operates differently. Executing the indexOf('a') operation will return −1, signifying the character 'a' does not appear in "JAVA IS FUN". Java is case-sensitive; a lowercase 'a' differs from an uppercase 'A'.

Visualizing a String object as a vending machine really helps us to keep in mind the object concept: state, behavior, and identity. To use strings in Java programs, we need to learn the notation used to create a string, and to ask it to provide one of its services. The declaration

```
String s = "Java lets us use objects.";
```

creates and initializes a String, which we refer to as s, while the method

```
s.length();
```

will return the number of characters in s, 25 in this example.

Note the object-oriented flavor of the invocation s.length(). We ask the object referred to by s to determine its length. Finding its length is one of the behaviors a string can exhibit.

FIGURE 2.4 Selected String methods

```
public char charAt(int index)              character at the specified index
public int compareTo(String anotherString) 0 if equal to anotherString
                                           negative if less
                                           positive if greater
public boolean equals(Object anObject)     true for equal Strings
public int indexOf(char ch)                index of first occurrence of ch
public int indexOf(char ch, int from)      index of first occurrence of ch starting
                                               at index from
public int indexOf(String str)             index of first occurrence of str
public int indexOf(String str, int from)   index of first occurrence of ch starting
                                               at index from
public int length()                        string length
public String substring                    new string with characters from
   (int beginIndex, int endIndex)              beginIndex to endIndex – 1
public String toLowerCase()                converts to lowercase
public String toUpperCase()                converts to uppercase
public String trim()                       removes leading and trailing
                                               whitespace
public static String valueOf(int i)        creates a string from an int
public static String valueOf(double d)     creates a string from a double
```

The syntax we use can serve as a reminder. We might think of the dot in the expression s.length() as a vending machine button. Calling the length method is like pressing a vending machine button. The length method is part of the String object, and its result depends on the state of that object. Just as it would not make any sense to push a candy button not connected to the vending machine, trying to call length() instead of s.length() would not connect with any string and would be meaningless.

Figure 2.4 shows some of the many String methods. Figure 2.5 presents examples of the use of these methods on the string given by

```
String s = "Java lets us use objects.   "
```

Each method in these examples refers to a specific String object. We call these methods **instance methods**, because each refers to a specific instance of a string. Just as selection buttons are part of a specific instance of a vending machine, instance methods are part of a specific instance of an object. In the expression s.length(), the length request goes to the string referred to by s, not to some other string. Not all the methods of Figure 2.4 are instance methods. We leave consideration of the valueOf methods to a later section.

FIGURE 2.5 Examples of String methods

Method	Return Value	Description of Return Value
s.charAt(6)	'e'	The character at position 6 is an 'e'.
s.compareTo("Toast")	negative integer	The string referred to by s is alphabetically less than "Toast" so the return value is a negative integer.
s.equals("Java is fun")	false	The string referred to by s does not have the same characters as "Java is fun".
s.indexOf('e')	6	The leftmost 'e' in the string occurs at index 6.
s.indexOf('e',8)	15	The first occurrence of 'e', starting from index 8, is at index 15.
s.indexOf("us")	10	The leftmost occurrence of "us" starts at index 10.
s.indexOf("us",11)	13	The first occurrence of "us", starting from index 11, begins at index 13.
s.indexOf("us",15)	–1	There is no occurrence of "us" starting at index 15.
s.length()	27	This string contains 27 characters.
s.substring(13,24)	A new String with characters "use objects"	The string "use objects" starts at index 13 and continues up to index 24.
s.toLowerCase()	A new String with characters "java lets us use objects. "	Returns a new string with all lowercase characters.
s.toUpperCase()	A new String with characters "JAVA LETS US USE OBJECTS. "	Returns a new string with all uppercase characters.
s.trim()	A new String with characters "Java lets us use objects."	Returns a new string with the blanks removed.

To illustrate the use of objects, Example 2.1 replaces every occurrence of the word "fish", in a string entered by the user, with the word "fowl". Our Io class, found on the disk included with this text, has a method, readString, to input a string from the keyboard. Example 2.1 demonstrates the use of String objects, but is not an object-oriented program in which objects cooperate to perform a task. We will see an object-oriented program in the next section.

The Big Picture

An object has state, behavior, and identity. We picture a vending machine in a given configuration which provides services and has an identity distinct from other vending machines. A String object hides its state, but provides many methods to access its services.

We ask an object to perform an operation using one of its instance methods. Such an instance method refers to that object. Thus s.length() returns the length of a specific String object referred to by s. Just as a selection button is part of a vending machine, an instance method is part of an object.

EXAMPLE 2.1   Replace.java

```java
/* Replaces every occurrence of "fish" with "fowl"
 */

import iopack.Io;
public class Replace {
    public static void main(String[] args) {
        String s = Io.readString("Enter a String which includes \"fish\" ");
        int length = s.length();
        int position = s.indexOf("fish");                              // Note 1
        while (position != -1) {
            s = s.substring(0,position) + "fowl"
                + s.substring(position+4, length);                     // Note 2
            position = s.indexOf("fish", position + 4);                // Note 3
        }
        System.out.println("The new string is:");
        System.out.println('\t' + s);
    }
}
```

Output—First Run
```
Enter a String which includes "fish" :  A fish is nice
The new string is:
        A fowl is nice
```

Output—Second Run
```
Enter a String which includes "fish" :  I like fish today and fish tomorrow.
The new string is:
        I like fowl today and fowl tomorrow.
```

Note 1: We initialize position with the index of the first occurrence of "fish", or −1 if it does not occur in s.

Note 2: When we find "fish" in the string s, we create a new string with that occurrence of "fish" replaced by "fowl". We have already used the string concatenation operator, +, in `println` statements.

Note 3: We continue the search from the character after "fowl".

Values vs. References

Java uses variables for objects such as strings differently than variables for basic data types such as integers. Let us review how we created variables to hold integers. The statement

```
int x = 4;
```

declares a variable, x, of type integer, and initializes it with the value of 4. Inside the computer, the variable x has a storage location and the computer places the value 4 into that location. Here the variable holds the value 4. A real world analogy might involve me asking at the dinner table to please pass the salt, and my wife graciously handing me the salt shaker. My wife can easily move the small salt shaker to my hand. My hand is like the variable x, and the salt is like the value 4.

By contrast, suppose I would like to stand next to the Grand Canyon. Even if I had a hundred helpers, they could not pass me the Grand Canyon. If I want to stand next to the Grand Canyon, I have to go there. In fact my wife might pass me a map showing how to get to there. When I hold that map in my hand I am holding a reference to the Grand Canyon, telling me where it is.

In Java, an object is like the Grand Canyon. An object may be large, so we do not want to pass it around. We put it in one place and tell users where it is. A variable for an object holds a reference to that object, telling where to find it in the computer's memory. If we declare

```
String myString = "We want a big car";
```

then the variable `myString` contains a reference to a `String` object, rather than the object itself. We leave the object in one place, and variables in our program refer to it.

When we declare and initialize an integer variable,

```
int x = 4;
```

we get a single entity that holds an integer value (see Figure 2.6a). When we declare and initialize a `String` variable,

```
String myString = "We want a big car";
```

FIGURE 2.6 Value vs. reference

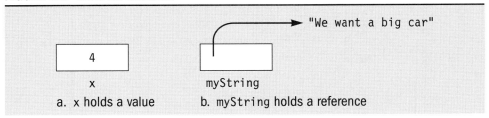

a. x holds a value b. myString holds a reference

Figure 2.6b shows we get two entities, an object representing "We want a big car" and a variable, myString, that refers to that String object. We show the reference as an arrow pointing to the object. The reference has a name, myString, while the string itself remains nameless.

A big difference between **int** variables which hold values and string variables which hold references occurs as a result of assignment. The assignment operation copies the contents of one variable to another. For primitive types such as **int**, assignment copies the integer value from one variable to another, as shown in Figure 2.7. The effect of the assignment statement y = x; is to copy the value 4, stored in x, to y, overwriting the previous value of 5 that y had.

FIGURE 2.7 Assignment of an integer

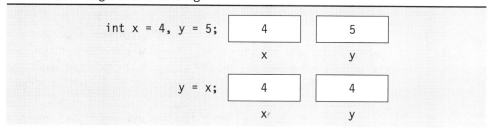

For object types such as String, assignment still copies the contents of one variable to another. A string variable holds a reference, so assignment copies that reference as Figure 2.8 shows. After the assignment, the variable t has the same reference as s. Both refer to "soup". The string "fish" has no references to it and is termed **garbage**. Java will automatically reclaim the memory used for "fish" with a process known as a **garbage collection**.

FIGURE 2.8 Assignment of a string

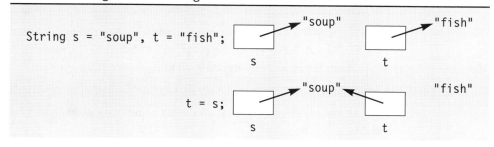

Comparing Figures 2.7 and 2.8, we see that, after the assignment, x and y each have their own copy of the value 4, but that s and t refer to the same string "soup". Figure 2.8 uses strings, but the same sharing would result from the assignment of any type of object, whereas no sharing results from assignment of primitive types.

We access objects through references to them. For example, to find the number of characters in "We want a big car" we invoke

myString.length()

accessing the string using the myString variable.

If we declare a string without initializing it, as in

```
String s;
```

then we have a variable, s, which does not refer to any string yet, as shown in Figure 2.9. Java uses **null** to represent the value of a reference before we create an object for it to refer to, so s is **null** until we give it a reference to a string.

FIGURE 2.9 An object declaration without object creation

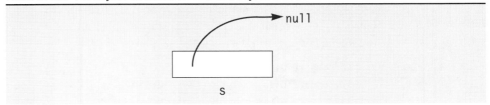

Comparing Strings

Often we need to check two strings for equality, meaning both strings have the same characters at each position. For example, for the strings s1, s2, and s3, given by

```
String s = "a houseboat";
String s1 = "house";
String s2 = s.substring(2,7);
String s3 = "horse";
String s4 = s1;
```

we would like s1 and s2 to be equal, because each has the same characters, but expect s1 to be unequal to s3, because these strings differ at index 2. Of course, s1 and s4 are equal, because they refer to the same string.

The equality operator, ==, which we have used many times to test the equality of primitive values, can mislead us when applied to strings. This operator compares the contents of each variable, returning **true** if both are the same. Figure 2.10 illustrates the problem.

FIGURE 2.10 s1 == s4 but s1 != s2

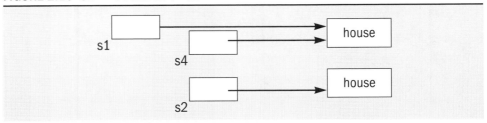

Object variables hold references to objects. The equality operator, ==, compares the contents of these variables, comparing the references, not the objects to which they refer. Both s1 and s4 refer to the same object, so s1 == s4 evaluates to **true**. The expression s1 == s2 will be **false** as these references are not equal; they refer to different objects. While these objects are not identical they are equal because they each contain the same sequence of characters. We need a replacement for the equality operator which compares the objects themselves, not the references to them.

The `equals` method solves our problem. It returns **true** if both strings have the same sequence of characters. Thus `s1.equals(s4)` and `s1.equals(s2)` will both return **true**. Figure 2.4 shows the argument to the `equals` method is of type `Object`. We will discuss the `Object` type in Section 3.1 on inheritance and will see a `String` is a type of `Object` so we can use `String` arguments such as `s4` and `s2`, as we did above.

In applications such as alphabetizing we need to determine whether one string is greater than another. Java provides a `compareTo` method which, comparing strings lexicographically, returns a negative number if the string is less than the string argument, returns zero if both strings are equal, and returns a positive number if the string is greater than the string argument. For our strings `s1`, `s2`, and `s3` above,

`s1.compareTo(s2)`	returns zero
`s1.compareTo(s3)`	returns a positive number
`s3.compareTo(s1)`	returns a negative number

The Big Picture

Variables declared with primitive types such as **int** hold values of those types. We can compare values using the == operator.

   String variables, and those with other object types, are references to objects which are themselves nameless. To compare `String` values we use the `equals` method.

TEST YOUR UNDERSTANDING

1. Given the `String` object `String s = "The three did feed the deer";` find

    a. `s.length()`   c. `s.indexOf('e')`   e. `s.substring(4,9)`
    b. `s.charAt(5)`   d. `s.indexOf("did")`   f. `s.toUpperCase()`

2. Given the `String` objects

    ```
    String s1 = "time";
    String s2 = "time";
    String s3 = s2;
    ```

    find

    a. `s1 == s2`   c. `s2 == s3`   e. `s1.equals(s3);`
    b. `s1 == s3`   d. `s1.equals(s2)`

3. Given the `String` objects

    ```
    String s1 = "Happy days";
    String s2 = "Hello world";
    String s3 = "77 Sunset Strip";
    ```

    determine whether the value returned by each of the following calls to the `compareTo` method returns zero, a negative value, or a positive value.

    a. `s1.compareTo(s2);`   b. `s2.compareTo(s3)`   c. `s1.compareTo("Happy days");`

2.2 ▪ Introduction to Object-Oriented Programming

Computer programmers must develop ever larger systems with more capabilities. They seek programming techniques that allow them to be highly productive, achieving quality programs. The procedural approach, used since the first high-level languages were introduced in the late 1950s, is being challenged by object-oriented programming, an approach first discussed in the 1960s, but which became popular in the late 1980s and the 1990s.

In fiction, each story has a point of view. A single person can tell the entire story from his or her point of view, with the thoughts of the other characters never appearing. Another approach is to have each character express his or her thoughts and feelings as well as show actions. Procedural programming is programming from a single point of view, using a controller that orchestrates all the action while the data acted on is passive. Object-oriented programming puts the data inside objects that actively manage that data by providing services to other objects. Rather than having a single controller manipulating passive data, an object-oriented program lets the objects interact with each other, each meeting its responsibilities. Object-oriented programming combines the data and its operations into objects.

Object-Oriented Design

Bertrand Meyer gives an example of a payroll program which produces paychecks from time-cards.[1] Management may later want to extend this program to produce statistics or tax information. The payroll function itself may need to be changed to produce weekly checks instead of biweekly checks, for example. The procedures used to implement the original payroll program would need to be changed to make any of these modifications. Meyer notes that any of these payroll programs will manipulate the same sort of data, employee records, company regulations, etc.

Focusing on the more stable aspect of such systems, Meyer states a principle

> "Ask not first what the system does:
> Ask WHAT it does it to!"[2]

and a definition

> "Object-oriented design is the method which leads to software architectures based on the objects every system or subsystem manipulates (rather than "the" function it is meant to ensure)."[3]

Use Cases and Scenarios

In this section we model a customer order at a fast food restaurant, showing how objects interact and call one another's services, leading to a true object-oriented simulation. We first need to identify the types of objects we will need, which we do by considering typical uses of the system, called **use cases**, and various **scenarios** which show, in a step-by-step manner, the interactions among objects that take place relating to that use case.

---

[1] Bertrand Meyer, Object-Oriented Software Construction, Second Edition, Prentice-Hall, 1997, p. 105.

[2] Ibid., p. 116.

[3] Ibid., p. 116.

When designing using objects, we may start by identifying the use cases. In our example, one use case would be a customer placing an order. To better understand each use of the system we describe typical scenarios. For each use case, we develop a primary scenario showing the normal interactions, and several secondary scenarios which list the interactions that take place in more unusual or error situations.

For our restaurant example, a normal scenario, shown in Figure 2.11, would involve a customer entering the restaurant, walking up to the counter, and placing an order.

FIGURE 2.11 A scenario for a fast food order

```
The customer orders a burger, soda, and fries from the waiter.
The waiter asks the cook to make a burger.
The waiter serves the soda to the customer.
The waiter asks the cook to make the fries.
The cook gives the waiter the burger to serve.
The cook gives the waiter the fries to serve.
The waiter asks the customer to pay.
```

From this scenario, we identify three objects, a customer, a waiter, and a cook. The scenario also shows the responsibilities of each object, and we can easily determine the information each object needs to maintain in its state. The customer has money to buy food, the waiter has cash received, and the cook has a supply of burgers and fries.

FIGURE 2.12 State and responsibilities for Customer, Waiter, and Cook

	Customer	**Waiter**	**Cook**
State	money	cash	burgers and fries
Responsibilities	place an order	take an order	make burgers
	pay	serve an item	make fries

An object-oriented program to simulate this use of a restaurant would create the customer, waiter, and cook objects and let them interact as the scenario indicates. In the last section we visualized objects as vending machines with buttons for their services. We depict them in Figure 2.13 more schematically as black boxes with a handle for each service. In design, a black box signifies we cannot see inside at the workings of the object but must use its services.

FIGURE 2.13 Objects showing only available services

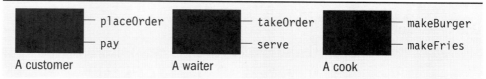

A customer   A waiter   A cook

Example 2.2 creates a customer, waiter, and cook and lets them interact.

EXAMPLE 2.2   **QuickFood.java**

```
/* Uses the Cook, Waiter, and Customer classes to be defined in
 * Section 2.4. A customer places an order.
 */
```

```java
public class QuickFood {
  public static void main(String[] args) {
    Cook aCook = new Cook();                          // Note 1
    Waiter aWaiter = new Waiter(cook);
    Customer aCustomer = new Customer(waiter);
    aCustomer.placeOrder();                            // Note 2
  }
}
```

Output

```
Waiter places BURGER order
Cook making BURGER
Waiter serves BURGER
Waiter serves SODA
Waiter places FRIES order
Cook making FRIES
Waiter serves FRIES
Customer pays $4.17
```

Note 1: The **new** operator creates a new object. We will discuss it in the next section.

Note 2: We start execution of this program by invoking the `placeOrder` behavior of a `Customer` object. The objects then collaborate to complete the scenario.

Example 2.2 illustrates, in a striking manner, how an object-oriented program leaves it to the objects to call one another's services. The `main` method creates three objects and starts the process with one line

```java
customer.placeOrder();
```

This one invocation gets the customer to communicate with the server who then communicates with the cook, and so on. The `main` method is not the master controller, but rather each object contains services called by other objects.

Where do we define these services and the items that make up the state of the objects? In Java, each object is an instance of a class which defines the state variables and service methods for that type of object. For example, a `Customer` class defines the state and behavior for `Customer` objects, as do the `Waiter` and `Cook` classes for `Waiter` and `Cook` objects. In the next section we go carefully over the coding of classes, and in Section 2.4 we write the `Customer`, `Waiter`, and `Cook` classes where we define the services and state for objects such as aCustomer, aWaiter, and aCook, that we used in Example 2.2. First we introduce another example of analysis and design to which we shall return several times in this text.

Scenarios for an ATM system

Before getting into the details of object-oriented programming in Java, we start another design, previewing the automated teller machine simulation we design in Section 3.5. The familiar situation is that a user inserts a card into the teller machine, the teller asks the user for a personal identification number (PIN), the user enters his or her PIN, and so on until the transaction is complete. Going through complete scenarios of typical uses of the system helps us to find the objects and identify their responsibilities. Figure 2.14 shows such a scenario for a deposit transaction.

FIGURE 2.14 A scenario for a successful deposit

```
The user asks the teller to accept an ATM card.
The teller asks the user to enter a PIN.
The user asks the teller to accept a PIN.
The teller asks the user to select a transaction type.
The user asks the teller to accept a deposit.
The teller asks the user to select an account type.
The user asks the teller to accept a savings account type.
The teller asks the bank to find the bank account of the chosen type for the
    user with the specified PIN.
The bank gives the teller a reference to the account.
The teller asks the user to specify an amount.
The user asks the teller to accept an amount.
The teller asks the account to deposit the specified amount.
The teller asks the user to select another transaction ...
```

Looking at the scenario of Figure 2.14, we identify four objects: user, teller, bank, and bank account. From this scenario we can identify the responsibilities of each object as shown in Figure 2.15.

FIGURE 2.15 Responsibilities derived from the scenario of Figure 2.14

User	Specify a PIN
	Select a transaction type
	Select an account type
	Specify an amount
Bank	Find a specified account
Account	Deposit an amount
Teller	Accept an ATM card
	Accept a PIN
	Accept a transaction type
	Accept an account type
	Accept an amount
	Accept an account

By writing scenarios for other types of transactions we would find other responsibilities for our objects, and perhaps other objects. For example, an account object will have the additional responsibilities of

```
Withdraw an amount
Get the account balance
```

We will look more closely at such an account object in the next section.

To implement the ATM system, we need to create User, Bank, Account, and Teller objects, that we identified using a use case and scenarios, and let them interact. The main method for the ATM system, like main in Example 2.2, will be very short, because rather than a single controller manipulating data, each object keeps its own data and invokes services of other objects.

Now we are ready to see how Java lets us define and create objects, which leads us to the classification process.

Classification

Many of our everyday concepts involve a classification of objects. When we sit on a chair, we are sitting on a specific instance of the Chair concept. The particular chair may be hard or soft, expensive or inexpensive. It provides a service, a place to sit, and has a state, occupied or unoccupied. A lamp, an instance of the lamp class, is not a chair. Its service is light and its state is on or off. We use language to categorize groups of objects. Natural language is imprecise, but we all generally agree on what we call a chair or a lamp.

In Java we must be more precise, but the idea is similar. Java uses a class to define the state and behavior of a type of object. The word *class* evokes the idea of classification. To understand how to build a class to define a type of object, we focus first on the BankAccount class.

A BankAccount class must describe the components of the state and behavior for the various BankAccount objects we create. In Java, variables save the state of an object, and methods implement its behaviors. Informally, Figure 2.16 shows how the BankAccount class implements the state and behaviors for each of its instances.

FIGURE 2.16 Implementing BankAccount state and behavior

balance	A decimal value representing the dollar amount of the current balance for that account.
deposit an amount	Add the amount to the current balance.
withdraw an amount	If the amount is not greater than the current balance, subtract the amount from the current balance.
getBalance	Get the current balance.

The Unified Modeling Language (UML) has become the standard for object-oriented modeling; we will use UML diagrams in this text.[4] Figure 2.17 represents the class of all bank accounts, showing their common structure.

FIGURE 2.17 The BankAccount class

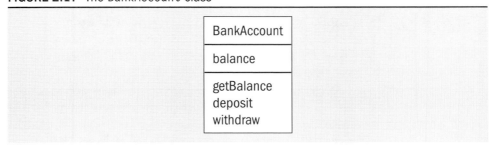

The top section in Figure 2.17 gives the class name, the middle section lists the variables used to represent the state of an account, and the bottom section lists each account's operations used to provide its services. We say each bank account object is an instance of the BankAccount

---

[4] See http://www.rational.com/uml/ for more information on the UML.

class. The class describes the state (data) and behavior (operations) each object instance possesses; it is like a pattern or template specifying the state and behavior of each of its instances.

Just as we may point to a favorite chair, confident it fully exemplifies the Chair concept, we create each BankAccount object to instantiate the BankAccount class. In the UML notation, objects, as pictured in Figure 2.18, also have three-part diagrams. In the top part we name the object and indicate the class that defines it, underlining both as in

```
myAccount : BankAccount
```

The middle section shows the balance with a specific value, for example

```
balance = 24.50
```

The BankAccount class specifies each account must have a balance. In the object itself, the balance has a specific value. This is analogous to the concept of a chair which specifies a seat and legs, contrasted with an actual chair object that has a hard seat and curved legs.

The third part of the object diagram lists the services the object provides. Each BankAccount object can deposit, withdraw, or get its balance.

The objects myAccount and yourAccount of Figure 2.18 are instances of the BankAccount class. In the next section we will write the Java code that implements the BankAccount class.

FIGURE 2.18 Two BankAccount objects

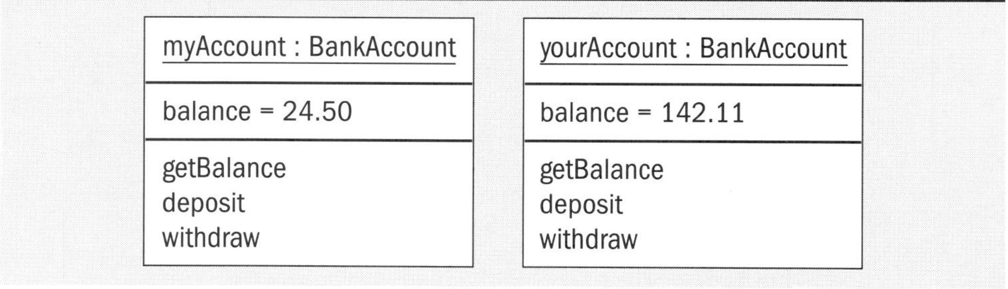

myAccount : BankAccount	yourAccount : BankAccount
balance = 24.50	balance = 142.11
getBalance deposit withdraw	getBalance deposit withdraw

The Big Picture

In designing an object-oriented program, we write use cases listing the desired uses of the system. For each use case, scenarios show the steps for successful and unsuccessful interactions with the system. We identify objects and their responsibilities from these scenarios.

Because there is no master controller, the main method of an object-oriented program is very short, just creating some objects and letting them invoke each other's services.

Just as a particular chair exemplifies the properties of the Chair concept, objects are instances of a class that defines their state and behavior. We define a class, to implement a type for each object, which specifies the data and implements operations for each object of that type.

The Unified Modeling Language (UML) is the standard notation for object-oriented design.

4. Write a scenario in which the customer orders a burger, fries, and a soda, but the drink machine is broken.

5. As a start for an object-oriented program, write a scenario describing the interactions when a customer rents a car. What objects, with what responsibilities, do you identify from this scenario?

2.3 ▪ Classes and Objects in Java

Our scenarios from the previous section showed objects invoking each other's services. One of the responsibilities of a bank account is to provide a deposit operation which the teller can invoke. Each bank account has a state given by the value of its balance. The BankAccount class defines the state and behavior of bank account objects. In this section we develop and test the BankAccount class, leaving the full development of the ATM system, which uses inheritance and other object-oriented concepts discussed in Chapter 3, until Section 3.5. In the next section we implement the classes for the QuickFood application of Example 2.2.

The Structure of the BankAccount Class

Our analysis in the last section used a BankAccount object in the scenarios for the ATM system. In discussing classification, we saw that, in Java, a class defines the state and behavior of its object instances. To use BankAccount objects we need to specify them in a BankAccount class. The BankAccount class we will develop in this section will have a structure shown in Figure 2.19.

FIGURE 2.19 The BankAccount class

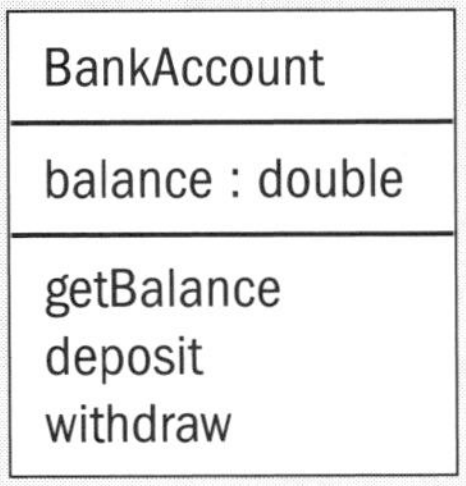

This BankAccount class specifies the account balance to represent the state of each account, and deposit, withdraw, and getBalance as an account's behavior. The UML notation allows us to specify the type of the state variables; here balance is of type **double**. For simplicity we do not show the arguments to the methods or their return types.

The overall structure of the BankAccount class in Java is

```
public class BankAccount {
    // put code to specify BankAccount objects here
}
```

A class is like a dictionary entry. The dictionary entry for *chair* defines what it means to be a chair. The BankAccount class defines what it means to be a bank account object. Each bank account has state, behavior, and identity. In Java, we use instance variables to hold the state of an object, and instance methods to implement the services needed to meet its responsibilities, which is what we call its behavior. We also have constructor operators to create bank accounts each with its own memory location distinct from other objects. Thinking of the BankAccount class as a definition of bank account objects, it will have the structure:

```
public class BankAccount {

    // instance variables go here    (for state)
    // constructors go here          (for identity)
    // instance methods go here      (for behavior)

}
```

We will now discuss each of the three parts of the BankAccount class: instance variables, instance methods, and constructors.

Instance Variables

An **instance variable**, also called a **field**, is a variable declared inside the class but outside of any method. The name *instance variable* reminds us these variables will be part of object instances. Analogously, the Chair concept includes mention of legs, but the concept does not have legs, only actual chairs do.

Our programs in Chapter 1 have used variables, but we have always declared these variables inside the main method. We call these variables **local variables** because they are declared inside one method, and cannot be used outside that method.

In Java, we declare instance variables to hold the state of an object. For a bank account we might declare

```
double balance;
```

to hold an account balance. The declaration (incomplete) for the bank account class shows how we declare the account balance outside of any method.

```
public class BankAccount  {
    private double balance;

    //  fill in the rest of the declaration here
}
```

In contrast to local variables, instance variables are available to all services of an object. Making a deposit to a bank account will increase its balance while making a withdrawal will reduce it.

Note the use of the modifier private. Using private signifies we want to keep the account balance data hidden within the object, accessible only by the services the object provides. The object appears as a black box to a user of its services. The user of an account can only inspect its balance by calling the object's getBalance method, and can only change its balance by making a deposit or a withdrawal.

Instance Methods

To specify an object's behavior we declare methods inside the class, such as a deposit method in the BankAccount class.

```
public class BankAccount {
    private double balance;

    public void deposit(double amount) {
        balance += amount;
    }

    // fill in the rest of the declaration here
}
```

The deposit method adds the amount passed in as a parameter to the balance. We call it an instance method because it is part of a specific bank account instance. Each bank account will have its own balance and its own deposit method. To put it simply, when I make a deposit into my account I want the money to increase my balance, not yours.

Our BankAccount class defines the concept of a bank account. Analyzing the ATM system with use cases and scenarios shows us each bank account should provide services to get its balance and to make a withdrawal, in addition to the deposit service. Thus we must add getBalance and withdraw operations to the BankAccount class.

The code for getBalance() simply returns the balance

```
public double getBalance() {
    return balance;
}
```

The withdraw method uses an if-else statement to check whether the account balance is large enough to make the withdrawal, printing an error message otherwise.

```
public void withdraw(double amount) {
    if (balance >= amount)              // check for sufficient funds
        balance -= amount;
    else
        System.out.println("Insufficient funds");
}
```

With the addition of the getBalance and withdraw methods the BankAccount class now has the structure

```
public class BankAccount {
    private double balance;

    public void deposit(double amount) {
        balance += amount;
    }
    public double getBalance() {
        return balance;
    }
    public void withdraw(double amount) {
        if (balance >= amount)
            balance -= amount;
        else
            System.out.println("Insufficient funds");
    }

    // fill in the rest of the declaration here
}
```

Constructors

Our BankAccount class now defines the concept we identified in the analysis of the ATM system. It provides an instance variable, balance, for the state, and getBalance, deposit, and withdraw methods for the services. However, we do need to add code to enable us to create and initialize BankAccount objects. These special methods, called **constructors**, always have the same name as the class, and never have a return value.

Every bank account has a balance. We can use the constructor to initialize the balance of a new account. The BankAccount constructor

```
public BankAccount () {
    balance = 0.0;
}
```

initializes the balance of a new BankAccount object to zero. We call a constructor with no arguments a **default constructor**.

We would also like a constructor to create a bank account with a specified initial balance. Using method overloading, which we discuss later in this section, we can use the constructor

```
public BankAccount(double initialAmount) {
    balance = initialAmount;
}
```

to create a bank account with a balance initialized to initialAmount. Figure 2.20 shows the class diagram for the BankAccount class as we developed it in this section.

FIGURE 2.20 The revised BankAccount class

BankAccount
balance : double
BankAccount() BankAccount(initialAmount : double) getBalance() : double deposit(amount : double) : void withdraw(amount : double) : void

We usually do not show this amount of detail in our class diagrams. The **UML** notation uses a different style than Java for specifying method parameters and return values. Whenever possible we omit parameters and return values from class diagrams to avoid confusion.

Example 2.3 shows the Java code for our BankAccount class. This class has no main method. We can compile it, but not execute it. Remember the BankAccount class defines BankAccount objects. We need to learn how to create and use BankAccount objects.

EXAMPLE 2.3 **BankAccount.java**

```
/* Declares a BankAccount class with an account balance,
 * two constructors, and getBalance, deposit, and
 * withdraw operations.
 */
```

```
import iopack.Io;
public class BankAccount {
    private double balance;                                      // Note 1

    public BankAccount()    {                                    // Note 2
        balance = 0;
    }
    public BankAccount(double initialAmount) {                   // Note 3
        balance = initialAmount;
    }
    public double getBalance() {                                 // Note 4
        return balance;
    }
    public void deposit(double amount) {                         // Note 5
        balance += amount;
    }
    public void withdraw(double amount) {
        if (balance >= amount)
            balance -= amount;
        else
            System.out.println("Insufficient funds");
    }
}
```

Note 1: We declare the variable balance outside of any method. Each object has its own balance variable which stores the balance for that specific account. We declare balance private so only BankAccount operations can use it.

Note 2: BankAccount() is a constructor. A constructor has no return value, not even **void**. It has the same name as the class, in this case, BankAccount. We need a BankAccount constructor to initialize the balance because balance is a private field, and can only be changed by BankAccount operations. A constructor is a special operation which we use when we create a new object as in the expression new BankAccount().

Note 3: This constructor overloads the name BankAccount, but it has a parameter giving the initial balance for the new account, so Java can tell the difference between it and the BankAccount constructor with no parameters. We discuss overloading later in this section.

Note 4: We include the getBalance method to tell us the account balance. Because balance is a private variable, we can only access it by using a method which is a member of the BankAccount class. The getBalance method has no parameters but returns a double value which is the balance.

Note 5: The deposit method refers to a specific BankAccount object. It adds the specified amount to the balance of that specific bank account. Depositing to my account will increase my account balance, while depositing to your account will increase yours.

☞ **STYLE**   Our BankAccount class declares the data first, then the constructors, and finally the other methods. Even though we cannot directly use the private data, we like to place it in an easy to spot location at the top of the class definition. Many programmers prefer a different style, which places the private data after the methods, which are usually public. The public methods provide the interface the programmer will use directly, and some feel that for this reason they should be at the top. For your programs choose one of these styles and use it consistently.

Using BankAccount Objects

Now that we have defined the BankAccount class, we can create BankAccount objects and invoke their services.

Creating Objects

For creating objects Java provides an operator, appropriately enough called **new**. We can declare a new bank account with the expression

```
new BankAccount();
```

which creates the bank account object, reserving space in memory, and calling the default BankAccount constructor to initialize it. This expression creates a new bank account, but does not tell anyone where it is. Figure 2.21 shows there are no references to this new account, and it has no name.

FIGURE 2.21 Result of new BankAccount()

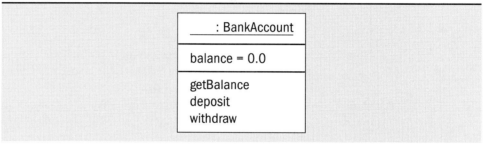

To use a BankAccount object we must declare a variable to refer to it, as in the statement

```
BankAccount myAccount = new BankAccount();
```

which declares the variable myAccount to have type BankAccount and to refer to a specific new BankAccount created with the **new** operator. The **new** operator allocates space in memory for a BankAccount object, calls the default constructor, and returns a reference to the object which the assignment operator stores in the variable myAccount. The myAccount variable has a reference to the newly created BankAccount, so it knows where to find that account in the computer's memory. We say the account is an **instance** of the BankAccount class. Figure 2.22 shows the variable myAccount referring to a newly created BankAccount instance.

FIGURE 2.22 myAccount refers to a new BankAccount

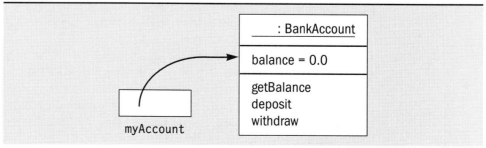

☛ **TIP**

We always use the **new** operator to create an object. For String objects, we used a special form, for example

```
String fruit = "apple";
```

which is equivalent to

```
String fruit = new String("apple");
```

Declaring a BankAccount variable without creating a BankAccount object for it to refer to, as in

```
BankAccount anAccount;
```

would cause anAccount, shown in Figure 2.23, to have the value **null**.

FIGURE 2.23 An object declaration without object creation

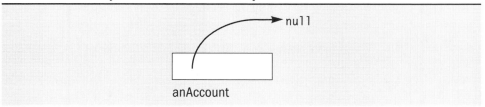

anAccount

Making an Object Behave

Once we have created a new BankAccount object we need to get it to perform its operations. Java uses a syntax that looks like we are commanding an object to perform a service. Informally, we might command myAccount to deposit $100 with

```
myAccount deposit $100
```

In Java, the code for myAccount to make a deposit is

```
myAccount.deposit(100.00);
```

The bank account object referred to by myAccount deposits 100 dollars. The deposit method adds the amount of the deposit to the balance field of the object referred to by myAccount. The code for this instance method is

```
public void deposit(double amount) {
    balance += amount;
}
```

Note we do not refer to the object explicitly in the code. We understand the balance field is the account balance of the specific BankAccount object that is getting the deposit, which in the statement myAccount.deposit(100.00) is the object referred to by myAccount.

☛ **A LITTLE EXTRA**

The **this** Reference

Java uses **this** to refer to the current object whose method we are invoking. We could write the deposit method as

```
public void deposit(double amount) {
    this.balance += amount;
}
```

which shows explicitly that we are adding the amount to the balance of this, referring to the current object, the account to which we are depositing.

When we invoke myAccount.deposit(100.00), the variable this refers to the current object, the BankAccount object referred to by myAccount. When we invoke yourAccount.deposit(100.00), the variable this refers to the BankAccount object referred to by yourAccount. English usage is similar. When I say "My book is on the table" the word *my* refers to my book, but when you say "My book is on the table" the word *my* refers to your book. When we use this inside the deposit method it refers to the account into which the deposit is being made.

The use of this is optional. We introduce it here to emphasize that instance variables and methods are always invoked by an object, even if that object is not explicitly mentioned.

As with the deposit method, each BankAccount object has its own getBalance method which we invoke using the dot notation as in

```
double money = myAccount.getBalance();
```

If the object referred to by myAccount has a balance of $24.50, the variable money will have the value 24.5.

Calling the withdraw method, as in

```
myAccount.withdraw(20.00);
```

will cause $20.00 to be deducted from the balance of the object referred to by myAccount if that balance is greater than or equal to $20.00, and will print a message otherwise.

Now that we have seen how to create a bank account and invoke its services, in Example 2.4 we create two accounts and use their services. Inside main we create a BankAccount object, as in:

```
BankAccount myAccount = new BankAccount(25.00);
```

and get its balance using the getBalance method, as in:

```
myAccount.getBalance();
```

which will return the $25.00 balance with which we initialized the account. In this example, we use the BankAccount class of Example 2.3.

EXAMPLE 2.4   **TestBankAccount.java**

```
/* Creates and uses some BankAccount objects.
 */

import iopack.Io;
public class TestBankAccount {                                      // Note 1
    public static void main (String [ ] args) {
        BankAccount myAccount = new BankAccount(25.00);
        System.out.print("My balance = ");                         // Note 2
        Io.println$(myAccount.getBalance());                       // Note 3
        myAccount.deposit(700.00);                                 // Note 4
        System.out.print("My balance = ");
        Io.println$(myAccount.getBalance());
```

```
        myAccount.withdraw(300.00);
        System.out.print("My balance = ");
        Io.println$(myAccount.getBalance());
        myAccount.withdraw(450.00);
        System.out.print("My balance = ");
        Io.println$(myAccount.getBalance());                    // Note 5
        BankAccount  yourAccount  = new BankAccount();
        yourAccount.deposit(1234.56);
        System.out.print("Your balance = ");
        Io.println$(yourAccount.getBalance());
    }
}
```

Output

```
My balance = $25.00
My balance = $725.00
My balance = $425.00
Insufficient funds
My balance = $425.00
Your balance = $1,234.56
```

Note 1: The BankAccount class defines a type of BankAccount object. The TestBankAccount class is a procedural program which we use to try out the BankAccount type. We could have included the main method in the BankAccount class itself, but it helps to differentiate between a class such as BankAccount, used to define a type, and a class such as TestBankAccount which contains a main method to test the type. Including a main method in a class such as BankAccount is the usual way to provide testing, but such mixing of functions in a class can be confusing when first creating class definitions.

Note 2: This print statement describes the output and stays on the same line so the next statement can display the balance. We could have used the standard System.out.println method to display the balance, but then the default formatting would have displayed the balance as 25.0, with only one place after the decimal point.

Note 3: We use the Io.println$ method to format the balance in dollars and cents, with a dollar sign.

Note 4: The variable myAccount invokes its deposit operation. Each object has certain operations that express its behavior. A BankAccount can deposit, and here myAccount deposits 700 dollars by adding that amount to its balance. We can think of deposit(700.00) as a message sent to myAccount asking it to handle this request according to its deposit method. The instance method deposit always refers to a specific BankAccount object.

Note 5: The balance remains the same because the previous withdrawal request was rejected for insufficient funds.

☞ **STYLE**

The objects referred to by myAccount and yourAccount are instances of the class BankAccount. Note the common style which uses capital letters to start class names and lowercase letters to start object names and method names.

Method Overloading

We use **method overloading** to define more than one method with the same name. Such overloaded methods must have differences in their arguments lists. For example, the String class contains overloaded methods. Two methods have the same name, indexOf, but one has a char parameter while the other has a parameter of type String. Their signatures are:

```
public int indexOf(char c);
public int indexOf(String s);
```

Programmers find it less cumbersome to use overloaded methods. For example if the indexOf method were not overloaded, we would have to use something like indexOfChar and indexOfString as the names for these two methods. Method overloading helps when we have methods which are similar except that they operate with different arguments.

When we use an overloaded method in a program, Java can determine which method to call by looking at the type of argument we pass to it. For example, in

```
String food = "potato";
int a     = food.indexOf('a');
int to  = food.indexOf("to");
```

Java will call the indexOf(char c) method to find the index of the first 'a' in "potato", because the argument 'a' passed in the call indexOf('a') has type char. However, Java will call indexOf(String s) to find the index of the first occurrence of "to", because the argument "to" has type String.

Another common use of method overloading is for constructors. We often overload constructors to provide a different way of creating objects. The BankAccount class of Example 2.3 has two constructors, one with no parameters and one with a single parameter specifying an initial balance. In Example 2.4, we used the constructor with no arguments to create yourAccount and the constructor with one argument to create myAccount.

When two methods have the same name, Java uses the argument types to determine which method to call. Thus Java will not let us define two methods with the same name and the same types of parameters; it could not determine which one to call when we invoke the method in our program.

Class Variables and Methods

Our BankAccount class of Example 2.3 has only instance variables, instance methods, and constructors. It defines a new type of object, a BankAccount. When we use BankAccount methods in Example 2.4, we first create a bank account using a constructor.

```
BankAccount myAccount = new BankAcccount(25.00);
```

Only then can we perform transactions such as

```
myAccount.deposit(700.00);
```

It would not make sense to invoke deposit(700.00) without prefixing it with the object name myAccount. The deposit method deposits into a specific account and can only be called as an operation of an account.

A class may also include **class variables** and **class methods**, which are associated with the class, rather than with a particular instance of the class. They are declared using the static modifier. The word *static* may remind us that class variables and methods stay with the class. The main method is always static, a class method, because Java calls it to start the program when there are no objects.

To illustrate class variables and methods, we modify the BankAccount class of Example 2.3 to count the total number of deposit, withdraw, and getBalance transactions successfully completed by all bank accounts created in a test program. We change the class name to Acct to avoid confusion with the unmodified BankAccount class. The Acct class has a class variable, transactions,

```
public static int transactions = 0;
```

which keeps count of the number of transactions. The Acct class has only one copy of the transactions class variable, whereas each Acct object has its own copy of the balance instance variable. Thus, if myAcct makes a deposit and yourAccount performs a withdrawal, the total number of transactions will increase by two. We modify the deposit, withdraw, and getBalance methods to increment the transactions variable.

We include a class method

```
public static int getTransactions() {
    return transactions;
}
```

which returns the total number of successful transactions by all Acct objects.

EXAMPLE 2.5   Acct.java

```
/* Modifies the BankAccount class to include a class variable
 * to store the total number of successful deposit, withdraw, and
 * getBalance operations by any Acct object.
 */

import iopack.Io;
public class Acct {
    private double balance;
    private static int transactions = 0;                     // Note 1

    public Acct()   {
        balance = 0;
    }
    public Acct(double initialAmount) {
        balance = initialAmount;
    }
    public void deposit(double amount) {
        balance += amount;
        transactions++;
    }
    public void withdraw(double amount) {
        if (balance >= amount) {
          balance -= amount;
          transactions++;                                     // Note 2
        }
```

continued

```
            else
                System.out.println("Insufficient funds");
        }
        public double getBalance() {
            transactions++;
            return balance;
        }
        public static int getTransactionCount() {                    // Note 3
            return transactions;
        }
}
```

Note 1: The `static` modifier signifies that `transactions` is a class variable. We initialize it to 0 because a test program starts with no transactions completed initially.

Note 2: We only increase the count of transactions when the withdrawal is successful.

Note 3: The `static` modifier signifies that `getTransactionCount` is a class method.

Example 2.6 tests the `Acct` class. We create `myAcct` and execute the `deposit`, `getBalance`, and `withdraw` methods. Because the last withdraw is unsuccessful, we get a transaction count of 3. Creating `yourAcct` and executing the `deposit` and `getBalance` methods, we find that the transaction count becomes 5. The `transactions` class variable is part of the class and is incremented by all instances. Of course each object only reads or writes its own `balance` instance variable.

When calling the `getTransactionCount` method, we prefix the class name, as in

`Acct.getTransactions`

reminding us that `getTransactionCount` is a class method, not part of any instance of the `Acct` class.

EXAMPLE 2.6 **TestAcct.java**

```
/* Creates some Acct objects and illustrates the use of
 * class variables and methods.
 */

import iopack.Io;
public class TestAcct {
    public static void main (String [ ] args) {
        Acct myAcct = new Acct(25.00);
        myAcct.deposit(700.00);
        myAcct.withdraw(300.00);
        myAcct.withdraw(450.00);
        System.out.print("My balance after completing transactions is ");
        Io.println$(myAcct.getBalance());
        System.out.println
            ("The number of transactions is " + Acct.getTransactionCount());
        Acct yourAcct = new Acct();
        yourAcct.deposit(1234.56);
        System.out.print("Your balance after completing transactions is ");
```

```
            Io.println$(yourAcct.getBalance());
            System.out.println
                ("The number of transactions is " +    Acct.getTransactionCount());
        }
    }
```

Output

```
Insufficient funds
My balance after completing transactions is $425.00
The number of transactions is 3
Your balance after completing transactions is $1,234.56
The number of transactions is 5
```

☛ TIP

In main we cannot write

deposit(100.00);

because deposit is a method of a specific Acct object, and main is a class method. We must write

myAcct.deposit(700.00);

where myAcct refers to an Acct object.

The Big Picture

The BankAccount class implements a programmer-defined type. It contains constructors, with the same name as the class, which initialize the private instance variable, balance, when a BankAccount object is created. The instance methods getBalance, deposit, and withdraw allow a BankAccount object to fulfill its responsibilities. Because Java supports method overloading we are able to include two constructors with the same name.

A user creates a BankAccount object with the **new** operator, and invokes its behavior with a message sending style,

myAccount.deposit(50.00);

Class variables and methods, declared with the static modifier, are part of the class but not part of any instance.

TEST YOUR UNDERSTANDING

6. Write the declaration for an integer account number instance variable in the BankAccount class. Restrict access to the account number to methods of the BankAccount class.

7. Where do we declare an instance variable of a class? Give an example of an instance variable in Example 2.3.

8. Rewrite the declaration BankAccount theAccount; so theAccount will refer to a newly created BankAccount.

9. What value does the variable `theAccount` have after the following declaration?

   ```
   BankAccount theAccount;
   ```

10. Given a `BankAccount`, `myAccount`, write Java statements to

    a. deposit $35.50
    b. get the current balance
    c. deposit $999

11. Which method does the **new** operator call in the following expression?

    ```
    new BankAccount();
    ```

TRY IT YOURSELF   12. In Example 2.4, add a line

   ```
   deposit(439.86);
   ```

   to the `main` method of the `TestBankAccount` class which tries to use the `deposit` instance method without referring to a specific `BankAccount` object. Try to compile this modified program and see what error you get.

TRY IT YOURSELF   13. Replace the first `myAccount.getBalance()` method call in the `TestBankAccount` class with a field access

   ```
   myAcccount.balance.
   ```

   This will create errors because the `balance` field is private and not accessible outside the `BankAccount` class. What errors do you get when you try to compile this modified version of Example 2.4?

14. We can overload methods other than constructors. Write another `withdraw` method, with no parameters, which will withdraw $40 if that amount is available. This method provides a quick withdrawal where the user does not have to specify any amount.

15. Suppose we want to add a third constructor to Example 2.3 which would take no parameters, but would set the initial balance to $25.00. We could code it as

    ```
    public BankAccount () {
        balance = 25.00;
    }
    ```

    Could we add this constructor to Example 2.3 or will Java not allow it? Explain.

TRY IT YOURSELF   16. Add the `withdraw` method written in question 14 to Example 2.3. Modify Example 2.4 to test the new `withdraw` method.

2.4 ■ The QuickFood Example

The `BankAccount` class we wrote in the last section defines the state and services for bank account objects. The `QuickFood` program of Example 2.2 uses `Customer`, `Waiter`, and `Cook` objects. We need to write the `Customer`, `Waiter`, and `Cook` classes which define the state and services for these objects. We introduce the UML class and sequence diagrams to illustrate the design.

Class Diagrams

We use the UML **class diagram** to show the associations between classes. An association represents a relationship between instances of the associated classes. For example, a customer places an order with a waiter, while a waiter asks the cook to make a burger (see Figure 2.24).

FIGURE 2.24 A class diagram

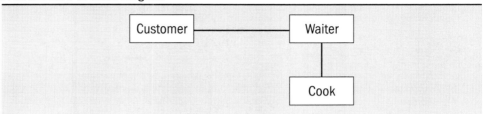

Sequence Diagrams

We can visualize the scenario of Figure 2.11 in a **sequence diagram**, another part of the UML, that shows object interactions arranged in time sequence. Each object appears at the top (see Figure 2.25), with a dashed line descending, called its **lifeline**. We represent each message from one object to another using a horizontal arrow.

FIGURE 2.25 A sequence diagram for a food order

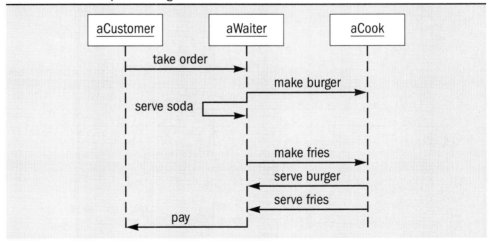

For ease of implementation of this introductory example, we change the model to serve each item as it is ordered. Figure 2.26 shows the new sequence diagram.

Our code for the Customer, Waiter, and Cook classes includes the state and responsibilities identified in Figure 2.12. For simplicity the placeOrder method just orders a burger, fries, and a soda. A more detailed implementation would provide a user interface for the customer to make selections. The pay method reduces the customer's money by the amount of the bill.

FIGURE 2.26 A revised sequence diagram for a food order

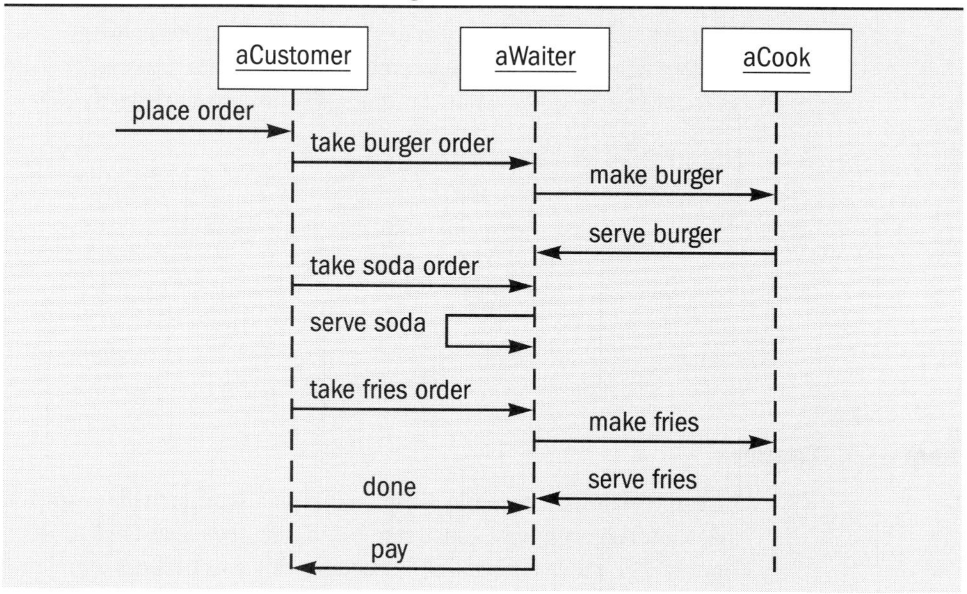

In implementing associations, we include a reference in one object to the object to which it is associated. The Customer constructor saves a reference to a Waiter.

```
public Customer(Waiter w) {
    waiter = w;
}
```

In taking an order, the waiter adds the price to the customer's bill and asks the cook to prepare the burger or fries, or serves the soda. The cook makes the burger or fries and asks the waiter to serve it. Example 2.7 shows the complete code for the simple fast food simulation.

EXAMPLE 2.7   **QuickFood.java**

```
/* Adds the Customer, Waiter, and Cook classes
 * to define the objects used in Example 2.2.
 */

import iopack.Io;

class Customer {                                          // Note 1
    private double money = 30.00;
    private Waiter waiter;

    public Customer(Waiter w) {
        waiter = w;
    }
    public void placeOrder() {
        waiter.takeOrder("Burger",this);
        waiter.takeOrder("Soda",this);
```

```
        waiter.takeOrder("Fries",this);
        waiter.takeOrder("Done",this);                    // Note 2
    }
    public double pay(double amount) {
        money -= amount;
        System.out.print("Customer pays ");
        Io.println$(amount);
        return amount;
    }
}

class Waiter {
    private double cash = 200.00;
    private Cook cook;
    private Customer customer;
    private double bill = 0;

    public Waiter(Cook c) {                                // Note 3
        cook = c;
    }
    public void takeOrder(String item, Customer c) {
        customer = c;
        if (item.toUpperCase().equals("BURGER")){          // Note 4
            System.out.println("Waiter places BURGER order");
            bill += 1.99;
            cook.makeBurger(this);
        }
        else if (item.toUpperCase().equals("FRIES")){
            System.out.println("Waiter places FRIES order");
            bill += 1.19;
            cook.makeFries(this);
        }
        else if (item.toUpperCase().equals("SODA")){
            serveFood("SODA");                             // Note 5
            bill += .99;
        }
        else if (item.toUpperCase().equals("DONE"))
            cash += customer.pay(bill);;
        }
    public void serveFood(String item) {
        System.out.println("Waiter serves " + item.toUpperCase());
    }
}
class Cook {
    private int burgers = 10;
    private int fries = 10;

    public void makeBurger(Waiter waiter) {                // Note 6
        if (burgers > 0) {
            System.out.println("Cook making BURGER");
            waiter.serveFood("Burger");
            burgers--;
        }
```

continued

```
        else
            System.out.println("Sorry -- No more BURGERS");
    }
    public void makeFries(Waiter waiter) {
        if (fries > 0) {
            System.out.println("Cook making FRIES");
            waiter.serveFood("Fries");
            fries--;
        }
        else
            System.out.println("Sorry -- No more FRIES");
    }
}
public class QuickFood {
    public static void main(String[] args) {
        Cook cook = new Cook();
        Waiter waiter = new Waiter(cook);
        Customer customer = new Customer(waiter);
        customer.placeOrder();
    }
}
```

Output (same as Example 2.2)
```
Waiter places BURGER order
Cook making BURGER
Waiter serves BURGER
Waiter serves SODA
Waiter places FRIES order
Cook making FRIES
Waiter serves FRIES
Customer pays $4.17
```

Note 1: We have omitted the public modifier on the Customer, Waiter, and Cook classes, because at most one public class may appear in any file, and for simplicity we wanted to include all the classes in the same file. Generally, classes define types that can be used in many applications, and should be declared using the **public** modifier and placed in separate files. We discuss the use of access modifiers in Section 3.4.

Note 2: The customer passes "DONE" to signal the order is complete.

Note 3: Reflecting the association between a waiter and a cook, we pass a reference to a cook when we construct a waiter.

Note 4: Converting the item to uppercase allows the comparison to be case insensitive, so "burger", when converted, would be equal to "BURGER".

Note 5: The waiter calls its own serveFood method, so the object is implicit. We could have used this.serveFood("SODA") to make the object explicit.

Note 6: `public void makeBurger(Waiter waiter) {`

We do not need an explicit constructor for the Cook class because the waiter passes a reference to itself to the cook when it calls the makeBurger method.

> **The Big Picture**
>
> A large part of an object-oriented program involves writing the classes which define the state and responsibilities of its object instances. In meeting its responsibilities an object uses the services of other objects.
>
> The UML class diagram shows the associations between classes, while the sequence diagram shown the sequence of interactions between objects, from the earliest at the top to the latest at the bottom.

TEST YOUR UNDERSTANDING

17. Redraw the sequence diagram of Figure 2.26 so the waiter asks the customer to take each item as it becomes available.

2.5 ▪ Object Composition

Our BankAccount class has an instance variable balance of type **double**. **Composition**, a powerful object-oriented design concept, builds objects which have other objects as data fields. Our Java objects can be composed of other objects, just as a computer, for example, is composed of a CPU, a keyboard, a monitor, a disk drive, and so on. Composition models the **HAS-A** relationship in which one object contains another. An automobile has tires and an engine, for example.

We shall build Name and Address objects that have fields defined as Strings, and Person objects that each have a field defined as a Name and another defined as an Address. Composition, along with inheritance, which we cover in Section 3.1, is one of the two ways of defining new classes using those previously defined. Composition models the whole-part relationship; the whole object is composed of its parts.

We define a class Person we can use in applications when we need data associated with a specific individual. We need many data items for each person, including first name, last name, street address, city, and so on. We choose to organize this data into coherent Name and Address classes rather than as an unorganized group of individual fields. Organizing our data will make our class easier to read. We can use the Name and Address classes in other applications.

Figure 2.27 lists the fields for our Name, Address, and Person classes. Each class in Figure 2.27 contains data fields that are objects. The Name class has two String objects and a char. Each Address has four String objects. The zip code, zip, uses digits, but because we do not do arithmetic on zip codes, we have no need to store it as an integer. Our Person class uses a String, a Name and an Address. As we did with the zip code, we treat the id, usually the social security number, as a String.

Figure 2.28 illustrates composition. A Person contains references to a String, a Name, and an Address. A Name contains two String references and a character, while an Address contains four String references. We are especially interested in the fields, so we omit the operations from the class diagrams.

FIGURE 2.27 Fields for the Name, Address, and Person classes

```
Name
    private String first
    private char initial
    private String last
    public  Name(String f, String l)
    public  Name(String f, char i, String l)
    public  String toString()

Address
    private String street
    private String city
    private String state
    private String zip
    public  Address(String st, String cy, String se, String zp)
    public  String toString()

Person
    private String id
    private Name name
    private Address address
    public  Person(String i, Name n, Address a)
    public  String getId()
    public  String toString()
```

FIGURE 2.28 Composition: The Person, Name, and Address classes

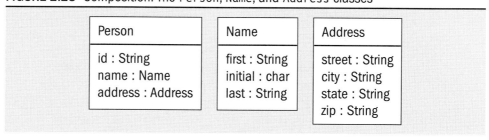

Each of the three classes has a constructor to initialize its fields (the Name class has two), and each has a toString method to provide a string representation for display purposes. Java provides a toString method for its library classes whose objects we need to display. The println method calls this toString method when asked to display an object, as in

```
System.out.println(aPerson);
```

which calls aPerson.toString() to get the string representation for the object aPerson of type Person.

☛ **TIP**

Define a toString method for each class whose objects you need to display. Then you will be able to display your objects using the println statement. When your object is a component of another object, its string representation will be part of the string representation of the containing object.

☞ STYLE	Do	group fields into classes such as Name and Address that give meaning to the fields, help to organize your data, and can be reused in other applications.
	Don't	build a class with a long list of unorganized fields.

None of the fields listed in Figure 2.27 are static. The Name, Address, and Person classes define the data and operations that will be part of each object of these types. Each object is an instance of its class type. For example the object given by

```
Name composer = new Name("Wolfgang", 'A', "Mozart");
```

shown in Figure 2.29, is an instance of the Name class, just as the String objects s1, s2, s3, and s4, defined on page 47, are instances of the String class.

FIGURE 2.29 An instance of Name

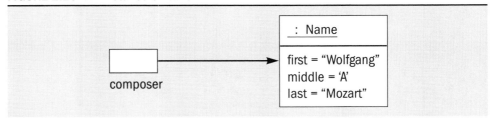

The composer object has **private** data, its first, initial, and last fields, and a method, toString, which returns its string representation. When we create an object instance of some type, we say we are **instantiating** an object of that type.

In contrast to the classes of many of our earlier examples, our classes in this chapter contain non-static fields which are instance variables meant to be part of each object instance.

Example 2.8 creates our Name, Address, and Person classes, which we will use in applications later in this chapter. To allow us to use each class in future applications, we make each public in its own file. The TestPerson class contains a main method to test the Person, Name, and Address classes, also illustrating some of the String methods of Section 2.1.

EXAMPLE 2.8   **Name.java**

```
/* Groups fields for a name.
 * Uses toString to display.
 */

public class Name {
    private String first;
    private char initial;
    private String last;

    public Name(String f, String l) {          // Note 1
        first = f;  last = l;
    }
```

continued

```
    public Name(String f, char i, String l) {
        this(f,l);                                          // Note 2
        initial = i;
    }
    public String toString() {
        if (initial == '\u0000')                            // Note 3
            return first + " " + last;
        else
            return first + " " + initial + " " + last;      // Note 4
    }
}
```

Address.java

```
/* Groups fields for an address.
 * Uses toString to display.
 */

public class Address {
    private String street;
    private String city;
    private String state;
    private String zip;
    public Address(String st, String cy, String se, String zp) {
        street = st; city = cy; state = se; zip = zp;
    }
    public String toString() {
        return street + "\n" + city + ", " + state + " " + zip;    // Note 5
    }
}
```

Person.java

```
/* Groups fields for a person.
 * Uses toString to display.
 */

import iopack.Io;
public class Person {
    private String id;
    private Name name;
    private Address address;
    public  Person(String i, Name n, Address a) {
        id = i; name = n; address = a;
    }
    public String getId() {                                 // Note 6
        return id;
    }
    public  String toString() {
        return name + "\n" + address;                       // Note 7
    }
}
```

TestPerson.java

/* Tests the Person, Name, and Address classes and uses String methods.
 */

```java
import iopack.Io;
public class TestPerson {
    public static void main (String [] args) {
        Name aName = new Name("Henry", "Johnson");
        Address anAddress =
            new Address("1512 Harbor Blvd.", "Long Beach", "CA", "99919");
        String anId = Io.readString("Enter an id string");      // Note 8
        Person aPerson = new Person(anId,aName,anAddress);
        System.out.println("Our person is ");
        System.out.println(aPerson);
        System.out.println("  with id " + aPerson.getId());
        System.out.println("\n And now some tests of string methods");
        String address = anAddress.toString();
        int i = address.indexOf("Harbor");                      // Note 9
        System.out.println("The index of Harbor in address is " + i);
        String z1 = String.valueOf(99919);                      // Note 10
        int l = address.length();                               // Note 11
        System.out.println("The length of address is " + l);
        String z2 = address.substring(l-5,l);                   // Note 12
        boolean same = z2.equals(z1);                           // Note 13
        System.out.println("These two zip codes are the same? " + same);
        int less = z1.compareTo("Harbor");                      // Note 14
        System.out.println("Compare returns the negative number " + less);
        String hat = "   hat   ";
        System.out.println(hat+"rack");
        System.out.println(hat.trim()+"rack");                  // Note 15
    }
}
```

Output

```
Enter an id string:   123456789
Our person is
Henry Johnson
1512 Harbor Blvd.
Long Beach, CA 99919
   with id 123456789

   And now some tests using string methods
The index of Harbor in address is 5
The length of address is 38
These two zip codes are the same? true
Compare returns the negative number -15
      hat    rack
hatrack
```

Note 1: We provide a constructor for a name without a middle initial. Java automatically initializes data fields with default values if none are supplied by the user. We do not supply a value for `initial` so Java initializes it with the default value for the type character which is the character with numerical code zero. We can write this character as either '\u0000', where the 'u' stands for Unicode, or '\000'.

Note 2: One constructor can call another constructor in the same class by using the name this which refers to the current object. Here we call the other Name constructor, passing it a name and an address, and then initialize the middle initial.

Note 3: We do not display the middle initial when it has the default value of '\u0000', the character code with value zero, which Java assigns to the initial field when none is specified in the constructor.

Note 4: The concatenation operator '+' converts any primitive type, such as char, to a string representation.

Note 5: We add the string "\n" to include a newline in the string, so the street will appear on a separate line from the city, state, and zip code.

Note 6: `public String getId() {`

We do not want to output the id every time we display the person's name and address. We can use this method to get the id when we need it.

Note 7: `return name + "\n" + address;`

The one string argument, "\n", tells Java the + is the string concatenation operator rather than an arithmetic addition. The string concatenation operator uses the toString methods we defined for Name and Address to get the string representations for the name and address fields, without having to write name.toString() or address.toString().

Note 8: `String anId = Io.readString("Enter an id string");`

The readString method can be found in the Io class on the disk included with this text.

Note 9: `int i = address.indexOf("Harbor");`

The indexOf method returns the position of the first occurrence of its argument in the string, or -1 if the argument is not found. Here "Harbor" occurs starting at position five.

Note 10: `String z1 = String.valueOf(99919);`

The static method valueOf makes a string from its argument. Java overloads valueOf to make strings from several primitive types.

Note 11: `int l = address.length();`

The length() method returns the length of the string.

Note 12: `String z2 = address.substring(l-5,l);`

The substring method returns a string made from a range of characters of the given String object. Here we make a string from the last five characters, those at position l-5 up to position l, which is just the zip code of the address.

Note 13: `boolean same = z2.equals(z1);`

The two strings, z1 and z2, are different objects but have the same characters, the zip code, so they are equal.

Note 14: `int less = z1.compareTo("Harbor");`

Digits come before letters in the ASCII and Unicode character orderings so a string of digits will be less than a string of letters. The `compareTo` method will return a negative number.

Note 15: `System.out.println(hat.trim()+"rack");`

The `trim` method removes the leading and trailing blanks from `hat`.

The Big Picture

Composition models the HAS-A relationship between objects in which one object contains another. We represent the contained object as an instance variable in the containing object.

TEST YOUR UNDERSTANDING

18. Declare and initialize a `Name` object using your own name.

19. Declare and initialize an `Address` object using your own address. List the four objects of which this `Address` is composed.

20. Declare and initialize a `Person` object using data of your choice. List the three objects of which this `Person` is composed.

Chapter 2 Summary

An object has state, behavior, and identity. In analogy with a vending machine, the services an object provides are like the buttons of the machine. Users access an object via its services which express the behavior of that object and meet its responsibilities. Using classes from the Java library, we can create objects and use the operations Java provides for them. Strings are a special class of great importance; Java provides a large library of `String` methods. A few, such as `valueOf`, are class methods which we invoke as if we were sending a message to the class, for example `String.valueOf(1234)`. Most, including `charAt`, `compareTo`, `equals`, `indexOf`, `length`, `substring`, `toUpperCase`, and `toLowerCase`, are instance methods which we invoke as if sending a message to a particular object, for example `myName.length()`, where `myName` is the string given by `String myName = "Art"`. The concatenation operator, +, not only concatenates strings, but converts primitive types to a string representation, and uses the `toString` method to get the string representation for objects. We can initialize a string using a string literal such as `"house"`.

Java implements primitive types such as integers and class types such as bank accounts differently. Primitive types have small fixed sizes, so Java variables hold their values. An integer variable stores the values of an integer. By contrast, objects may be quite large and have varying sizes, so it is easier for an object variable to store a reference to an object rather than trying to hold it directly.

Procedural programming focuses mainly on the function performed by the program and only incidentally on the data. The function performed is the aspect most likely to change, so a more stable and maintainable system results from object-oriented programming which focuses on the objects that comprise the system. These objects have data representing their state and operations representing their behavior. The operations give each object an active role in the system. Objects communicate with each other by invoking operations that allow an object to meet its responsibilities. We can use scenarios for typical uses of the system to find the relevant objects and identify their responsibilities.

In Java, we use a class to define a type of object. Each object has state, data which we represent using instance variables which are non-static variables declared inside the class but outside of any method. Our BankAccount class defines an instance variable, balance, to hold the account balance. By contrast, local variables are declared inside methods and available only inside the method in which they are declared.

We add methods to our class to implement the behavior of its type of objects. In our BankAccount class we add instance methods to deposit an amount, to withdraw an amount, and to get the account balance. These methods are not static, signifying they will be operations of each BankAccount object, representing that object's behaviors. Static methods such as the amount method of Example 1.6 are sometimes called class methods because they are class tools, not associated with any object.

We access objects using methods which are operations representing their behavior. Java uses the period to separate the name of the object from the operation it is invoking. This notation emphasizes that an object such as myAccount is performing one of its operations such as deposit. Usually, we make the object's data fields private, requiring that users of our objects access fields only via the object's operations.

A special kind of method, called a constructor, helps the user to construct an object. Constructors have the same name as the class and no return value. We define a BankAccount constructor which creates a BankAccount object with an initial balance of zero. We create objects with the new operator which allocates space for the object, calls the object's constructor, and returns a reference to the object. Java uses **null** to represent an uncreated object.

Java supports method overloading, where two methods have the same name but different parameters. One important use of overloading is to provide multiple constructors for objects of a given class. We add another constructor to our BankAccount class to create a BankAccount with a specified initial balance.

We use composition, a powerful design tool, for building objects composed of other objects. Example 2.8 builds a Name and an Address from strings, and a Person class from a name and an address.

Build Your Own Glossary

Find the uses, in this chapter, of the terms below. Enter a definition of each in the glossary.txt file on the disk included with this text.

class diagram	constructor	garbage collection
class method	default constructor	HAS-A
class variable	field	instance
composition	garbage	instance method

instance variable	method overloading	scenario
instantiate	null	sequence diagram
lifeline	object	use case
local variable	reference	value

Skill Builder Exercises

1. Match the concept name on the left with its function on the right.

   a. instance variable   i. stores a value within a method

   b. local variable   ii. represents a behavior of an object

   c. class method   iii. represents an attribute of an object

   d. instance method   iv. used by the class as a whole

2. Fill in the blanks in the following:

   A variable of a primitive type holds a _____, while a variable of a class type holds a _____.

3. What will be the output from the following?

```
String s = "hat";
String t = s + " rack";
System.out.println(s.substring(0,1) + t.substring(5,8));
```

Critical Thinking Exercises

4. Declaring a bank account as `BankAccount acct;` and making a deposit using `acct.deposit(500.00);` will have the following result:

   a. The compiler will generate an error message.

   b. The account `acct` will have a balance of $500.00.

   c. The account `acct` will have its previous balance increased by $500.

   d. none of the above.

5. A constructor

   a. must have the same name as the class it is declared within.

   b. is used to create objects.

   c. may be overloaded.

   d. b and c above

   e. all of the above

6. Which of the following are never part of a class definition?

   a. instance variables

   b. static methods

   c. instance methods

   d. constructors

   e. none of the above

7. Suppose `acct1` refers to a `BankAccount` with a balance of $300, while `acct2` refers to a `BankAccount` with a balance of $200. After the assignment

```
acct1 = acct2;
```

Which of the following is true?

a. Withdrawing $100 from acct1 will leave its balance at $200.
b. Withdrawing $100 from acct2 and then withdrawing $100 from acct1 will leave the balance of acct1 at $100.
c. Withdrawing $100 from acct1 and then withdrawing $100 from acct2 will leave the balance of acct1 at $0.00.
d. none of the above
e. all of the above

Debugging Exercise

8. The following isEqual method, for the Name class of Example 2.8, attempts to check two names for equality. They should be equal if they have the same first name, middle initial, and last name. Find and correct any errors.

```
public boolean isEqual(Name name) {
    boolean result = false;
    if (first == name.first &&
            initial == name.initial  &&
            last == name.last)
        result = true;
    return result;
}
```

Program Modification Exercises

9. Add a compareTo method to the Name class of Example 2.8. The compareTo method should return a negative integer when the object is less than the argument, zero when equal, and a positive number when greater. Compare last names first. If the last names are equal, compare first names. If the first names are equal, compare the middle initial.

10. Revise the BankAccount class of Example 2.3 to overload the withdraw method. Include a withdraw method with no parameters, which will withdraw $40 if available, and display a message otherwise. Add tests of this new method to the main method.

11. Modify Example 2.7 to add a pickUpOrder method to the Customer class, and let the waiter invoke this method when each item is ready.

12. Modify Example 2.3 to include an account holder of type Person in the BankAccount class.

Program Design Exercises

13. Write a Java program which illustrates the use of each of the String methods in Figure 2.4.

14. Write a class for soccer game scoring. Provide a constructor which starts each team with a score of zero. Include instance variables to keep the score for both teams. Include a method to add one to the score of the first team and a method to add one to the score of the second team. Include a method which displays the score of both

teams, and the main method to test, creating two different soccer games. Score points so the first game is 3–2 and the second game is 0–1. Display the scores of each game.

15. Write a class which uses the soccer game class of Exercise 14. In the main method create some games, score points, and display the results.

16. Write a class for a warehouse which hold radios, televisions, and computers. Provide a constructor which starts a warehouse with no items. Include instance variables to store the quantity of each item in the warehouse. Include methods to add to the stock of each item, and a method to display the contents of the warehouse. Test in a main method, creating two warehouses. Add items to each and display the final contents of each warehouse.

17. Write a class to use the warehouse class of Exercise 16. In the main method create some warehouses, add some items to each and display the contents of each warehouse.

18. Write a class to keep track of the movement of a cat. Include three instance variables to hold the x, y, and z positions of the cat. Include a method for the cat to walk to another position. This method has two parameters specifying the change in x and the change in y. If the cat is at (3,4,5) and we ask it to walk(1,4) then it will be at (4,8,5). (Walking is a horizontal action here.) Include a method for the cat to jump to another position. This method has one parameter specifying the change in the cat's vertical position. If the cat is at (3,4,5) and we ask it to jump(5) it will be at (3,4,10). Include a method to display a cat's position. Test in a main method, creating a few cats and have them walk and jump. Display their final positions.

19. Write a class to use the cat tracking class of Exercise 18. In the main method, create some cats, make them walk and jump, and display their final positions.

20. For the Warehouse class of Exercise 16 overload the Warehouse constructor by adding another constructor to create a warehouse with specified initial quantities of radios, televisions, and computers. Revise the main method to include tests of this new constructor.

21. For the Warehouse class of Exercise 16, add methods to remove a specified quantity of each item. If the quantity specified is greater than the amount of that item in the warehouse then no items are removed. Add tests of the remove methods to the main method.

22. Write a coffee vending machine class. Include fields giving the number of cups of coffee available, the cost of one cup of coffee, and the total amount of money inserted by the user. This machine requires exact change. Include one constructor which stocks the machine with a quantity and price of coffee specified as parameters, and another with no parameters which stocks the machine with 10 cups of coffee at fifty cents each. Include the following methods:

```
menu()     // displays the quantity and price of coffee
insert(int quarters, int dimes, int nickels)   // inserts the given amount
select()   // dispenses a cup of coffee if user has inserted enough
           // money and coffee is available, otherwise displays a message
refund()   // returns the money inserted
```

Write a main method to create some vending machines and test their operation.

23. Create a Fraction class to provide a data type for rational numbers. Each Fraction will have an integer numerator and denominator. Include a constructor with two integer arguments to initialize the numerator and denominator. Include methods to add, multiply, subtract, and divide, each having a Fraction argument and returning a Fraction result. Use the main method to test the Fraction class.

3

Object-Oriented Concepts

Java is an object-oriented programming language. Object-oriented programming is especially useful in managing the complexity of very large software systems. It focuses on the objects which are the most stable aspect of these systems. Designing with objects is natural for us and allows us to reuse designs in new contexts, making software less expensive to produce and easier to maintain. To get these benefits we need to use the full range of object-oriented techniques.

Chapter 2 introduced us to object concepts and the use of objects. In this chapter, we present the key features of object-oriented programming: inheritance, polymorphism, abstract classes, and interfaces, leading to a case study illustrating the flavor of the object-oriented programming paradigm using the object-oriented design techniques of use cases and scenarios. This case study is a program-driven version of the AtmScreen applet shown in Figure 3.1 prompting the user to select an account type. These account types, Savings and Checking, use inheritance to extend the basic Account type.

In the Atm case study of Example 3.6 we use all the object-oriented concepts covered in this chapter to provide almost the same functionality as the AtmScreen applet without the graphics and user interface components that we will cover in later chapters.

OBJECTIVES

◆ Use inheritance to relate classes.

◆ Use polymorphism to improve design.

◆ Use abstract classes to achieve implementation independence.

◆   Use interfaces to view an object from different perspectives.
◆   Understand the use of modifiers in specifying access.
◆   Develop an object-oriented design from use cases and scenarios.

FIGURE 3.1 The `AtmScreen` applet: select an account type

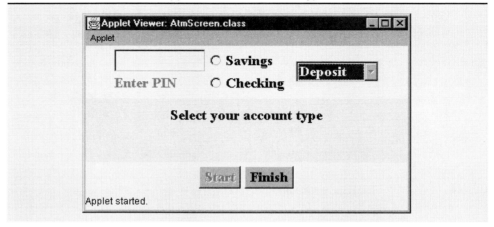

3.1 ▪ Subclasses and Inheritance

Inheritance lets us do in Java what we do when classifying natural objects, which is to group common properties and behavior into a higher-level superclass. We can talk about objects generally on a higher level or more specifically with more lower-level details.

Classification Revisited

Classification organizes knowledge. We divide living things into plant and animal categories. Among animals we differentiate reptiles from mammals, and among mammals we can tell cats from bats. Finally, we recognize individual cats, Tabby and Tom.

Using these categories we can refer to the `Animal` class for behavior, such as movement, that Tabby and Tom share just by being animals. The `Mammal` class stores the common property of having hair, and all `Cat` instances are carnivorous. Finally some properties differ from cat to cat. We have thin cats and fat cats, feisty cats and fraidy cats.

Classification reflects the **IS-A** relationship in that every `Mammal` is an `Animal`, and every `Cat` is a `Mammal`.

Class Hierarchies

Java lets us organize our classes into a hierarchy in which a class can have several **subclasses**, each of these subclasses can itself have subclasses, and so on. For example, our `BankAccount` class could specify the state and behavior common to all bank accounts. A `SavingsAccount` subclass might add an interest rate variable and a method to post the interest. A `CheckingAccount` could have a per check charge, a minimum balance necessary to waive the service charge, and a method to process a check which deducts the service charge, if any.

A TimedAccount subclass of SavingsAccount may not permit withdrawals and may have a limit on the number of deposits.

To create a subclass, we **extend** a class, inheriting all the attributes and behavior of that class. A declaration for a SavingsAccount subclass would have the pattern

```
public class SavingsAccount extends BankAccount {
    . . .
}
```

where the keyword *extends* tells us that SavingsAccount is a subclass of the BankAccount class.

The SavingsAccount class inherits the data fields and methods of the BankAccount class. Every SavingsAccount is a BankAccount. Our BankAccount class of Example 2.1 has an instance variable, balance, to hold the balance, and instance methods, getBalance, deposit, and withdraw. The SavingsAccount class extends BankAccount, so it will inherit the balance field, and the getBalance, deposit, and withdraw methods. We say that the SavingsAccount class is a subclass of the BankAccount class, and that the BankAccount class is a **superclass** or parent class of the SavingsAcccount class.

A subclass can add data fields and methods to those it inherits from its superclass. The SavingsAccount class needs an interestRate attribute to store the interest rate for an account, and a postInterest method to compute the interest, adding it to the account balance.

Subclasses do not inherit the constructors of their superclasses. Even though SavingsAccount does not inherit the BankAccount constructors, it is a BankAccount and, as part of its construction, must use one of the BankAccount constructors to initialize the field it inherits. A subclass constructor calls a constructor for its superclass by calling the super method in the first statement, passing it any arguments that the superclass constructor needs. The SavingsAccount constructor calls the BankAccount constructor using the statement

```
super(amount);
```

where amount is the desired initial balance for the account. The SavingsAccount constructor then initializes the interestRate field. Figure 3.2 shows the code for the SavingsAccount constructor.

FIGURE 3.2 A SavingsAccount constructor

```
public SavingsAccount(double amount, double rate) {
    super(amount);
    interestRate = rate;
}
```

Objects exhibit their unique behavior. Each type of account can handle a withdraw request, for example, in its own way. A subclass automatically inherits the public methods of its superclass, but it may choose to **override** some of them to implement its own specific behavior.

The CheckingAccount class will inherit the getBalance and deposit methods from BankAccount, but will override the withdraw method to call a processCheck method as this type of checking account only permits withdrawals by check. A CheckingAccount is a kind of BankAccount, but it handles withdrawals in its own way. Figure 3.3 contains the code for the CheckingAccount withdraw method which overrides the BankAccount withdraw method.

FIGURE 3.3 Withdraw method overrides BankAccount withdraw

```
public void withdraw(double amount) {
    processCheck(amount);
}
```

☛ TIP

Do not confuse overriding with overloading. An overloaded method has the same name as the original method, but different parameters. The BankAccount class of Example 2.1 has overloaded constructors, one with no parameters and the other with a single parameter of type **double**. A user of the BankAccount class may call either constructor.

The CheckingAccount withdraw method overrides the withdraw method of the BankAccount class, and each has one parameter of type **double.** We could overload the withdraw method by declaring a CheckingAccount withdraw method of the form

```
public void withdraw(int amount)
```

where the parameter has type **int**. The CheckingAccount class would then have two withdraw methods, one inherited from BankAccount, and one added with an **int** parameter.

If you intend to override a method, make sure to use the same parameters that are used in the method you are overriding.

The processCheck method of the CheckingAccount class withdraws the requested amount if the balance is above the minimum for free checks, but withdraws the amount requested plus the service charge if the balance is below the minimum. In either case it calls

FIGURE 3.4 Inheritance

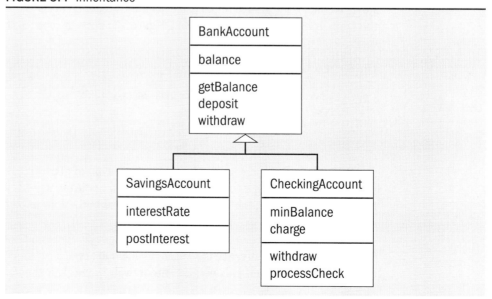

the `withdraw` method of the `BankAccount` class to do the actual withdrawal of funds. The `CheckingAccount` class overrides the `BankAccount` `withdraw` method (see Figure 3.3), so it needs to refer to `BankAccount` `withdraw` as

```
super.withdraw(amount);
```

which calls the `withdraw` method of the superclass, `BankAccount`.

Figure 3.4 shows the inheritance relationship between the `BankAccount` class and its `SavingsAccount` and `CheckingAccount` subclasses. The UML uses the unfilled arrow to denote the inheritance relationship. The subclasses inherit from their parent superclass.

In Example 3.1, we illustrate inheritance with the code for subclasses `SavingsAccount` and `CheckingAccount` of the `BankAccount` class. An `Inherit` class uses the different accounts to show how inheritance works.

EXAMPLE 3.1   **Inherit.java**

```
/* Defines SavingsAccount and CheckingAccount
 * of BankAccount.  The Inherit class uses these
 * classes to illustrate inheritance.
 */

import iopack.Io;
class SavingsAccount extends BankAccount  {                  // Note 1
    private double interestRate;     // % interest
    public SavingsAccount(double amount, double rate) {      // Note 2
        super(amount);                                       // Note 3
        interestRate = rate;                                 // Note 4
    }
    public void postInterest()  {
        double balance = getBalance();
        double interest = interestRate/100*balance;          // Note 5
        setBalance(balance + interest);                      // Note 6
    }
}

class CheckingAccount extends BankAccount {
    private double minBalance;             // Balance needed to avoid charge
    private double charge;                 // Per check charge
    public CheckingAccount(double minAmount, double charge) {  // Note 7
        super();                                             // Note 8
        minBalance = minAmount;
        this.charge = charge;
    }
    public void processCheck(double amount)  {
        if (getBalance() >= minBalance)
            super.withdraw(amount);                          // Note 9
        else
            super.withdraw(amount + charge);                 // Note 10
    }
    public void withdraw(double amount) {                    // Note 11
        processCheck(amount);
    }
}
```

continued

```
public class Inherit {
    public static void main(String [] args) {
        SavingsAccount s =  new SavingsAccount(500.00, 4.5);
        CheckingAccount c = new CheckingAccount(2500.00, .50);    // Note 12
        s.deposit(135.22);                                        // Note 13
        s.postInterest();
        s.withdraw(50);
        System.out.print("The balance of SavingsAccount s is ");
        Io.println$(s.getBalance());
        c.deposit(1000.00);                                       // Note 14
        c.processCheck(200.00);                                   // Note 15
        c.withdraw(100.00);                                       // Note 16
        System.out.print("The balance of CheckingAccount c is ");
        Io.println$(c.getBalance());
    }
}
```

Output

```
The balance of SavingsAccount s is $613.80
The balance of CheckingAccount c is $699.00
```

Note 1: We omit the public modifier because there can be at most one public class in a file. We discuss the use of modifiers in Section 3.4.

Note 2: For the sake of simplicity, we only define one constructor, instead of overloading it as we did in the BankAccount class.

Note 3: We call the BankAccount constructor, passing it the amount of the initial balance to assign to the balance field.

Note 4: We initialize the interest rate.

Note 5: Divide the interest rate specified as a percent by 100 to get the decimal equivalent.

Note 6: setBalance(balance + interest);

Although a SavingsAccount is a BankAccount and has a balance field, that field is private to the BankAccount class; we cannot manipulate it directly from a subclass. We revise our BankAccount class of Example 2.1 to include a public setBalance method. In Section 3.4, we will see an alternate approach using no modifier or the protected modifier, to access fields of a superclass.

Note 7: public CheckingAccount (double minAmount, double charge) {

We only define one CheckingAccount constructor. We do not pass an initial balance to the CheckingAccount constructor, but do pass it to the constructor for the SavingsAccount class to see the contrast.

Note 8: super();

Because we do not specify an initial balance, we call the default constructor (the one with no arguments which sets the balance to zero) for the BankAccount superclass. If we omit this line, Java will add it anyway, because a subclass constructor must always call some superclass constructor to correctly initialize the part of the object inherited from the superclass.

Note 9: `super.withdraw(amount);`

To call the `BankAccount` `withdraw` method, we use the prefix `super`. Because the balance is above the minimum needed for free checks, we withdraw the amount requested. We do not have to check if the amount requested is available, since the `withdraw` method performs that check.

Note 10: `super.withdraw(amount + charge);`

The balance is below the minimum so we withdraw the amount requested plus the service charge for the check.

Note 11: `public void withdraw(double amount) {`

We override the `BankAccount` `withdraw` method to permit withdrawal only by check.

Note 12: `CheckingAccount c = new CheckingAccount(2500.00, .50);`

The minimum balance for free checking is $2500 and the service charge per check if the balance is below $2500 is $.50.

Note 13: `s.deposit(135.22);`

We try all the `SavingsAccount` methods.

Note 14: `c.deposit(1000.00);`

Because the initial account balance is zero, we first make a deposit.

Note 15: `c.processCheck(200.00);`

Because the balance is below $2500, we will withdraw $200.50 to cover the amount requested and the service charge.

Note 16: `c.withdraw(100.00);`

Because the balance is below $2500, we will withdraw $100.50 to cover the amount requested and the service charge.

The Big Picture

The superclass contains data and operations common to all subclasses. Rather than duplicating these fields in each subclass, we let the subclasses inherit them from the superclass. A subclass can override inherited methods and/or add additional methods or data fields.

Inheritance supports the IS-A relationship in which a subclass object is also a type of superclass object.

The Object Class

The `Object` class, in the `java.lang` package, is a superclass of every Java class. A class, such as `Person` from Example 2.8, that does not explicitly extend any class, implicitly extends `Object`. Java treats the `Person` class as if we had declared it as

```
public class Person extends Object {// same as Example 2.8}
```

Our `SavingsAccount` class from Example 3.1 explicitly extends `BankAccount`, while `BankAccount` implicitly extends `Object`. The Java library classes all directly or indirectly extend `Object`.

The `Object` class provides a default implementation of the `toString` method which returns an empty string. Because every class inherits from `Object`, every class can invoke its `toString` method. For example, a `BankAccount` object, `myAccount`, is also an `Object`, so we can invoke the `toString` method, explicitly as in

```
myAccount.toString();
```

or implicitly, as in

```
System.out.println(myAccount);
```

either of which will display an empty string. Any class that wants to have a string representation should override the `toString` method as we did in Example 2.8 for the `Name`, `Address`, and `Person` classes. Figure 3.5 shows the `BankAccount` class inheriting the `toString` method from `Object`, while the `Person` class overrides it.

FIGURE 3.5 Inheriting from `Object`

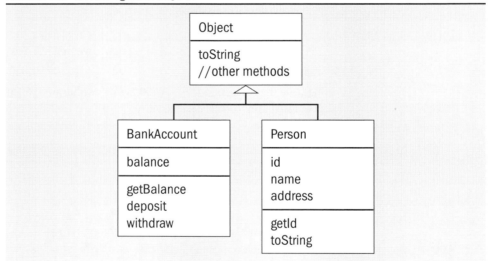

TEST YOUR UNDERSTANDING

1. Each class, except `Object`, extends another class either implicitly or explicitly. For each of the following classes, which class does it extend?

    a. `SavingsAccount`   c. `BankAccount`   e. `CheckingAccount`
    b. `String`   d. `Address`

2. Declare and initialize a `SavingsAccount`, `s`, with an initial balance of $50.00 and an interest rate of 3.5 percent.

3. Declare and initialize a `CheckingAccount`, `c`, which requires a minimum balance of $1500 to avoid the $.35 service charge per check.

4. Describe the output from each of the following, where s is the SavingsAccount defined in question 2, and

```
Name president = new Name("Abraham", "Lincoln");
String s3 = "horse";
```

   a. `System.out.println(s);`
   b. `System.out.println(s3);`
   c. `System.out.println(president);`

5. Suppose we add a withdraw method to the SavingsAccount class, using the pattern

```
public void withdraw(double amount, boolean receipt)
    { ... },
```

   which would perform the withdrawal giving a receipt if the **boolean** variable receipt is **true**. Would we be overriding the withdraw method of the BankAccount super-class, or overloading it? Explain.

3.2 ▪ Polymorphism and Abstract Classes

With **polymorphism** we get the benefit of letting each type of object define its own behavior. Our programs ask objects to do something, but each does that operation in its own way. Our programs are easier to create and maintain because we leave to the objects the details of how to perform their operations.

Abstract classes let us talk about operations in general deferring to concrete subclasses which the implementation details.

Polymorphic Operations

When we say "that animal is eating", it could mean that a lion is tearing flesh from its kill or that a giraffe is munching tree leaves. Each subclass of Animal has its own way of eating. The eating operation is **polymorphic**, meaning that it has many structures. A lion implements it one way, while a giraffe does something quite different, but they both eat.

Most of us do not talk to the animals, but if we did, we could command animal1 eat, animal2 eat, and so on. Each animal knows how to eat, and will do so in its own way. We do not even have to know what kind of animals we are talking to, or how they eat. Our command is like a program that will execute correctly, even if other animals come along later that we had not known about. We say animal902 eat, and it eats because it knows how to eat.

Our command program would not be nearly as flexible if we had to know the type of each animal. We could start commanding lion eat, giraffe eat, but this program only works for a lion followed by a giraffe, and does not allow for some other type of animal. Our earlier command was more flexible because we used the superclass type, Animal, in our command, rather than the subclass types, Lion and Giraffe.

A program fragment

```
animal.eat();
```

is much easier to code and maintain than the fragment shown at the top of the following page.

```
if (animal is a giraffe)
    animal.eatLikeAGiraffe();
else if (animal is a frog)
    animal.eatLikeAFrog();
else if (animal is a lion)
    animal.eatLikeALion();
```

To deal with a duck, the first version requires no change, but the second needs another test added, namely,

```
else if (animal is a duck)
    animal.eatLikeADuck();
```

In the first version, `animal.eat()`, Java calls the correct version of the polymorphic `eat` method, depending on what type of object `animal` refers.

The Big Picture

Polymorphism is one of the key benefits of object-oriented programming. A statement such as `animal.eat()` will apply to objects of any subclass of animal, such as lions and giraffes, and will apply to those objects yet to be defined, such as ducks. Such code needs less modification because it is independent of the details of how each type of animal eats, which are left to the subclasses to implement.

Java lets us use polymorphic operations for our objects. We cannot make our bank accounts eat, but we can make withdrawals. To be flexible we want to refer to all our accounts using the superclass type `BankAccount`. After all, every `SavingsAccount` or `CheckingAccount` is a `BankAccount`, just as every lion or giraffe is an animal.

Suppose we have several `BankAccount` objects, b1, b2, and b3. We can command them to withdraw $50 each with the statements

```
b1.withdraw(50.00); b2.withdraw(50.00);
b3.withdraw(50.00);
```

We don't have to know what kind of bank accounts b1, b2, and b3 are, because every bank account knows how to withdraw, just as every animal knows how to eat. This is the beauty of object-oriented programming—each object implements its own behavior.

If bank account b1 happens to be a checking account, it will deduct a service charge if the balance is below the minimum for free checking; we just ask it to withdraw, and trust b1 to know how to process a `withdraw` request. Our little program is quite flexible. We can apply it at some future time when we have written a `TimedAccount` subclass of `BankAccount` which overrides the `withdraw` method to prohibit withdrawals. If b2 happens to refer to a `TimedAccount` then Java will execute the `withdraw` method defined in the `TimedAccount` class.

The `withdraw` operation has many structures, depending upon which subclass of `BankAccount` is processing the withdrawal. As we saw with animals, to get the flexibility of polymorphism, we need to refer to objects by their superclass type. We start by creating two `BankAccount` objects (see Figure 3.6)

FIGURE 3.6 Two BankAccount objects

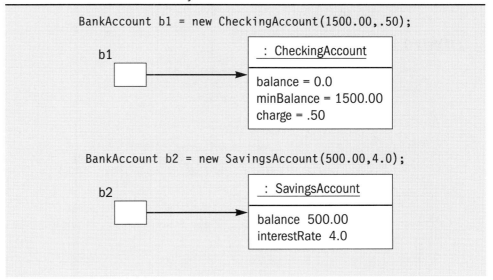

```
BankAccount b1 = new CheckingAccount(1500.00,.50);
```

b1

: CheckingAccount

balance = 0.0
minBalance = 1500.00
charge = .50

```
BankAccount b2 = new SavingsAccount(500.00,4.0);
```

b2

: SavingsAccount

balance 500.00
interestRate 4.0

We declare b1 with type BankAccount, but actually assign it a value of type CheckingAccount which is a subtype of BankAccount. Similarly, we declare b2 with type BankAccount, but assign it a SavingsAccount. If we make a deposit and withdrawals

```
b1.deposit(400.00);
b1.withdraw(50.00);
b2.withdraw(50.00);
```

then b1 will process the withdraw using CheckingAccount withdraw, deducting a service charge, while b2 will process the withdraw using SavingsAccount withdraw (inherited from BankAccount), not deducting a service charge.

If we now change b2 to refer to the CheckingAccount to which b1 refers, and then do a withdrawal, as in

```
b2 = b1;
b2.withdraw(50.00);
```

the account b2 will now process the withdrawal using CheckingAccount withdraw, and will deduct a service charge, because as shown in Figure 3.7, b2 refers to a checking account.

FIGURE 3.7 Variable b2 refers to a checking account

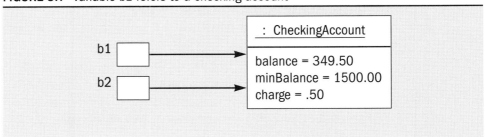

b1

b2

: CheckingAccount

balance = 349.50
minBalance = 1500.00
charge = .50

As the program is running, Java determines to which type object the variable b2 refers, and invokes the withdraw method for that type.

EXAMPLE 3.2   **Withdraw.java**

```
/*  Uses the withdraw operation of
 *  BankAccount and its subclasses
 *  to illustrate polymorphism.
 */

public class Withdraw {
    public static void main(String [] args) {
        BankAccount b1 = new CheckingAccount(1500.00,.50);          // Note 1
        BankAccount b2 = new SavingsAccount(500.00, 4.0);
        b1.deposit(400.00);
        b1.withdraw(50.00);                                         // Note 2
        System.out.print
            ("The balance of the BankAccount to which b1 refers is ");
        Io.println$(b1.getBalance());
        b2.withdraw(50.00);                                         // Note 3
        System.out.print
            ("The balance of the BankAccount to which b2 refers is ");
        Io.println$(b2.getBalance());
        b2 = b1;                                                    // Note 4
        b2.withdraw(50.00);                                         // Note 5
        System.out.print
            ("The balance of the BankAccount to which b2 refers is "); // Note 6
        Io.println$(b2.getBalance());
    }
}
```

Output
```
The balance of the BankAccount to which b1 refers is $349.50
The balance of the BankAccount to which b2 refers is $450.00
The balance of the BankAccount to which b2 refers is $299.00
```

Note 1: The key to using polymorphism is to declare objects of a general superclass type but assign instances of various subclasses. We do that here, declaring b1 to have type BankAccount, but assigning it an object of type CheckingAccount.

Note 2: Because b1 refers to a CheckingAccount, the withdraw deducts a service charge.

Note 3: Because b2 currently refers to a SavingsAccount, the withdraw does not deduct a service charge.

Note 4: We change b2 to refer to a CheckingAccount.

Note 5: Because b2 now refers to a CheckingAccount, the withdraw deducts a service charge.

Note 6: "The balance of the BankAccount to which b2 refers is"

The variable b2 refers to some bank account object, either a SavingsAccount or a CheckingAccount. This example shows that we can change this reference during the course of the program among various subclasses of BankAccount which is the declared type of b2.

We illustrated polymorphism with animals eating and withdrawals from bank accounts. Every bank account can process a withdrawal and every animal can eat. Suppose we try another command, `animal1 brushTeeth`, `animal2 brushTeeth`, and so on. If `animal1` is our brother, he might diligently execute this command, but if `animal1` is a lion, he might show us his teeth, but he would not be brushing them. Not every animal brushes its teeth. Lion command specialists would advise us not to try to command a lion to brush its teeth, as we may get an error (called a runtime error) while we are running our command program (not enough time to run).

Just as we cannot expect every animal to brush its teeth, we cannot expect every bank account to process a check. Only a `CheckingAccount` has a `processCheck` method. If we create a `SavingsAccount` and try to process a check, as in

```
BankAccount b = new SavingsAccount(750.00, 5.0);
b.processCheck(50.00);        // rejected
```

the Java command specialist will reject the `processCheck` statement. We declared b to have type `BankAccount`, which cannot always process a check. At runtime b might actually refer to a `CheckingAccount` which can process a check, but then again it might refer to a `SavingsAccount` which cannot. The Java compiler stops us from making a fatal runtime error, allowing only a method, like `deposit` or `withdraw`, that every account can execute.

If we want to use methods that apply only to a certain type of account, then we should declare our objects to be of that type. For example, declaring c to be a `CheckingAccount` will allow us to execute `c.processCheck`, as we saw in Example 3.1.

Abstract Classes

Nothing is just an `Animal`. Every animal is a member of some subclass such as `Lion` or `Giraffe`. The class `Animal` has no instances, in contrast to the class `Lion` which has various, perhaps ferocious, instances.

In retrospect, now that we have `SavingsAccount` and `CheckingAccount` classes, our `BankAccount` class might be better defined to have no instances. We never want an object that is just a `BankAccount`; every account is either a `CheckingAccount` or a `SavingsAccount`. Java uses the modifier `abstract` to denote that a class cannot have any instances. Declaring `BankAccount` using the pattern

```
public abstract class BankAccount { // same as before }
```

will cause the Java compiler to reject any attempts to instantiate `BankAccount` objects, as for example,

```
BankAccount b = new BankAccount(1000.00); // rejected now
```

By using the `abstract` modifier we can make `BankAccount` an abstract class. Nevertheless, `BankAccount` implements the `deposit`, `withdraw`, `getBalance`, and `setBalance` methods, which its subclasses inherit. The `BankAccount` class implements all its methods. An abstract class can implement only some of its methods and leave others to be implemented by subclasses as we will see when we develop the `Shape` class next.

A Shape Class

Nothing is just a Shape, but Shape has subclasses such as Line and Circle that have instances. We define every Shape to have an instance variable, center, which is a Point object, and instance methods toString, draw, and move. The toString method provides a string representation for a shape.

A shape has no particular form, so we make the draw operation abstract, to be implemented only in subclasses of Shape. Java allows us to do this by using the abstract modifier, as in

```
public abstract void draw();
```

which declares, but does not implement, the draw method.

☞ **TIP**   Any class that has an abstract method must be declared abstract using the abstract modifier.

The move method has two arguments—the distances to move in the x and y directions. Every Shape moves by moving its center by the specified distances in each direction. Figure 3.8 shows the diagram for the abstract Shape class. Using the UML notation, we italicize the class name and the abstract draw method, and also designate Shape as abstract, using {abstract}.

FIGURE 3.8 The abstract Shape class

Shape {abstract}
center : Point
draw toString move

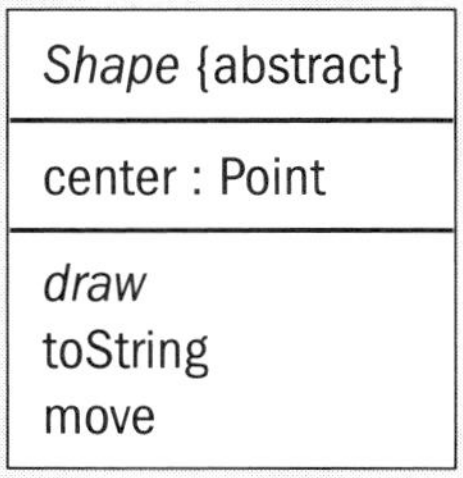

The Line class extends Shape, adding an instance variable, end, of type Point to represent the other end of the line (the center point of the parent shape represents one end of the line). The Line class implements its own toString and move methods. The move method first calls the move method of the Shape superclass to move the center point and then moves its own endpoint. When we learn to use Java graphics we can draw a line, but for now we just draw a line by displaying its string representation.

The Circle class extends Shape, adding an instance variable, radius, of type **int** to represent the radius. (The center point of the parent shape represents the center of the circle.) The Circle class implements its own toString method, and its draw method just calls the toString method for now. The Circle class is happy to inherit the move method from the Shape class, which moves the center by the specified amounts.

The java.awt package, that we discuss later in the text, has a Point class that we can use in our shapes. We use a constructor

```
public Point(int x, int y);
```

that creates a Point, given its x and y coordinates, a method

```
public void translate(int x, int y);
```

that translates a Point by the specified x and y values, and the toString method. Translating a point p at $(3,4)$ by 2 in the x-direction and 5 in the y-direction, as in:

```
Point p = new Point(3,4);
p.translate(2,5);
```

would move the point p to $(5,9)$.

Example 3.3 contains the Shape, Line, and Circle classes. We add a main method to the Shape class to test our classes. We create two shapes, assigning a line to the first and a circle to the second. Because the Shape class declares draw and move methods, the Java compiler lets each of our shapes invoke them. Java, at runtime, finds the right version of each method. If we move a shape which is a line, then Java executes the move defined in the Line class, but if we move a shape which is a circle, then Java, realizing that circles inherit the move operation from the Shape class, calls the move method defined in the Shape class.

EXAMPLE 3.3   **Shape.java**

```
/*   Shape is an abstract class with
 *   subclasses Line and Circle.
 *   Shape has an abstract draw method
 *   which the subclasses implement.
 *   The test uses shapes polymorphically.
 */

import java.awt.Point;                                    // Note 1

public abstract class Shape {                             // Note 2
    Point center;                                         // Note 3
    public Shape() {                                      // Note 4
      center =  new Point(0,0);
    }
    public Shape(Point p) {
      center = p;
    }
    public abstract void draw();                          // Note 5
    public String toString(){
      return "Shape with center " + center;               // Note 6
    }
    public void move(int xamount, int yamount) {
      center.translate(xamount,yamount);                  // Note 7
    }
    public static void main(String[] argv) {
        Shape s0 = new Line(2,5,6,7);                     // Note 8
        Shape s1 = new Circle( new Point(3,4) ,5);
        s0.move(3,-1);                                    // Note 9
        s0.draw();                                        // Note 10
        s1.move(3,-1);                                    // Note 11
        s1.draw();                                        // Note 12
    }
}
```
continued

```
class Line extends Shape {
    Point end;
    public Line(int x1, int y1, int x2, int y2) {        // Note 13
        super(new Point(x1,y1));                          // Note 14
        end = new Point(x2,y2);
    }
    public void draw() {                                 // Note 15
      System.out.println(toString());
    }
    public String toString() {
        return  "Line from "+center + " to " + end;
    }
    public void move(int xamount, int yamount) {
        super.move(xamount,yamount);
        end.translate(xamount,yamount);
    }
}
class Circle extends Shape {
    int radius;
    public Circle(Point p, int r) {
        super(p);
        radius = r;
    }
    public void draw() {
        System.out.println(toString());
    }
    public String toString() {
        return "Circle at "+center+" with radius "+radius;
    }
}
```

Output
```
Line from java.awt.Point[x=5,y=4] to java.awt.Point[x=9,y=6]
Circle at java.awt.Point[x=6,y=3] with radius 5
```

Note 1: We discuss the `java.awt` package in Chapter 5.

Note 2: The `abstract` modifier indicates that Shape will have no instances.

Note 3: We omit the `public` modifier to make the center `Point` visible to other classes in the package. We discuss modifiers and access in Section 3.4. We could have used the `private` modifier and provided a public method, `getCenter`, to access the center point.

Note 4: The default constructor sets `center` to the origin, (0,0).

Note 5: We declare the `draw` method as abstract. Any class with an abstract method is an abstract class with no instances. Any subclass that we want to instantiate must implement the `draw` method.

Note 6: `return "Shape with center " + center;`

Java invokes the `center.toString` method to get the string representation for the center.

Note 7: `center.translate(xamount,yamount);`

The translate method of the `Point` class moves the `center` point by the specified x and y values.

Note 8: `Shape s0 =new Line(2,5,6,7);`

The line has coordinates (2,5) and (6,7).

Note 9: `s0.move(3,-1);`

Java determines at runtime that `s0` refers to a `Line` and calls the correct version of move from the `Line` class.

Note 10: `s0.draw();`

Because `s0` refers to a `Line`, Java calls the draw method from the `Line` class, which is no longer abstract.

Note 11: `s1.move(3,-1);`

Because `s1` refers to a `Circle`, Java uses the move method inherited from `Shape`.

Note 12: `s1.draw();`

Because `s1` refers to a `Circle`, Java calls the draw method defined in `Circle`.

Note 13: `public Line(int x1, int y1, int x2, int y2) {`

We pass the x and y coordinates of the two endpoints of the line. Alternatively, we could have defined the constructor to accept two `Point` arguments.

Note 14: `super(new Point(x1,y1));`

We pass one `Point` to the `Shape` constructor to initialize the `center` point.

Note 15: `public void draw() {`

Until we cover graphics in Chapter 5 we implement `Line` and `Circle` draw by just displaying the string representation of each object.

The Big Picture

By declaring a reference, s, as an abstract `Shape`, we can ask it to draw or move,

`s.draw();`   `s.move(3,4);`

leaving the details of how a concrete shape such as a line moves or draws itself to the details of the implementation of the subclass. The same operations will apply to any subclass of Shape, even those, such as Rectangle, Polygon, or Ellipse, which we have not yet defined.

Our program will depend on the abstract class, which is less likely to change than are the details of the concrete subclasses that extend it.

TEST YOUR UNDERSTANDING

6. If we declare b as in

   `BankAccount b = new CheckingAccount(1500.00,.50);`

which of the following will generate a compiler error?

a. `b.deposit(100.00);`   d. `b.getBalance(); ;`

b. `b.processCheck(25.25)`   e. `b.withdraw(75.00);`

c. `b.postInterest();`

7. If we declare b as in

   `CheckingAccount b = new CheckingAccount(1500.00,.50);`

   which of the method calls of question 6 will generate a compiler error?

TRY IT YOURSELF

8. What will be the result if in the `Shape` class of Example 3.3 we declare it without the `abstract` modifier, declaring it as

   `public class Shape { // same as Example 3.3 };?`

   Revise Example 3.3 to omit the modifier at that position. Does the compiler allow this change?

9. Will the `Shape` class of Example 3.3 still be abstract if we revise that example to implement the `draw` method in the `Shape` class to return the string representation of a `Shape`? Explain.

3.3 Interfaces

An **interface** is like an abstract class but an interface only contains abstract methods and constants, while an abstract class may implement some or all of its methods. We will see that we can use an interface to provide a view of objects from a certain perspective, showing some but not all of their behaviors. By using more than one interface we can view a single object from multiple perspectives. An interface for a component tells what services it supports, independently of any implementation. By programming to an interface we can replace one component by another as long as it supports the same interface; our program will not have to change.

An interface specifies abstract methods and constants only. A class can implement an interface by providing implementations of each method in the interface. For example, Figure 3.9 shows an interface, `Scaleable`, which specifies a scale method. In the diagram we use the <<interface>> designation for interfaces. We italicize the interface name and its methods since they are always abstract. The empty middle row shows that an interface has no data fields.

FIGURE 3.9 The `Scaleable` interface

```
public interface Scaleable {
    public void scale (int factor);
}
```

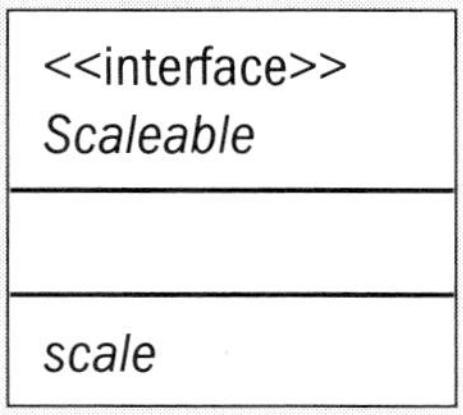

<<interface>>
Scaleable
scale

Any class that wants to scale the size of its objects can implement the `Scaleable` interface, providing an implementation for the scale method. For example, Figure 3.10 shows the

ScaleableCircle class which extends Circle to provide the usual Circle methods and data, and implements Scaleable to be able to scale the circle by multiplying its radius by a scale factor.

FIGURE 3.10 Implementing an interface

```
class ScaleableCircle extends Circle implements Scaleable    {
    public ScaleableCircle(Point p, int r) {
        super(p,r);
    }
    public void scale(int factor) {
        radius *= factor;
    }
}
```

The keyword implements signifies that the ScaleableCircle class will implement the scale method specified in the Scaleable interface. Because interfaces do not contain data fields or method implementations which could cause ambiguities, Java permits a class to implement more than one interface.

We can include interface definitions along with class definitions in our programs. An interface cannot have non-constant data fields, and cannot implement any methods. All methods of an interface are abstract. We have the option to use the modifier abstract in interface method declarations, but do not need to, and typically do not.

For our example program we use an interface to notify businesses and investors interested in responding to interest rate changes. We will continue this example in Section 4.6 after we introduce arrays.

The RateChangeListener interface will specify methods rateRaised and rateLowered.

```
interface RateChangeListener {
    public void rateRaised(double amount);
    public void rateLowered(double amount);
}
```

The Investor and Business classes each implement the RateChangeListener interface to take appropriate action when the interest rate changes. In the main method, we call the rateRaised method to inform the investor and the business that the we raised the interest rate.

The declarations

```
RateChangeListener investor = new Investor();
RateChangeListener business = new Business();
```

create the investor and business objects using the interface type RateChangeListener. Since we are only dealing with these objects regarding interest rate changes we do not need to know what other behavior each has. The Investor class has a countMoney operation which we cannot access using the investor object. In this example, we are using an interface to view an object from a particular perspective, ignoring its other behavior. Similarly, we cannot access the doPayroll operation of the Business class because we are viewing the business object only as a RateChangeListener. By using interfaces we can allow an application to see only those aspects of an object that it needs.

EXAMPLE 3.4 SendRateChange.java

```java
/* Any object that implements the RateChangeListener
 * interface can be notified about interest rate
 * changes.  We notify investor and business objects
 * which implement RateChangeListener to take action
 * when the rate change occurs.
 */

interface RateChangeListener {
    public void rateRaised(double amount);
    public void rateLowered(double amount);
}
class Investor implements RateChangeListener {
    public void rateRaised(double amount) {
        System.out.println("   Investor sells stocks");
    }
    public void rateLowered(double amount) {
        System.out.println("   Investor buys stocks");
    }
    public void countMoney() {
        System.out.println("Counting money");
    }
}
class Business implements RateChangeListener {
    public void rateRaised(double amount) {
        System.out.println("   Business reduces debt");
    }
    public void rateLowered(double amount) {
        System.out.println("   Business takes a loan");
    }
    public void doPayroll() {
        System.out.println("Doing payroll");
    }
}
public class SendRateChange {
    public static void main(String [] args) {
        RateChangeListener investor = new Investor();      // Note 1
        RateChangeListener business = new Business();
        System.out.println("Raising interest rates");
        investor.rateRaised(.50);                          // Note 2
        business.rateRaised(.50);
        System.out.println("Lowering interest rates");
        investor.rateLowered(.25);
        business.rateLowered(.25);
    }
}
```

Output
```
Raising interest rates
   Investor sells stocks
   Business reduces debt
Lowering interest rates
   Investor buys stocks
   Business takes a loan
```

Note 1: We declare the investor object as a RateChangeListener. An interface is abstract, so we must initialize the investor object with a class which implements the RateChangeListener interface. By declaring investor as a RateChangeListener we limit the operations we can perform on it to the two methods, rateRaised and rateLowered, contained in that interface. We view the Investor object from the perspective of interest rate changes, and have no access to any other behavior, such as countMoney, that it might have.

Note 2: We know that the investor implements the RateChangeListener interface, so we can call the rateRaised method.

The Big Picture

Interfaces allow us to view objects from a certain perspective, showing part, but not necessarily all, of their behaviors.

Interfaces permit more flexibility for polymorphism than can be attained using inheritance from a superclass. In Example 3.3, we attain polymorphism when asking shapes to move, but each object must be a subclass of Shape. By contrast, in Example 3.4 we achieve polymorphism by calling the rateRaised method for objects that implement the RateChangeListener interface. These objects, like business and investor, do not have to be subclasses of a common superclass.

TEST YOUR UNDERSTANDING

10. Explain the difference between an interface and an abstract class.

11. What interface would a Consumer class need to implement to be notified about interest rate changes in Example 3.4? Which methods?

TRY IT YOURSELF

12. Modify the Circle class of Example 3.3 to implement the Scaleable interface, which will make all circles scaleable.

3.4 ▪ Modifiers and Access

In a library, we may use reference books such as encyclopedias but not take them home, whereas we are encouraged to borrow non-reference books to read at home. Rare books may be restricted to scholars with special credentials.

Just as access to library books varies, so can access to Java classes, data fields, and methods vary. Java uses the modifiers public, private, and protected to specify the type of access. Before we consider modifiers and access we need to take a second look at packages, as package access is one of the topics we will discuss.

A Second Look at Packages[1]

Java groups related code into packages. Figure 3.11 shows the Java packages that we use in this text.

---

[1] Section 1.6 contains a first look at packages.

FIGURE 3.11 Java packages used in this text

Package Name	Description
java.applet	applets
java.awt	abstract windowing toolkit
java.awt.event	event handling in the AWT
java.beans	Java Beans
java.io	input and output
java.lang	classes central to Java
java.lang.reflect	reflection
java.math	big numbers
java.net	networking
java.rmi	Remote Method Invocation
java.rmi.server	Remote Method Invocation
java.sql	database access
java.text	formatting and internationalization
java.util	utility classes

We can put our own code in a package and would certainly do so if our code was meant to be used by others or as part of a larger system. To put code in a named package, we use a `package` statement as the first statement in the file. For example, our Io class, containing methods like `readInt` that we often use, is in a package named `iopack`. The file `Io.java` has as its first statement

```
package iopack;
```

All files that are part of a package must be in a directory with the same name as that of the package. Because the Io class is in the package `iopack`, the file `Io.class` must be in a directory named `iopack`, the same name as the package. Java has to find the classes that we use, and it uses the directory structure on the host machine to locate these classes.

If we want to use the Io class from the package `iopack` in a program, we have to tell Java which package it is in. We can do this in two ways. One way is to use the fully qualified name of the class, as in

```
iopack.Io.readInt("Enter an integer");
```

which tells Java to look in the directory `iopack` to find the Io class.[2] If we use methods from the Io class often this will result in many longer method names. The other way is to use an `import` statement in our program, as in

```
import iopack.Io;
```

which also tells Java to look in the `iopack` directory for the Io class. In this case, when we use static methods from the class, we can refer to them just using the class name, as in

```
Io.readInt("Enter an integer");
```

---

[2] Java also needs to find the `iopack` directory. See Appendix E for instructions on making packages available using various development environments.

To use the `Name`, `Address`, and `Person` classes from Example 2.8 in other applications we put them in a package `personData` by adding the statement

```
package personData;
```

as the first line in each of these programs and placing the files `Name.class`, `Address.class`, and `Person.class` in a directory named `personData`. We use the `personData` package in the `AtmScreen` applet of Example 1.5 and in the `Atm` application of Example 3.6.

Now that we have seen how to put classes in a package, we return to the topic of modifiers and access.

Class (and Interface) Visibility

(Replace 'class' everywhere in this section by 'class or interface'; the visibility rules are the same for both.)

We can declare a class using the `public` modifier, as in

```
public class SendRateChange { ... }
```

or without any modifier, as in

```
class Investor { ... }
```

We can access a `public` class from other packages, but we can only access a class that lacks the `public` modifier from the package it is defined in. We say that a class declared without the `public` modifier has package visibility, meaning that it is visible only in its own package.

Our small examples are self-contained; each would perform the same with or without the `public` modifier for classes. The significance of the `public` modifier only shows up if we want to use a class outside of its package. For example, our `Io` class is public in the package `iopack`. Leaving off the `public` modifier would make the `Io` class visible only within the package `iopack`, defeating its purpose of providing methods to use in classes outside of the `iopack` package.

Looking at the source code provided with the Java Development Kit, in the `src` directory, shows that the classes we use, such as the `Math` class from the `java.lang` package are public. If they were not, we could not have used them. We can put at most one public class on any file. Again looking at the JDK source code shows that each public class is in a separate file which has the same name as that class.

Data Field and Method Visibility

We can declare data fields and methods using at most one of the modifiers `public`, `private`, or `protected` to specify the type of access we want for that data field or method. Figure 3.12 lists these modifiers with the least restrictive at the top and the most restrictive at the bottom.

Access to Data Fields

In our examples, we usually make data fields private to hide the data, allowing users to access it only by means of the public methods of the class.

When we define a class which we expect to be a superclass of various subclasses, we have to decide how we would like these subclasses to have access to data of this superclass. Had we

FIGURE 3.12 Access modifiers for data fields and methods

public	Accessible anywhere the class name is accessible.
protected	Accessible in the package that contains the class in which the data field or method is declared, and in any subclass of that class.
(no modifier)	Accessible in the package that contains the class in which the data field or method is declared.
private	Accessible only in the class in which the data field or method is declared.

declared the center `Point` as private in the `Shape` class of Example 3.3, then the `toString` methods of subclasses `Line` and `Circle` would not have had access to the `center` variable.

In Example 3.3, we declared the center `Point` without using any access modifier, meaning that the `center` variable is accessible in the package containing the `Shape` class, which in this example is the current directory. Another choice would be to declare `center` with private access, and include a method, `getCenter`, as in,

```
public Point getCenter() {
    return center;
}
```

which would allow all users of shapes to get the value of `center`. The `toString` method of the `Circle` class could use `getCenter` to find the center point, as in

```
public String toString() {
    return "Circle at " + getCenter() + " with radius " + radius;
}
```

Comparing these two choices, using no modifier lets any class in the same package as `Shape` use the `center` variable directly. Typically the classes in a package have a common purpose; the package developer makes the compiled classes in the package available for use. Only the package developer would be using the data directly and could make desired changes without affecting users of the package.

Making the data private ensures that no other class can use the data directly. The developer can change the data without affecting any users as long as the public methods remain the same. This approach requires a method such as `getCenter()` if users need to read the data, and another method such as `setCenter()` if users need to change the data.

The other two modifiers, `public` and `protected`, have drawbacks when applied to data fields. Public data can be modified by any user of the class. We would like our classes to provide a carefully chosen interface using public operations, following the model of primitive types which we access using operations such as + and *, never using the representation of the primitive data.

Using the `protected` modifier makes a data field accessible in the current package, and in any subclass of the class in which the field is declared. Because a class has no way to know which classes may have extended it, it has no way to communicate a change in its data representation to subclasses who have access to protected data fields. For example, had we declared the center point as

```
protected Point center;
```

and then later change the representation to use x- and y-coordinates instead of a `Point`, all subclasses of `Shape` would have to be revised.

☞ TIP   Avoiding public or protected data fields will allow you to change the representation of the data without affecting users of your class. Provide methods if users need to access the data.

Access to Methods

Most often we make methods `public`. An object's methods provide its behavior for others to use. Our public `draw`, `move`, and `toString` methods allow users to perform these operations on any shape. Class developers might use a `private` method to help implement the `public` methods.

☞ TIP   Make any methods private that are not intended for users of the class. Such a method might use an algorithm to help implement the public methods. You can change a private method later to use a better algorithm, knowing that no users of the class will be affected.

Protected methods also have their uses. For example, if we declare the `center` instance variable of the `Shape` class of Example 3.3 using the `private` modifier, we could declare `getCenter` and `setCenter` methods using the protected modifier, as in

```
protected Point getCenter() {
    return center;
}
protected void setCenter(Point p) {
    center = p;
}
```

which would limit access to these methods to the package containing `Shape` and to subclasses of `Shape`.

Example 3.5 illustrates the use of the access modifiers for classes, data fields, and methods. We put classes A, B, and C in a package named `visibility`, which means that they must be in a directory with the same name. Class D extends A but is not in the same package as A, while class E is another unrelated class. Inside A we declare data fields and methods with each of the four kinds of access, and show in which classes these data fields and methods will be visible. We demonstrate the difference in visibility between the public class A and the non-public class B.

The UML notation allows us to indicate the access (visibility) of each field in a class diagram. Figure 3.13, for class A of Example 3.5, shows these symbols, which are useful when we wish to specify more details.

When we do not put our class in a named package by using a package statement at the beginning of the program, that class is in a default package consisting of all Java classes in the current directory. Figure 3.14 diagrams the two packages in Example 3.5, the package visibility and the default package. We do not show the class B in the diagram because access to it is limited to the classes in the visibility package.

FIGURE 3.13 Indicating access in a class diagram

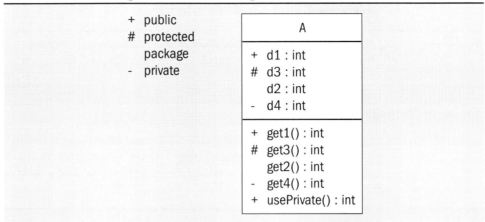

FIGURE 3.14 The packages of Example 3.5

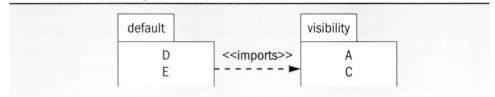

EXAMPLE 3.5   **A.java**

```
/* Public class A declares data fields and methods
 * with each of the four access modifiers
 * (including none) to illustrate their use.
 * Class B, not public, will be visible only in the
 * package visibility in which it is declared.
 */

package visibility;                                      // Note 1

public class A {                                         // Note 2
    public int d1=1;                                     // Note 3
    int d2=2;
    protected int d3=3;
    private int d4=4;

    public int get1() {
        return d1;
    }
    int get2() {
        return d2;
    }
    protected int get3() {
        return d3;
    }
```

```
        private int get4() {
            return d4;
        }
        public int usePrivate(){                          // Note 4
            return d4 + get4();
        }
    }

    class B {                                             // Note 5
        public int d5=5;
    }
```

C.java

```
/*  Class C in the same package as A and B
 *  can use A, B and all data fields and methods
 *  except the private d4 and getd4().
 */

package visibility;

public class C {                                          // Note 6
    public static void main(String [] args) {
        A a = new A();
        B b = new B();
        int i, j;
        i = a.d1 + a.d2 + a.d3 + b.d5 + a.usePrivate();
        j = a.get1() + a.get2() + a.get3();
        System.out.println("i is " + i + " and j is " + j);
    }
}
```

D.java

```
/* Class D extends A but is not in the same package.
 * D cannot use d2, get2(), or the class B which are
 * visible only in the package in which they are declared.
 * D can use d3 and getd3(), but only from an object of
 * type D, not from an object of type A.  Class E can
 * only use public data and methods.
 */

import visibility.*;                                      // Note 7

public class D extends A {                                // Note 8
    public static void main(String [] args) {
        A a = new A();
        D d = new D();
        int i,j;
        i = a.d1 + d.d3 + a.usePrivate();                 // Note 9
        j = a.get1() + d.get3();
        System.out.println("i is " + i + " and j is " + j);
    }
}
```

continued

```
class E {                                              // Note 10
    public static void main(String [] args) {
        A a = new A();
        D d = new D();
        int i,j;
        i = a.d1 + d.usePrivate();                     // Note 11
        j = d.get1();
        System.out.println("i is " + i + " and j is " + j);
    }
}
```

Output—C

i is 19 and j is 6

Output—D

i is 12 and j is 4

Output—E

i is 9 and j is 1

Note 1: We put classes A, B, and C in a package to illustrate the effect of access modifiers.

Note 2: The public class A will be visible everywhere.

Note 3: We declare four variables and four methods using each of the three modifiers and no modifier. We will see which we can access in classes C, D, and E.

Note 4: We can directly access the private d4 and get4 only in A, but we can access a public method that uses them anywhere.

Note 5: Declared with no modifier, the class B is visible only in its package, visibility. The public d5 is visible anywhere B is.

Note 6: `public class C {`

Class C, in the same package as A and B, can use both A and B, and all but the private d4 and get4.

Note 7: `import visibility.*;`

This `import` statement tells Java to look in the `visibility` package for any classes it cannot otherwise find. The star * matches every class. We could have used two `import` statements to name the classes explicitly, as in:

```
import visibility.A;
import visibility.C;
```

Note 8: `public class D extends A {`

Class D extends A, but is not in the same package as A. D cannot use B or its data, or d2 and get2 all of which are visible only in the package in which they are declared.

Note 9: `i = a.d1 + d.d3 + a.usePrivate();`

A subclass object d, of type D, can use the protected d3 and get3, but an object a, of type A, cannot access d3 and get3.

Note 10: `class E {`

The class E, which is not in the same package as A, and not a subclass of A, can only access the public data fields and methods of A.

Note 11: `i = a.d1 + d.usePrivate();`

Either a, of type A, or d, of the subtype D of A, can access the public fields of A.

The steps to compile and run programs with named packages depend upon the development environment used.[3]

The Big Picture

Groups of related classes belong in a named package with each public class on a separate file, contained in a directory with the same name as the package. Java looks in the classpath to find the directory. Classes declared without using the public modifier may only be used inside the package.

Instance variables and methods can have four levels of access. Usually data is private; the default is access within the same package. Methods representing the behavior of the object are public, but the other access modes have their uses.

TEST YOUR UNDERSTANDING

13. What must be the first statement in a file whose contents we wish to include in a package named `stuff`?

14. Suppose you wish to use a class, `GoodStuff`, from the package `stuff`, in your program. Write the `import` statement that would allow Java to find the `GoodStuff` class.

15. Suppose you want to call the `static` method `doStuff()` from the class `GoodStuff` in the package `stuff`. How could you do it without using an `import` statement?

16. What must be the directory name which contains all files that are part of the package `stuff`?

17. Example 3.3 uses the `Point` class from the `java.awt` package. Without looking at the source, can you tell whether or not `Point` is declared using the `public` modifier? Why or why not?

18. In Example 3.5, what is the difference in accessibility to d2, declared with no modifier, and accessibility to d3, which is protected?

---

[3] Using the JDK, execute the following commands from the directory which contains the `visibility` directory (on other than Windows systems replace the backslash with a forward slash)

```
javac visibility\A.java
javac visibility\C.java
javac D.java
java visibility.C
java D
java E
```

19. In Example 3.5, what is the difference in accessibility to `get2()`, which is declared with no modifier, and accessibility to `get4()`, which is private?

20. Explain why the field d5, declared public in Example 3.5, is not visible in either class D or E.

ENHANCEMENT

3.5 ▪ Object-Oriented Design with Use Cases and Scenarios

In object-oriented programming, we solve our problem by identifying objects, each having certain responsibilities, and let these objects use each others' services. Each object's methods provide that object's services which allow it to meet its responsibilities.

To identify the objects we analyze the system using use cases and scenarios. Each use case describes one function that the system should provide. For each use case, we develop several scenarios, which are step by step listings of the interactions among the user and other parts of the system to provide the function described by that use case. Usually for each use case there will be a primary scenario which represents the interactions for a successful use, and several secondary scenarios representing the various errors that might occur.

Defining the Problem

For our example case study we will develop a simple automatic teller application. A user of the system should be able to choose an account, either savings or checking, and make deposits to, withdrawals from, or get the balance of that account. For simplicity, we assume that each user has at most one account of each type. (The `AtmScreen` applet of Example 1.5 which exemplifies the topics covered in the first six chapters is a solution of this problem based on the analysis presented in this section but with a graphical user interface.)

Designing a Solution

Object-Oriented Design—Developing Scenarios

For the automatic teller system our use cases consist of the deposit, withdrawal, and get balance transactions that the user can perform. To discover the objects we need, we can look at scenarios which represent each use case, first looking at scenarios where everything goes well, and then looking at some processing failures. Figure 3.15 describes a scenario for a successful deposit.

The scenario in Figure 3.15 involves four objects, `user`, `teller`, `bank`, and `account`. In a real system, the user would be an actor who inserts an ATM card and pushes various buttons to respond to menus. This actor would interact with, but not be part of, the system. To enhance the object-oriented flavor of this example, we will use a surrogate user object. Our surrogate in the system could use random numbers to make choices or, as we do here, could ask for guidance from the real user via keyboard input.

For every successful scenario there are usually several scenarios where something goes wrong; Figure 3.16 shows one of them, when no account exists of the type specified by the user.

FIGURE 3.15 A scenario for a successful deposit

```
The user asks the teller to accept an ATM card.
The teller asks the user to enter a PIN.
The user asks the teller to accept a PIN.
The teller asks the user to select a transaction type.
The user asks the teller to accept a deposit.
The teller asks the user to select an account type.
The user asks the teller to accept a savings account type.
The teller asks the bank to find the account of the chosen type for the user
  with the specified PIN.
The bank gives the teller a reference to the account.
The teller asks the user to specify an amount.
The user asks the teller to accept an amount.
The teller asks the account to deposit the specified amount.
The teller asks the user to select another transaction, ...
```

FIGURE 3.16 A scenario when the specified account does not exist

```
The user asks the teller to accept an ATM card.
The teller asks the user to enter a PIN.
The user asks the teller to accept a PIN.
The teller asks the user to select a transaction type.
The user asks the teller to accept a get balance transaction.
The teller asks the user to select an account type.
The user asks the teller to accept a checking account type.
The teller asks the bank to find the account of the chosen type for the user
  with the specified PIN.
The bank gives the teller a null account.
The teller asks the user to select another transaction, ...
```

We leave to the exercises the writing of other scenarios to explore possible uses (and misuses) of the automatic teller system.

Designing a Solution

Object-Oriented Design—Assigning Responsibilities

Using scenarios gives us an idea of the responsibilities for each object. Figure 3.17 shows these responsibilities.

Figure 3.18 shows the relationships between classes we have identified. Each line represents an association between the two classes it connects. These associations are evident from the scenarios in which one class makes a request of another.

Designing a Solution

Object-Oriented Design—Defining the Classes

We can define our classes and create methods to handle each of these responsibilities. The User class contains a Person object of Example 2.8 to store the personal information for that user. Because the user object is going to get input from the human user, we implement the user methods to select a transaction and an account type by using a menu.

FIGURE 3.17 Objects and their responsibilities

User	Enter a PIN
	Select a transaction type
	Select an account type
	Specify an amount
Bank	Find a specified account
Account	Deposit
	Withdraw
	Get balance
Teller	Accept an ATM card
	Accept a PIN
	Accept a transaction type
	Accept an account type
	Accept an amount
	Accept an account

FIGURE 3.18 Class relationships

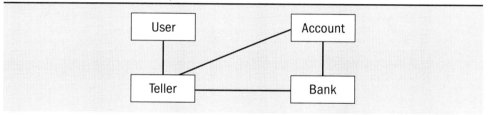

The `Teller` methods accept information from the user and the bank, and use the information to initiate the next step of the transaction. When the teller accepts the account from the bank, it can ask the account to execute a `getBalance` transaction, but needs to ask the user for the amount to deposit or withdraw before asking the account to execute one of these transactions. When a transaction is complete, the teller asks the user to select another transaction, until the user cancels the session.

The `Account` class revises the `BankAccount` class of Example 2.3 by making `Account` abstract and adding a `Person` data field for the account holder. The `Savings` class revises class `SavingsAccount` of Example 3.1 to inherit from `Account` and accept a `Person` in its constructor. The `Checking` class revises class `CheckingAccount` of Example 3.1 to inherit from `Account` and accept a `Person` in its constructor.

Large systems use databases to hold account data. We will cover files in Chapter 8, and Java database access in Chapter 11. Here we simply create three accounts in the `Bank` class. When asked to find an account, the bank checks each account to see if the account id matches the PIN specified by the user and if its type matches the type specified by the user.

We purposely avoid complicating the system. Obviously a real ATM system would be orders of magnitude more complex. We omit any security considerations for entering PINs and greatly simplify the interactions with the bank. Normally the user would have an id such as a social security number and a separate PIN for a bank account. We use only one identification number which we store as the id field in the `Person` class. Each account has a holder field of type `Person` which allows the bank to compare the PIN entered by the user to the id of the account holder.

Completing the Java Code

The object-oriented programming style really becomes evident when we look at the `main` method in the `Atm` class. We just create the objects, a `bank` (and its `accounts`), a `teller`, and two `users`, and let the objects interact to fulfill the scenarios which express the desired behavior of the system. Each user starts the processing by asking the teller to accept an ATM card.

We find modeling with objects natural, because it mirrors our world of objects, even though objects in our programs can also model abstractions not based upon real-world analogies. The process of developing scenarios modifies the stepwise refinement process to apply to interacting objects. Traditional stepwise refinement is essential in implementing those methods of a class which use more complex algorithms.

EXAMPLE 3.6   **Atm.java**

```java
/*  Illustrates object-oriented programming, identifying
 *  objects from scenarios for use cases, and their methods
 *  from the responsibilities identified in the scenarios.
 *  The main method just constructs objects, which interact,
 *  providing the behavior for which they are responsible.
 */
import personData.*;
import iopack.Io;
class User {
    private Person me;
    private Teller teller;
    public User(Person p, Teller t) {                        // Note 1
        me = p;
        teller = t;
        System.out.println("\nUser is \n" + p + "\n");
        teller.acceptCard(this);
    }
    public void enterPIN() {
        String pin = Io.readString("Enter your PIN");
        teller.acceptPIN(pin);
    }
    public void selectTransaction() {                        // Note 2
        int choice;
        do {
            System.out.println();
            System.out.println("Choose a transaction");
            System.out.println("1.  Deposit");
            System.out.println("2.  Withdraw");
            System.out.println("3.  Cancel");
            System.out.println("4.  Get balance");
            choice = Io.readInt("Enter your choice, 1, 2, 3, or 4");
        } while (choice != 1 && choice != 2 && choice != 3 && choice != 4);
        teller.acceptTransaction(choice);
    }
```

continued

```java
    public void selectType() {
        int choice;
        do {
            System.out.println();
            System.out.println("Choose an account type");
            System.out.println("1.  Savings");
            System.out.println("2.  Checking");
            System.out.println("3.  Cancel");
            choice = Io.readInt("Enter your choice, 1, 2 or 3");
        } while (choice != 1 && choice != 2 && choice != 3);
        teller.acceptType(choice);
    }
    public void specifyAmount() {
        double d = Io.readDouble("Enter an amount");
        teller.acceptAmount(d);
    }
}

class Teller {
    public static final int DEPOSIT =1;                    // Note 3
    public static final int WITHDRAW = 2;
    public static final int CANCEL = 3;
    public static final int BALANCE = 4;
    String id;
    int transType;
    int  acctType;
    User user;
    Bank bank;
    Account account;

    public Teller(Bank b) {
        bank = b;                                          // Note 4
    }
    public void acceptCard(User u) {
        user = u;                                          // Note 5
        user.enterPIN();
    }
    public void acceptPIN(String s) {
        id = s;
        user.selectTransaction();
    }
    public void acceptTransaction(int trans) {
        transType = trans;
        if (transType != CANCEL)                           // Note 6
            user.selectType();
    }
    public void acceptType(int type) {                     // Note 7
        acctType = type;
        if (acctType != CANCEL)
          bank.find(id,acctType,this);
    }
```

```
    public void acceptAccount(Account a) {
      account = a;
      if (account != null)                                      // Note 8
        if (transType == BALANCE){
          System.out.print("The balance is ");
          Io.println$(account.getBalance());
          user.selectTransaction();                             // Note 9
        }
        else {
          if (transType == DEPOSIT || transType == WITHDRAW)
            user.specifyAmount();
        }
      else {
        System.out.println("No such account");
        user.selectTransaction();
      }
    }
    public void acceptAmount(double amount) {
        switch(transType) {
          case DEPOSIT :
              account.deposit(amount);
              break;
          case WITHDRAW:
              account.withdraw(amount);
              break;
        }
        user.selectTransaction();
    }
}

class Bank {
    public static final int SAVINGS = 1;
    public static final int CHECKING = 2;
    Name n1 = new Name("John","Venn");
    Address a1 = new Address( "123 Main St.", "Tyler","WY", "45654");
    Person p1 = new Person("123123123",n1,a1);
    Name n2 = new Name("Mabel","Venn");
    Person p2 = new Person("456456456",n2,a1);
    Account p1Savings = new Savings(1500.00,p1,4.0);            // Note 10
    Account p1Checking = new Checking(p1,2500.00,.50);
    Account p2Savings = new Savings(1000.00,p2,3.5);

    public void find(String id, int acctType, Teller teller) {
      if (p1Savings.getId().equals(id) && acctType==SAVINGS)   // Note 11
        teller.acceptAccount(p1Savings);
      else if (p1Checking.getId().equals(id)&& acctType==CHECKING)
        teller.acceptAccount(p1Checking);
      else if (p2Savings.getId().equals(id) && acctType==SAVINGS)
        teller.acceptAccount(p2Savings);
      else
        teller.acceptAccount(null);                            // Note 12
    }
}
```

continued

```java
public class Atm {
    public static void main(String [] args) {                    // Note 13
        Bank bank = new Bank();
        Teller teller = new Teller(bank);
        User user1 = new User(bank.p1,teller);
        User user2 = new User(bank.p2,teller);
    }
}

abstract class Account {                                          // Note 14
    private double balance;
    private Person holder;

    public Account(Person p)     {
        this(0,p);
    }
    public Account(double initialAmount, Person p) {
        balance = initialAmount;
        holder = p;
    }
    public String getId() {
        return holder.getId();
    }
    public void deposit(double amount) {
        balance += amount;
    }
    public void withdraw(double amount) {
        if (balance >= amount)
            balance -= amount;
        else
            System.out.println("Insufficient funds");
    }
    public double getBalance() {
        return balance;
    }
    public void setBalance(double amount) {
        balance = amount;
    }
}

class Checking extends Account {
    private double minBalance;
    private double charge;
    public Checking(Person p, double minAmount, double charge) {
        super(p);
        minBalance = minAmount;
        this.charge = charge;
    }
    public void processCheck(double amount)    {
        if (getBalance() >= minBalance)
            super.withdraw(amount);
        else
            super.withdraw(amount + charge);
    }
```

```
        public void withdraw(double amount) {
            processCheck(amount);
        }
    }

    class Savings extends Account  {
        private double interestRate;
        public Savings(double amount, Person p, double rate) {
            super(amount,p);
            interestRate = rate;
        }
        public void postInterest()  {
            double balance = getBalance();
            double interest = interestRate/100*balance;
            setBalance(balance + interest);
        }
    }
```

Output

```
User is
John Venn
123 Main St.
Tyler, WY 45654

Enter your PIN:   123123123

Choose a transaction
1.  Deposit
2.  Withdraw
3.  Cancel
4.  Get balance
Enter your choice, 1, 2, 3, or 4:   1

Choose an account type
1.  Savings
2.  Checking
3.  Cancel
Enter your choice, 1, 2 or 3:   1
Enter an amount:   50.00

Choose a transaction
1.  Deposit
2.  Withdraw
3.  Cancel
4.  Get balance
Enter your choice, 1, 2, 3, or 4:   3

User is
Mabel Venn
123 Main St.
Tyler, WY 45654

Enter your PIN:   456456456
```

continued

```
Choose a transaction
1.  Deposit
2.  Withdraw
3.  Cancel
4.  Get balance
Enter your choice, 1, 2, 3, or 4:   4

Choose an account type
1.  Savings
2.  Checking
3.  Cancel
Enter your choice, 1, 2 or 3:   2
No such account

Choose a transaction
1.  Deposit
2.  Withdraw
3.  Cancel
4.  Get balance
Enter your choice, 1, 2, 3, or 4:   3
```

Note 1: `public User(Person p, Teller t) {`

We construct a user with a person to hold its personal data, and a teller. The user starts the transaction by sending the teller its id, and a reference to itself.

Note 2: `public void selectTransaction() {`

The methods to select a transaction type and an account type use menus to make the selections and send them to the teller.

Note 3: `public static final int DEPOSIT =1;`

Named constants make the program easier to maintain and modify. We use the C-C++ style, naming constants with uppercase identifiers. The `final` modifier indicates that the variable cannot change so that it becomes a constant.

Note 4: `bank = b;`

The teller saves the reference to the bank so that it can find accounts.

Note 5: `user = u;`

The teller saves the reference to the user, to continue communication during the execution of the transactions for that user.

Note 6: `if (transType != CANCEL)`

The transaction will end if the user cancels when requested to choose a transaction type.

Note 7: `public void acceptType(int type) {`

If the user does not cancel when selecting an account type, the teller asks the bank to find the user's account of the type selected, savings or checking.

Note 8: `if (account != null)`

The bank sends a **null** account to the teller if it cannot find an account of the specified type for that user.

Note 9: `user.selectTransaction();`

After each transaction is complete the teller asks the user to select another transaction.

Note 10: `Account p1Savings = new Savings(1500.00,p1,4.0);`

We create three accounts to test our program.

Note 11: `if (p1Savings.getId().equals(id) && acctType==SAVINGS)`

The `find` method checks each account to see if it matches the id and account type specified by the user. To check a large number of accounts we would use an array which we discuss in Chapter 4.

Note 12: `teller.acceptAccount(null);`

If none of the accounts match the id and account type specified by the user, the bank sends the value **null** back to the teller.

Note 13: `public static void main(String [] args) {`

The `main` method simply creates the objects and lets them execute their responsibilities determined from the scenarios.

Note 14: `public abstract class Account {`

The account classes modify those of Examples 2.3 and 3.1.

Testing the Code

We test the code with the scenarios of Figures 3.14 and 3.15, leaving other tests for the exercises. The first user, John Venn, makes a deposit of $50 to a savings account. This transaction succeeds because this user has a savings account. The second user, Mabel Venn, attempts to get the balance of a checking account, but this transaction fails because this user does not have a checking account.

The use cases and scenarios express the intended functionality of the system, so testing based on these scenarios will help ensure that our system correctly provides the required behavior.

The Big Picture

Use cases and scenarios help us to identify the objects and their responsibilities. A class implementing an object type includes methods to enable objects of that type to meet their responsibilities. Rather than following a recipe, as in a procedural program, the `main` method merely creates some objects. The program proceeds with objects interacting with one another.

TEST YOUR UNDERSTANDING

21. Write a scenario for a successful withdrawal of $100 from a checking account.

22. Write a scenario for a user who chooses to cancel rather than input a transaction type.

23. Write a scenario for a user who cancels a deposit while specifying the account type.

24. Write a scenario for a successful `getBalance` transaction from a savings account.

TRY IT YOURSELF   25. Run Example 3.6 testing the scenarios of questions 21–24.

Chapter 3 Summary

Inheritance, often contrasted with composition, is another way to relate objects. A class that extends another class, said to be a subclass of the class it extends, can add data fields and methods to those of its parent class; it can also override methods, tailoring their implementations to express its own behavior. In our bank account example, a savings account adds an interest rate field, and a method to post the interest. A checking account overrides the `withdraw` method of its bank account superclass to deduct a service charge if the balance is below the minimum needed for free checking. Every class, directly or indirectly, extends the `Object` class which provides a default implementation for the `toString` method that returns an empty string.

Polymorphism, the many forms that operations can have, distinguishes object-oriented programming from other paradigms. A superclass variable can refer to objects of any of its subclasses. A bank account variable might refer to a savings account or it might refer to a checking account; its withdraw behavior will be different in each case. As programmers, we just ask the object to execute its withdraw behavior. Our programs can have broad applicability, leaving each object to implement the specifics of its behavior. We use the same name, `withdraw`, but the result depends on the type of object whose `withdraw` method we invoke.

Often the superclasses we use polymorphically are abstract, not having any instances. In Example 3.3 the abstract `Shape` class implements its `move` method, but not its `draw` method. Subclasses `Line` and `Circle` must override `draw`, but may either override or inherit the `move` method from `Shape`. When we draw or move shapes, each instance will draw or move itself behaving according to its subclass type.

An interface is similar to an abstract class with no data fields, and with all methods abstract. Java allows a class to extend only one other class, but it may implement many interfaces. When implementing multiple interfaces, there is no danger of ambiguity as to which field to use or method implementation to invoke, because interfaces have neither. Declaring an object using an interface type that it implements allows that object to present itself from the perspective of that interface, hiding its other behavior. Using an interface allows a very flexible polymorphism because classes that implement an interface do not have to be subclasses of a common superclass.

Java uses modifiers to specify the type of access for classes, interfaces, data fields, and methods. We can declare a class or interface public, making it visible everywhere, or without any modifier, restricting its visibility to the package it is declared in. We can declare data fields and methods with the `private`, `protected`, or `public` modifiers or without any modifier. Using the `private` modifier restricts access to the methods of the class in which the data field or method is declared. This is the most restrictive modifier. Using no modifier restricts access to the package containing the data field or method. We can use the `package` statement to put our code in a package.

Using the `protected` modifier restricts access to the containing package or to any subclasses of the containing class. Finally, the `public` modifier makes the data field or method visible anywhere its containing class is visible. Generally speaking, we declare data fields `private` or with no modifier. Methods are most often `public`, but the other access types can be useful.

Our case study demonstrates the object-oriented programming methodology. Thinking in objects fits in with our normal experiences. To develop a system, we look at the uses of that system, giving scenarios for each use which step through the interactions needed to accomplish that use. We construct normal scenarios describing successful outcomes, and scenarios for the many cases when something goes wrong. From these scenarios we identify the objects of the system and their responsibilities. We implement methods to allow objects to satisfy these responsibilities. The system runs by creating objects and letting them interact, serving each other with their methods that execute their responsibilities. Our automatic teller application is quite simple, but illustrates this approach nicely.

Build Your Own Glossary

Find the uses, in this chapter, of the terms below. Enter a definition of each in the `glossary.txt` file on the disk that comes with this text.

abstract class	inheritance	polymorphism
classification	interface	subclass
extends	IS-A	superclass
implements	modifier	
import	override	

Skill Builder Exercises

1. Fill in the blanks in the following:

    A class can extend one _____ but may implement more than one _____.
    An _____ class has no instances, but it can implement some of its
    _____. Redefining an operation in a subclass is called _____.
    The behavior expressed by that operation depends on the _____ of the
    object that invokes it.

2. Fill in the modifier needed to provide the desired access for the following declarations
    in a public class C which is defined in a package P.

    a. _____ `int a;`   access only in the package P
    b. _____ `double d;`   access in P and in subclasses of C
    c. _____ `String s;`   access everywhere C is accessible
    d. _____ `char c;`   access only within C

3. What will be the output from the following program?

    ```
    class H {
        int a;
        public H() {
            a = 0;
        }
        public H(int i) {
            a = i;
        }
        public void display() {
            System.out.println(a);
        }
    }
    ```

continued

```
class K extends H {
    public K(int i) {
        a = i*i;
    }
}
public class HK {
    public static void main(String[] args) {
        H first = new H(4);
        H second  = new K(5);
        first.display();
        first = second;
        first.display();
    }
}
```

Critical Thinking Exercises

4. Using the Name class of Example 2.8 and the SavingsAccount class of Example 3.1, given

    ```
    Name name = new Name("Ben","Franklin");
    SavingsAccount savings = new SavingsAccount(500.00,4.0);
    String s = "The account of " + name + " is " + savings;
    ```

    choose the correct value of the string s.

    a. "The account of Ben Franklin is $500 at 4% interest"
    b. "The account of Ben Franklin is "
    c. "The account of Franklin, Ben is $500 at 4% interest"
    d. None of the above.

5. A subclass can override a method of its superclass or overload it. Choose the correct description from the following:

    a. Overriding replaces the method of the superclass, while overloading adds another method with the same name.
    b. Overloading replaces the method of the superclass, while overriding adds another method with the same name.
    c. Both overriding and overloading replace the method of the superclass.
    d. None of the above.

6. Which of the following best describes the relation between an abstract class and an interface?

    a. An interface is another name for an abstract class.
    b. An abstract class must be totally abstract, but an interface may be partially implemented.
    c. Every abstract class implements an interface.
    d. None of the above.

7. Given

    ```
    BankAccount b = new SavingsAccount(500.00,4.0);
    ```

referring to the classes of Example 3.1, choose which of the following is correct.

a. The statement b.postInterest() is not valid.

b. Later in the program, the variable b may refer to a CheckingAccount object.

c. The statement b.withdraw(100.00) will invoke the withdraw method from the BankAccount class.

d. All of the above.

Debugging Exercise

8. The following class, Check, might be useful for validating account inquiries by telephone. It attempts to check the four digits submitted by the inquirer against the last four digits of the account id. It uses the Account class of Example 3.1, and the Name and Address classes of Example 2.8. Find and fix any errors in the code below.

```java
public class Check {
    Account account;
    public Check(Account a){
        account = a;
    }
    public boolean checkId(String id){
        String s = account.getId();
        int length = s.length();
        boolean result = false;
        if (s.substring(length-4,length)==id)
            result = true;
        return result;
    }
    public static void main (String[] args) {
        Name name = new Name("Java", 'A', "Student");
        Address address =
                new Address("76 Applet Way","Web City","FL","44444");
        Person person = new Person("123456789",name,address);
        Account a = new Savings(1500.00,person,4.0);
        Check c = new Check(a);
        System.out.println(c.checkId("6789"));
    }
}
```

Program Modification Exercises

9. Modify Example 3.4 to add a Consumer class, which will implement the RateChangeListener interface so consumers can be notified about interest rate changes.

10. Modify Example 3.1 to add a TimedAccount subclass of SavingsAccount, which has an instance variable, fundsAvailable, to indicate that part of the balance is available for withdrawal. Override the withdraw method to check that the amount requested does not exceed the funds available for withdrawal. Override the deposit method to permit, at most, three deposits during the life of the account. Override the postInterest method to add the interest to fundsAvailable. Use an instance variable to hold the number of deposits made.

11 Modify Example 3.1 to add a readAccount method to the BankAccount class which will return a BankAccount constructed from data input from the keyboard. Override readAccount in SavingsAccount to return an account which refers to a SavingsAccount that you construct, again initializing it with data from the keyboard. Similarly implement readAccount in the CheckingAccount class.

12. Modify the Line, Circle, and Shape classes of Example 3.3 to make all data fields private and add public methods, such as getCenter, to access those fields.

13. Add a Rectangle subclass of Shape to Example 3.3. Let the center point of the shape represent the upper-left corner, and add another point which represents the lower-right corner of the rectangle.

14. Modify Example 3.3, rewriting the main method to pass the two shapes to a method which draws and moves each of them.

Program Design Exercises

15. Implement the ScaleableCircle class of Figure 3.10. Run a test program which scales, moves, and draws scaleable circles.

16. Define and implement a ScaleableLine class which implements the Scaleable interface. Scale the line by changing its end point so the length of the new line is f times the length of the old line, where f is the scale factor.

17. Implement a set of classes for dining out which will demonstrate polymorphism. An abstract Restaurant class will have abstract methods such as getMenu, getBill, orderFood, payBill, and so on. Implement an eatOut method by calling the abstract methods in the order they would occur in a typical restaurant scenario. Implement FastFood, CoffeeShop, and Fancy subclasses of Restaurant, which each implement all the abstract methods using stubs to print messages describing what would happen in that type of restaurant. For example the orderFood method in the FastFood class might describe talking to a machine from a car window in a drive through line, while the orderFood method in the Fancy class might call a method to order wine to go with the meal.

   To demonstrate the polymorphism, declare several restaurant objects, which refer to the various subclasses. Calling the eatOut method for each restaurant will show that each behaves in its own way, appropriate to that type of restaurant. A subclass may override the eatOut method if the order of method calls defined in the eatOut method in the Restaurant class is not appropriate for that subclass.

18. Implement a set of classes for accommodations which will demonstrate polymorphism. An abstract Accommodation class will have abstract methods such as reserve, checkIn, tipStaff, payBill, and so on. Implement a sleepOut method by calling the abstract methods in the order they would occur in a typical accommodation scenario. Implement LuxuryHotel, Motel, and Campground subclasses of Accommodation, which each implement all the abstract methods using stubs to print messages describing what would happen in that type of accommodation. For example the checkIn method in the Campground class might describe pitching a tent, while

the checkIn method in the LuxuryHotel class might call a method to have the luggage carried to the room.

To demonstrate the polymorphism, declare several accommodation objects, which refer to the various subclasses. Calling the sleepOut method for each accommodation will show that each behaves in its own way, appropriate to that type of accommodation. A subclass may override the sleepOut method if the order of method calls defined in the sleepOut method in the Accommodation class is not appropriate for that subclass.

19. Identify use cases and develop scenarios for a car rental system in which users may reserve a car in advance, cancel a reservation, and pick up or return a car. The company has different types of cars, such as compact cars and luxury cars. Make a list of the responsibilities of each object you identify from the scenarios. Implement the rental system using an object-oriented approach, developing methods to implement an object's responsibilities, and letting the objects interact. There is no "right" answer for this exercise. Good solutions will differ in many respects from one another.

20. a. Create an interface, Bendable, with one method, bend. Create two classes, Spoon and Arm, which implement Bendable. Spoon will also have an eat method, and Arm will also have a raise method. Each of these methods prints a message indicating its function. Write another class with a main method, declaring two objects of type Bendable, one a spoon and another an arm.

    b. Put the Bendable interface and the Arm and Spoon classes in a named package. The test class will not be in that package. Put it in a directory that does not contain the package. Show how you configure your system to provide the test class access to the package.

4 Arrays and Vectors

Each variable whose type is a primitive type holds a single value. Using an integer variable score to hold a single test score, we can write a program to add a list of test scores. Once we add a score to the total we no longer need that value and can read the next test score, saving it in the same variable, score.

Suppose we want to arrange the scores from highest to lowest and display an ordered list of all scores. To sort the scores, they must all be available so we can compare one score with another. Using variables of type **int** would hardly be feasible. For 50 scores we need variables score1, score2, score3, and so on up to score50. We might manage the burdensome task of typing 50 variable declarations, but suppose we had 500 scores, or 5000. Fortunately, the **array** concept solves this problem. An array is a sequence of values.

Figure 4.1 shows the AtmScreen applet of Example 1.5 when the user tries to deposit to a non-existent checking account. To find that the account does not exist, the bank searched an array of its accounts. Arrays are a useful data structure in many applications.

We look carefully at the declaration and creation of Java arrays, as there are some differences from array usage in C or C++. In this context, as with objects, we need to fully understand the distinction between array variables that hold references and primitive type variables that hold values.

FIGURE 4.1 The AtmScreen applet: No such account

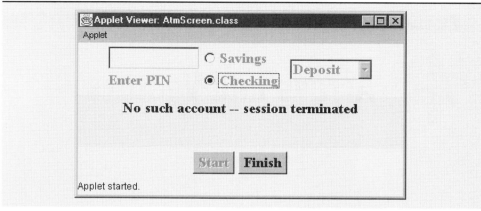

After presenting basic array concepts, we include several sections which illustrate array usage and introduce important ideas and techniques. **Arrays of arrays** extend the array concept to multidimensional sets of data. Our case study takes up the problem of sorting, which arranges data in order. We introduce insertion sort, developing a complete sorting program. As an extra topic, we show how to add timing statements to determine the efficiency of this sorting algorithm.

Arrays have a fixed size, but vectors can grow as we add more elements. We show the most useful vector methods, and in the last section illustrate using an interface to support callbacks, when one object registers with another to be notified (called back) when an event occurs. This use of interfaces previews the technique Java uses to signal user interface events that we will study in Chapter 6.

OBJECTIVES

◆ Know how to create and use Java arrays.
◆ Understand that array variables hold references.
◆ Understand memory allocation for arrays, and arrays of objects.
◆ Copy arrays and pass array arguments.
◆ Use arrays of arrays.
◆ Implement insertion sort using an array to hold the data.
◆ Estimate efficiency using timing statements.
◆ Use the Vector class for more flexibility.
◆ Use an interface and a vector of objects to provide callbacks, previewing the Java event model.

OPTIONS

◆ Omit 4.3 Arrays of Arrays (Used in 5.6 and 7.5)

4.1 ■ Java Arrays

It is not feasible to have 500 variables, score1, score2, ..., score500, to hold 500 test scores. The Java array lets us use one variable to refer to a collection of elements. Like object variables, Java array variables are references. We can declare an array using C-like syntax

```
int score[ ];
```

or as

```
int [ ] score;
```

The latter form indicates more clearly that the type, **int [ ]**, of the variable `score` is an array of integers; we use it exclusively in this text. Because array variables are references, the `score` variable has the value **null** until we refer it to an actual array. C and C++ programmers specify the size of the array in the declaration, but in Java we do not because such a declaration creates a reference to an array, but not the array itself.

We show two ways to create an array of scores. We can initialize `score` with a sequence of values, as for example

```
int[ ] score = {74, 38, 92};
```

which is a short form of

```
int[ ] score = new int[ ]{74, 38, 92};
```

in which we use the **new** operator to allocate space for the array and initialize the array elements to the given values. Figure 4.2 shows the `score` array.

FIGURE 4.2 The `score` array

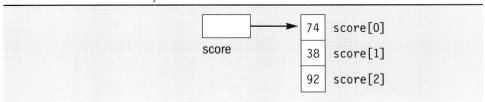

We use an integer **index** to access an **array element**. An array element is one of the sequence of values in the array. The index specifies its position in the array. As in C and C++, array indices always start with 0, so we denote the three elements of the `score` array as `score[0]`, `score[1]`, and `score[2]`. Using an index other than 0, 1, or 2 for the `score` array will cause an error.[1] An array has a `length` attribute, so for the `score` array, the field `score.length` has the value 3. We use this field in the test condition in **for** loops involving arrays.

We can use the **new** operator

```
new int[3]
```

to allocate an array with space for three elements. Figure 4.3 shows the memory after execution of the statement `anArray = new int[3];`

FIGURE 4.3 Memory allocation using the operator **new**

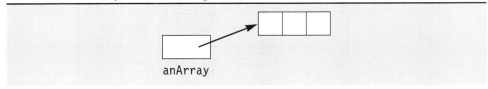

---

[1] We cover array index out of bounds errors, which generate exceptions, in Section 8.1.

After allocating the memory for the array elements, we can assign values to them, as in

```
anArray[0] = 17;
anArray[1] =  3 + x;
anArray[2] = Io.readInt("Enter a integer");
```

The first statement assigns a constant value, the second assigns the value of an expression, while the third assigns a value input by the user. Figure 4.4 diagrams the memory, if the variable x has the value 7, and the user inputs 22.

FIGURE 4.4 Memory configuration for anArray

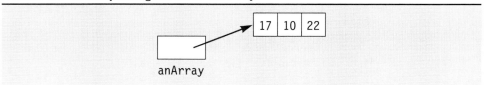

We declared anArray to have three elements. By using a variable, size, in the declaration instead of the constant, 3, as in:

```
int[] myArray = new int[size];
```

we can declare an array myArray whose size depends on the value of the variable size. By allowing the user to input the value of size before making the declaration, as in:

```
int size = Io.readInt("Enter the size of the array");
int[]  myArray = new int[size];
```

we can create the myArray at runtime to have the desired size.

In Example 4.1, we use random numbers to simulate the tossing of two dice. Assuming each die has an equal chance of landing so any one of its six faces, numbered 1 through 6, will be showing, we tabulate the frequency of each sum, 2 through 12, of the numbers showing on each die.

We write a method, roll, that simulates the roll of two dice, returning the sum of the numbers showing on each face. The Math.random() method returns a **double** between 0 and 1. Multiplying that value by 6 gives a value ranging from 0 up to 6. Casting that value to an **int** type gives an integer from 0 through 5. Adding one gives an integer from 1 through 6, and each of these should occur with equal frequency because the random numbers are distributed evenly. Writing these steps in a formula to generate a random number from 1 to 6, we have

```
(int)(6*Math.random())+ 1
```

for the simulation of tossing one die.[2] We toss two and return the sum.

The method tossResults tabulates the frequencies of each outcome, 2 through 12, in an array named result, so result[i] is the number of times the sum on the two dice was i. We use the array elements result[2] through result[12], leaving result[0] and result[1] unused. We simply store the outcome of the roll method in the result array by incrementing the value in the array. For example, if the outcome is 6 we execute result[6]++.

---

[2] If using the Java™ 2 platform, we can import the java.util.Random class, create Random random =new Random(), and use the method random.nextInt(6) to get a random integer from 0 through 5.

Example 4.1 shows the code for this dice simulation. To find the expected frequencies, make a 6x6 grid with row labels 1–6, and column labels 1-6, and enter the sum of the row and column labels in the grid as shown in Figure 4.5.

FIGURE 4.5 Outcomes when tossing two dice

	1	2	3	4	5	6
1	2	3	4	5	6	7
2	3	4	5	6	7	8
3	4	5	6	7	8	9
4	5	6	7	8	9	10
5	6	7	8	9	10	11
6	7	8	9	10	11	12

Tabulating the 36 outcomes from Figure 4.5 gives

Sum	2	3	4	5	6	7	8	9	10	11	12
Frequency	1	2	3	4	5	6	5	4	3	2	1

showing, for example, that 6 of the 36 outcomes are 7 while 3 are 10. If the number of rolls is a simple multiple of 36, such as 36, 360, 3600, and so on, we can easily see how closely the simulation matches the prediction. For example, for 360 tosses we expect, on average, to get a distribution like

Sum	2	3	4	5	6	7	8	9	10	11	12
Frequency	10	20	30	40	50	60	50	40	30	20	10

EXAMPLE 4.1   **Dice.java**

```
/* Simulates the rolling of two dice.
 * The user inputs the number of tosses.
 * Uses arrays to tabulate the results.
 */

import iopack.Io;
public class Dice {
    public static int roll() {
        int die1 = (int)(6*Math.random())+ 1;              // Note 1
        int die2 = (int)(6*Math.random())+ 1;
        return die1 + die2;
    }
    public static int[] tossResults(int number) {
        int[] result = {0,0,0,0,0,0,0,0,0,0,0,0,0};         // Note 2
        for (int i=0; i<number; i++)
            result[roll()]++;                               // Note 3
        return result;
    }
    public static void main(String[] args) {
        int[] diceThrows;        // Elements 2-12 store frequencies
        int numberOfRolls;       // Number of rolls of the dice
```

continued

```
                numberOfRolls = Io.readInt("Enter the number of rolls of the dice");
                diceThrows = tossResults(numberOfRolls);
                System.out.println("Sum\tFrequency");
                for (int i=2; i<=12; i++)
                        System.out.println(i + "\t" +diceThrows[i]);
        }
}
```

Output
```
Enter the number of rolls of the dice:   3600
Sum     Frequency
2       101
3       209
4       307
5       402
6       498
7       590
8       481
9       403
10      308
11      213
12      88
```

Note 1: This formula produces values 1 through 6 with equal expected frequency, as we explained in the previous text.

Note 2: We initialize the array with 13 components, all zero, but only use the last 11, with `result[i]` holding the number of times the sum on the two dice was `i`.

Note 3: The call to `roll()` returns a value from 2 through 12. We increment the number stored in the array element `result[roll()]` to indicate the occurrence of the value returned by `roll`. For example, if `roll()` returns five, we increment `result[5]`.

The Big Picture

In Java, we may declare an array variable as **int[ ] x** or **int x[ ]**, which creates an array variable x to refer to an array. The variable x has the value **null** until we refer it to an array. We can initialize an array with a sequence of values, or use the **new** operator to allocate space for the elements and later assign the elements, perhaps to values input by the user.

TEST YOUR UNDERSTANDING

1. Declare and initialize an array with values 37, 44, 68, and −12. Replace the second element with 55.

2. Declare and initialize an array with values −4.3, 6.8, 32.12, −11.4, and 16.88. Copy the element in the fourth location into the first.

3.   Declare and initialize an array with values 's', 'y', 't', 'c', 'v', and 'w'.

4.   Write a **for** loop to find the sum of the elements of the array in question 1.

5.   Write a Java expression which generates a random integer greater than or equal to 50 but less than 500.

6.   If in Example 4.1 we were to roll three dice instead of two how many elements are needed in the roll array? Which elements would be unused?

4.2 ▪ Copying and Passing Arrays

To use arrays effectively we need to understand how array assignment works and how to pass array arguments to methods and return array values from them.

Copying References vs. Copying Values

Assigning one array to another copies the reference, not the array value.

For example, Figure 4.6a diagrams the memory Java uses for an array variable x initialized to refer to an array of five integers, and an uninitialized array variable y. Figure 4.6b shows the memory usage after we assign x to y.

FIGURE 4.6 Memory usage for an array assignment

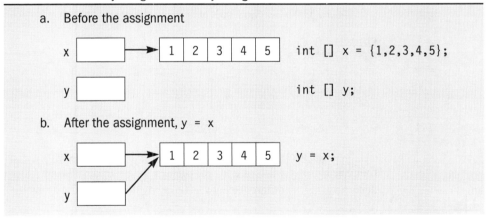

a.   Before the assignment

x [ ] → | 1 | 2 | 3 | 4 | 5 | `int [] x = {1,2,3,4,5};`

y [ ]   `int [] y;`

b.   After the assignment, y = x

x [ ] → | 1 | 2 | 3 | 4 | 5 | `y = x;`

y [ ] →

We see that the assignment copies the reference from the variable x into the variable y. After the assignment, both variables refer to the same array. Copying a reference is more efficient than copying the whole array, which can be quite large. It takes time to copy the array values, and space to hold them.

Because the variables x and y in Figure 4.6b refer to the same array, any changes made using x will affect y, and vice versa. For example, if we execute

```
y[2] = -38;
System.out.println(x[2]);
```

we will see that x[2] has the value −38; x and y refer to the same array so their elements must be the same. Figure 4.7 shows the effect of the assignment to y[2].

FIGURE 4.7 Memory usage after the assignment y[2] = -38

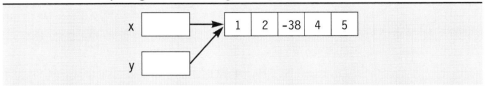

The assignment in Figure 4.6b copies the reference in the variable x, resulting in two variables, x and y, referring to the same array. If we want to copy the array elements, not the reference, then we need to use the **new** operator to allocate space for a second array, as in

```
int[] y = new int[x.length];
```

and write a loop to copy each element from the old array to the new array, as in

```
for (int i=0; i<y.length; i++)
  y[i] = x[i];
```

We can also use the arraycopy method to copy an array. The statement

```
System.arraycopy(x, 0, y, 0, y.length);
```

will copy the entire array x to the array y. The five arguments to the arraycopy method are the source array, the starting index in the source, the target array, the starting index in the target, and the number of elements to copy.

Figure 4.8 shows the result of copying the array x to the array y.

FIGURE 4.8 Copying array elements

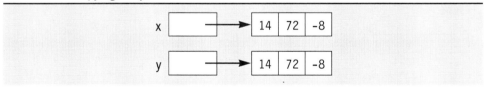

With the arrays in Figure 4.8, if we change the x array, say by the assignment x[1] = 11, then the y array will not change, since y refers to a different array than x. Example 4.2 demonstrates the difference between copying a reference and copying the array itself.

EXAMPLE 4.2 **ArrayCopy.java**

```
/*  Illustrates the difference between array
 *  assignment, which copies a reference to an array,
 *  and making a copy of one array in a new array.
 */

public class ArrayCopy {
    public static void main(String [] args) {
        int [] x = {4,5,6};
        int [] z = x;                          // Note 1
        int [] y = new int[x.length];          // Note 2
        for (int i=0; i<y.length; i++)         // Note 3
            y[i] = x[i];
```

```
        x[1] = 7;                                              // Note 4
        System.out.println                                     // Note 5
            ("The x array is now {"
                + x[0]+ "," + x[1]+ "," + x[2] + "}");
        System.out.println                                     // Note 6
            ("The y array, after changing x, is {"
                + y[0] + "," + y[1]+ "," + y[2] + "}");
        System.out.println                                     // Note 7
            ("The z array, after changing x, is {"
                + z[0]+ "," + z[1]+ "," + z[2] + "}");
    }
}
```

Output
```
The x array is now {4,7,6}
The y array, after changing x, is {4,5,6}
The z array, after changing x, is {4,7,6}
```

Note 1: Assigning x to z copies the reference in x, so z refers to the same array as x.

Note 2: We allocate space for a new array with the same number of elements as the x array.

Note 3: The **for** loop copies the elements from the x array to the y array.

Note 4: This assignment changes the x array to see how it affects the y and z arrays.

Note 5: This code displays the changed x array. For brevity we write a simple `println` statement for the three array values. Better code would use a **for** loop. Because we have to display three arrays, we opted for the less general, but briefer form. In the next example, we will see how to use array arguments so we can write a method to display an array.

Note 6: Because the array variable y refers to a true copy of the x array, changing x has no effect on the y array; it retains its original values.

Note 7: Because the array variable z refers to the same array as x does, changing the x array also changes the z array.

Passing an Array Argument

The code to display array values in Example 4.2 is repetitive and not suitable for large arrays. By writing a method to display an array we can display larger arrays and do not need to repeat the same code. Figure 4.9 shows the `display` method, which has a formal parameter, anArray, of type **int[ ]**, an array of integers.

FIGURE 4.9 The `display` method

```
public static void display(int [] anArray) {
    System.out.print("{");
    for (int i=0; i<anArray.length; i++) {
        if (i!=0) System.out.print(",");
        System.out.print(anArray[i]);
    }
    System.out.println("}");
}
```

To call the `display` method we pass it an array argument, as in the code fragment

```
int[] score = {40, 50, 60};
display(score);
```

which displays the array {40, 50, 60} on the screen.

The actual argument, `score`, refers to the array {40,50,60}. When we call the method `display(score)`, Java copies `score` to the formal parameter `anArray`; Figure 4.10 displays the memory configuration.

FIGURE 4.10 Passing the `score` reference to the `anArray` parameter

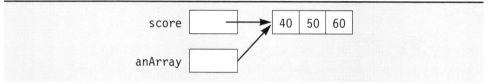

The code inside the `display` method uses the reference given by the `anArray` variable to display each element of the array. We can use the `display` method again with a different argument. In Example 4.2 we displayed three arrays. In Example 4.3 we revise Example 4.2 to use the `display` method to display the arrays. We only have to write the code once, in the `display` method, but call it three times. Each time we call the `display` method, the parameter, `anArray`, refers to the array that we pass in as the argument.

The `display` method of Example 4.3 has an array parameter, but does not return any value. However, methods can return array values. We define a method, `readIntArray`, to input an array from the user. This method will not have any parameters but it will have a return value which will be the array that the user inputs. Inside the `readIntArray`, we first ask the user to input the size of the array, then use the operator **new** to allocate memory for an array of that size. Finally, we ask the user to input each element of the array, and use the `return` statement to return that array to the caller of the method. Figure 4.11 contains the code for the `readIntArray` method.

FIGURE 4.11 The `readIntArray` method

```
public static int[] readIntArray() {
    int size = Io.readInt("Enter the array size");
    int [] anArray = new int[size];
    for (int i=0; i<size; i++)
        anArray[i] = Io.readInt("Enter anArray["+i+"]");
    return anArray;
}
```

In the `readIntArray` method, we specify the return type as **int[ ]**, an array of integers. We use the `return` statement to return the array, `anArray`, that we get from the user. To use the `readIntArray` method, we can declare an array in our program and initialize that array with the value returned by the `readIntArray` method, as in

```
int[] score = readIntArray();
```

where the variable `score` will refer to the array returned by the `readIntArray` method.

We put the code in a package, `arrayIo`, to make it easier to use these methods in later programs.

EXAMPLE 4.3   **ArrayMethods.java**

```java
/*  Revises Example 4.2 to use a display
 *  method to output an array and the readIntArray
 *  method to input an array.
 */

package arrayIo;

import iopack.Io;
public class ArrayMethods  {
    public static void display(int [] anArray) {                    // Note 1
        System.out.print("{");
        for (int i=0; i<anArray.length; i++) {
            if (i!=0) System.out.print(",");                        // Note 2
            System.out.print(anArray[i]);
        }
        System.out.println("}");
    }
    public static int[] readIntArray() {
        int size = Io.readInt("Enter the array size");
        int [] anArray = new int[size];                             // Note 3
        for (int i=0; i<size; i++)
            anArray[i] = Io.readInt("Enter element " + i);
        return anArray;
    }
    public static void main(String [] args) {
        int [] x = readIntArray();
        int [] z = x;
        int [] y = new int[x.length];
        System.arraycopy(x,0,y,0,x.length);
        x[0] = 7;
        System.out.print("The x array is now ");
        display(x);                                                 // Note 4
        System.out.print("The y array, after changing x, is ");
        display(y);                                                 // Note 5
        System.out.print("The z array, after changing x, is ");
        display(z);                                                 // Note 6
    }
}
```

Output
```
Enter the array size:  8
Enter element 0:  4
Enter element 1:  5
Enter element 2:  6
Enter element 3:  7
Enter element 4:  8
Enter element 5:  9
```

continued

```
Enter element 6:  10
Enter element 7:  11
The x array is now {7,5,6,7,8,9,10,11}
The y array, after changing x, is {4,5,6,7,8,9,10,11}
The z array, after changing x, is {7,5,6,7,8,9,10,11}
```

Note 1: The display method outputs the values of the array argument passed to it.

Note 2: We print a comma before every element except the first.

Note 3: We create the array inside the readIntArray method.

Note 4: We pass the display method the array we wish to display. This is the first of three calls to the display method. In this call the formal parameter, anArray, refers to the array to which x refers.

Note 5: In this call the formal parameter, anArray, refers to the array to which y refers.

Note 6: display(z);

In this call the formal parameter, anArray, refers to the array to which z refers.

☛ A LITTLE EXTRA **Comparing int[ ] and int arguments**

When we pass an array to a method, we pass a reference. As shown in Figure 4.10, we get two references to the same array, one from the actual argument passed in score, in this example, and one from the formal parameter, anArray. Although we did not do it in the display method, an assignment to anArray such as

```
anArray[o] = 75;
```

would also change the score array that we pass to the display method.

By contrast, when we pass a primitive type to a method, the formal parameter is a copy of the actual argument and changing it has no effect on the actual argument. To illustrate this, consider the assign4 method given by

```
public static void assign4(int someNumber) {
    someNumber = 4;
}
```

which we might call with the code

```
int x = 27;
assign4(x);
```

When we pass the variable x to assign4, Java copies its value, 27, to the formal parameter someNumber. Figure 4.12 shows the result.

FIGURE 4.12 Memory configuration after passing x to assign4

| 27 | 27 |
| x | someNumber |

When Java executes the assign4 method, the formal parameter changes its value to 4, but the actual argument x is unchanged, as Figure 4.13 shows.

FIGURE 4.13 Effect of the assign4 method

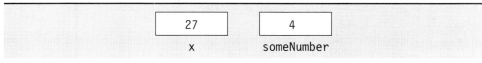

> **The Big Picture**
>
> Array variables, like object variables, are references. Assignment copies one reference to another and both variables refer to the same array. We can use the array-copy statement to copy one array to another. We can pass array arguments to and return array values from methods. The formal array parameter, used in the body of the method, refers to the array passed from the calling program, whereas for primitive types, changes to the formal parameter do not affect the value passed from the caller.

TEST YOUR UNDERSTANDING

7. Diagram the memory usage for each of the following:
   a. `int[] intArray = {2,-4,5,9,-1};`
   b. `char[] charArray = new char[8];`
   c. `double[] doubleArray;`

8. Diagram the memory usage resulting from the execution of

   ```
   int[] a = {36, -2, 44, 55};
   int[] b = a;
   ```

9. Write Java code to make a copy of the array a of question 8.

10. Diagram the memory configuration resulting from the variables and arrays of question 9.

11. Write a statement to call the display method of Figure 4.9 to display the array `myArray = {52, 63, 74, 85};`

12. Diagram the memory usage for each of the following:
    a. `int[] intArray;`
    b. `char[] charArray = {'a','b','c'};`
    c. `double[] doubleArray = new double[6];`

13. The display method of Figure 4.9 uses a formal parameter, anArray, to refer to the array. Suppose we call the display method with the argument `myArray = {52, 63, 74, 85}`. Diagram the memory configuration, showing both anArray and myArray.

4.3 ▪ Arrays of Arrays

Arrays let us access a collection of data. An array `score` might have 10 elements, or 100, which we access using an array index, as in `score[9]`. An instructor might have an array of scores for each student in the class. For example,

```
student1 has scores {52, 76, 65}
student2 has scores {98, 87, 93}
student3 has scores {43, 77, 62}
student4 has scores {72, 73, 74}
```

and so on for the 30 students (or is it 300?) in the class.

We do not want to declare 30 or 300 variables. We have the same problem that we faced with one set of scores, where we had to declare variables `score1`, `score2`, `score3`, and so on, until we learned to declare an array variable score of type **int[ ]**. The type **int[ ]** has two parts, the type of each element, **int**, followed by the square brackets, [] , to indicate an array.

What we need here is an array with one element for each student, where the type of each element is **int[ ]**, an array of integers. This array will hold the student's test scores. Again we write the type declaration in two parts, the type of each element, **int[ ]**, followed by the square brackets, [], to indicate an array of an array of integers. For a class of 30 students with three scores each, we would declare

```
int[][] s =  new int[30][3];
```

We initialize the array, as we did for arrays of integers, by giving a list of elements. Here each element of the array is itself an array of integers. To shorten the example, we declare an array of four students with three scores each

```
int[][] student =
    {{52,76,65},{98,87,93},{43,77,62},{72,73,74}};
```

Each element of the `student` array is itself an array of integers. In fact,

```
student[0] is {52, 76, 65}
student[1] is {98, 87, 93}
student[2] is {43, 77, 62}
student[3] is {72, 73, 74}
```

Figure 4.14 shows the `student` array of array of scores for each student.

FIGURE 4.14 The `student` array

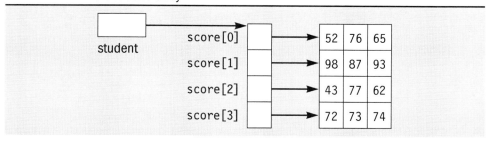

We can get individual scores by referring to components of these arrays, as in

```
student[1][0] is 98
student[2][2] is 62
```

The length of the student array, given by student.length, is four. Each element of the student array is itself an array. The length of element i, student[i], is given by student[i].length, which in this example is three.

Example 4.4 computes the average score for each student. To compute the average we use nested **for** statements. The outer loop uses an index i to refer to student i. The inner loop uses an index j to refer to the grades of student i.

EXAMPLE 4.4   **StudentScore.java**

```java
/* Uses an array of arrays.  Computes the average
 * score of each student.
 */

import iopack.Io;
public class StudentScore {
    public static void main(String [] args) {
        int [] [] student = {{52, 76, 65},{98, 87, 93},{43, 77, 62},{72, 73, 74}};
        double sum;                                             // Note 1
            for (int i=0; i<student.length; i++) {             // Note 2
                sum = 0;                                       // Note 3
            for (int j=0; j<student[i].length; j++)            // Note 4
                sum += student[i][j];                          // Note 5
            System.out.print("The average score for student " + i + " is ");
            Io.println(sum/student[i].length,1);               // Note 6
        }
    }
}
```

Output
```
The average score for student 0 is 64.3
The average score for student 1 is 92.7
The average score for student 2 is 60.7
The average score for student 3 is 73.0
```

Note 1: We declare sum, representing the sum of the scores for each student, as type **double** so that when we compute the average it will be of type **double**.

Note 2: The index i runs through the students. The field student.length gives the length of the array of score arrays, which is 4 in this example.

Note 3: We must initialize sum for each student, so this statement must occur inside the outer loop. Forgetting to initialize sum here would cause the second student's scores to be added to the those of the first, giving an incorrect result.

Note 4: This loop runs through the array of scores for student i. The length of this array is student[i].length.

Note 5: We add score j of student i to the sum of scores for student i.

Note 6: Io.println(sum/student[i].length,1);

We output the average, which is the sum of the scores divided by the number of scores. Here the number of scores is the length of the array of scores for student i.

In Example 4.4 each student had the same number of test scores, but we could have defined an array whose rows have irregular sizes, as, for example,

```
int[ ] scores = {{54, 76}, {67, 83, 44}};
```

instead of the array in Example 4.4.

Instead of initializing the array, we can enter the scores at the keyboard, assuming each student has the same number of scores, using

```
int numStudents = Io.readInt("Enter the number of students");
int numScores = Io.readInt("How many test scores for each?");
int[ ][ ] scores = new int[numStudents] [numScores];
for (int i = 0; i < scores.length; i++)
    for (int j = 0; j < scores[i].length; j++)
        scores[i][j] = Io.readInt("Enter score[" + i + " , " + j + "]' ");
```

In Example 4.5, we show how to compute average scores if each student may have a different number of scores which we enter from the keyboard. We use the readIntArray and display methods from the arrayIo package of Example 4.3.

EXAMPLE 4.5 **EnterScores.java**

```
/*  Revises Example 4.4 to allow each student to
 *  have a different number of scores.  Enters
 *  scores from the keyboard.
 */

import iopack.Io;
import arrayIo.ArrayMethods;

public class EnterScores  {
    public static void main(String [] args) {
        double sum;
        int numStudents = Io.readInt("Enter the number of students");
        int[ ][ ] scores = new int[numStudents] [ ];                    // Note 1
        for (int i = 0; i < scores.length; i++)
            scores[i] = ArrayMethods.readIntArray
                            ("How many scores for student " + i);       // Note 2
        for (int i = 0; i < scores.length; i++){
            sum = 0.0;
            for (int j = 0; j < scores[i].length; j++)
                sum += scores[i][j];
            System.out.println("The scores for student " + i + " are:");
            System.out.print('\t');
            ArrayMethods.display(scores[i]);
            System.out.print("\twith average  ");
            Io.println(sum/scores[i].length,1);
        }
    }
}
```

Output
```
Enter the number of students:  2
How many scores for student 0:  2
Enter element 0:  66
Enter element 1:  99
How many scores for student 1:  3
Enter element 0:  44
Enter element 1:  67
Enter element 2:  65
The scores for student 0 are:
        {66,99}
        with average  82.5
The scores for student 1 are:
        {44,67,65}
        with average  58.7
```

Note 1: We must always specify the size of the array in the first bracket, but may leave the second empty to allow us, in this example, to create arrays of different sizes to hold the scores for each student.

Note 2: The readIntArray method creates and returns the array of scores for each student.

The Big Picture

An array of arrays is an array whose elements are also arrays. These element arrays do not have to be the same size. When declaring a variable for an array of arrays, we can either initialize it to refer to a specific array of arrays, or we can use the **new** operator. Declaring int [ ] [ ] x = new int [4] [3] will create an array x whose four elements are each arrays of three elements. Declaring int [ ] [ ] y = new int [4] [ ] will create an array y whose four elements may have different sizes. Executing y[0] = new int [5] will give the first element of the y array five elements.

TEST YOUR UNDERSTANDING

TRY IT YOURSELF

14. Rerun Example 4.4 after changing the student array to have a different number of scores for each student. Does the program work properly without any other changes?

15. Declare and initialize an array of arrays of scores for two students. The first student has scores 55, 66, 87, and 76, while the second's scores are: 86, 92, 88, and 95.

16. Declare and initialize an array of arrays of batting averages for five baseball players for each of the last three years. Batting averages are typically computed to three decimal places and range from .150 to .400.

17. Given

    ```
    char[][] letter ={{'a','b','c'}, {'x','y','z','w'}}
    ```

    find

    a. letter[0][1]   b. letter[1][0]
    c. letter[1][3]   d. letter[0][2]

4.4 ▪ Insertion Sort

In this section we develop a program to sort an array, arranging the elements in order, using **insertion sort**. Sorting has many uses; various algorithms have been developed to solve this important problem. Insertion sort is useful for smaller data sets, while other methods such as quicksort or merge sort work much more efficiently for larger sets of data.

Defining the Problem

To understand how insertion sort works, we start with an example. Given the values

    54   23   78   42   26   12   41   64

the sorted array should be

    12   23   26   41   42   54   64   78.

Insertion sort takes each element from the second element to the last element and inserts it in its proper place with respect to the preceding elements. Starting with the second element, which we underline,

    54   <u>23</u>   78   42   26   12   41   64

we see that 23 < 54, so we move 54 to the right one position in the array, and insert 23 in the beginning of the array giving

    23   54   78   42   26   12   41   64

Next we insert 78.

    23   54   <u>78</u>   42   26   12   41   64

Because 78 > 23, we compare 78 to 54. Since 78 > 54, we know that 78 is greater than all its predecessors, so we leave it where it is in position three, and the array is unchanged. At this point we have the first three elements in order.

Next we insert the fourth element, 42.

    23   54   78   <u>42</u>   26   12   41   64

Because 42 > 23 and 42 < 54, we move both 78 and 54 to the right by one position and insert 42 in the second position giving

    23   42   54   78   26   12   41   64

Figure 4.15 shows the array at this point, with its first four elements in order. In Chapter 6 we will use Java to draw such a diagram.

The fifth element, 26, is greater than 23 and less than 42, so we move the elements 42, 54, and 78 to the right and insert 26 in the second position giving

    23   26   42   54   78   12   41   64

FIGURE 4.15 Partially sorted array

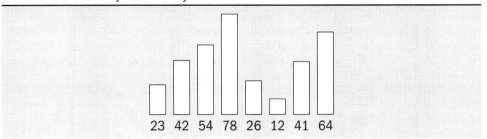

23 42 54 78 26 12 41 64

Because 12 is smaller than 23, we move the first five elements to the right, and insert 12 at the start of the array, giving

12   23   26   42   54   78   41   64

The seventh element, 41, is greater than the first three elements, but less than 42, so we move 42, 54, and 78 to the right and insert 41 as the fourth element of the array, giving

12   23   26   41   42   54   78   64

Finally, the last element, 64, is greater than all elements except 78, so we move 78 to the right, and insert 64 as the seventh element and the array is fully sorted.

Designing a Solution

Developing the Pseudocode

To design a program for insertion sorting we use **stepwise refinement** to develop **pseudocode**, a less formal and more understandable statement of the solution process in English. In step-wise refinement we start with the overall solution and refine each step with more details, until we reach a level of detail we can translate readily into Java code. Figure 4.16 expresses the top-level solution in pseudocode.

FIGURE 4.16 Insertion sort: Top-level pseudocode

```
Get the data to sort;
Display the data to sort;
Insert each item in the correct
    position in its predecessors;
Display the sorted data;
```

To get the data, we can allow the user to input data, or we can generate random numbers to test without requiring user input. Figure 4.17 shows this refinement.

FIGURE 4.17 Refinement: Get the data to sort

```
Ask if the user wants to enter data;
if (yes) Get data from the user;
else Generate random data;
```

When getting data from the user, one approach is to ask the user each time if there is more data to input, which does not require the user to know the size of the data set, but does require an extra response to input the next item. Alternatively, we choose to have the user input the size of the data first, as shown in the refinement in Figure 4.18, where we use **loop** to represent a loop that we will implement using one of the Java repetition statements.

FIGURE 4.18 Refinement: Get data from the user

```
Input the size of the data;
loop
    Get the next item;
```

To generate random data, we first ask the user to input the size of the data and then use a loop to generate the required number of random values. At this point we see that no matter how we get the data, we will ask the user to input the size of the data. Rather than coding this request twice, we can get the data size before deciding whether to input from the user or generate random numbers. Figure 4.19 shows this revision of the refinement in Figure 4.17, including the refinements for the subproblems.

FIGURE 4.19 Revised refinement: Get the data to sort

```
Input the size of the data;
Ask if the user wants to enter the data;
if (yes)
    loop
        Get the next item;
else
    loop
        Generate the next random item;
```

To sort the data we insert one element at a time, starting with the second element, in the correct position in its predecessors. Figure 4.20 shows this loop.

FIGURE 4.20 Refinement: Insert items

```
loop, from second item to last
    Insert item i in the correct position
        in its predecessors;
```

From our example, we see that to insert the item at index i we first find its correct position, say j, then move the elements at positions j through $i-1$ to the right by one position. Lastly we insert the item in the correct position. Figure 4.21 shows this refinement.

FIGURE 4.21 Refinement: Insert item i

```
Find the correct position, say j, for item i;
Move elements at j to i-1 one position to the right;
Insert item i at position j.
```

To find the correct position for item i, we start at the leftmost element of the array, with j initialized to zero, and while item i is greater than item j, we increment j. In our example above, when we insert item 6, which is 41, we do the following steps:

41 > 12 so increment j to 1
41 > 23 so increment j to 2
41 > 26 so increment j to 3
41 < 42 so stop, finding that index 3 is the correct position

Figure 4.22 shows the refinement for finding the correct position.

FIGURE 4.22 Refinement: Finding the correct position for item i

```
j = 0;
while (item i > item j) j++;
```

We have to be careful to move the items starting at the right, so we do not overwrite any items before they are moved, and to save item i before writing over it. In our example above, we found that index 3 is the correct position for 41, so we need to move items 3 to 5 to the right. The steps are:

save the 41 that we want to insert
move 78 to the right, storing it as the item at index 6
move 54 to the right, storing it as the item at index 5
move 42 to the right, storing it as the item at index 4

Figure 4.23 shows the **for** loop we need to move the elements at indices from j to i-1 to the right by one position.

FIGURE 4.23 Refinement: Move elements j to i-1 to the right

```
Save item i;
for (int k=i; k>j; k--)
    item[k] = item[k-1];
```

To insert item i at position j we use the simple assignment

```
item[j] = item[i];
```

To display the data we can use the display method we wrote in Example 4.3; we can leave that refinement until we write the Java code. We will also postpone the decision as to which random numbers to use until we write the code. We have refined all but the simplest subproblems and can put the complete pseudocode together in Figure 4.24 at the top of the next page.

Designing a Solution

Alternatives

After completing the pseudocode for the insertion sort, we might think about other possible solutions. One alternative would allow us to combine the steps of finding the correct position and moving the elements to the right. For this alternative, when we insert element i we start

FIGURE 4.24 Pseudocode for insertion sort

```
Input the size of the data;
Ask if the user wants to enter the data;
if (yes)
    loop
       Get the next item;
else
    loop
        Generate the next random item;
Display the data to sort;
loop, from second item to last {
   j = 0;
   while (item i > item j) j++;
   Save item i;
   for (int k=i; k>j; k--)
       item[k] = item[k-1];
   item[j] = item[i];
}
Display the sorted data;
```

checking at element i-1 instead of at element 0. We leave the development of this alternative solution to the exercises.

Completing the Java Code

The pseudocode in Figure 4.24 gives us a good basis for writing the Java program.

EXAMPLE 4.6 **InsertionSort.java**

```
/* Sorts an array of data using the
 * insertion sort algorithm.  Uses
 * data from the user or random data.
 */

import iopack.Io;
import arrayIo.ArrayMethods;

public class InsertionSort {
    public static void main(String [] args)  {
        int size = Io.readInt("Enter the number of data items");
        int[] item = new int[size];       // allocate array to hold data
        char enter = Io.readChar("Enter 'Y' to enter data, 'N' for random data");
        if (enter=='Y' || enter=='y')
           for (int i=0; i<size; i++)
               item[i] = Io.readInt("Enter item[" + i + "]");
        else
           for (int i=0; i<size; i++)
               item[i] = (int)(100*Math.random());          // Note 1
        System.out.print("The data to sort is ");
        ArrayMethods.display(item);                          // Note 2
```

```
        for (int i=1; i<size; i++) {
            int current = item[i];                      // Note 3
            int j = 0;
            while (current > item[j]) j++;              // Note 4
            for (int k=i; k>j; k—)
                    item[k] = item[k-1];
            item[j] = current;
        }
    System.out.print("The sorted data is ");
    ArrayMethods.display(item);
    }
}
```

Output—First Run
```
Enter the number of data items:  10
Enter 'Y' to enter data, 'N' for random data:  n
The data to sort is {51,70,41,62,95,89,63,78,80,5}
The sorted data is {5,41,51,62,63,70,78,80,89,95}
```

Output—Second Run
```
Enter the number of data items:  5
Enter 'Y' to enter data, 'N' for random data:  y
Enter item[0]:   88
Enter item[1]:   77
Enter item[2]:   66
Enter item[3]:   55
Enter item[4]:   44
The data to sort is {88,77,66,55,44}
The sorted data is {44,55,66,77,88}
```

Note 1: To make the display shorter, we restrict our random numbers to the range 0 to 99. The Math.random method returns a random number of type **double** between 0 and 1. Multiplying it by 100 produces a value between 0 and 100. Casting that value to type **int** produces an integer from 0 to 99.

Note 2: We use the display method from the ArrayMethods class of Example 4.e.

Note 3: We store the element at index i before we write over it. This store occurs in the pseudocode just before the inner **for** loop when we are about to write over item i. We refer to item i earlier in the code, so in the **while** statement, we decided to store it earlier, before the **while** statement. This has the small advantage that, in the **while** statement, we can use the value of the variable current which is a little easier to look up than the array element, item[i]. It is like the difference between looking up a number in your personal address book or the big phone book. Either way you look up a value, but one way is a little easier. For beginning programmers, we rank clarity and ease of understanding higher than efficiency concerns, but good programmers need to be aware of performance. See the A Little Extra section which follows for a more important discussion of efficiency.

Note 4: Such a short loop body fits easily on the same line. It might make it stand out more to put it on its own line.

Testing the Code

We tested `InsertionSort` with two cases, one using random numbers and one with user supplied data. The user supplied the data in reverse order which is a special case worthy of testing. Realistically we should provide many more tests, but because of space limitations we defer that to the exercises.

☞ A LITTLE EXTRA   **Efficiency**

Polishing your code to improve efficiency is important because getting results fast is always a selling point in commercial projects. However, the best performance increases will come from using the best **algorithm** for the task. Data structures and algorithm analysis courses show how to analyze algorithms for efficiency, and present good algorithms for important computing tasks. As an introduction to this area, we will investigate the efficiency of the `InsertionSort` algorithm of Example 4.6.

The `System` class has a method, `currentTimeMillis()`, which returns a **long** value containing the current time in milliseconds from a starting point of January 1, 1970. We call this method, just before the loop to do the insertion sort, to get the start time, and just after the loop to get the stop time. Computing `stoptime-starttime` tells us the time spent doing the insertion sort.

Example 4.7 revises Example 4.6 to add these timing statements.[3] We omit the output of the array because we want to run the program for very large arrays. For the same reason, we use only random number input, omitting the choice to let the user input the data.

The most interesting statistic about an algorithm is its **growth rate**. We want to know not just how much time insertion sort will take for a single array, but how does that time grow as the size of the array grows. In our test, sorting an array of size 1000 takes 210 milliseconds,

FIGURE 4.25 Rate of growth of insertion sort

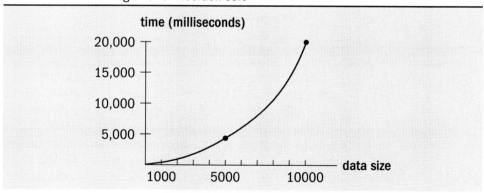

---

[3] In running this example, we disabled the JIT (Just-in-time compiler) which compiles the byte code during runtime to improve performance. The results will be similar with the JIT enabled, but we need to use larger data sizes, because the processing is so fast. For example, one run took 1292 milliseconds to sort 10000 items, and 5097 to sort 20000. To disable the JIT in Java 1.1 use the command `java -nojit InsertionSortTiming`. To disable the JIT in Java 2 (1.2), use the command `java -Djava.compiler=NONE InsertionSortTiming`.

while sorting an array of size 10000 takes 20149 milliseconds. If we repeat with the same size arrays, we will get similar results. The size of the data increased 10-fold from 1000 to 10000, while the time increased about 100-fold from 210 to 20149 milliseconds.

We say that the time needed by insertion sort has a rate of growth which varies as the square of the size of the data. Based on this growth rate, we would predict that running this program with an array of size 2000 should take about 800 milliseconds. Try it and see how good our prediction is. Figure 4.25 shows the number of steps plotted as a function of the size of the data.

EXAMPLE 4.7 **InsertionSortTiming.java**

```java
/* Sorts random numbers using the insertion sort algorithm. Uses
 * random data. Outputs the milliseconds taken to sort. Use this
 * program to estimate the efficiency of insertion sort.
 */

import iopack.Io;

public class InsertionSortTiming {
    public static void main(String [] args)  {
        int size = Io.readInt("Enter the number of data items");
        int[] item = new int[size];       // allocate array to hold data
        for (int i=0; i<size; i++)
            item[i] = (int)(100*Math.random());
        long starttime = System.currentTimeMillis();            // Note 1
        for (int i=1; i<size; i++) {
            int current = item[i];
            int j = 0;
            while (current > item[j]) j++;
            for (int k=i; k>j; k-)
               item[k] = item[k-1];
            item[j] = current;
        }
        long stoptime = System.currentTimeMillis();             // Note 2
        System.out.println("The time used in milliseconds is "
                          + (stoptime-starttime));
    }
}
```

Output—First Run

```
Enter the number of data items:   1000
The time used in milliseconds is 210
```

Output—Second Run

```
Enter the number of data items:   10000
The time used in milliseconds is 20149
```

Note 1: We save the time just before the loop to do the sorting.

Note 2: We save the time just after the sorting loop, so subtracting the start time from the stop time will give the time used by the sort.

The Big Picture

Carefully developed pseudocode makes it much easier to code correctly in Java. Insertion sort is one of many algorithms that use data stored in arrays. By adding timing statements we can estimate the efficiency of the code. For insertion sort, we find that the time used grows proportionally to the square of the size of the data.

TEST YOUR UNDERSTANDING

18. Show the stages of insertion sort by starting with the array 52, 38, 6, 97, 3, 41, 67, 44, 15 and showing that array after each insertion of an element in the correct position in its predecessors.

19. Change the insertion sort algorithm by providing a different refinement for the `insert item i` subproblem of Figure 4.21. In this solution, compare item `i` to item `i-1`, exchanging the two if item `i` is less than item `i-1`. Repeat this process of moving the original item `i` to the left until either that item is greater than or equal to its predecessor or there are no more predecessors.

TRY IT YOURSELF

20. Test Example 4.6 carefully with a range of test data.

21. Using the data of question 18, show each change in that array following the revised insertion sort algorithm of question 19.

TRY IT YOURSELF –A LITTLE EXTRA

22. Add timing statements before and after the call to the `tossResults` method in Example 4.1. Try test cases to estimate the rate of growth of time to toss the dice as the number of tosses increases. If the number of tosses doubles, by what factor does the time increase?

4.5 ▪ Vectors and Enumerations

A **vector** is like an array, but it can grow in size. The cost of this flexibility is a decrease in performance compared to arrays. Vectors are useful in multithreaded applications as they are designed for safe access from concurrent threads.[4]

The `Vector` class, in the `java.util` package, has three constructors. We can specify the initial size of the vector and the amount to increase its size when it becomes full. Using the default

```
new Vector();
```

will give us a vector of capacity 10 which doubles in size when more space is needed. The constructor

```
new Vector(20);
```

creates a vector with the capacity to hold 20 elements, which doubles in size when necessary.

---

[4] Threads, which we introduce in Chapter 9, allow different parts of the program to proceed in parallel, sharing the processor.

Finally,

```
new Vector(15,5);
```

starts out with a capacity of 15 which increases by 5 when necessary.

To add an element to the end of a vector, we use the addElement method which has a parameter of type Object. Because every class is, directly or indirectly, a subclass of Object, we can add any object to a vector. The code

```
Vector v = new Vector();
String s = "Happy days";
v.addElement(s);
```

creates a vector and a string and adds the string to the vector. We can only add objects to a vector. To add primitive types we must use the wrapper classes discussed in Section 6.2.

Using the addElement method, as in

```
v.addElement("A big car");
v.addElement("Less is more");
```

adds the strings at the end of the vector

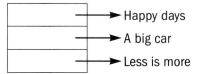

We could use the insertElementAt method to insert an item at a given index in the vector, but that is less efficient than adding at the end. For example,

```
v.insertElementAt("Candy and cake",1);
```

changes v to

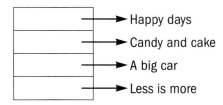

requiring the moving of two elements. In a large vector, insertion might require moving a large number of elements.

The elementAt method lets us get the element at a given index. For example,

```
String atTwo = (String)v.elementAt(2);
```

assigns "A big car" to atTwo. Because elementAt returns a value of type Object, we must cast it to a string to assign it to a string variable. The vector v contains only string elements so we know casting to a string will not cause an error.

The Vector class provides several methods to locate elements in a vector. The contains method returns **true** if its argument is an element of the vector and **false** otherwise. Thus

```
v.contains("A big car");
```

returns **true**, while

```
v.contains("Sweet dreams");
```

returns **false**. If we need the exact location of an element, the indexOf method returns the index of its first occurrence in the vector, or −1 if it does not occur. For example,

```
v.indexOf("A big car");
```

would return 2, while

```
v.indexOf("Sweet dreams");
```

returns −1. The call

```
v.indexOf("Happy days",1);
```

returns −1, because there is no occurrence, in v, of "Happy days" starting at index 1.

The capacity method returns the number of elements allocated for the vector, while the size method returns the number of its elements. Thus

```
v.capacity();
```

returns 10, while

```
v.size();
```

returns 4. Either

```
v.removeElement("Candy and cake");
```

or

```
v.removeElementAt(1);
```

would remove the element at index 1 from v.

The elements method returns an enumeration which allows us to get all the elements of a vector. The Enumeration interface is a general facility for retrieving the elements of a container. It has two methods, nextElement, which returns the nextElement in an arbitrary order, and hasMoreElements which returns **true** if there are more elements not yet returned. We can use the method

```
public void listAll(Enumeration e) {
    while(e.hasMoreElements())
        System.out.println(e.nextElement());
}
```

to list the elements of any enumeration. This listAll method applies to any container that has an enumeration, completely separating the details of the container type, Vector, Stack, List, or other container, from the listing process. We could list the elements of v with

```
Enumeration e = v.elements();
listAll(e);
```

To illustrate Vector objects, we create a vector of the first 1000 Fibonacci numbers. The Fibonacci sequence starts with its first two elements, $f_1 = f_2 = 1$, and the remaining computed by

$$f_{i+1} = f_i + f_{i-1}$$

so the first 10 Fibonacci numbers are 1, 1, 2, 3, 5, 8, 13, 21, 34, and 55. The Fibonacci numbers have useful applications in numerical analysis and occur in nature, but here we use them solely to illustrate vectors.

We use the `BigInteger` class, in the `java.math` package, to handle large Fibonacci numbers. We create big integers from strings as, for example,

```
BigInteger twentyDigits = new BigInteger("12345678909876543210");
```

and add them using the `add` method, as in

```
BigInteger stillTwentyDigits = twentyDigits.add(twentyDigits);
```

EXAMPLE 4.8   **Fibonacci.java**

```
/* Uses the Fibonacci sequence to
 * illustrate vectors.
 */

import java.util.*;
import java.math.BigInteger;

public class Fibonacci {
    public static void main(String[] args) {
        Vector fib = new Vector(1000);
        BigInteger previous = new BigInteger("1");
        BigInteger current = previous;
        fib.addElement(previous);
        fib.addElement(current);
        BigInteger temp;
        for(int i=2; i<fib.capacity(); i++) {                        // Note 1
            temp = current;
            current = previous.add(current);
            previous = temp;
            fib.addElement(current);
        }
        System.out.println
            ("The fifth Fibonacci number is " + fib.elementAt(4));   // Note 2
        System.out.println("The one-thousandth Fibonacci number is "
                                    + fib.elementAt(999));
        Vector prime = new Vector();
        for (int i = 0; i < 100; i++) {
            BigInteger aFib = (BigInteger)fib.elementAt(i);
            if (aFib.isProbablePrime(10))                            // Note 3
                prime.addElement(aFib);
        }
        System.out.println
            ("The probable primes in the first 100 Fibonacci numbers are:");
        Enumeration e = prime.elements();                           // Note 4
        while(e.hasMoreElements())
            System.out.println("\t" + e.nextElement());
        System.out.println
            ("Vector prime's capacity is " + prime.capacity());     // Note 5
        System.out.println("Vector prime's size is " + prime.size());
        System.out.println("The ninth probable prime is the "
                        + (fib.indexOf(new BigInteger("514229")) + 1) // Note 6
                        + "th Fibonacci number");
        int count = 0;
        BigInteger random100;                                   continued
```

```
            do {
                random100 = new BigInteger
                    (String.valueOf((int)(100*Math.random()) + 1));     // Note 7
                count++;
            }while (!fib.contains(random100));                          // Note 8
            System.out.println("It took " + count
                    + " tries to find a Fibonacci number randomly");
        }
    }
```

Output

```
The fifth Fibonacci number is 5
The one-thousandth Fibonacci number is
    4346655768693745643568852767504062580256466051737178040248l7
    2908953655541794905189040387984007925516929592259308032263471
    7520968962323987332247116164299644090653318793829896964992851
    6003704476137795166849228875
The probable primes in the first 100 Fibonacci numbers are:
        2
        3
        5
        13
        89
        233
        1597
        28657
        514229
        433494437
        2971215073
        99194853094755497
Vector prime's capacity is 20
Vector prime's size is 12
The ninth probable prime is the 29th Fibonacci number
It took 12 tries to find a Fibonacci number randomly
```

Note 1: We fill fib to its capacity of 1000. We use two variables, previous and current to represent the last two Fibonacci numbers computed. Each time through the loop we save the current number, add it to the previous number to get the updated current, and then copy the saved old current to get the new previous.

Note 2: It is always a good practice to check a computation with a known value, which we do here, checking that the fifth Fibonacci number is 5.

Note 3: We create a new vector containing those of the first 100 Fibonacci numbers that are probably prime. Determining whether a number is prime (has no divisors other than itself and 1) can be time consuming for large numbers. The BigInteger class has a method, isProbablyPrime, that determines with a certain probability that a number is prime. Making the probability higher makes the computation longer. This method uses the probability $1 - (1/2)^n$ where n is its argument. We pass the argument 10, so it uses the probability $1 - 1/1024$, which means there is greater than a 99.9% chance that the number is prime.

Note 4: Because we are using the `Vector` class, from the `java.util` package, we can use the `elements` method to get an enumeration. When creating our own container such as the `List` of Section 8.6, we can implement an enumeration by implementing the `hasMoreElements` and `nextElement` methods.

Note 5: There turned out to be 12 probable primes in the first 100 Fibonaaci numbers, so the vector, `prime`, automatically grew to capacity 20 from its initial capacity of 10.

Note 6: We check which Fibonacci number happens to be the ninth probable prime.

Note 7: We get a random number between 1 and 100 and then convert it to a string, using the `valueOf` method, so we can construct a `BigInteger` from that random number.

Note 8: We keep computing random numbers from 1 to 100 as long as they are not Fibonacci numbers. Because there are 11 Fibonacci numbers between 1 and 100, we expect about `100/11` = `9.09` trials, on the average, until we find a Fibonacci number.

The Big Picture

A vector grows automatically to accommodate more data. We add values of type `Object`, or any of its subtypes, to a vector. An enumeration, with `hasMoreElements` and `nextElement` methods, lets us iterate through the elements of a vector.

TEST YOUR UNDERSTANDING

23. Declare a vector which initially can hold 25 elements, and grows by seven when it becomes full.

24. Explain the difference between the `capacity` and the `size` methods for the `Vector` class.

4.6 ▪ Using a Vector: Interfaces and Callbacks

Vectors conveniently hold items without us having to specify how many. We use a vector to hold listeners who asked to be notified when the federal bank changes interest rates.

As we shall see when we introduce the Java event model in Chapter 6, Java often uses interfaces to provide callbacks. The term **callback** describes itself. We often leave a message, with our telephone number, for someone to call us back. In our example, an object, which implements an interface, registers with an object which generates an event, asking that object to call it back using one of the methods of that interface. The object receiving these callback requests saves them in an array, making the calls to each object in the array when an event of interest to them occurs.

For example, the economy is quite sensitive to interest rate changes. We can use an interface to provide a callback mechanism to notify interested parties about interest rate changes. The `RateChangeListener` interface will specify methods `rateRaised` and `rateLowered`.

The `FederalBank` class will have a method, `addRateChangeListener`, which interested parties can invoke to ask the bank to notify them when the rate changes. (This is analogous to someone leaving a message asking to be called back.) The bank keeps these listeners in an array.

Each interested party implements the two methods of the `RateChangeListener` interface to do whatever they need to do when the interest rate changes. When the bank raises the interest rate, it invokes the `rateRaised` method for each listener in its array, and when it lowers the rate it invokes the `rateLowered` method for each listener. (This is the callback.) In Figure 4.25 the investor and the business show the `rateRaised` and `rateLowered` methods of the `RateChangeListener` interface to the bank who will notify them when an interest rate change occurs. The bank shows its `addRateChangeListener` method so the investor and business can register for the callbacks.

FIGURE 4.26 Callback communication

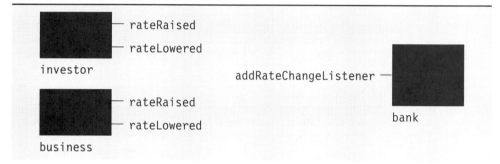

An investor and a business will be the parties who need to know about interest rate changes. Example 4.9 illustrates this use of an interface to provide callbacks when interest rates change. Figure 4.27 shows the interactions of the objects in this example.

FIGURE 4.27 Interactions of the investor, business, and bank

```
An investor asks the bank to notify her about a rate change.
The bank adds the investor to its array of listeners.
A business asks the bank to notify it about a rate change.
The bank adds the business to its array of listeners.
The bank decides to raise the interest rate.
The bank calls every listener in its array, using the rateRaised method that
    each listener implements to handle an interest rate increase.
The bank decides to lower the interest rate.
The bank calls every listener on its list, using the rateLowered method that
    each listener implements to handle an interest rate decrease.
```

EXAMPLE 4.9   **RateChange.java**

```java
/* Uses the RateChangeListener interface
 * to provide callbacks to investor and
 * business objects from the federal bank
 * when the rate changes.
 */

import java.util.*;
```

```
interface RateChangeListener {                                    // Note 1
   public void rateRaised(double amount);
   public void rateLowered(double amount);
}

class Investor implements RateChangeListener {                    // Note 2
   private FederalBank bank;                                      // Note 3
   public Investor(FederalBank f) {
      bank = f;
      bank.addRateChangeListener(this);                           // Note 4
   }
   public void rateRaised(double amount) {                        // Note 5
      System.out.println("   Investor sells stocks");
   }
   public void rateLowered(double amount) {
      System.out.println("   Investor buys stocks");
   }
}

class Business implements RateChangeListener {
   private FederalBank bank;
   public Business(FederalBank f) {
      bank = f;
      bank.addRateChangeListener(this);
   }
   public void rateRaised(double amount) {
      System.out.println("   Business reduces debt");
   }
   public void rateLowered(double amount) {
      System.out.println("   Business takes a loan");
   }
}

class FederalBank {
   private Vector listener = new Vector();
   public void addRateChangeListener(RateChangeListener r) {      // Note 6
      listener.addElement(r);
   }
   public void raiseRate(double amount) {                         // Note 7
      Enumeration e = listener.elements();
      while (e.hasMoreElements())
         ((RateChangeListener)(e.nextElement())).rateRaised(amount);// Note 8
   }
   public void lowerRate(double amount) {                         // Note 9
      Enumeration e = listener.elements();
      while (e.hasMoreElements())
         ((RateChangeListener)(e.nextElement())).rateLowered(amount);
   }
}
public class RateChange {
   public static void main(String [] args) {
      FederalBank bank = new FederalBank();                       // Note 10
      RateChangeListener investor = new Investor(bank);
      RateChangeListener business = new Business(bank);
      System.out.println("Fed raises rates");
      bank.raiseRate(.50);                                        // Note 11
      System.out.println("Fed lowers rates");
      bank.lowerRate(.25);
   }
}
```

continued

Output
```
Fed raises rates
    Investor sells stocks
    Business reduces debt
Fed lowers rates
    Investor buys stocks
    Business takes a loan
```

Note 1: Any class that wants its objects to be notified about rate changes implements this interface.

Note 2: The `Investor` class declares that it implements the `RateChangeListener` interface. This requires it to implement the two methods in the interface.

Note 3: We save the `bank` object in case we need to communicate with it later, say to remove ourselves as listeners. We do not use the `bank` further in this example.

Note 4: The investor tells the bank to add it to the array of listeners waiting for news about interest rate changes. The `investor` object refers to itself using the `this` variable.

Note 5: The bank will call this `rateRaised` method when it raises the interest rate. The investor can take appropriate action based on that news. We just print a message here to show how callbacks work.

Note 6: `public void addRateChangeListener (RateChangeListener r) {`

We add a `RateChangeListener` to the array of listeners. Notice how we use the interface as a type name for the formal parameter. Each actual argument that we pass to this method will be an object which is an instance of a class that implements the `RateChangeListener` interface.

Note 7: `public void raiseRate(double amount) {`

When the bank raises the interest rate, it calls this method to notify all the listeners. The bank knows that each listener implements the `RateChangeListener` interface, so it calls the `rateRaised` method of each listener, which the listener has implemented to take action when the rate is raised.

Note 8: `((RateChangeListener)(e.nextElement())) .rateRaised(amount);`

The `nextElement` method returns type `Object` which we cast to a `RateChangeListener` as that is the item type we put into the vector. This is the callback from the federal bank to the listeners.

Note 9: `public void lowerRate(double amount) {`

When the bank lowers the interest rate, it calls this method to notify all the listeners.

Note 10: `FederalBank bank = new FederalBank();`

To illustrate callbacks, we create a bank, an investor, and a business. The investor and the business, in their constructors, register themselves with the bank as rate change listeners.

Note 11: `bank.raiseRate(.50);`

The bank raises the interest rate, notifying all the listeners.

> **The Big Picture**
> ....
> Callbacks are an important mechanism for event handling. To illustrate, a federal bank keeps a vector of rate change listeners, notifying (calling back) each when it raises or lowers interest rates. The listeners implement the methods of the `RateChangeListener` interface which specify what they want to do when the rate changes.

Chapter 4 Summary

Arrays allow us to conveniently refer to large collections of elements. An array variable refers to the array; we use an index to refer to a specific element such as `myArray[2]`.

To indicate an array type we add the square brackets, `[]`, to the element type. Thus **int[ ]** denotes the type of an array whose elements have type **int**, and **int[ ][ ]** denotes the type of an array whose elements have type **int[ ]**.

The statement

```
int[] myArray = {4,5,6};
```

declares an array variable, allocates space for three elements, and initializes the three elements to have values four, five, and six. Array indices start at zero, so that `myArray[0]` has the value four.

Array variables refer to an array of elements. The statement

```
int[] myArray;
```

declares an array variable, but does not allocate space for elements to which it can refer. The statement

```
int[] myArray = new int[3];
```

uses the operator **new** to allocate space for an array of three integers to which the variable `myArray` refers. We still need to initialize this array with desired values.

We can pass array arguments to methods and return array values from methods. A formal parameter of an array type has a copy of the reference to an array passed as the actual argument to the method, so inside the method we can change the actual array argument passed from the caller.

An array of arrays is a collection of elements which are themselves arrays. Our test score example uses an array of arrays of test scores, where each array of test scores contains the scores for a single student. We first assume each student has the same number of scores, and then revise the example to allow students to have differing numbers of scores.

Many applications involve sorting. The insertion sort algorithm arranges the elements of an array in order, by inserting each element in the correct position in its predecessors. It is useful for small arrays, but gives way to more efficient methods for larger arrays. We can add timing statements to determine the time taken by an algorithm as a function of the size of the data; we call this the growth rate of the algorithm. For insertion sort, we found that the time increased as the square of the size of the data.

A vector is like an array, but it can grow in size. The addElement method adds an element, of type Object or a subtype, to the end of a vector, causing the vector to grow in size to accommodate it, if necessary. We get an element using the elementAt method. Other methods help us find an element in a vector. The elements method returns an Enumeration object with which we can list the elements of an array using the hasMoreElements and nextElement methods. Enumerations are useful for listing objects from many types of containers.

Java often uses interfaces to provide callbacks. Example 4.9 illustrates the use of an interface to provide a callback with the RateChangeListener interface, which specifies two methods, rateRaised and rateLowered. A class implements the RateChangeListener interface by implementing both of these methods to provide the actions to take when the interest rate rises or falls. The class registers as a listener by calling the federal bank's addRateChangeListener method. This registration is like leaving a message asking to be called back. When the bank raises the rate it calls the rateRaised method for each of the registered listeners, and when it lowers the rate it calls their rateLowered method. This is the callback. Our example uses investor and business objects interested in receiving callbacks when the interest rate changes.

Build Your Own Glossary

Find the uses, in this chapter, of the terms below. Enter a definition of each in the glossary.txt file on the disk that comes with this text.

algorithm   callback   pseudocode
array   growth rate   stepwise refinement
array element   index   vector
array of arrays   insertion sort

Skill Builder Exercises

1. Given the array

    ```
    static int[] nums = {45,23,67,12,11,88,3,77};
    ```

    what value does split(0,7) return, and how does the nums array change given the following Java code?

    ```
    static void interchange (int a, int b){
        int temp = nums[a];
        nums[a] = nums[b];
        nums[b] = temp;
    }
    static int split (int first, int last) {
        int x, splitPoint;
        x = nums[first];
        splitPoint = first;
        for (int i = first;i <= last; i++)
            if (nums[i] < x) {
                splitPoint++;
    ```

```
            interchange (splitPoint,i);
        }
    interchange(first,splitPoint);
    return splitPoint;
}
```

2. Find the array that results from the execution of f(3,4,21) where the code for the method f is:

```
public static int[] [] f(int n, int m, int value) {
    int[] [] x = new int[n] [m];
    for (int i=0; i < x.length; i++)
        for (int j=0; j < x[i].length; j++)
            x[i] [j] = value;
    return x;
}
```

3. Find the array that results from the execution of g(5,34) where the code for the method g is:

```
public static int[] [] g(int n, int value) {
    int[] [] x = new int[n] [];
    for (int i=0; i < x.length; i++) {
        x[i] = new int[i+1];
        for (int j=0; j < x[i].length; j++)
            x[i] [j] = value;
    }
    return x;
}
```

Critical Thinking Exercises

4. For each statement below, choose one of the following which best describes it:
    i.   declares an array only
    ii. declares an array, and allocates space for array elements
    iii. declares an array, allocates space for and initializes its elements with values supplied by the user
    iv. is incorrectly formed

    a. `int[] x;`   c. `int[] z = new int[5];`
    b. `int y = {32, 41};`   d. `int[] w = {5, 6};`

5. Consider the code

```
int[] x = {5,6,7,8,9};
int[] y = x;
y[2] = 3;
```

    Which of the following is correct?

    a. x[2] has the value 7.
    b. x[2] has the value 6.
    c. x[2] has the value 3.
    d. y[3] has the value 7.

6. Choose one of the following which best describes the result of the statement

```
int[] myArray = display(score);
```

where `display` is the method of Example 4.3.

    a.   The array `myArray` will refer to the same array as `score`.
    b.   The array `myArray` will refer to a different array than `score`.
    c.   This statement is incorrectly formed.
    d.   None of the above.

7. The statement

```
import iopack.Io;
```

as used in Example 4.1, tells the Java compiler that the code for

    a.   the file `Io.class` must be in the same directory as the file `Dice.class`.
    b.   the file `Io.class` must be in a directory named `iopack` which is a subdirectory of a directory in the classpath.
    c.   the file `Dice.class` must be in a directory named `iopack.Io`.
    d.   None of the above.

Debugging Exercise

8. A word is a palindrome if it reads the same backward and forward. For example, dad and otto are palindromes while hat and boat are not. The following program attempts to find if a word is a palindrome. The user enters the number of characters in the word, and then enters each character, using only lower-case characters. Find and correct any errors in this program.

```
public class Pal {
    public static void main(String [] args) {
        int size = Io.readInt("How many characters?");
        char [] a = new char[size];
        for (int i=0; i<a.length; i++)
            a[i] = Io.readChar("Enter next character");
        System.out.print("The word ");
        for (int i=0; i<a.length; i++)
            System.out.print(a[i]);
        for (int i=1; i<a.length/2; i++)
            if (a[i] != a[a.length-i]) {
                System.out.println(" is not a palindrome");
                System.exit(0);
            }
        System.out.println(" is a palindrome");
    }
}
```

Program Modification Exercises

9. Modify Example 4.4 to use a method to return the average score for each student.

10. Modify Example 4.4 to compute the class average for each test, rather than the average of the test scores for each student.

11. a. Modify the insertion sort program of Example 4.6 to use the insertion method described in the pseudocode of Test Your Understanding question 19.

A LITTLE EXTRA

    b. Add timing statements to the code to estimate the growth rate for the insertion algorithm (See Example 4.7).

12. Modify the insertion sort algorithm of Example 4.6 to sort an array of `String` objects. Use the `compareTo` method to replace the less than operator, `<`, used for integers. Enter the strings to sort using the `Io.readString` method.

13. Modify the `FederalBank` class of Example 4.9 to include a `removeRateChange-Listener` method, so rate change listeners who previously registered with the bank can request that they no longer be notified of interest rate changes.

Program Design Exercises

14. Write a program to search for a string in a `String` array. Return its position in the array if found and return –1 otherwise.

15. Write a program to reverse an array of `String` objects.

16. a. Generate an array of 20 random integers from zero to nine, using the formula

    `(int)(10*Math.random())`

    to generate each number. Search for the first occurrence, if any, of the number seven, and report its position in the array.

    b. Repeat the computation of 16a 1000 times and for each position in the array, report the number of times the first occurrence of a seven in the array is at that position.

17. Generate an array of 10,000 random numbers from zero to four, using the formula

    `(int)(5*Math.random())`

to generate each number. Report the percentage of each number, zero, one, two, three, and four, in the array.

18. The standard deviation is a measure of the spread of the data with respect to the mean (average). Data with a small standard deviation will be clustered close to the mean and data with a larger standard deviation will be more spread out. To compute the standard deviation, find the mean, find the difference of each item from the mean, square those differences, find the average of those squares, and, finally, find the square root of that average which is the standard deviation. For example, given the data 10, 20, and 30,

mean	$(10+20+30)/3 = 20$
differences	$(10–20) = –10\ (20\text{-}20) = 0\ (30–20) = 10$
squares of differences	100, 0, 100
average of the squares	$(100+0+100)/3 = 66.7$
square root of the average	8.2

Write a Java program to compute the mean and standard deviation of the elements in an array with elements of type **double**.

19. A company has five stores. Input the weekly sales for each store. Find the store with the maximum sales, the one with the minimum sales, and find the average weekly sales for the five stores.

20. A company has three regions with five stores in the first region, three in the second, and two in the third. Input the weekly sales for each store. Find the average weekly sales for each region, and for the whole company.

21. A company has five stores. Input the weekly sales for each store. Determine which stores have sales in the top half of the sales range. To find the range of sales, first find the maximum and minimum sales. The range is the maximum minus the minimum.

22. a. Write a program to partition an array. Read in n values to an array, and a test value x. Rearrange the array, so the elements up to and including index p are less than or equal to x and the elements from p+1 to n are greater than x. Elements may be repeated. The test value, x, may be larger than all values or smaller than all values or in between somewhere. You may only visit each element once, and may not copy it to another array. For example, given

    28   26   25   11   16   12   24   29   6   10

    with test 17 the result might be

    1   0   6   12   11   16   25   24   29   26   28

    with partition index 4.

      An outline of an algorithm is:

    ```
    Start with markers at each end. Move markers toward each other until you
    find a wrongly placed pair. Allow for x being outside the range of array
    values.

    While the two markers have not crossed over:
        exchange the wrongly placed pair and move both markers inward by one.
        move the left marker to the right while elements are less than or
            equal to x.
        move the right marker to the left while elements are greater than x.
    ```

A LITTLE EXTRA   b. Add timing statements to the code of 22a to estimate the growth rate of the array partitioning algorithm (See Example 4.7).

23. a. Write a Java program to perform a selection sort of an array of integers. In a selection sort, we find the smallest element of the array and interchange it with the first element. We repeat this process, finding the smallest element of the remaining elements and exchanging it with the first of the remaining elements. At each repetition the number of elements remaining decreases by one, until the whole array is sorted.

A LITTLE EXTRA   b. Add timing statements to the code of 23a to estimate the growth rate of the selection sort algorithm (See Example 4.7).

24. Write a Matrix class to operate with NxN matrices. For example, we represent a 3x3 matrix as

    {{2.3,4.1,-1.7}, {12.4,15.0,1.2},{2.0,3.0,4.0}}

Provide the addition, subtraction, and scalar multiplication operations. Adding or subtracting two matrices x and y produces a matrix with each element the sum or difference of the corresponding elements of x and y. In formulas,

z[i][j] = x[i][j] + y[i][j] for the sum
z[i][j] = x[i][j] - y[i][j] for the difference

The scalar multiplication of a matrix x by a number n produces a matrix with each element the product of n times x[i][j]. For example, 2.0 times the 3x3 matrix above produces the matrix

{{4.6,8.2,-3.4},{24.8,30.0,2.4},{4.0,6.0,8.0}}.

Include a constructor with parameters to specify the dimension n of the matrix, and the array of arrays for its initial value. Use the main method to test, applying matrix operations to several Matrix objects.

PUTTING IT ALL TOGETHER

25. Add a Picture subclass of Shape to Example 3.3. A picture contains an array of shapes, which may themselves be pictures, or any other shapes. Implement the draw, move, and toString methods for pictures. Implement an add method which will add shapes to the picture. The center of a shape will be the center of the picture. When drawing a picture, draw its shapes relative to the center of the picture. For example, if a picture has center (2,3) and contains a circle with center (1,3) and radius two, then draw that circle at center (3,6) with radius two. (Hint: One approach is to add a Point, base, as a static field in the Shape class, which will represent the absolute position of the current picture. When drawing a picture, temporarily augment the base to the center of the picture, and reset it after the drawing is done.)

5 Applets and Graphics

In this chapter we discuss event-driven programming. As we learned in Section 1.3, event-driven programs will respond to external events generated by the user or the operating system. By contrast, in our earlier programs, the Java interpreter executed our main method, which controlled the flow of the computation without interruption.

Figure 5.1 shows the AtmScreen applet of Example 1.5 after the user, John Venn in this case, has entered his name, entered a PIN, and chosen to withdraw from his savings account. The AtmScreen prompts Mr. Venn to specify the amount of the withdrawal, and waits for an event. In this simple example, Mr. Venn has two choices: either enter an amount or press the Finish button. The AtmScreen passively waits for one of these events to occur.

We will see another big change in our programs, involving the style of input and output. Previously our programs used the character mode, inputting characters from the keyboard and outputting characters to the screen. Now we will use graphical input and output which uses pixels (picture elements), dots on the screen used to create graphic images. A screen may show 25 columns of 80 characters each, for a total of 2000 characters. At a resolution of 1024 by 768, the screen displays 786,432 pixels, while at a resolution of 640 by 480, the screen displays 307,200 pixels.

Figure 5.1 shows this graphical style. Notice that we drew the text in a bigger font than that used for character output, and centered the prompt "Specify the amount" horizontally

FIGURE 5.1 The `AtmScreen`: specify the amount

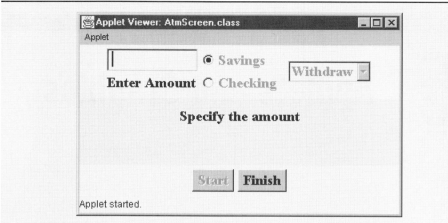

in the window. Running the applet will show the yellow canvas on which we drew the prompt. We will cover the use of text and color as well as the drawing of shapes in this chapter.

In this text we introduce event-driven programming in three chapters graded according to complexity. In this chapter we present only the events which cause the screen to be drawn or redrawn. These events are coordinated by the operating system and our program only has to specify what to draw. In Chapter 6 we cover high-level events associated with a component of a user interface such as a button. The button deals with the mouse directly and just tells us when the user presses it. In Chapter 7 we discuss the low-level mouse, key, and window events.

So far all our Java programs have been standalone applications. In this chapter we introduce applets, which as we recall from Section 1.2 are Java programs executed from a browser or an applet viewer. The `AtmScreen` shown in Figure 5.1 is an applet. We wrote the `AtmScreen` example to show all the different aspects of Java we cover in the first six chapters are important pieces that fit together in a complete program. However, it shows, at a simple level, how a bank might serve customers using the Internet. A customer could use a browser to connect to the bank's web site and download a Java applet to perform banking transactions.

With applets we enter the world of distributed computing, rapidly growing in importance, in which computers at different locations coordinate their efforts. We start with an introduction to the World Wide Web.[1]

OBJECTIVES
- Learn the uses of protocols.
- Learn basic HTML tags.
- Write and test applets.
- Understand the idea of event-handling code.
- Respond to paint events.
- Know AWT classes needed for drawing.

---

[1] For a thought provoking article on the significance of interaction in computing see "Why Interaction is More Powerful than Algorithms," Peter Wegner, *Communications of the ACM,* May 1997, pp. 80–91.

◆ Draw shapes and text.
◆ Draw in color.
◆ Set up the tangram puzzle.

OPTIONS
◆ Omit 5.6: Drawing the Tangram Puzzle (Used in 7.5)

5.1 ■ The World Wide Web and Applets

The Internet includes many applications of which the most used is email. The rapidly growing **World Wide Web (WWW)** allows computers all over the world to explore the enormous web of links from one site to another for educational, commercial, and recreational purposes. We will explain a little about how the World Wide Web works, and then show how Java applets enhance its capabilities.

Protocols

Diplomats and heads-of-state follow protocol, special rules of etiquette governing their formal interactions. Each knows what is expected in a particular situation. Similarly computers have rules of interaction, called **protocols**, specifying how they interact. To send email, a computer uses **SMTP** (Simple Mail Transfer Protocol). To receive email, a computer uses **POP3** (Post Office Protocol-version 3). Computers can transfer files using **FTP** (File Transfer Protocol). The `java.net` package that we introduce in Chapter 10 makes it easy to use these protocols from a Java program.

For web programs, we most often use **HTTP** (Hypertext Transfer Protocol). Hypertext, text augmented with links to other files, images, and other resources, makes the Web a web of links connecting one document to another document on the same computer or on one perhaps halfway around the world. As Java programmers, we are not required to know the details of HTTP.

Clients and Servers

A salesperson serves a customer who is sometimes called a client. Computers follow this analogy. A client can ask a server for service. The server may be a program that will give a client its email if both client and server send each other messages using POP3. The server may be ready to serve hypertext files to a client, both using HTTP to communicate. We call the client getting hypertext files a browser, while the server sending hypertext files is called a web server. Popular browsers include Communicator from Netscape, Internet Explorer from Microsoft, and HotJava from Sun.

The URL

We use a **URL** (Uniform Resource Locator) to specify an Internet resource. For example, the URL

```
http://java.sun.com/applets/index.html
```

specifies the page on Sun Microsystems' web site that provides applets. The URL has three parts,

`http`	specifies the protocol, in this case HTTP
`java.sun.com`	specifies the domain name of the server (a unique address for that computer)
`applets/index.html`	the path to the resource

If we enter this URL in the location field of a browser, then the browser will connect to Sun's computer, and using HTTP, ask the web server for the `index.html` file in the specified path. Sometimes we do have a reference, like this one, to a specific file that we request; more often we go to a home page and click on links from that home page to find the information we want. For example, Sun's Java home page has the URL

`http://java.sun.com/`

Clicking on the word Applets on that page gets the page given by the first URL above. When the URL does not include a path to the resource, then the server returns a file with a default file name, usually `index.html`.

HTML

Notice that the page providing applets has the `.html` extension. We use **HTML** (Hypertext Markup Language) to create the hypertext files found on the Web. This markup language adds tags to specify the formatting of the text. For example the tag `<br>` causes a break to a new line. The browser interprets these tags, formatting the page for the client. Using tags allows browsers of different capabilities to interpret the tags differently. For example, the tag `<em>`, requesting emphasis for the text that follows, might cause one browser to display the text in italics, but another browser, without the capability to use italics, might underline that text for emphasis.

The World Wide Web must adapt itself to many computers with differing capabilities. By using **HTML tags**, web documents can be displayed by a variety of browsers including those on terminals without graphics capabilities.

We will soon see how to include Java applets in web pages. Although HTML is not hard to learn to use, we do not really need to design web pages to learn how to use Java applets. Nevertheless, a brief introduction to HTML here will remove some of the mystery from the Web, and provide a foundation for further study for those who may wish to integrate Java into web applications.

To get the flavor of HTML we list a few tags in Figure 5.2 and use them to write a rudimentary web page. Tags are not case sensitive; the tag `<BR>` is the same as `<br>`.

FIGURE 5.2 Some HTML tags

` `	break to the next line
`<p>`	new paragraph (after a blank line)
`<em>` ... `</em>`	emphasize the text
`<strong>` ... `</strong>`	strongly emphasize the text
`<title>` ... `</title>`	title, displayed separately from text
`<h1>` ... `</h1>`	top-level header
`<h3>` ... `</h3>`	third-level header, (lowest is sixth)
`<ul>` ... `</ul>`	an unordered list
`<li>`	element of a list
`<a>` ... `</a>`	an anchor, a hypertext link
`<img>`	an image
`<applet>` ... `</applet>`	a Java applet

We can insert an empty tag like <br> anywhere to cause a line break. Non-empty tags like <em> have a closing form using the forward slash that marks the end of the text covered by that tag. Thus

```
<em> Java is fun. </em>
```

would emphasize the text *Java is fun.* The six levels of header tags specify the importance of the header, with h1 being the most important, and h6 the least. Browsers will try to make the more important headers larger and more impressive. An unordered list includes, between its starting and ending tags, various list elements with tags <li>.

Some tags use attributes embedded in the tag to provide information needed to interpret that tag. The **anchor tag** uses the href attribute to specify the URL of a hypertext link. For example, to link to Sun's Java home page we can use the anchor

```
<a href = "http://java.sun.com/"> Sun's home page. </a>
```

The href attribute gives the URL for Sun's Java home page. The text *Sun's home page.* will usually appear underlined and in blue, indicating that a mouse click will cause the browser to request, using HTTP, the Sun server to serve up its home page HTML file, which the browser then interprets, displaying Sun's home page.

The client must be connected to the Internet to link to other computers. Anchors can also link to files on the same machine using a relative URL. For example, to link to a file funStuff.html in the same directory, we could use the anchor

```
<a href = "funStuff.html"> some fun stuff </a>
```

Use the <img> tag to display an image, with the src attribute which gives the URL of the source of the picture. For example, to display a picture of the author of the text, found in the same directory as the web page itself, use

```
<img src="gittleman.gif">
```

A browser that cannot display graphics will fill the space with text such as [IMAGE].

The **applet tag** will, in Java enabled browsers, cause the Java interpreter to execute a Java applet. The code attribute specifies the class file for the applet, while the width and height attributes give the size of the applet in **pixels**. For example,

```
<applet code=Sort.class width=300 height=200>
If you see this, your browser is not Java enabled.
</applet>
```

will cause Java enabled browsers to execute the Sort applet of Example 1.3 and other browsers to display the message between the <applet> and </applet> tags.

The World Wide Web really is worldwide. In our browser we access web pages developed on a variety of machines running various operating systems. Java bytecode is platform-independent, so the server and client can use different hardware and software, but the client can still execute the applet developed on the server. This platform independence makes Java an ideal language for network applications, including the WWW.

Example 5.1 shows an HTML file for a very simple web page, displayed in Figure 5.3, which uses some of the tags from Figure 5.2.

FIGURE 5.3 Displaying `WebPage.html` in a browser

EXAMPLE 5.1   **WebPage.html**

```
<!-- Illustrates some html tags in                              // Note 1
-- a simple web page.
-->

<title> Let's try HTML </title>                                 // Note 2

<h1> Java is fun  </h1>                                          // Note 3
<p>
<h3> With <em>Java</em> we can </h3>                             // Note 4
<ul> <li> Do object-oriented programming                        // Note 5
        <li> Create nifty graphics
```

```
            <li> Add applets to our web pages
            <li> Network to remote computers
        </ul><p>
        Download Java from
            <a href = "http://java.sun.com/"> Sun's home page. </a>   // Note 6
        <br>
        <h2> Get ready -- Here comes the prof
        <img src=gittleman.gif><br>                                   // Note 7
            who wrote this applet </h2><br>
 Press the <strong> Sort </strong> button to display 10 random numbers.<br>
Then press the <strong> Next </strong> button to insert the next value in
 order.<br>
<applet code=Sort.class width=300 height=200>                         // Note 8
        If you see this, your browser is not Java enabled.
</applet>
```

Note 1: Comments in HTML documents start with `<!--` and end with `-->`.

Note 2: The title displays at the top of the frame, not in the document itself. Web search engines use the title in their searches.

Note 3: The text between the `h1` tags has the largest size.

Note 4: The em tag causes the text to be displayed in italics.

Note 5: Each item of an unordered list is preceded by a bullet.

Note 6: `<a href = "http://java.sun.com/"> Sun's home page. </a>`

The URL of the anchor does not show up; the blue text is underlined.

Note 7: `<img src=gittleman.gif><br>`

The image is a `.gif` file, a graphics format.

Note 8: `<applet code=Sort.class width=300 height=200>`

`Sort.class` was compiled from the `Sort.java` program containing the Java code for the applet of Example 1.3 that displays 10 random numbers in a bar chart when the user presses the Sort button and inserts the next bar in order with respect to its predecessors when the user presses the Next button. In this chapter we will learn how to write applets; we will study the code for a more complex version of this applet in Section 6.6.

Use a browser to see this page. In Netscape Communicator, click on File, click on Open Page, and click on Choose File to locate the `WebPage.html` file. In Microsoft Internet Explorer, click on File, click on Open, and click on Browse to locate `WebPage.html`.[2] The URL is a file URL, using the `file` protocol. The domain name of the server is just the local host, which can be omitted, so the URL looks like

`file:///path/WebPage.html`

where `path` is the path on the local machine to the `WebPage.html` file.

---

[2] The versions of these browsers or other browsers used to view applets in this text must be Java 1.1 enabled. Netscape Communicator 4.5 and above and Microsoft Internet Explorer 4.0 and above will run the applets in this text.

We created a very simple web page in Example 5.1, but to illustrate Java applets we could have made it a lot simpler, leaving out everything but the applet tags, as in the following code for the HTML file Sort.html.

```
<applet code=Sort.class width=300 height=200>
</applet>
```

Loading Sort.html in a Java 1.1 enabled browser will cause the Sort applet to execute; the page will just contain the bar chart and the button we use to sort the bars. To test applets without using a browser, development environments provide an applet viewer that executes only the applets in an HTML file, and ignores everything else.[3] Using an applet viewer, both Example 5.1 and Sort.html would produce the same result.

The Big Picture

Using a browser, we can connect to sites anywhere in the World Wide Web. Web pages, written using HTML, the Hypertext Markup Language, may include applets, which are Java programs that the browser downloads and executes on our machine. A browser displays the full web page, while an applet viewer just tests any applets on the page, ignoring other HTML tags.

TEST YOUR UNDERSTANDING

1. Which protocol does the browser use to download web pages?

2. Given the URL

   ```
   http://developer.javasoft.com/developer/readAboutJava/jpg/ball.html
   ```

   a. What is the protocol?
   b. What is the domain name of the server?
   c. What is the path to the resource?

3. What language do we use to write web pages?

4. What are the three required attributes in the <applet> HTML tag? What does each specify?

5. For what purpose is an HTML anchor tag used?

6. Which header tag, h2 or h5, will most likely cause a more prominent display of the text to which it applies?

5.2 ▪ Event-Driven Programming and a First Look at the AWT

Redrawing a window gives a gentle introduction to event-driven programming. The operating system manages its windows, notifying Java when a Java window needs refreshing. In our program we write the code to execute when our applet window needs to be redrawn.

---

[3] The JDK provides the applet viewer which we use from the console window, as for example,

```
appletviewer WebPage.html
```

See Appendix E for instructions on creating applets with other development environments.

However, we do not call this code, but rather we wait for an event such as a user resizing the window to cause Java to call our `paint` method.

Responding to Paint Events

Example 5.1 showed an applet executed from a web page, but we will not see until Section 6.6 how the code for that `Sort` applet works. Let us look at the simple applet in Example 5.2 to start event-driven programming.

EXAMPLE 5.2   **HelloCount.java**

```java
/* Displays the message "Hello World!" in
 * a browser or applet viewer window.
 */

import java.applet.Applet;
import java.awt.Graphics;
public class HelloCount extends Applet {
  private int count = 1;       // counts paint calls
  public void paint(Graphics g) {
    g.drawString("Hello " + count++,30,30);                    // Note 1
  }
}
```

Note 1: We increment `count` each time we call `paint` so we can see how events that the user creates cause calls to the `paint` method to redraw the applet.

We will discuss the code further after we see how `HelloCount` works. Notice that the `HelloCount` class has no `main` method. We always run applets using a browser or applet viewer which initializes and starts them. The big difference between this program and the Java programs of earlier chapters is that our program is no longer in control. We implement the `paint` method, but we do not call it. The Java system calls our `paint` method when our applet needs to be redrawn. It needs to draw the window when the applet starts running, and again in response to external events such as a user minimizing the window, resizing it, or just changing to another web page and then returning back to this one. Figure 5.4 shows the applet when we first start it.

The `paint` method increments the `count` variable by one every time Java calls it so we can see how the user's actions cause Java to call `paint`. Figure 5.5 shows the applet window after the user resizes it with the mouse, making it smaller. The `count` goes up to 2, showing that Java called the `paint` method to redraw the applet when the user resized it.

Covering the applet with another window and then uncovering it will cause Java to call our `paint` method to redraw the applet. Figure 5.6 shows the applet partially covered, while Figure 5.7 shows it uncovered again, now showing a count of three indicating another call to `paint`.

Paint events are simpler than those we study in Chapters 6 and 7 because the operating system coordinates these events, keeping all its windows properly painted, not just our Java programs. All our Java program has to do is to implement the `paint` method.

FIGURE 5.4 HelloCount just starting

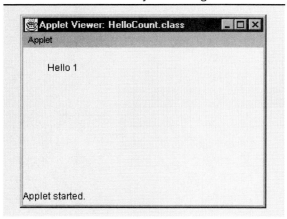

FIGURE 5.5 HelloCount after the user resizes it

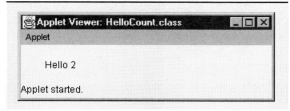

FIGURE 5.6 HelloCount partially covered

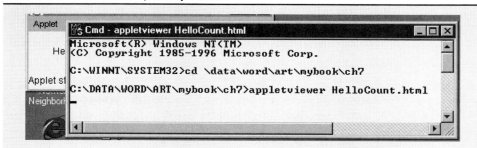

FIGURE 5.7 HelloCount uncovered again

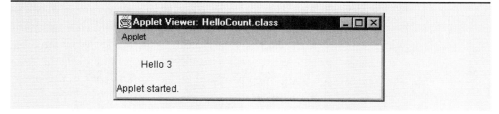

Our code in an event-driven program such as HelloCount.java implements the paint method which is a type of event handler. It contains the code to draw the applet window when an event causes the window to need to be drawn again. Although this example is an applet, we will see in Section 7.4 that we can create standalone event-driven programs. Now that we have seen how HelloCount works, responding to paint events, we will look at more detail at some other aspects of that code.

Introducing the AWT

The **AWT** (Abstract Windowing Toolkit) provides a large library of classes to build graphical user interfaces, programs that use the computer's graphics capabilities to make applications easier to use. The java.awt package contains the graphics classes for drawing shapes and text that we cover in this chapter, and the user interface controls such as buttons, checkboxes, text fields, and so on that we cover in Chapter 6. The java.awt.event package contains the classes needed for event handling which we use in Chapters 6 and 7.

☞ TIP

Even though the package name java.awt.event makes it look like this package is part of the java.awt package, it is a separate package contained in the directory java/awt/event.

Every applet must extend the Applet class in the package java.applet, which provides an interface between applets and the browser or applet viewer which runs them. We do not refer to any methods of the Applet class in this example.

The Applet class extends the class Panel found in the AWT. A panel is a container that can hold components such as buttons, and is itself contained in another container such as a browser window. In Chapter 6 we will see we can indeed fill an applet with various components.

An applet looks much like any other Java program. The HelloCount class can have data fields and methods just like other Java classes. In this example, HelloCount implements just one method, paint, which overrides the paint method it inherits from the Container class.

The paint method has a formal parameter g of type Graphics, a class of the AWT. Java has a reason to call the AWT abstract. The same Java code runs on vastly different systems. Windows, UNIX, and Macintosh systems each have their own windowing libraries. A Java program that used graphics would be very difficult to write if it had to refer specifically to each of the many windowing libraries. The Graphics package provides an abstract interface to these windowing libraries. When we run a Java program that uses graphics, the local system provides an implementation of the Graphics class that uses the windowing library available on that system. As Java programmers, we just use the methods of the Graphics class for all systems, and let each system give us an implementation which works on that system.

The drawString method of the Graphics class allows us to draw a string. Its first argument is the string to draw; the second and third arguments are the (x,y) coordinates, in pixels, where the string is to be drawn. In a window or panel, the origin, (0,0), of the coordinate system is the upper-left corner of the screen. The x-coordinate increases toward the right, while the y-coordinate increases toward the bottom of the screen. We indicate the size in pixels for our applet's panel in the width and height attributes of the applet tag in the HTML file. Figure 5.8 shows a panel with a width of 400 and height 300.

Running the HelloCount applet of Example 5.2 displays the string Hello n starting at position (20,30) which is near the upper-left corner of the screen. Here n is the number of times Java has called the paint method. In the rest of this chapter, we will learn how to do more interesting drawing in the paint method, but the function of our program will remain the same, to tell the Java system what to do when our applet needs to be painted. In Chapter 6 we will provide code to handle button presses, text field entries, and other user interface events. In Chapter 7 we will learn how to tell the Java system what to do in response to mouse clicks and key presses.

To run the HelloCount applet, we compile that code, and write an HTML file,

```
<applet code=HelloApplet.class width=300 height=200>
</applet>
```

which we load in the browser or applet viewer.

FIGURE 5.8 Coordinates, in pixels, for 400 by 300 panel

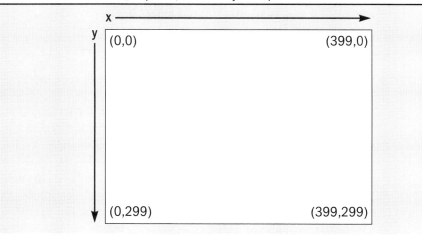

The Big Picture

We write code in an event-driven program that is called by the system when an event occurs. Our `paint` method indicates how to redraw the window when the user resizes it or uncovers it from behind another window. The Abstract Windowing Toolkit (AWT) includes the `Graphics` class whose methods let us draw shapes and text in a platform-independent manner.

TEST YOUR UNDERSTANDING

7. What class does every applet extend? In what package is that class found?

8. Arrange the following classes in order, from left to right, so that each class, except the rightmost, extends the class on its right.

   `Component, Applet, Container, Object, Panel`

9. Which classes from the `java.awt` package did we mention is this section?

10. Explain how the `Graphics` class helps provide platform independence for Java.

11. We use coordinates to represent each pixel in a window or panel. What is the position of the origin `(0,0)`?

12. Give some examples of events that would cause the `paint` method of the `HelloCount` applet of Example 5.2 to be called.

5.3 ▪ Drawing Shapes

We introduce the methods of the `Graphics` class for drawing lines, rectangles, ovals, and arcs.

Graphics Drawing

The `Graphics` class contains a number of methods for drawing various shapes. We will draw lines, rectangles, ovals, polygons, and arcs. In each case, we will do our drawing in the `paint` method which Java calls when events affecting our window cause it to need repainting.

The `drawLine` method has four **int** parameters, the x- and y-coordinates of each endpoint of that line segment. For example

```
g.drawLine(70,80,200,300)
```

draws the line between (70,80) and (200,300).

The `drawRect` method

```
g.drawRect(int x, int y, int w, int h)
```

draws the rectangle with upper-left corner (x,y), width w, and height h. Java draws an oval which just touches the center points of its bounding rectangle. The arguments specify the corner (x,y) , width, and height of the bounding rectangle. For example, Figure 5.9 shows the oval resulting from

```
g.drawOval(50,50,200,100)
```

FIGURE 5.9 An oval with its bounding rectangle

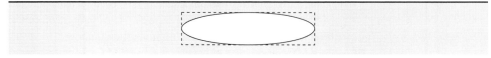

To draw a circle, use a bounding rectangle which is a square.

An arc is a piece of an oval. The `drawArc` method has six parameters; the first four specify the bounding rectangle while the fifth gives the start angle for the arc and the sixth gives the angle swept out by the arc. Referring to the face of a clock, the zero-degree angle is at three o'clock, and angles increase counterclockwise with 360 degrees in the full circle, as Figure 5.10 shows.

FIGURE 5.10 Degree measure around a circle

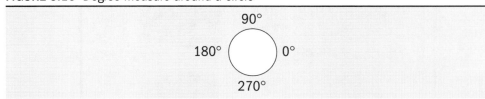

Figure 5.11 shows the part of the oval of Figure 5.9 that starts at 45° and sweeps out an arc of 90°, drawn by the method

```
g.drawArc(50,50,200,100,45,90);
```

FIGURE 5.11 An arc from the oval of Figure 5.9

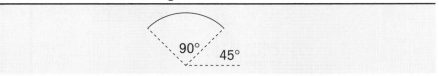

Positive sweep angles sweep out an arc from the starting angle in the counterclockwise direction.

Java lets us draw a rounded rectangle, which is a rectangle with the corners rounded. For example,

```
drawRoundRect(50,50,200,100,30,30);
```

will draw a rounded rectangle of width 200 and height 100 with each corner replaced by an arc of an oval of diameter 30 in each direction.

To draw a polygon we can use two arrays, one for the x-coordinates of the polygon's point, and one for the y-coordinates. To draw the triangle below at right we create the arrays

```
int [] x = {0,50,100};
int [] y = {86,0,86};
```

and use the method

```
g.drawPolygon(x,y,3);
```

(50,0)

(0,86)   (100,86)

where the last argument gives the number of points. The `drawPolygon` method will automatically connect the last point with the first, closing the polygon. Use the `drawPolyline` method to draw without connecting the last point to the first.

To draw filled ovals, arcs, rectangles, rounded rectangles, or polygons, use the `fillOval`, `fillArc`, `fillRect`, `fillRoundRect`, and `fillPolygon` methods which have the same arguments as the corresponding draw methods shown above.

Example 5.3 demonstrates these `draw` and `fill` methods. The `paint` method of this applet draws various shapes when the Java system calls it in response to events requiring the applet to be painted.

FIGURE 5.12 The `DrawFill` applet of Example 5.3

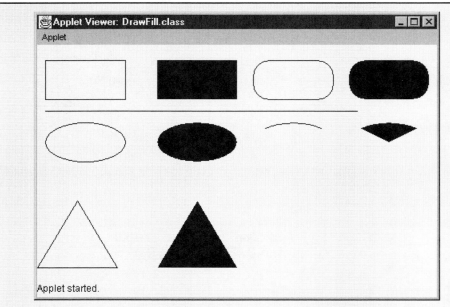

Figure 5.12 shows the applet viewer running the applet of Example 5.3 using DrawFill.html given by

```
<applet code=DrawFill.class width=500 height=400>
</applet>
```

EXAMPLE 5.3

DrawFill.java

....

```java
/* Uses Graphics draw and fill
 * methods for various shapes.
 * The Java system calls the paint
 * method when the applet needs repainting.
 */

import java.awt.Graphics;
import java.applet.Applet;

public class DrawFill extends Applet {
  public void paint(Graphics g) {
      g.drawRect(10,20,100,50);                    // Note 1
      g.fillRect(150,20,100,50);
      g.drawRoundRect(270,20,100,50,40,40);        // Note 2
      g.fillRoundRect(390,20,100,50,40,40);
      g.drawLine(10,85,400,85);                    // Note 3
      g.drawOval(10,100,100,50);                   // Note 4
      g.fillOval(150,100,100,50);
      g.drawArc(270,100,100,50,45,90);             // Note 5
      g.fillArc(390,100,100,50,45,90);
      int [] x = {0,50,100};
      int [] y = {286,200,286};
      g.drawPolygon(x,y,3);                        // Note 6
      x[0]=150; x[1]=200; x[2]=250;
      g.fillPolygon(x,y,3);
  }
}
```

Note 1: The rectangle has upper left corner (10,20), width 100, and height 50.

Note 2: Rounds the corner with an arc of a circle of diameter 40.

Note 3: Draws the line from (10,85) to (400,85).

Note 4: Draws the oval whose bounding rectangle has corner (10,100), width 100 and height 50.

Note 5: Draws the arc, starting from a 45 degree angle and sweeping out a 90 degree angle counterclockwise, which is part of the oval whose bounding rectangle has upper left corner (270,100), width 100, and height 50.

Note 6: g.drawPolygon(x,y,3);

Draws the polygon from point (0,286) to (50,200) to (100,286) to (0,286).

☞ **TIP**

Test applets using the applet viewer rather than a browser. Browsers store your class file in a cache. When you make changes and recompile, they load your Java class from the cache, using the old file, not the changed version. Even pressing the Reload button does not help as that just reloads the HTML file. To load the new version of the Java class you have to clear the cache or exit the browser and start it again.

We are not developing web pages, so we can just use an applet viewer to test our applets. Even when we include an applet in a web page, we can still test it with the applet viewer, using the browser only when we have completed the applet development.

In Example 5.3 we passed numbers as arguments to the drawing methods, drawing the filled rounded rectangle, for example, starting from the point (390,20). We gave the applet a width of 500 in the HTML file, so there was space in the applet to draw this rectangle. Had we given the applet a width of 350, then the filled rounded rectangle would not have appeared. Our program would be more flexible if we drew the figures relative to the size of the applet.

Drawing Relative to the Screen

We can get the size of the applet using the `getSize` method of the `Component` class which the `Applet` class inherits because `Applet` is a subclass of `Panel` which is a subclass of `Container` which is a subclass of `Component`. The `getSize` method returns an object of type `Dimension`, a class in the `java.awt` package, which has public data fields, `width` and `height`, that tell us the width and height of the component, in this case an applet.

Drawing relative to the size of the applet takes more effort to implement, but will work correctly when we change the applet's width and height in the HTML file. For example, Figure 5.13's code would draw a rectangle whose width is one-third the width of the applet, whose height is one-third the height of the applet, and which appears at the upper-right of the applet.

FIGURE 5.13 Drawing a rectangle relative to the applet's size

```
Dimension d = getSize();
int w = d.width;
int h = d.height;
g.drawRect(2*w/3,0,w/3,h/3);
```

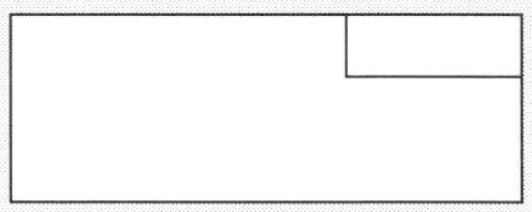

No matter what size the applet has, the rectangle of Figure 5.13 has its corner at a point two-thirds of the width from the left. Its width and height are always one-third the width and height of the applet. To use the `Dimension` class, we must import `java.awt.Dimension`. We will leave the revision of Example 5.3 to draw figures relative to the size of the applet to the exercises, but we will use this technique in the next section when we draw text.

The Big Picture

Given a `Graphics` object for our platform, we draw lines, rectangles, rounded rectangles, ovals, arcs, and polygons and fill all but the lines. When drawing ovals and arcs, we specify the dimensions of their bounding rectangles. We can draw relative to the size of the applet so the figures will resize when the applet does.

13. Write a statement to draw a horizontal line of length 12 whose left endpoint is (3,5).

14. Write a statement to draw a rectangle whose opposite corners are (10,10) and (100,200).

15. Write a statement to draw a square of side 50 whose upper left corner is (30,60).

16. Write a statement to draw an oval which touches the left side of its bounding rectangle at (100,100) and the top of its bounding rectangle at (200,50).

17. Write a statement to draw an arc of the oval of question 16 which starts at 90 degrees and sweeps out an angle of 120 degrees.

18. Write Java code to draw a closed polygon connecting the points (0,0), (200,50), (100,100), and (0,0) in that order.

5.4 ▪ Drawing Text

We have often used the System.out.println method to display characters in the console window. Using characters gives little or no choice of fonts, styles, and sizes, much less colors. In this section, we will draw text in graphics mode where we can choose different fonts, sizes, and styles for our text. In the next section we will use color for both text and shapes.

Fonts

The Font class in the java.awt package lets us create fonts of different types, with various sizes, and a choice of four styles. Java uses generic font names, so that it can more easily handle different languages which use a range of Unicode characters. Java generic font names are

Serif	The letters have hooks at the ends of the strokes.
SansSerif	The letters do not have hooks.
Monospaced	Each character has the same width.
Dialog	Used in dialog boxes.
DialogInput	Used for dialog input.

These generic font names are mapped to actual fonts available on the user's machine. For example, Windows systems use the Times New Roman font for Serif, the Arial font for SansSerif and Dialog, and the Courier font for Monospaced and DialogInput. Do not use the font names, such as Helvetica, that were introduced in version 1.0 of Java, which refer to specific fonts. These names may not have a meaning for non-Latin characters.

Java provides four styles, Font.PLAIN, Font.ITALIC, Font.BOLD, and Font.ITALIC + Font.BOLD, which are constants of the Font class. We measure the character size in points; one inch is equal to 72 points.

The Font constructor takes three arguments: the name, style, and size.

```
Font serifBold24 = new Font("Serif",Font.BOLD,24)
```

Notice that the three arguments are all constants. We really only need to construct the font serifBold24 once, but if we put the call to the constructor in the paint method, we will construct the same font again each time Java calls our paint method. Constructing this font should be part of the applet's initialization, done only once when the applet is first loaded.

Initializing an Applet

The `Applet` class has just the method we need. The browser calls the applet's `init` method, which has the form

```
public void init() {
   // do initialization here
}
```

once when it loads the applet. The default implementation of `init` in the `Applet` class does nothing, so we override it to provide any needed initialization for our applet, such as the creation of a font.

☛ **TIP**   Use the `init` method to compute values that will not change during the life of the applet. Do not compute something that might change, such as the applet's width, in the `init` method, because `init` is called only once when the applet is loaded.

We can specify the font for the applet in the `init` method using the `setFont` method from the `Component` class, which `Applet` inherits, as in

```
setFont(serifBold24);
```

which would cause all the strings in the applet to appear in the font `serifBold24`. If we want to change the font in the `paint` method, then we can use the `setFont` method of the `Graphics` class:

```
g.setFont(serifBold24);
```

We will draw all our text at positions defined in terms of the dimensions of the applet, so if the size of our applet changes, then our text will still be displayed at the same relative positions. As we saw in Figure 5.13, we can use the `getSize` method to return a `Dimension` object which gives us the height, h, and width, w, of our applet.

Suppose we want to center a string in the applet, horizontally. If we know the number of pixels, n, that a string will use when displayed in a given font, we can easily center it horizontally. If the applet's width is w, then w-n gives the number of pixels on both sides of the string. Dividing w-n in half gives the number of pixels on each side of the string needed to make it centered, as shown in Figure 5.14.

FIGURE 5.14 Centering a string

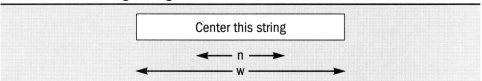

Font Metrics

To find the width of a string drawn in our applet using a certain font, we can get the `FontMetrics` object that will give us the data. We get a `FontMetrics` object using the `getFontMetrics` method inherited from `Component`, as in:

```
FontMetrics metrics = getFontMetrics(serifBold24);
```

which gives us the `metrics` object for the `serifBold24` font as it will appear in this applet. To get the width of a string, we use the `stringWidth` method:

```
n = metrics.stringWidth("Center this string");
```

We can get the `metrics` object and find the width of a string in the `init` method because these values do not change, even if the applet changes size. However, we need to find the height and width of the applet in the `paint` method. Java calls our `paint` method whenever events require our applet to be redrawn. Many of these events, including maximizing, minimizing, and other resizing operations, change the size of our applet, so we need to compute the applet's size in the `paint` method, which will be called in response to such changes in its size.

In Example 5.4 we will draw five lines of text, each one using a different font name, and illustrating the four styles with different point sizes. For the text of each line, we draw the generic font name given by the `getName` method, as in

```
serifBold24.getName()
```

which will return the string `Serif`.

We will center the first line horizontally, and start each of the other four lines one-fourth of the way in from the left. Vertically, we will position the five lines so they always divide the applet into six equal parts. We put as much of the computation as possible in the `init` method which is called just once, rather than in `paint` which is called whenever the applet needs to be redrawn.

Figure 5.15 shows the `Text` applet of Example 5.4 running in the applet viewer.

Figure 5.15 The `Text` applet of Example 5.4

EXAMPLE 5.4 **Text.java**

```
/* Draws the different fonts, trying all the
 * styles, and using various point sizes.  Draws
 * all text relative to the applet's size.  Uses
 * the init method to compute quantities that
 * do not change.
 */

import java.awt.*;                                          // Note 1
import java.applet.Applet;

public class Text extends Applet {
    private Font serifBold24;                               // Note 2
    private Font sansSerifItalic14;
    private Font monoPlain18;
    private Font dialogBI20;        // Dialog, bold and italic, 20 point
    private Font DIPlain18;         // DialogInput, plain, 18 point
    private String serif;           // The string we want to center
    private int serifStart;         // The distance in for the centered string
    private int serifWide;          // The width of the centered string

    public void init() {
        serifBold24 = new Font("Serif",Font.BOLD,24);       // Note 3
        sansSerifItalic14 = new Font("SansSerif",Font.ITALIC,14);
        monoPlain18 = new Font("Monospaced",Font.PLAIN,18);
        dialogBI20 = new Font("Dialog",Font.BOLD+Font.ITALIC,20);
        DIPlain18 = new Font("DialogInput",Font.PLAIN,18);
        setFont(serifBold24);                               // Note 4
        FontMetrics metrics = getFontMetrics(serifBold24);  // Note 5
        serif = serifBold24.getName();                      // Note 6
        serifWide = metrics.stringWidth(serif);             // Note 7
    }
    public void paint(Graphics g) {
        Dimension d = getSize();                            // Note 8
        int w = d.width;                    // The width of the applet
        int h = d.height;                   // The height of the applet
        serifStart = (w-serifWide)/2;                       // Note 9
        int otherStart = w/4;                               // Note 10
        g.drawString(serif,serifStart,h/6);                 // Note 11
        g.setFont(sansSerifItalic14);                       // Note 12
        g.drawString(sansSerifItalic14.getName(),otherStart,2*h/6);  // Note 13
        g.setFont(monoPlain18);
        g.drawString(monoPlain18.getName(),otherStart,3*h/6);
        g.setFont(dialogBI20);
        g.drawString(dialogBI20.getName(),otherStart,4*h/6);
        g.setFont(DIPlain18);
        g.drawString(DIPlain18.getName(),otherStart,5*h/6);
    }
}
```

Note 1: Because we have to use Graphics, Font, and FontMetrics from java.awt, it is easier just to use the * which stands for any class in the java.awt package, rather than to import each class individually.

Note 2: We will create this font in the `init` method, but we declare it here because we need to use it in both the `init` and the `paint` methods.

Note 3: The three arguments to the `Font` constructor are the font name, the style, and the point size. Note that styles are constants of the `Font` class.

Note 4: Using the `setFont` method from the `Component` class sets the font for the applet. Setting the font here does not help us in this example, because we are going to change the font several times in the `paint` method, but if our applet used one font, then this would be the best way to set it.

Note 5: We get the font metrics for the `serifBold24` font as it will appear in this applet. The `FontMetrics` class has many methods to provide data about the font; we use just one in this example. Each font has its own `FontMetrics` object; we illustrate just this one.

Note 6: `serif = serifBold24.getName();`

The `getName` method returns the name of the font, which in this case is Serif.

Note 7: `serifWide = metrics.stringWidth(serif);`

The `stringWidth` method of the `FontMetrics` class returns the width in pixels of its string argument as drawn in this font.

Note 8: `Dimension d = getSize();`

The `getSize` method returns the size of the applet as a `Dimension` object which we can use to get the height and width of the applet. We get the applet's size in `paint` which is called when the size changes; we need to find the new size before drawing the strings again.

Note 9: `serifStart = (w-serifWide)/2;`

We divide the whitespace into two equal parts so the first string will be centered as in Figure 5.14.

Note 10: `int otherStart = w/4;`

The other four strings will start at a distance w/4 from the left of the applet, where w is the applet's width.

Note 11: `g.drawString(serif,serifStart,h/6);`

We draw the first string with the x-coordinate computed so as to center the string horizontally and the y-coordinate at h/6 which will always be one-sixth of the way down from the top, no matter what height, h, the applet has.

Note 12: `g.setFont(sansSerifItalic14);`

Inside the `paint` method, we use the `setFont` method of the Graphics class to change the font.

Note 13: `g.drawString(sansSerifItalic14.getName(),otherStart,2*h/6);`

We draw each of the other strings starting at a horizontal position one-fourth of the way across from the left. We position the strings vertically using the applet's height to divide it into six equally spaced parts each separated by one of our strings.

> **The Big Picture**
> ....
> Using the graphics mode we can draw text in different fonts, with italics, boldface, and in different sizes. Using a `FontMetrics` object, we can find properties of a particular font, which we can use to position text.
>
> We override the `init` method of the Applet class to include code that needs to be executed only once when the applet is initialized.

TEST YOUR UNDERSTANDING

19. Declare and initialize a monospaced, italic, 30-point font.

20. Declare and initialize a 12-point sans serif font which is both bold and italic.

21. Write a statement to draw a string s of width 50 centered horizontally in an applet whose width is 400 and height is 200.

22. Write a statement to draw a string s of width 100 centered both horizontally and vertically in an applet of width 300 and height 200. The font height is 20.

5.5 ▪ Using Color

The `Color` class, found in the `java.awt` package, defines the color name constants

Color.black	Color.gray	Color.magenta	Color.red
Color.blue	Color.green	Color.orange	Color.white
Color.cyan	Color.lightGray	Color.pink	Color.yellow
Color.darkGray			

We can construct other colors using their red, green, and blue components. One `Color` constructor uses values between 0 and 255 for each component. Thus

```
Color itsRed = new Color(255,0,0);
```

constructs a color, `itsRed`, that is equal to the `Color.red`, while

```
Color itsGreen = new Color(0,255,0);
```

constructs another, `Color.green`.

We can find the components of a color by using the `getRed`, `getGreen`, and `getBlue` methods. For example, the methods

```
Color.pink.getRed();
Color.pink.getGreen();
Color.pink.getBlue();
```

will return the values 255, 175, and 175 for the red, green, and blue components of `Color.pink`.

Another way to specify colors uses float values between 0.0 and 1.0. Using the constructor which takes three arguments of type **float**,

```
new Color(1.0f, 0.0f, 0.0f)        would be red,
new Color(0.0f, 1.0f, 0.0f)        would be green, and
new Color(1.0f, 175f/255, 175f/255)  would be pink.
```

To initialize the applet's background and foreground colors we can use the `setBackground` and `setForeground` methods inherited from the `Component` class:

```
setBackground(Color.yellow);
setForeground(new Color(100,150,200));
```

Inside the `paint` method we can change the drawing color using the `setColor` method of the `Graphics` class:

```
g.setColor(Color.blue);
```

If we need to get any of these colors, perhaps to save them to use later, we can use the `getBackground`, `getForeground`, and `g.getColor` methods.

Example 5.5 uses colors, making a rendition of this text that resizes itself when the applet is resized. Using the foreground color, we draw a green rectangle for the front of the book, then we draw light gray lines to represent the top and side views of the pages. Black lines outline the book, which is drawn on a pink background. The text's title uses a reddish brown color, while the author's name is dark blue. We do not simplify the expressions which compute the locations of the points needed to do the drawing in order to make it easier to see how they were obtained. We leave this simplification, which would make the program more efficient, to the exercises.

Figure 5.16 shows a black and white version of this applet. Run it to get the effect of the color. Be sure to resize it to see how the book and text resize along with the applet.

FIGURE 5.16 The applet of Example 5.5

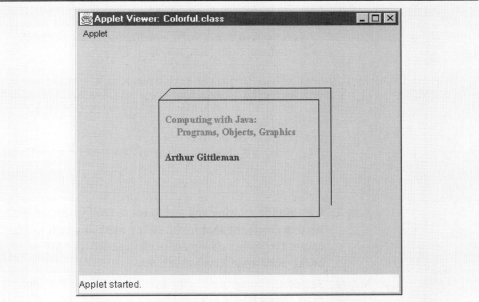

EXAMPLE 5.5 Colorful.java

```
/* Uses colors, drawing a shape and
 * text relative to the size of
 * the applet.
 */
```

continued

```
import java.awt.*;
import java.applet.Applet;

public class Colorful extends Applet{
    public void init() {                                        // Note 1
        setBackground(Color.pink);
        setForeground(Color.green);                             // Note 2
    }
    public void paint(Graphics g) {
        Dimension d = getSize();
        int w = d.width;
        int h = d.height;
        Font serifBold = new Font("Serif",Font.BOLD,h/20);      // Note 3
        g.setFont(serifBold);
        FontMetrics metrics = getFontMetrics(serifBold);
        g.fillRect(w/4,h/4,w/2,h/2);                            // Note 4
        g.setColor(Color.lightGray);
        for (int i=0; i<h/20; i++){
            g.drawLine(w/4+i,h/4-i, w/4+i+w/2, h/4-i);          // Note 5
            g.drawLine(w/4+i+w/2,h/4-i,w/4+i+w/2,h/4-i+h/2);    // Note 6
        }
        g.setColor(Color.black);
        g.drawRect(w/4,h/4,w/2,h/2);                            // Note 7
        g.drawLine(w/4,h/4,w/4+h/20,h/4-h/20);                  // Note 8
        g.drawLine(w/4+h/20,h/4-h/20,w/4+h/20+w/2,h/4-h/20);    // Note 9
        g.drawLine
            (w/4+h/20+w/2,h/4-h/20,w/4+h/20+w/2,h/4-h/20+h/2);  // Note 10
        g.setColor(new Color(100,100,0));                       // Note 11

        g.drawString("Computing with Java:", w/4 + w/50, h/4 + h/10);  // Note 12
        g.drawString("    Programs, Objects, Graphics",w/4 + w/50,
                            h/4 + h/10 + metrics.getHeight());
        g.setColor(new Color(0,0,125));                         // Note 13
        g.drawString("Arthur Gittleman",
                w/4 + w/50, h/4 + h/10 + 3*metrics.getHeight()); // Note 14
    }
}
```

Note 1: We want the drawing and text to resize itself whenever the applet is resized, so we need to create the font in the `paint` method which will be called when an event causes the size of the applet to change. We use the `init` method only to set the background and foreground colors.

Note 2: Setting the foreground color would be useful if the applet used just one color to draw. We set it here just to illustrate the use of this method; after drawing the front cover of the book in the foreground color, we set other colors in the `paint` method to do the rest of the drawing.

Note 3: We define the point size as h/20, so when the height h of the applet changes, the point size will adjust. We chose a size such that the two strings fit within the boundaries of the book.

Note 4: This centers a rectangle whose dimensions are half those of the applet.

Note 5: The top line of the front cover of the book has endpoints (w/4,h/4) and (w/4+w/2,h/4). In the loop we draw horizontal lines representing the top view of the pages of the book, in increments of one pixel. We add the number of pixels i to the x-coordinate, but subtract it from the y-coordinate, because the positive y direction is down. The ith horizontal line has endpoints (w/4+i,h/4-i) and (w/4+i+w/2,h/4-i). The y-coordinate of both endpoints is the same because the line is horizontal.

Note 6: `g.drawLine(w/4+i+w/2,h/4-i,w/4+i+w/2,h/4-i+h/2);`

The right border line of the front cover of the book has endpoints (w/4+w/2,h/4) and (w/4+w/2,h/4+h/2). Here the x-coordinates are the same because the line is vertical. In the loop, we draw vertical lines representing the front view of the pages of the book, in increments of one pixel. We add the number of pixels i to the x-coordinate, but subtract it from the y-coordinate, because the positive y direction is down. The ith vertical line has endpoints (w/4+i+w/2,h/4-i) and (w/4+i+w/2,h/4-i+h/2).

Note 7: `g.drawRect(w/4,h/4,w/2,h/2);`

This rectangle provides a black border around the front of the book.

Note 8: `g.drawLine(w/4,h/4,w/4+h/20,h/4-h/20);`

This line is the top of the binding of the book.

Note 9: `g.drawLine(w/4+h/20,h/4-h/20,w/4+h/20+w/2,h/4-h/20);`

This line is the top of the back cover of the book.

Note 10: `g.drawLine(w/4+h/20+w/2,h/4-h/20,w/4+h/20+w/2,h/4-h/20+h/2)`

This line is the front edge of the back cover of the book. We drew all these boundary lines after drawing the gray pages; drawing them before drawing the pages would cause the pages to draw over the borders.

Note 11: `g.setColor(new Color(100,100,0));`

This color appears as a reddish brown.

Note 12: `g.drawString("Computing with Java:", w/4 + w/50, h/4 + h/10);`

We position the title inside the book's front cover.

Note 13: `g.setColor(new Color(0,0,125));`

This color appears as a dull blue.

Note 14: `g.drawString("Arthur Gittleman",`
`                w/4 + w/50, h/4 + h/10 + 3*metrics.getHeight());`

Because the font changes size as the applet changes size, we have to get the FontMetrics object inside the paint method. We use the getHeight method to get the height of the font, adding three times the font height to the y-coordinate of the previous line, so that we draw the author's name line below the title.

The Graphics class has two drawing methods, draw3DRect and fill3DRect, that we did not mention in Section 5.3 because we need to set a different color than the default black to see these effects, which brighten and darken the drawing color to provide a very thin shading to give a slight three-dimensional effect. These methods have a fifth parameter of type **boolean** which we set to **true** to show the rectangle raised and to **false** to show it recessed. Thus the statement

```
g.fill3DRect(150,20,100,50,true);
```

draws a raised, filled rectangle with upper-left corner (150,20), width 100, and height 50.

The Color class has methods, brighter and darker, which modify the color, so

```
Color.orange.brighter()
```

will brighten Color.orange. We can call the method twice to achieve a greater effect, as in

```
Color.orange.darker().darker()
```

which darkens the darkened Color.orange.darker(). Figure 5.17 shows the applet of Example 5.6 which displays these effects.

FIGURE 5.17 The applet of Example 5.6

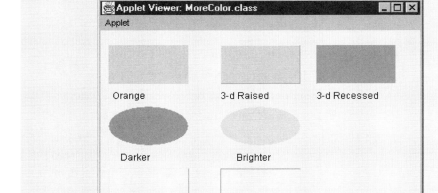

EXAMPLE 5.6 **MoreColor.java**

```
/*  Demonstrates drawing and filling 3D
 *  rectangles, and brightening and darkening
 *  colors.
 */

import java.awt.*;
import java.applet.Applet;

public class MoreColor extends Applet {
    public void paint(Graphics g) {
```

```
        g.setColor(Color.orange);
        g.fillRect(10,20,100,50);
        g.fill3DRect(150,20,100,50,true);              // Note 1
        g.fill3DRect(270,20,100,50,false);
        g.setColor(Color.orange.darker());
        g.fillOval(10,100,100,50);
        g.setColor(Color.orange.brighter());
        g.fillOval(150,100,100,50);
        g.setColor(Color.orange);
        g.draw3DRect(10,180,100,50,true);
        g.draw3DRect(150,180,100,50,false);            // Note 2
        g.setColor(Color.black);                       // Note 3
        g.drawString("Orange",15,90);                  // Note 4
        g.drawString("3-d Raised",150,90);
        g.drawString("3-d Recessed",270,90);
        g.drawString("Darker",25,170);
        g.drawString("Brighter",170,170);
        g.drawString("3-d Raised",20,250);
        g.drawString("3-d Recessed",160,250);
    }
}
```

Note 1: The argument **true** specifies a raised rectangle.

Note 2: The argument **false** specifies a recessed rectangle.

Note 3: We want the text in black. We do not have to draw the text right after the figure it describes, which would have forced us to keep changing colors between black and orange. We can set the color to black after drawing all the figures, and then draw all the text.

Note 4: To draw the text we use the default font name, style, and point size.

The Big Picture

....

The Color class has 13 predefined colors. We can define others using red, green, and blue values, either integers from 0 to 255, or floats from 0.0f to 1.0f. We can set the foreground and background colors for any component and set the drawing color used by a Graphics object.

TEST YOUR UNDERSTANDING

23. Construct a color which is the same as Color.blue, using the constructor that takes three integer arguments.

24. Construct a color which is the same as Color.blue, using the constructor that takes three float arguments.

25. Find the red, green, and blue components for Color.black.

26. Find the red, green, and blue components for Color.white.

ENHANCEMENT

5.6 ▪ Drawing the Tangram Puzzle

The traditional tangram puzzle uses seven plastic or wooden pieces that fit nicely into a square shape, but which can be moved to create other figures. In this section we begin to computerize the tangram puzzle.

Defining the Problem

In this section we begin to experiment with the ancient Chinese tangram puzzle, in which a square is cut into five triangles, a square, and a parallelogram. The object of the puzzle is to reassemble the pieces into various fanciful shapes. Figure 5.18 shows the seven pieces in the square, and Figure 5.19 shows a cat into which they can be arranged.

FIGURE 5.18 The seven tangram pieces

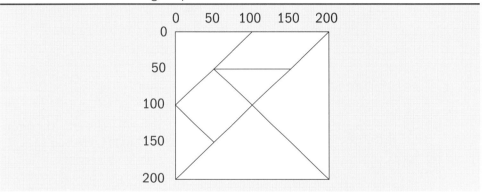

To solve the puzzle it would be nice if we could move the pieces around. In Chapter 7 we can return to this puzzle when we learn to drag figures with the mouse or move them by pressing keys. At this point, we can create polygons to represent each of the pieces, draw the original square, and translate the pieces to a new location. This will not only prepare us to solve the puzzle later, but will give us an opportunity to review arrays from Chapter 4.

Designing a Solution

Starting with One Polygon

To begin to build our tangram solver, we first work with a single polygon, constructing, translating, and drawing it to prepare ourselves to manipulate the puzzle with its seven polygons.

In Section 5.3 we drew a polygon, passing the drawPolygon method arrays of x- and y-coordinates, and the number of points. We can also draw a polygon by passing the drawPolygon method a Polygon object. The Polygon class, part of the AWT, has some methods that might help us solve this puzzle problem. For example, if p is a polygon, then

```
p.contains(x,y)
```

tells us if the point (x,y) is inside p.

FIGURE 5.19 The Cat

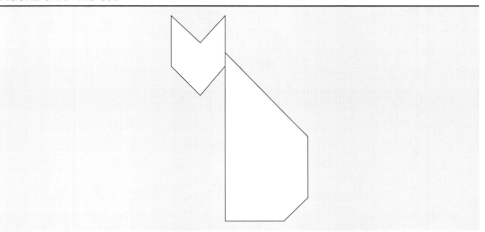

We will use the `translate` method of the `Polygon` class, as in:

```
p.translate(x,y)
```

which will add x to the x-coordinate of every point in p, and add y to the y-coordinate of every point in p.

To construct the polygons for each of the seven pieces, we will use the constructor

```
Polygon(int[] xpoints, int[] ypoints, int npoints)
```

Before we try to construct all seven polygons, let us get started by working with the triangle in the upper-left corner of Figure 5.18. From that figure we learn the coordinates of its three corners are (0,0), (100,0), and (0,100). To construct this triangle in Java we need the arrays of the x- and y-coordinates for these points, which are

```
int[] x = {0,100,0};   // x-coordinates
int[] y = {0,0,100};    //  y-coordinates
```

The number of points to connect to form the triangle is 3. We use the `Polygon` constructor to construct this triangle:

```
Polygon triangle = new Polygon(x,y,3);
```

If we translate this triangle 100 pixels across to the right and 50 pixels down, as in:

```
triangle.translate(100,50);
```

we will move the triangle to the position shown in Figure 5.20.

To draw the triangle we use the `drawPolygon` method, as in:

```
g.drawPolygon(triangle);
```

which will display the triangle in its original position if we call it before we translate the triangle, and in its new position if we call it after we translate the triangle.

Translating the triangle moves it to its new location. If we want to show the original position of the triangle and its new position in the same figure, we need to construct two

FIGURE 5.20 The translated triangle

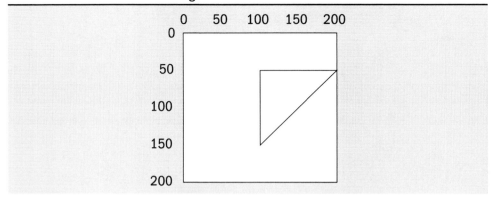

triangles, one which remains in its original position, and one which we translate to the new position. The following code will construct and display both triangles:

```
Polygon stay = new Polygon(x,y,3);
Polygon go = new Polygon(x,y,3);
go.translate(100,50);
g.drawPolygon(stay);
g.drawPolygon(go);
```

Having constructed one polygon, translated it, and drawn it, we can begin to solve the tangram puzzle containing seven polygons.

Designing a Solution

Solving a Simpler Problem

We still will not be able to solve the tangram puzzle, but we can solve the simpler problem of drawing the puzzle in its original configuration, moving all seven polygons to the right, and drawing them with different colors in the translated position.

In Figure 5.18, we see a 200x200 square. Using the numerical values shown, we can get the array of points for each polygon. For example, the big triangle at the bottom of the square connects the three points (0,200), (100,100), and (200,200). The array, xpoints, of x-coordinates, that we need is {0,100,200}, while the array, ypoints, of y-coordinates, is {200,100,200}.

Because we have seven polygons to construct, we have to create 14 arrays and call seven constructors. To save our fingers from this excessive amount of typing, we use an array of arrays[4] and write a loop to create the seven polygons. From Figure 5.18 we can get the coordinates of all the points and define the arrays for each polygon, as we did above for the big triangle at the bottom.

We put the seven arrays of x-coordinates into an array, x,

```
int[][] x =
    {{0,100,200},{100,200,200},{0,100,0},{0,50,0},
    {50,150,100},{0,50,100,50},{100,200,150,50}};
```

---

[4] See Section 4.3, "Arrays of Arrays."

and the seven arrays of y-coordinates into an array, y,

```
int[][] y =
    {{200,100,200},{100,0,200},{0,0,100},{100,150,200},
     {50,50,100},{100,50,100,150},{0,0,50,50}};
```

The order of the polygons in the arrays is as follows:

Array Index	Polygon
0	big triangle at bottom
1	big triangle at right
2	medium triangle at upper left
3	small triangle at lower left
4	small triangle in center
5	square
6	parallelogram

To construct the seven polygons, we use the loop

```
for (int i=0; i<x.length; i++)
    polygons[i] = new Polygon(x[i],y[i],x[i].length);
```

where x[i], the ith element of the x array, is itself an array of the x-coordinates for the ith polygon, y[i] is the array of the y-coordinates for the ith polygon, and x[i].length is the number of points in the ith polygon.

To draw these polygons we could use a loop

```
for (int i=0; i<x.length; i++)
    g.drawPolygon(polygons[i]);
```

in the paint method, but we would like to be more flexible, choosing a drawing color, and choosing whether or not to fill the polygon, so we write a showPolygon method. We pass a polygon, a color, a graphics object, and a **boolean** flag, indicating whether or not to fill the polygon, to the showPolygon method. The pseudocode for showPolygon is:

```
Save the current drawing color;
Set the color to the desired color;
if (fill flag is true)
    Fill the polygon;
else
    Draw the polygon;
Restore the saved drawing color.
```

For our initial tangram display, we would like to draw the polygons in their original configuration, as shown in Figure 5.18, on the left of the applet, and show them filled with different colors on the right of the applet. The translate operation moves the polygon to a new position. If we just translate the polygons on the left, then we will get polygons on the right, but nothing on the left.

We need two sets of polygons, one to draw on the left, and one to show, filled and colored, on the right. We have all the coordinates for the polygons in the x and y arrays, so we can create another array of seven polygons, and translate each to the right.

Completing the Java Code

Example 5.7 will draw the original configuration on the left and the colored, filled configuration on the right. We use an array to specify the colors for the filled polygons. Figure 5.21 shows this `Tangram` applet. In Chapter 7 we will expand this example to try to solve the puzzle.

FIGURE 5.21 The applet of Example 5.7

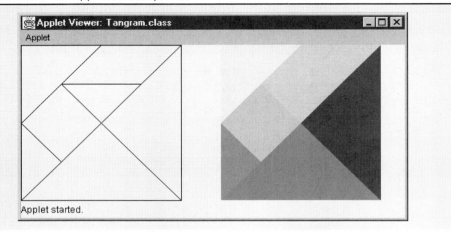

EXAMPLE 5.7   **Tangram.java**

```java
/* Draws the seven polygons of the
 * tangram puzzle on the left, and, filled
 * and colored, on the right.
 */

import java.awt.*;
import java.applet.Applet;

public class Tangram extends Applet {
    private int [][] x = {{0,100,200},{100,200,200},{0,100,0},{0,50,0},
                         {50,150,100},{0,50,100,50},{100,200,150,50}};
    private int [][] y = {{200,100,200},{100,0,200},{0,0,100},{100,150,200},
                         {50,50,100},{100,50,100,150},{0,0,50,50}};
    private Color [] colors = {Color.red,Color.blue,Color.yellow,
                Color.magenta,Color.cyan,Color.pink,Color.orange};
    private Polygon [] polygons = new Polygon[7];
    private Polygon [] translates = new Polygon[7];

    public void init() {
        for (int i=0; i<x.length; i++) {
            polygons[i] = new Polygon(x[i],y[i],x[i].length);
            translates[i]= new Polygon(x[i],y[i],x[i].length);     // Note 1
            translates[i].translate(250,0);                        // Note 2
        }
    }
}
```

```
    public void showPolygon
                   (Polygon p, Color c, Graphics g, boolean fill) {
        Color oldColor = g.getColor();                            // Note 3
        g.setColor(c);
        if (fill)
           g.fillPolygon(p);
        else
           g.drawPolygon(p);
        g.setColor(oldColor);                                     // Note 4
    }
    public void paint(Graphics g) {
        for (int i=0; i<polygons.length; i++) {
           showPolygon(polygons[i],Color.black,g,false);          // Note 5
           showPolygon(translates[i],colors[i],g,true);           // Note 6
        }
    }
}
```

Note 1: We construct a second array of polygons having the same points as the first.

Note 2: Each polygon moves 250 pixels to the right.

Note 3: We save the current drawing color, so we can restore it later.

Note 4: After drawing the polygon in the desired color, we restore the previous drawing color.

Note 5: We set the `fill` argument to **false** to draw the polygons unfilled.

Note 6: `showPolygon(translates[i],colors[i],g,true);`

We choose a color from the color array, and set the `fill` argument to **true** to fill the polygon with the chosen color.

Testing the Code

We cannot test the tangram solver until we complete the code for it in Section 7.5, where we use the mouse and the keyboard to drag and rotate the polygons to solve tangram puzzles.

The Big Picture

By constructing polygons and passing them to the `drawPolygon` method, we can use polygon methods such as `translate` to move polygons to new locations. Had we drawn the polygons directly from the arrays of the x- and y-coordinates of their vertices, we would not have had access to the `translate` method.

Chapter 5 Summary

Chapter 5 introduces graphics, event-driven programming, and applets. Web pages may include Java applets, which we download and execute. Our computer communicates with a Web server using HTTP, one of several networking protocols. HTML uses tags to markup text, images, and applets for display in a browser. We can write web pages using just a few tags, but web page design is beyond the scope of this Java text. The `<applet>` tag allows Java enabled browsers to run Java applets. This tag requires us to specify code, width, and height attributes of the applet.

The AWT (Abstract Windowing Toolkit) has many classes for graphical user interface (GUI) design. The `Applet` class is part of the `java.applet` package, but it inherits from `Panel`, a container that is itself part of another container such as a browser window. The `Panel` class inherits from `Container` which inherits from `Component`, a class representing the various components of a user interface, some of which we will study in Chapter 6.

Applets do use some of the methods they inherit from the `Component` class, including the `paint` method which has a `Graphics` object, `g`, as its parameter. The `Graphics` class provides abstract methods for drawing, so we can write a Java applet or application that runs on any platform without knowing the specific windowing library used on that system. We write our Java programs using the abstract methods of the `Graphics` class, and the system gives us a graphics object which implements these methods on that system.

In the `paint` method we include statements describing what text and shapes we want to draw, but our program does not call the `paint` method. The Java system calls our `paint` method when events cause our applet to need repainting. For example, if the user resizes the applet, making it larger or smaller, or transfers back to the applet from another web page, then the applet needs to be redrawn. The operating system handles the events from the user, passing them to Java which calls our `paint` method to redraw our applet. In this chapter our programs only respond to paint events, but in the next two chapters we will learn to respond to mouse, key, button, checkbox, and other user-interface generated events.

The `Graphics` class has methods to draw lines, rectangles, ovals, arcs, rounded rectangles, 3-d rectangles, and polygons, and to fill any of them (except, of course, the line). We specify

- a line with the coordinates of its endpoints,
- a rectangle with its upper-left corner, width, and height,
- an oval with the arguments for its bounding rectangle,
- an arc with the arguments for an oval, a start angle, and a sweep angle,
- a rounded rectangle with the arguments for a rectangle, and the width and height of each arc rounding a corner,
- a 3-d rectangle with the arguments for a rectangle, and a raised flag,
- a polygon with either a `Polygon` object, or arrays of x- and y-coordinates and the number of its points.

The `drawString` method of the `Graphics` class has three parameters, the string to draw, and the x and y position at which to draw it. We can construct a font, specifying a font name, style, and point size. Java uses several generic font names to allow implementation of diverse character sets. The style can be plain, bold, italic, or bold and italic.

The `Applet` class has an `init` method which the browser calls when it first loads the applet. The default `init` method does nothing, but we can override it to initialize our applet, for example, creating a font. It is best to put code that needs to be executed only once in the `init` method rather than in `paint` which will cause that code to be executed every time Java calls the `paint` method.

The Color class provides 13 predefined colors and allows us to construct other colors specifying their red, green, and blue components either as integers from 0 to 255 or as float values from 0.0 to 1.0. We can set the foreground and background colors for the whole applet and set drawing colors using a Graphics object, usually in the paint method. The Color class has methods to brighten and darken a color.

The tangram puzzle provides an opportunity to construct and draw polygons as well as to review the use of arrays.

Build Your Own Glossary

Find the uses, in this chapter, of the terms below. Enter a definition of each in the glossary.txt file on the disk that comes with this text.

anchor tag	HTML tag	SMTP
applet tag	HTTP	URL
AWT	pixel	World Wide Web (WWW)
FTP	POP3	
HTML	protocol	

Skill Builder Exercises

1. Match the protocol names on the left with their functions on the right.

   a. POP3   i.   send email
   b. FTP   ii. not a protocol
   c. SMTP   iii. web file server
   d. HTTP   iv. receive email
   e. HTML   v.   transfer files

2. Describe what the following applet will display.

```java
import java.awt.*;
import java.applet.Applet;

public class Skill extends Applet {
    public void paint(Graphics g) {
        Dimension d = getSize();
        int w = d.width;
        int h = d.height;
        g.drawRect(w/4,h/4,w/2,h/2);
    }
}
```

3. A student was unsuccessful in trying to run the DrawFill applet using the following HTML file. Identify and correct the problem.

```html
<applet code=DrawFill.class, width=500, height=320>
</applet>
```

Critical Thinking Exercises

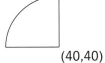

(40,40)

4. Which of the following will draw the arc shown to the left with center at (40,40)and radius 40?

   a. `g.drawArc(0,0,40,40,0,90);`   d. `g.drawArc(0,0,80,80,90,180);`
   b. `g.drawArc(40,40,0,0,0,90);`   e. `g.drawArc(40,40,0,0,90,90);`
   c. `g.drawArc(0,0,80,80,90,90);`

5. Which of the following colors is most like the color given by

   `new Color(51,85,85)?`

   a. `new Color(.20f,.33f,.33f)`   c. `new Color(.51f,.85f,.85f)`
   b. `Color.pink`   d. `new Color(53,87,87)`

6. In an applet shown in a 500 x 300 area, where is the point (50,150) relative to the point (50,50)?

   a. 100 pixels above it   c. 100 pixels below it
   b. 100 pixels to the right of it   d. 100 pixels to the left of it

7. Which of the following HTML tags does not require an end tag?

   a. `<h1>`
   b. `<ul>`
   c. `<em>`
   d. `<li>`
   e. `<title>`

Debugging Exercise

8. The following applet attempts to center a string horizontally. Find and fix any errors in it.

```java
import java.awt.*;
import java.applet.Applet;
public class Center extends Applet {
    Font f;
    String s;
    int x, w;
    public void init() {
        f = new Font("Monospaced",Font.PLAIN,18);
        setFont(f);
        FontMetrics metrics = getFontMetrics(f);
        s = f.getName();
        w = metrics.stringWidth(s);
    }
    public void paint(Graphics g) {
        Dimension d = getSize();
        int x = w-d.height;
        g.drawString(s,x,50);
    }
}
```

Program Modification Exercises

9. Modify Example 5.4 so that each line of text is centered horizontally.

10. Modify Example 5.3 to draw each figure relative to the size of the applet. When the applet is resized, the figures will be resized proportionally.

11. Modify Example 5.5 to make the program more efficient by reducing the number of division operations performed. Simplify by storing frequently computed values in variables, using the variable names to replace repeated computations.

12. Modify Example 5.7 to use random numbers to create the colors to draw each polygon. Get three numbers to specify the red, green, and blue components for a color.

Program Design Exercises

13. Write a method, `getPointSize`, which has formal parameters, a string, a font, an applet width, and an applet height, and which returns the largest point size such that drawing the string in that point size just fits within the applet. Write an applet which draws the string using that point size.

14. Write an HTML file which provides links to each of the Java example programs of this chapter. Test it using a browser.

15. Write an applet which centers your name both vertically and horizontally. (The `getHeight` method of the `FontMetrics` class gives the height of a font.)

16. The Tomato Soup Company wants a new logo on their web site. Write an applet containing a design for the logo which includes a nice red tomato with a green stem. Include the name of the company in the applet.

17 Write an applet which draws a happy face of your design. Use colors as desired.

18. Draw the happy face of Exercise 17 so it resizes when the applet changes size.

19. Write an applet which draws a clock (the old-fashioned kind with an hour hand, a minute hand, and numerals for each hour.)

20. Write an applet that draws a maze of your own design.

21. Write an applet which draws a pentagon inscribed in a circle. Connect the diagonals of the pentagon forming a star (which will contain a smaller pentagon in its center). Use the `sine` and `cosine` methods from the `Math` class with arguments 0, 2pi/5, 4pi/5, 6pi/5, and 8pi/5 to get the coordinates of the points as (`r*cos x`, `r*sin x`) where `r` is the radius of the circle, and `x` is one of the five angles listed.

22. Write an applet to draw a chessboard which has 64 squares alternating in color.

23. Write an applet to display an array of five integers in a bar graph, using one bar for each value. Scale the graph so the bar for the largest value just fits in the applet. Draw the graph relative to the size of the applet, so it will resize as the applet does.

24. Write a Java applet which displays a store's daily sales over the period of one week in a bar graph. It should look like Figure 5.22 on page 210.

FIGURE 5.22 A bar graph

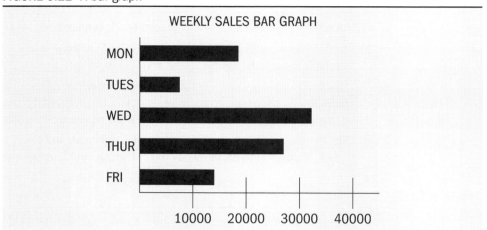

6

Graphical User Interfaces

Graphical user interface components such as buttons and text fields let the user communicate directly with our program. We introduce graphical user interfaces (or GUI's as they are often called) with our AtmScreen of Figure 6.1 which shows the applet of Example 1.5 just after the user has finished a transaction. The applet is still active, waiting for the next user to press the Start button to begin another transaction. The user has finished a transaction; in this chapter we finish presenting the topics needed to understand the AtmScreen code.

The AtmScreen presents a graphical interface to the user, with

◆ a text field to enter the user's name and PIN, and the amount of the transaction,
◆ two checkboxes for selecting the type of account,
◆ a choice box, which will pop up with three choices for the type of transaction: deposit, withdraw, or balance,
◆ a canvas to display prompts and transaction results, and
◆ two buttons for the user to start and finish a transaction.

When the user presses the Start button, that event causes the applet to take some actions:

◆ Enable the text field to permit the entry of the user name
◆ Enable the Enter Name label, making it dark instead of gray
◆ Direct key presses to the text field
◆ Disable the Start button

FIGURE 6.1 The AtmScreen ready to start another transaction

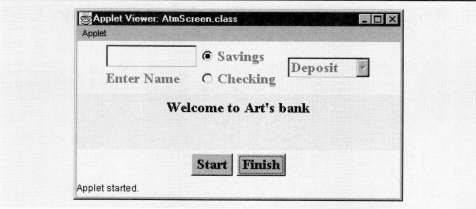

The Start button is the source of this event, while the applet is the event handler. The text field, checkboxes, and choice box also generate events which the applet handles. The events a user might generate using the AtmScreen interface of Figure 6.1 are:

Event source	User behavior	Event generated
text field	Enter text and press ENTER key	action event
checkbox	Select a checkbox	item event
choice box	Select an item	item event
button	Press the button	action event

We previously stated that when the user presses the Start button, the applet takes four actions, but how does the applet know that the user pressed the button? The Java 1.1 event model uses interfaces to support callbacks[1] to communicate from the event source to the event handler. The applet registers with the button, requesting to be notified when the user presses it. When such an event occurs the button calls the applet which can take the desired actions.

☞ **TIP**

Event-handling for redrawing the screen is simpler than for user interface events. Redrawing the screen involves windows from the environment outside our Java window. For example, the user can obscure our Java applet with a window from a different program which will cause our applet to need to be redrawn when the user removes the obstruction. The window manager part of the operating system coordinates this painting of its windows, so all we have to do is to handle the event generated by writing a paint method with the code needed to draw our Java component.[2] In contrast, the Start button is part of our Java program and we have full responsibility for communicating the occurrence of a button press from the source button to the event handler wishing to take appropriate action.

---

[1] See Section 4.6 where we used interfaces to support callbacks by the bank to interested listeners for interest rate changes.

[2] We considered this simpler case of event-driven programming in Chapter 5.

An even more basic question about the user interface of Figure 6.1 is how do we position the components? Looking at Figure 6.2 shows us what happens to this interface when the user resizes the applet.

FIGURE 6.2 The AtmScreen resized

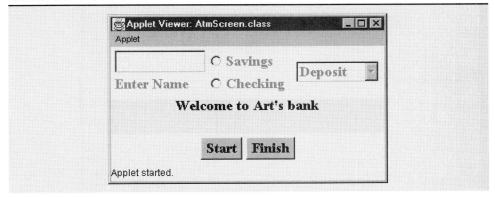

We see the canvas is thinner while the text field, choice box, and buttons remain the same size. The buttons are still centered so the Start button is closer to the left edge of the applet than it was in Figure 6.1. The programmer could not easily anticipate all the possible resizings and changes to the layout needed in each case. Java uses layout managers, objects that know how to layout components to figure out the new layout when necessary.

We have a lot to cover in this chapter, first introducing the various user interface components, next showing how the Java 1.1 event model allows button, text field, checkbox, and choice box event sources to communicate with event handlers, and then exploring layout managers. Along the way we introduce panels, labels, number formatting, wrapper classes, and inner classes.

In our small case study, we do some exploratory programming to develop a graphical user interface for insertion sorting, a problem we solved in Chapter 4. With a GUI we will be able to see each step of the sorting as it happens.

We show the displays from our examples and the code, but running these highly interactive programs will give the best sense of how they work.

OBJECTIVES

◆ Know important Component subclasses.
◆ Understand the purpose of layout managers.
◆ Use flow, border, grid, and gridbag layouts.
◆ Draw on a canvas.
◆ Understand the Java event model.
◆ Use buttons, text fields, and labels.
◆ Use inner classes.
◆ Use checkboxes and choice boxes.
◆ Design a GUI for insertion sorting.
◆ Estimate efficiency using timing statements.

OPTIONS
◆ Omit GridBag Layout (Used in 11.6)

6.1 ■ Introducing Components

After presenting the user interface components that we will study, we demonstrate the role of a layout manager in displaying components in a container. We illustrate to the use of a canvas component for drawing.

The Component Hierarchy

The Component class is the parent class for the graphical objects on the screen that provide the interface to the user. Figure 6.3 shows the subclasses of Component that we will treat in this text.

FIGURE 6.3 Some subclasses of Component

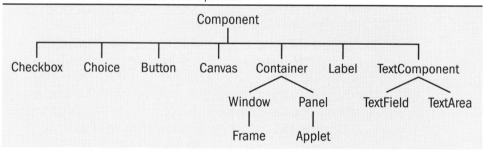

Checkbox, Choice, Button, Label, TextField, and TextArea objects provide the actual graphical components with which the user interacts. The Container class is the parent of components that contain other components. The Window and Frame containers are used by standalone applications.[3] The **panel** defines containers that are part of another container which might be a panel, or a frame, or a browser window. Our applets in Chapter 5 inherit from the Applet class. We see from Figure 6.3 that an applet is a container, and we will soon learn how to add components to it.

Adding Buttons to a Container

Java places components in a container which can adjust the positions of the components, as needed, when a component or the container itself is resized. For example, Figure 6.4 shows a container with five buttons arranged in a row. When the user resizes the container, there may only be room for three buttons in one row, so Java will put the other two buttons in a second row below the first as shown in Figure 6.5.

Of course, the button positions can only adjust if they are not specified with fixed numbers. For example if the rightmost button in Figure 6.4 was fixed at position (400,20), there would be no place for it in the 300 x 200 container of Figure 6.5. The position (400,20) would

---

[3] We cover standalone graphics application in Section 7.4.

FIGURE 6.4 Five buttons in a row

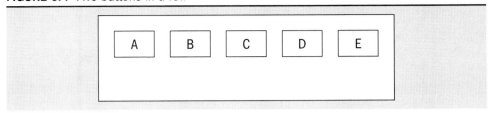

FIGURE 6.5 The five buttons after resizing Figure 6.4

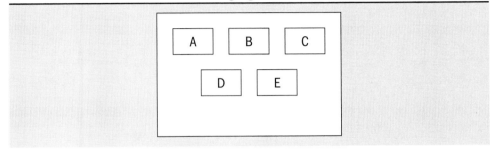

be too far to the right. For this reason, Java does not assign fixed positions to components, but uses a **layout manager** to arrange the components in the container.

Every time a container is resized, the layout manager repositions its components. A layout manager does not always succeed. If a user constricts the container of Figure 6.5 to have size 20x200, then none of the buttons will fit in such a thin container. But in most cases the layout manager can reposition the components to fit within the newly dimensioned container.

Flow Layout

Java has five layout managers, each useful in certain situations, and allows us to create custom layout managers if desired. The **flow layout** is the default layout manager for applets. The flow layout manager arranges components in a left-to-right flow until no more fit in a row, and centers any remaining components in the last row.

We can add components to an applet in its `init` method. For example, we can declare a button

```
Button button = new Button("Press me");
```

and add it to an applet using the `add` method of the `Container` class

```
add(button);
```

The flow layout manager will add the button in the next available space according to its flow layout method.

In Example 6.1 we try adding five buttons to an applet. By changing the size of the applet in the HTML file, we can see how the layout manager adjusts the placement of the buttons. Of course the buttons do not have any effect yet; we will learn how to respond to button presses in the next section.

FIGURE 6.6 Layout of five buttons in different size containers

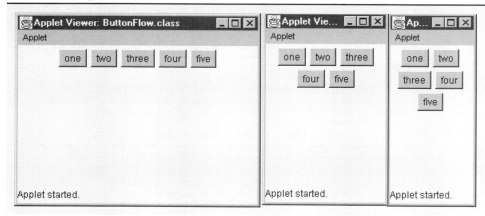

300 x 200   150 x 200   100 x 200

Figure 6.6 shows the layout using the ButtonFlow code from Example 6.1 with each of the following three HTML files:

- `<applet code = ButtonFlow.class width=300 height=200> </applet>`
- `<applet code = ButtonFlow.class width=150 height=200> </applet>`
- `<applet code = ButtonFlow.class width=100 height=200> </applet>`

EXAMPLE 6.1   ButtonFlow.java

```
/* Adds five buttons to an applet.
 * Tries different sizes for the applet
 * to see how the flow layout manager
 * adjusts the placement of the buttons.
 */

import java.awt.*;
import java.applet.Applet;

public class ButtonFlow extends Applet {
    public void init() {                          // Note 1
        add(new Button("one"));                   // Note 2
        add(new Button("two"));
        add(new Button("three"));
        add(new Button("four"));
        add(new Button("five"));
    }
}
```

Note 1: We do not even override the paint method in this example, because we do not draw anything. We just want to show how the flow layout manager arranges the buttons.

Note 2: We add buttons without specifying precisely where to position them. The flow layout manager handles the positioning of these components.

Drawing on a Canvas

In Chapter 5 we did not add any components to our applets, and did all our drawing in the applet itself by overriding its `paint` method. As we add components, such as buttons, to make a user interface, we want to be careful where we draw so we are not trying to draw on top of these components.

Java provides the `Canvas` class, a component on which we can draw text and shapes. Figure 6.7 shows the applet of Example 6.2 in which we use two buttons and a canvas on which to draw so as not to interfere with the buttons.

FIGURE 6.7 The applet of Example 6.2

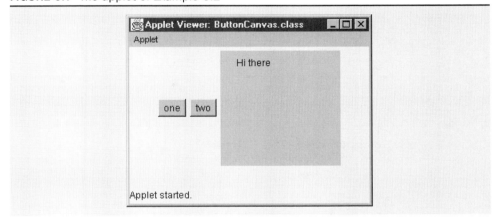

To use a canvas to do our drawing, we must define a subclass in which we override the `paint` method, as in

```
class DrawOn extends Canvas {
    public void paint(Graphics g) {
        g.drawString("Hi there",20,20);
    }
}
```

We create a new canvas and add it to our applet, setting its background color to pink to make it visible:

```
DrawOn canvas = new DrawOn();
add(canvas);
canvas.setBackground(Color.pink);
```

Java creates each component with a preferred size. The flow layout manager uses these preferred sizes for each component when it lays out the applet. Java made 0 x 0 the preferred size for a canvas, so we must set the size of the canvas, as in

```
canvas.setSize(150,150);
```

EXAMPLE 6.2 **ButtonCanvas.java**

```
/* Adds two buttons and a canvas to an applet.
 * Sets size of canvas, which by default is 0x0.
 * Sets background to pink to see the canvas.
 * Overrides paint to draw on the canvas.
 */

import java.awt.*;
import java.applet.Applet;

public class ButtonCanvas extends Applet {
    public void init() {
        add(new Button("one"));
        add(new Button("two"));
        DrawOn canvas = new DrawOn();
        add(canvas);
        canvas.setBackground(Color.pink);              // Note 1
        canvas.setSize(150,150);                       // Note 2
    }
}
class DrawOn extends Canvas {
    public void paint(Graphics g) {
        g.drawString("Hi there",20,20);                // Note 3
    }
}
```

Note 1: White is the default background color for both the canvas and the applet. Leaving the canvas white would blend it in with the applet. Even though our drawing would appear, we would not see the borders of our canvas. Setting the background to pink makes the canvas stand out.

Note 2: If we do not set its size, the canvas will have the default 0x0 size, so we will not see any of our drawing, which will always be too large to fit.

Note 3: We are drawing on the canvas component, so the coordinates are relative to the canvas. We position the string 20 pixels in each direction from the upper-left corner of the canvas, not the applet.

☛ A LITTLE EXTRA **Getting the Preferred Size**

The flow layout manager finds the preferred size of a component by calling its getPreferredSize method. We can override this method for our subclass of Canvas to give our canvas a preferred size of our choosing, rather than the 0x0 default. Overriding this method, as in

```
public Dimension getPreferredSize() {
    return new Dimension(150,150);
}
```

would remove the need for the applet to set the size of the canvas.

The Big Picture
....

We add components to make a user interface. Layout managers place components in relation to the dimensions of the window. When the user resizes the window, the layout manager rearranges the components. The flow layout manager adds components one after another until it fills a row. We draw on a canvas to separate our drawing from other user interface components such as buttons.

---

1. Describe the way the flow layout manager positions components in an applet.

2. Which of the classes shown in Figure 6.3 can be instantiated to create components we can add to an applet, and are which poses abstract without instances? (The JDK documentation will show which classes are abstract.)

TRY IT YOURSELF 3. Remove the `setSize` statement from Example 6.2, rerun that example, and explain the result.

....

6.2 ▪ Button and TextField
....

In Examples 6.1 and 6.2 we added buttons to an applet, but pressing these buttons had no effect. In this section we learn how to respond to button presses. We introduce the `TextField`, a component for entering text, in this section as both text fields and buttons generate action events.

The Java Event Model[4] and Action Events

We began event-driven programming in Chapter 5. Our applets in that chapter provided a `paint` method to be called when events such as resizing caused our applet to need to be redrawn. The applet did not have to ask to be redrawn; a windowing operating system keeps its windows properly drawn as part of its normal functioning. The operating system listens for events that require a window to be redrawn, passing these events to the Java event handler which calls the appropriate method to actually do the redrawing.

To make buttons work, Java follows a similar approach. The user pressing the button generates an event, called an **action event** because it will cause some action. The action event object, of type `ActionEvent`, contains information describing the event. The button keeps a list of any objects (listeners) that want to be notified when the button is pressed. When the user presses a button, the button notifies each object that is listening, passing it the action event that describes the button press. The listeners implement the actions they wish to take as a result of the button press.

This process of registering to listen for button presses is like an investor and a business registering with the federal bank to be notified when the bank changes the interest rate.[5] The

---

[4] Starting with version 1.1, Java introduced a new event model. We do not use the older 1.0 event model in this text.

[5] See Section 4.6 for a discussion of the `RateChangeListener` interface and callbacks.

investor and the business implement the `RateChangeListener` interface to provide methods to handle changes in the interest rate. The federal bank calls these methods when it changes the rate.

The `RateChangeListener` interface allows the investor and business to register with the federal bank and be notified when the rate changes. For button presses and other action events, Java provides the `ActionListener` interface which has one method

```
public void actionPerformed(ActionEvent event);
```

To be notified about a button press, an object registers as an action listener with the button, and implements the `actionPerformed` method to define what it will do when the button notifies it that an action event representing the button press has occurred.

Making Buttons Work

The button, like the federal bank of Example 4.9, keeps a list of listeners, and when the user presses the button, passes an action event as the argument to the `actionPerfomed` method of each object that registered as an action listener.

To illustrate this process, Example 6.3 will add two buttons to an applet. A Print button will display a message, and a Clear button will erase the message. We let the applet listen for button presses, and implement the `actionPerformed` method to display a message when the user presses the Print button, and to erase the message when the Clear button is pressed. One scenario might proceed as follows:

The applet registers with the Print button as an action listener.
The user presses the Print button.
The Print button sends an action event as an argument to the applet's
`actionPerformed` method.
The applet's `actionPerformed` method displays a message `Hi there`.

To be an action listener, the applet must declare that it implements the `ActionListener` interface, using the keyword `implements`, as in:

```
public class ButtonPress extends Applet implements ActionListener
```

To register with the button, the applet calls the button's `addActionListener` method, passing itself as the listener, as in

```
print.addActionListener(this);
```

where `print` refers to the Print button. The `this` argument is the applet itself which waits for notification about button presses.

To handle button presses, when it is notified that one occurred, the applet implements the `actionPerformed` method. We use two buttons in this example, so the applet, in the `actionPerformed` method, calls `event.getSource()` which will return the object that generated the event, either the Print button or the Clear button.

Using Labels

To display a message, we use a `Label` component. We generally use a label object to label a text field or other component, but it serves here to illustrate the action of a button to change the text of a label. Figure 6.8 shows the applet initially, and just after the Print button was pressed. The button just pressed has a border around it. Run this applet to see how it responds to button presses.

FIGURE 6.8 The applet of Example 6.3

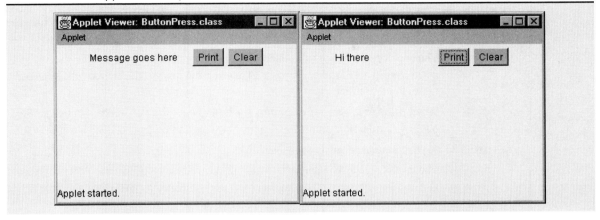

EXAMPLE 6.3   ButtonPress.java
....

```java
/* A Print button prints a message.
 * A Clear button erases the message.
 * We use a Label component to hold
 * the message.
 */

import java.awt.*;
import java.awt.event.*;                                    // Note 1
import java.applet.Applet;

public class ButtonPress extends Applet
                  implements ActionListener {               // Note 2
    private Button print = new Button("Print");
    private Button clear = new Button("Clear");
    private Label message = new Label("Message goes here"); // Note 3

    public void init() {
        add(message);
        add(print);
        add(clear);
        print.addActionListener(this);                      // Note 4
        clear.addActionListener(this);
    }
    public void actionPerformed(ActionEvent event) {        // Note 5
        Object source = event.getSource();                  // Note 6
        if (source == print)
            message.setText("Hi there");                    // Note 7
        else if (source == clear)
            message.setText("");                            // Note 8
    }
}
```

Note 1: We must import the `java.awt.event` package to use the event classes such as `ActionEvent`.

Note 2: The applet declares it will implement the `ActionListener` interface, meaning it must implement the `actionPerformed` method to receive action events from components with which it registers.

Note 3: Java allocates space for the label based on the size of the initial message. Using the default constructor, `Label()`, may not allow enough space for the text we use later.

Note 4: The applet registers with the Print button as an action listener, passing itself (`this`) as an argument to the `addActionListener` method, so the button will know who needs to be notified when the user presses this button.

Note 5: When the user presses either button, the button calls the `actionPerformed` method, passing it an action event that describes the button press.

Note 6: `Object source = event.getSource();`

The applet registered itself with two buttons and will be notified if either is pressed. The `getSource` method returns the object that generated the action event, so the applet can determine if it was the Print button or the Clear button.

Note 7: `if (source == print) message.setText("Hi there");`

If the user pressed the Print button, then the applet sets the text of the label to Hi there.

Note 8: `else if (source == clear) message.setText("");`

If the user pressed the Clear button, then the applet sets the text of the label to the empty string.

We want to revise Example 6.3 to use a canvas and do the drawing in the `paint` method of that canvas, as we did in Example 6.2. The canvas will do the drawing, so the canvas should register with the button to listen for button presses. When the user presses a button, the canvas will find out which button the user pressed and save the button's label, so when Java calls the `paint` method it will know whether to display a message or to erase the message.

Remember that the `paint` method is also event-driven, activated by users who resize the applet or return to it after browsing another page. Rather than waiting for one of these events to happen, the canvas asks Java to schedule a call to the `paint` method by calling `repaint()`. A scenario for this process is

The canvas registers with the Print button as an action listener.
The user presses the Print button.
The Print button sends an action event as an argument to the canvas'
  `actionPerformed` method.
The canvas' `actionPerformed` method saves the label of the button which the user
  pressed.
The canvas' `actionPerformed` method calls the `repaint` method, which asks Java to
  schedule a call the `paint` method to redraw the canvas.
Java calls the `paint` method of the canvas.
The canvas' `paint` method uses the button's label to determine that the Print button
  was pressed, and displays the message.

☛ TIP

For simplicity, we left out one important step involved when the applet calls repaint(). Java actually calls the update method which the applet inherits from Component. The update method first clears the component and then calls the paint method. Because the update method clears the component by default, we do not have to write code to clear it.

In Example 6.3, the listener used the getSource method to find the object that generated the action event. In Example 6.4 the listener, our canvas, uses the getActionCommand method to determine the name of the action that caused the event. By default, Java uses the button's label to name the action, so getActionCommand will return the label of the button that was pressed.

Figure 6.9 shows the applet just after the user pressed the Print button.

FIGURE 6.9 The applet of Example 6.4

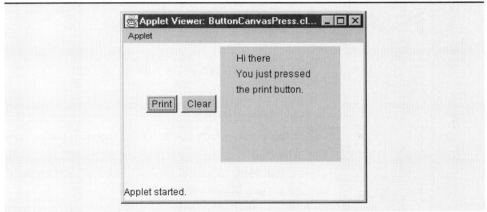

EXAMPLE 6.4 **ButtonCanvasPress.java**

```
/* Adds two buttons and a canvas to an applet.
 * The canvas registers with each button to
 * listen for button presses.  Its actionPerfomed
 * method saves the button label, and asks Java
 * to redraw the canvas (calling update which calls
 * paint).  Paint displays a message if Print was
 * pressed but clears the canvas if Clear was pressed.
 */

import java.awt.*;
import java.awt.event.*;
import java.applet.Applet;

public class ButtonCanvasPress extends Applet {
    private Button print = new Button("Print");
    private Button clear = new Button("Clear");
    private Draw canvas = new Draw();
```
                                                                    continued

```
public void init() {
    add(print);
    add(clear);
    print.addActionListener(canvas);                    // Note 1
    clear.addActionListener(canvas);
    add(canvas);
    canvas.setBackground(Color.pink);
    canvas.setSize(150,150);
  }
}
class Draw extends Canvas implements ActionListener {    // Note 2
  String command = "";                                  // Note 3

  public void actionPerformed(ActionEvent event) {
    command = event.getActionCommand();                 // Note 4
    repaint();                                          // Note 5
  }
  public void paint(Graphics g) {
    if (command.equals("Print")){                       // Note 6
      g.drawString("Hi there",20,20);
      g.drawString("You just pressed",20,40);
      g.drawString("the print button.",20,60);
    }
  }
}
```

Note 1: The canvas registers with the buttons to listen for button presses.

Note 2: The Draw class declares that it implements the ActionListener interface, which requires it implement the actionPerformed method to specify what it will do when notified about a button press.

Note 3: We need to initialize the command string, because the paint method needs to refer to it when the applet is initialized which is before it gets the label of the button which the user pressed.

Note 4: The getActionCommand method returns the name of the command for the action that generated the event. By default, the name of the action command for a button is its label.

Note 5: Java calls the paint method in response to events. When we want to refresh the screen to show the changes we made, we call repaint() to ask Java to schedule such a call to paint (via update which clears the canvas before calling paint). The call to update will clear the canvas, so the paint method does not need to supply any code for the case when the user presses the Clear button.

Note 6: if (command.equals("Print")){

For strings, we want to check the equality of the characters even if they are in different objects, so we use the equals method rather than the equality operator, ==, which checks the equality of references.

Number Formatting

In Example 6.4 we display text, but in Example 6.5 we will need to display decimal numbers or currency values in a graphical user interface.

The `java.text` package has a `NumberFormat` class[6] which allows us to output decimal and currency values using the default format for the user's locale. Decimal outputs in the US will use the dot, ., for the decimal point, and currency outputs in the US will appear as dollars and cents with a dollar sign preceding the amount.

To format decimal numbers, we use the `getInstance` method to get an instance of a `NumberFormat` class, and use the `setMaximumFractionDigits` method to specify the number of places to the right of the decimal point. For example, the code

```
NumberFormat n = NumberFormat.getInstance();
n.setMaximumFractionDigits(3);
System.out.println(n.format(Math.PI));
```

will output the value 3.142.

To format a currency value, we use the `getCurrencyInstance` method, as in:

```
NumberFormat nf = NumberFormat.getCurrencyInstance();
```

to get a `NumberFormat` object `nf` and then use that object to display **double** values in a currency format, using the `format` method, as in:

```
g.drawString(nf.format(cost), 20, 30);
```

which would display $2.10 if the variable `cost` had the value 2.1.

Wrapper Classes

In graphics mode, we would like to enter data in a text field. If those entries represent numerical values we need to convert them from the strings that the user inputs. We use wrapper classes to do these conversions. Java provides **wrapper classes**: `Integer`, `Double`, `Byte`, `Float`, `Long`, `Short`, `Boolean`, and `Character`, corresponding to each primitive type.

We also use wrapper classes to include primitive types within the class hierarchy. Java treats primitive types differently than class types. Every class extends, directly or indirectly, the `Object` class. If we declare an array of elements of type `Object`, as in:

```
Object[] anArray = new Object[10];
```

that array can contain objects of any subclass of `Object` which include every class type, such as `BigInteger` or `SavingsAccount`, but not primitive types such as **int** and **double**. To include primitive values in the class hierarchy, Java defines a wrapper class for each primitive type, which wraps a value of that type as a data field of an object.

An object of type `Integer` contains a data field of type **int**. The code

```
Integer wrapsFive = new Integer(5);
```

creates an object, `wrapsFive`, which contains an **int** field with value 5. We cannot place an **int** in `anArray`, but we can place an `Integer`, as in:

```
anArray[0] = wrapsFive;
```

We can use the `intValue()` method to retrieve the **int** value of 5 which is wrapped inside the `wrapsFive` object.

---

[6] We used the `NumberFormat` class to implement the `Io.print` and `Io.print$` methods introduced in Section 1.6.

To convert a string such as "345" to the integer 345, we can use

```
int converted = new Integer("345").intValue();
```

which creates a new `Integer` and uses the `intValue` method to retrieve the **int** value it contains. We could also use the static `parseInt` method, as in

```
int alsoConverted = Integer.parseInt("345");
```

to do the conversion.

Similarly the class `Double` wraps a **double** value inside an object of type `Double`, and the `doubleValue()` method retrieves it, as in:

```
Double wrapsPI = new Double(Math.PI);
double itsPI = wrapsPI.doubleValue();
```

To convert from a string to a **double** value, we use the `doubleValue` method, as, for example,

```
double converted = new Double("345.67").doubleValue();
```

There is no analogue for type **double** of the `parseInt` method.

Entering Data in Text Fields

A button enables a user to indicate an action, but does not allow the user to input any data. A text field provides a line for the user to input data. When the user presses the ENTER key on the keyboard, after entering data in a text field, Java sends an action event to all listeners who registered with that text field. Another way to use a text field is to get its contents when the user presses a button.

We can specify the number of columns for the text field in the constructor, as in

```
TextField text = new TextField(5);
```

which gives us a text field five columns wide. We use the `getText` method to retrieve the text that the user enters in the text field. This text is always of type `String` even if it is meant to represent a number such as 100 or 52.7.

In Example 6.5 we add a text field for the user to input **double** values. If the user presses the ENTER key or the Enter Price button, we add the value in the text field to the sum of the numbers entered so far. After we get the string entered by the user we set the text in the text field to the empty string using the `setText` method, as in:

```
text.setText("");
```

so that the user does not have to erase the old value before entering the next. If the user presses the Average button, we output, in a label, the average of the numbers entered. Figure 6.10 shows the applet of Example 6.5 just after the user pressed the Average button.

Because the text in a text field is of type `String` we need to convert it to a **double** which we do using a wrapper class, `Double`. The expression

```
new Double(text.getText()).doubleValue()
```

converts the string, `text.getText()`, that the user entered in the text field to a value of type **double**.

FIGURE 6.10 The applet of Example 6.5

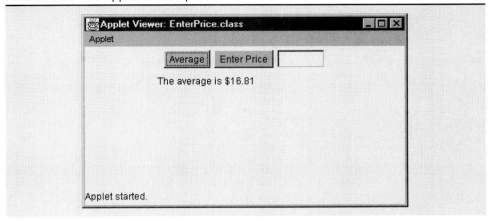

The applet registers itself as a listener with the text field, so the applet will be notified when the user hits the ENTER key. The applet is also listening for button presses, and will be notified when the user presses the Enter Price button, another way to signify the entry of a number in the text field. The Average button will notify the applet when the user presses it requesting the average, which the applet will display in a label.

EXAMPLE 6.5 **EnterPrice.java**

```java
/* Converts the price the user enters in the text field to a
 * double and adds it to the sum when the user hits the
 * Enter key or presses the Enter button.  Displays the
 * average in a label when the user presses the Average button.
 */

import java.awt.*;
import java.awt.event.*;
import java.applet.Applet;
import java.text.NumberFormat;

public class EnterPrice extends Applet
                        implements ActionListener {
    private Button average = new Button("Average");
    private Button enter = new Button("Enter Price");
    private TextField text = new TextField(5);                    // Note 1
    private Label answer = new Label("Enter prices -- the average goes here");
    private double sum = 0.0;
    private int count = 0;
    private NumberFormat nf;

    public void init() {
        add(average);
        add(enter);
        add(text);                                               // Note 2
        add(answer);
```

continued

```
            average.addActionListener(this);
            enter.addActionListener(this);
            text.addActionListener(this);                        // Note 3
            nf = NumberFormat.getCurrencyInstance();
        }
    public void actionPerformed(ActionEvent event) {
        Object source = event.getSource();
        if (source == text || source == enter){                  // Note 4
            sum += new Double(text.getText()).doubleValue();     // Note 5
            count++;
            text.setText("");                                    // Note 6
        }
        else if (source == average) {
            answer.setText("The average is "+ nf.format(sum/count));  // Note 7
        }
    }
}
```

Note 1: The text field shows five columns, but the user can enter more digits that will not all show.

Note 2: We add the text field component to the applet.

Note 3: The applet registers as a listener to be notified whenever the user hits the ⎡ENTER⎤ key.

Note 4: Hitting the ⎡ENTER⎤ key and pressing the Enter Price button generates action events. We could have used one or the other but wanted to show both approaches are possible.

Note 5: We use the `getText` method to get the string from the text field. The expression `new Double(text.getText()).doubleValue()` converts the string to a **double** which we then add to the sum.

Note 6: `text.setText("");`

Clears the text field of its previous contents.

Note 7: `answer.setText("The average is "+ nf.format(sum/count));`

When the user presses the Average button, we use the `format` method to display the average as a string with a currency format in dollars and cents.

The Big Picture

In the Java event model, listeners, who want to handle an event, register with the event source. When the event occurs, the source notifies the registered listeners. Buttons and text fields both generate action events. An action listener registers with a button, and when the user presses that button it notifies listeners by calling their `actionPerformed` methods, which each action listener must have implemented to make the button work as desired.

The `NumberFormat` class helps us format decimal numbers and currency values, using any locale specific variants. Wrapper classes include primitive types in objects. We use them here to convert strings to numbers.

4. Write a scenario showing the interactions in the applet of Example 6.3 for the case when the user presses the Clear button after pressing the Print button.

5. Write a scenario showing the interactions in the applet of Example 6.4 for the case when the user presses the Clear button after pressing the Print button.

6. Write a scenario showing the interactions in the applet of Example 6.5 for the case when the user enters two values, hitting the ENTER key after entering each, and then presses the Average button.

6.3 ▪ Checkbox and Choice

Checkboxes and **choice boxes** allow us to make selections. They generate an **item event** when the user makes a selection, but we can also use them with a button to signify that the user has finished making selections. We show both approaches in this section. First we introduce inner classes which we illustrate in our checkbox and choice box examples.

Inner Classes

An **inner class** can be defined as a member of another class or within a block of code. It has access to the fields of the outer class which simplifies code, and keeps its name nested within the outer class, avoiding conflict with the same name used elsewhere. We often use inner classes for event handlers. Inner classes add some complexity to Java; we only introduce them here, omitting some of the details we do not use.

The following code fragments illustrate the differences.

```
class A {
    String s = "potato";
}
class B extends Canvas {
    A other;
    public B(A a) {
        other = a;
    }
    public void paint(Graphics g) {
        g.drawString(other.s, 20, 20);
    }
}
```

```
class C {
    String s = "tomato";
    class D extends Canvas {
        public void paint(Graphics g) {
            g.drawString(s, 20, 20);
        }
    }
}
```

In the fragment on the left we need to pass an object of class A to class B to draw its instance variable s. Compiling a program with this structure will produce the class files A.class and B.class, so the class named B may be confused with other classes of the same name.

In the fragment on the right, the inner class D, declared as a member of C, has access to the instance variable, s, of C. Compiling a program with this structure will produce the class files C.class and C$D.class, so the class named D only appears prefixed by C$ and will not be confused with another class named D.

We use inner classes in the following examples and in other examples throughout the text. In Section 12.3 we illustrate anonymous inner classes showing if such an inner class is to be used only once we do not have to give it a name.

Checkbox and Choice with a Button

Checkboxes and choice boxes allow us to make selections, rather than initiate actions. Figure 6.11 shows two checkboxes, for selecting bold or italic fonts. The user can select none, either, or both styles.

Figure 6.12 shows two checkboxes, for selecting large or small size. We put these in a **checkbox group** so exactly one is selected. Checkboxes in a checkbox group are often called radio buttons, referring to the use of buttons on older radios used to select stations.

FIGURE 6.11 Checkboxes with both selected   **FIGURE 6.12** Checkboxes in a checkbox group

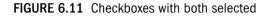

We can construct a checkbox in several ways, by passing a string to label the checkbox

```
Checkbox bold = new Checkbox("Bold");
```

or by passing a string and a **boolean** value indicating whether that item is selected initially

```
Checkbox italics = new Checkbox("Italics", true);
```

or we can pass a string, a **boolean**, and a CheckboxGroup object

```
Checkbox large = new Checkbox("Large",false,size);
```

where size is a checkbox group and is declared as:

```
CheckboxGroup size = new CheckboxGroup();
```

The declaration for large puts it in the checkbox group size. Each checkbox has a getState method which returns **true** if that checkbox is selected, and **false** otherwise.

A choice box shows one item, but allows the user to pop up other choices. Figure 6.13 shows a choice box containing the items Red, Green, and Blue to allow the user to choose a color. The item Red shows, but clicking on the button at the right will cause all three choices to pop up.

FIGURE 6.13 A choice box

The choice box has a simple constructor

```
Choice color = new Choice();
```

We use the add method to add each choice item to the choice box:

```
color.add("Red");
```

We can get the selected item using `getSelectedItem()` or get the index of the selected item using the `getSelectedIndex` method. We can initialize a choice box by selecting an item using its index

```
color.select(0);
```

or using its label

```
color.select("Red");
```

We can use checkboxes and choice boxes in two ways. In the first approach we add a button to the applet, and when the user presses the button signifying the selections have been made, we find those selections, taking whatever action is appropriate. In this approach the selections only take effect when the user presses a button. The only event the program responds to is the user's button press.

In the second approach, the program responds to each selection as it is made. Both the checkbox and the choice box generate an item event when the user makes a selection, and send that item event, which describes the item selected, to all objects that registered as item listeners with that checkbox or choice box.

To illustrate these two ways of using checkboxes and choice boxes we program the same example using both approaches. In each program we include a choice box to select one of three colors, red, green, or blue, and two checkboxes, one to select a square and the other to select a circle. We put the checkboxes in a checkbox group so that we must select exactly one shape, a square or a circle, but not both.

In Example 6.6, illustrating the first approach, we add a Draw button. When the user presses the Draw button, we draw the selected shape in the selected color. The canvas, an instance of the `DrawOn` class, listens for button presses. We declare the `DrawOn` class inside the applet class, as an inner class. Inner classes have access to the fields of the classes in which they are defined, so the `DrawOn` class will have access to the fields of the applet.

Figure 6.14 shows the applet initially, displaying a red circle, and then, after the user presses the Draw button, displaying the green square that the user selected.

FIGURE 6.14 The applet of Example 6.6

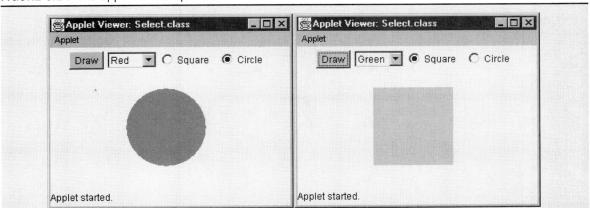

EXAMPLE 6.6 Select.java

```
/* Uses a choice box to choose a color, and
 * two checkboxes to choose a shape.  Uses a Draw button
 * to request drawing the selected shape in the
 * selected color.  Declares the DrawOn class as an
 * inner class, inside the class Select.
 */

import java.awt.*;
import java.awt.event.*;
import java.applet.Applet;

public class Select extends Applet {
    private Button draw = new Button("Draw");
    private DrawOn canvas = new DrawOn();
    private Choice color = new Choice();
    private CheckboxGroup shapes = new CheckboxGroup();
    private Checkbox square =
                    new Checkbox("Square",false,shapes);          // Note 1
    private Checkbox circle = new Checkbox("Circle",true,shapes);
    private String [] colorName = {"Red","Green","Blue"};         // Note 2
    private Color [] theColor = {Color.red,Color.green,Color.blue};  // Note 3

    public void init() {
        add(draw);                                                // Note 4
        add(color);
        add(square);
        add(circle);
        add(canvas);
        for (int i=0; i<colorName.length; i++)                    // Note 5
            color.add(colorName[i]);
        color.select(0);                                          // Note 6
        canvas.setSize(150,150);
        draw.addActionListener(canvas);                           // Note 7
    }

    class DrawOn extends Canvas implements ActionListener {       // Note 8
        public void actionPerformed(ActionEvent event) {         // Note 9
            repaint();                                            // Note 10
        }
        public void paint(Graphics g) {
            g.setColor(theColor[color.getSelectedIndex()]);      // Note 11
            if (circle.getState())                               // Note 12
                g.fillOval(20,20,100,100);
            else
                g.fillRect(20,20,100,100);
        }
    }
}
```

Note 1: The argument **false** specifies that checkbox `square` will not be selected initially. The third argument, `shapes`, names the checkbox group to which this checkbox belongs. We must select exactly one checkbox of a group.

Note 2: We use an array of strings to add items to the choice box. We could easily increase the number of choices by adding more elements to this array.

Note 3: The items in the choice box are strings, but we need color values to draw. We put the `Color` constants in this array at the same positions as their corresponding names in the `colorName` array. When we draw a shape, we find the index of the color name the user selected in the choice box, and use that same index to find the corresponding color in the `colorName` array to set as the drawing color.

Note 4: We add five components to the applet.

Note 5: This loop adds each color name to the choice box.

Note 6: `color.select(0);`

We select the first color, `Color.red`, as the initial drawing color, used when the applet starts running, before the user has a chance to select a color.

Note 7: `draw.addActionListener(canvas);`

The canvas registers with the button to listen for button presses, which indicate the user has selected a color and a shape and wants us to draw it. The choice box and the checkboxes generate item events, but in this program no one registers to listen to them. In Example 6.7, we listen for item events rather than the action events created by button presses.

Note 8: `class DrawOn extends Canvas implements ActionListener {`

We declare `DrawOn` as an inner class, meaning it is defined inside the class `Select` of our applet. As an inner class it can access the fields of the `Select` class. The `DrawOn` class declares that it implements the `ActionListener` interface, which means it must implement the `actionPerformed` method.

Note 9: `public void actionPerformed(ActionEvent event) {`

This is the method that the source of an action event, such as a button, calls to notify a listener when an action event occurs.

Note 10: `repaint();`

When the user presses the Draw button, the button will call the canvas' `actionPerformed` method. The canvas wants to be redrawn to implement the user's selections, so it calls its `repaint` method which requests Java to schedule a call to its `paint` method to redraw the canvas.

Note 11: `g.setColor(theColor[color.getSelectedIndex()]);`

The `getSelectedIndex` method gives the index of the user's choice of a color. We cannot use the choice items directly because they are strings such as `Green`, not colors such as `Color.green`. Once we know the index of the string the user selected, we use that index to find the color in the `theColor` array. Because `DrawOn` is an inner class it can access the fields `color` and `theColor` of the `Select` class containing it.

Note 12:`if (circle.getState()) g.fillOval(20,20,100,100);`

The `getState` method returns **true** if the user selected that checkbox. If the user selected circle, then we draw a circle, otherwise the user must have selected square, so we draw a square. Because `DrawOn` is an inner class it can access the circle data field of the `Select` class.

Checkbox and Choice Item Events

To illustrate the second approach, we respond to the item events generated by the checkboxes and the choice box. When the user selects a color we immediately redraw the current shape in that color. When the user selects a shape, we immediately draw that shape in the current color. The canvas, an instance of the `DrawOn` class, listens for checkbox or choice box selections.

In the second approach the `DrawOn` class listens for item events rather than action events. A class that wants to listen to item events must implement the `ItemListener` interface, which requires the class to implement the `itemStateChanged` method.

In our example, the `DrawOn` class will implement the `ItemListener` interface. The canvas will register as an item listener to listen for item events generated by the choice box when the user chooses a color, or generated by a checkbox when the user selects a square or a circle. When the user selects the color green, for example, the choice box calls the `itemStateChanged` method of all registered listeners, passing an `ItemEvent` object describing the selection to each listener. The `itemStateChanged` method calls the `repaint` method to ask Java to redraw the selected shape in green. A scenario for a choice box selection is:

The canvas registers with the choice box as an item listener.

The user chooses the color green.

The choice box sends an item event describing this selection to the canvas' `itemStateChanged` method.

The canvas' `itemStateChanged` method asks Java to schedule a call to the `paint` method to redraw the canvas.

Java calls the canvas' `paint` method (via `update`).

The `paint` method draws the currently selected shape in the selected color.

Choice boxes and checkboxes generate item events when the user selects an item or when the user deselects an item. We only want to draw the shape in the color that the user selects, and do not care about the shapes or colors that were deselected. An item event has a method, `getStateChange`, which returns either `ItemEvent.SELECTED` or `ItemEvent.DESELECTED`. We can use this method to allow us to handle only item events that return `ItemEvent.SELECTED`, ignoring the deselection events.

☛ **TIP**

In our example, when the user selects an item, the previous item is automatically deselected, but in cases where the checkboxes are not in a checkbox group, it is possible to select an item without a corresponding deselection of the previous item. It is also possible to deselect an item without selecting another one.

An item event has a `getItem` method that returns an `Object` which represents the selected item. If a choice box selection generated the item event, then the `getItem` method returns the string selected. If a checkbox generated the item event, then the getItem method returns the label of that checkbox. For example if the event evt represents the selection of green in a choice box, then

```
(String)evt.getItem()
```

returns the string Green. Java declares the return type of getItem to be Object, so we need to explicitly cast the return value to a string.

Example 6.7 uses much of the same code as Example 6.6, but because the canvas is an item listener rather than an action listener, the selections take effect immediately, not when the user presses a button as in Example 6.6. Figure 6.15 shows the SelectItem applet after the user selects the square shape.

FIGURE 6.15 The applet of Example 6.7

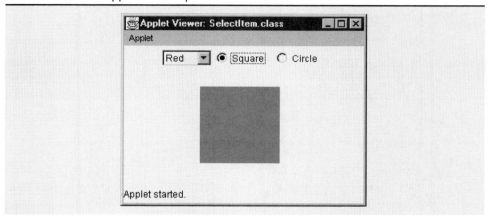

EXAMPLE 6.7 **SelectItem.java**

```java
/* Revises Example 6.6 to listen for item
 * events rather than action events.
 */

import java.awt.*;
import java.awt.event.*;
import java.applet.Applet;

public class SelectItem extends Applet {
    private DrawOn canvas = new DrawOn();
    private Choice color = new Choice();
    private CheckboxGroup shapes = new CheckboxGroup();
    private Checkbox square = new Checkbox("Square",false,shapes);
    private Checkbox circle = new Checkbox("Circle",true,shapes);
    private Color [] theColor = {Color.red,Color.green,Color.blue};
    private String [] colorName = {"Red","Green","Blue"};

    public void init() {
        add(color);
```

continued

```
                add(square);
                add(circle);
                add(canvas);
                for (int i=0; i<colorName.length; i++)
                   color.add(colorName[i]);
                color.select(0);
                canvas.setSize(150,150);
                color.addItemListener(canvas);                           // Note 1
                square.addItemListener(canvas);
                circle.addItemListener(canvas);
             }

         class DrawOn extends Canvas implements ItemListener {            // Note 2
             public void itemStateChanged(ItemEvent event) {             // Note 3
                 if (event.getStateChange() == ItemEvent.SELECTED)       // Note 4
                     repaint();
             }
             public void paint(Graphics g) {
                 g.setColor(theColor[color.getSelectedIndex()]);
                 if (circle.getState())
                     g.fillOval(20,20,100,100);
                 else
                     g.fillRect(20,20,100,100);
             }
         }
}
```

Note 1: We register the canvas to listen for item events generated by the choice box and the two checkboxes.

Note 2: We make the DrawOn class an inner class, as in Example 6.6. The DrawOn class declares that it implements the ItemListener interface which requires that it implement the itemStateChanged method to handle item events.

Note 3: An item event source, such as a choice box, calls the itemStateChanged method of all registered listeners, passing them an item event object containing information about that event.

Note 4: The getStateChange method tells us whether the item event was a selection or a deselection. We ignore deselection events, just repainting according to the user's selection.

The Big Picture

When using checkboxes and choice boxes we can respond to the item events generated when the user makes a selection, which will cause the selection to take effect immediately, or we can have the selection take effect when the user presses a button.

Inner classes, nested within another class, often simplify code, allowing access to the outer class fields.

7. Write a scenario showing the interactions in the applet of Example 6.6, for the case when the user selects a green square and presses the Draw button.

8. Write a scenario showing the interactions in the applet of Example 6.7, for the case when the user selects a square.

TRY IT YOURSELF 9. Modify Example 6.6 to add four additional colors. Rerun the applet to try out the new colors.

6.4 ■ Border and Grid Layouts

Thus far in Chapter 6 we have used only the flow layout, which is the default for applets. Java provides other layouts to give some other options when designing a user interface. In this section we look at the border layout and the grid layout. We also show how to use panels to group components nested inside another container, an approach that often enables us to achieve a better design.

The Border Layout

The **border layout** divides the container into five regions, North, South, East, West, and Center, as shown in Figure 6.16.

FIGURE 6.16 The border layout

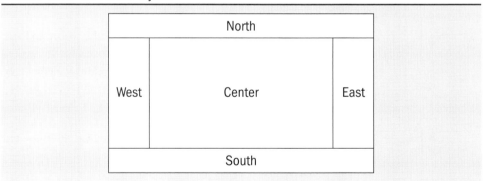

To use a border layout for an applet, we need to explicitly call the `setLayout` method to override the default

```
setLayout(new BorderLayout());
```

To add a component, we specify the region along with the component, as in

```
add(draw,"North");
```

where `draw` is a component such as a button.

Each component has a preferred size, and the flow layout respects that size, but the border layout does not. For example, the preferred size of a button just covers the text of its label, as, for example, in Figure 6.14 which shows a Draw button. The flow layout, used in Example 6.6, displays the Draw button in its preferred size. If we add this button to the North or South regions of a border layout, the border layout manager will stretch it horizontally until it fills the entire region. Adding the Draw button to the East or West regions will cause the border layout to stretch it vertically, and adding it to the Center will cause it to be stretched in both directions.

To see these effects we revise Example 6.6 to use a border layout. Figure 6.17 shows the unsatisfactory result.

FIGURE 6.17 The applet of Example 6.8

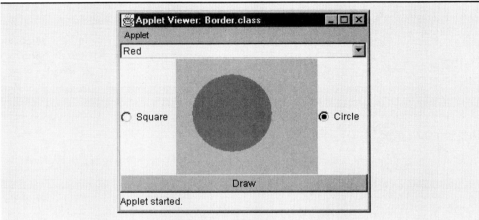

The border layout stretched the choice box in the North and the button in the South horizontally, and stretched the canvas in both directions to fill the Center region. While this is not what we want for the button, it improves the canvas. Remember that the default preferred size of the canvas is 0x0, so using the flow layout we had to set the size of the canvas. Because the border layout stretches the component in the center we do not need to set the size of the canvas.

Example 6.8 shows the code for the `init` method (which is the only part that differs) of this border layout revision of Example 6.6. We will show in Example 6.9 how to use a panel to make a better border layout for this example.

EXAMPLE 6.8 **Border.java**

```
/* Revises Example 6.6 to use a
 * border layout.
 */

public class Border extends Applet{
    //  the rest is the same as Example 6.6

    public void init() {
        setLayout(new BorderLayout());                          // Note 1
```

```
        add(draw,"South");                              // Note 2
        add(color,"North");                             // Note 3
        add(square,"West");
        add(circle,"East");
        add(canvas,"Center");                           // Note 4
         for (int i=0; i<colorName.length; i++)
            color.add(colorName[i]);
         color.select(0);
         canvas.setBackground(Color.pink);              // Note 5
         draw.addActionListener(canvas);
    }
}
```

Note 1: To use any layout for an applet (or any panel) except the flow layout, we need to set it explicitly, in this case passing a new `BorderLayout` object to the `setLayout` method.

Note 2: Directly adding a button to a border layout stretches it so it no longer looks like a normal button.

Note 3: Directly adding a choice box to a border layout stretches it, making it too wide, in this example.

Note 4: Adding a canvas to a border layout stretches it to fill the region, which means we no longer have to set its size.

Note 5: Setting the canvas color to pink allows us to see its extent.

Panels

To use the border layout more effectively, we can put the button, the choice box, and the two checkboxes into a panel, and then add that panel to the North region. The panel, which uses a flow layout by default, shows these components in their preferred sizes, if possible. We put the canvas in the Center region which will take up the rest of the applet's space because we do not put anything in the East, West, or South regions.

To use a panel, we first create it,

```
Panel p = new Panel();
```

then add components to it, as in

```
p.add(draw);
```

which adds the Draw button. Finally we use the statement

```
add(p,"North");
```

to add the panel p to the North region. Figure 6.18 shows the improved layout given by the applet of Example 6.9.

FIGURE 6.18 The applet of Example 6.9

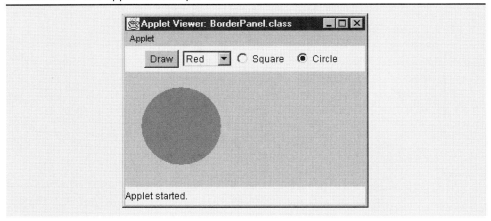

EXAMPLE 6.9 **BorderPanel.java**

```
/* Revises Example 6.8 to put the button,
 * the choice box, and the two checkboxes into
 * a panel.  Adds the panel to the North region,
 * and the canvas in the Center.
 */

public class BorderPanel extends Applet {
    // the rest is the same as Example 6.8

    public void init() {
        setLayout(new BorderLayout());
        Panel bar = new Panel();                    // Note 1
        bar.add(draw);                              // Note 2
        bar.add(color);
        bar.add(square);
        bar.add(circle);
        add(bar,"North");                           // Note 3
        add(canvas,"Center");                       // Note 4
        ...
    }
}
```

Note 1: We create a new panel which is a container that is itself contained in a container, in this case the applet.

Note 2: The panel uses the flow layout; adding the button, the choice box, and the two checkboxes to the panel keeps these components at their preferred sizes.

Note 3: We add the panel to the North region.

Note 4: We add the canvas to the Center region, which takes up the remaining space, as none of the other regions contain any components.

The Grid Layout

The grid layout divides the container into a rectangular grid. Figure 6.19 shows a 2 x 3 grid layout.

FIGURE 6.19 A 2 x 3 grid layout

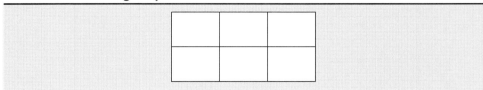

The grid layout stretches a component to fill the grid cell to which it has been added. This might be useful for a calculator applet in which each grid cell contains a button. To keep a component at its preferred size, we could add it to a panel, and then add the panel to a grid cell.

We can specify a 2x3 grid layout using the statement

```
setLayout(new GridLayout(2,3));
```

and can add a button using the statement

```
add(draw);
```

which adds the Draw button. Example 6.10 creates a 2x3 grid layout and adds a button to each of the six cells. We do not make the buttons do anything other than illustrate the appearance of the grid layout.

Figure 6.20 shows the applet of Example 6.10 in which the each button fills its grid cell.

FIGURE 6.20 The applet of Example 6.10

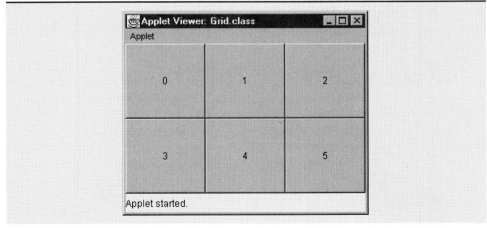

EXAMPLE 6.10 **Grid.java**

```java
/* Adds six buttons to illustrate the use
 * of the grid layout. The buttons do not
 * do anything
 */

import java.awt.*;
import java.applet.Applet;

public class Grid extends Applet {
    public void init() {
        setLayout(new GridLayout(2,3));
        for (int i=0; i<6; i++)
            add(new Button(String.valueOf(i)));          // Note 1
    }
}
```

Note 1: We convert the index i to a string to use as the label for the button. Note that the grid layout manager adds the buttons across the row first, and when the first row is full, goes to the second row.

The Big Picture

The border layout provides five regions to add components. It does not respect the preferred sizes of its components, but we can nest components in a panel, which uses the flow layout, and then add the panel to one of the border layout regions. Using a panel allows us to arrange components within one of the regions of the grid layout, which also does not respect preferred sizes. The grid layout divides the window into an m x n grid of rows and columns.

TEST YOUR UNDERSTANDING

10. For each region of a border layout, specify the dimensions of a component, width or height, or both, that the border layout ignores when laying out the component in that region.

11. Compare the flow layout with the grid layout with respect to how they use the preferred sizes of their components.

TRY IT YOURSELF

12. Modify Example 6.10 to display the buttons at their preferred size, rather than expanded to fill the grid cell, while still using the grid layout.

6.5 ▪ Gridbag Layout

The gridbag layout provides a grid of variable-sized cells, giving more flexibility than the flow, border, or grid layouts, but is more complex and difficult to learn. The best way to understand its features is through experimentation with simple examples. Figure 6.21 shows a form designed using a gridbag layout whose code appears as Example 6.16.

FIGURE 6.21 A form with a gridbag layout

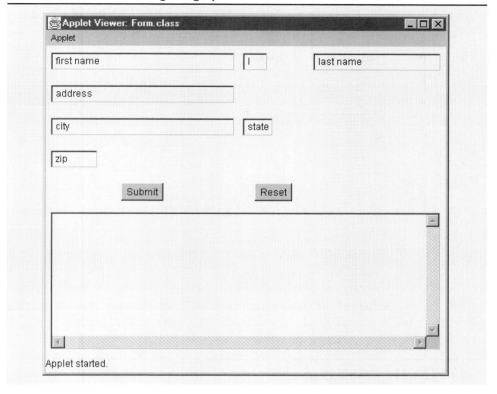

Default GridBag Constraints

A GridBagConstraints object holds the values which customize the location and appearance of each component. Setting the values of these constraint variables determines how the layout manager will display a component. The GridBagConstraints variables, with their default values and uses, are:

Variable	Default Value	Use
gridx, gridy	RELATIVE	Upper-left corner of display area
gridwidth	1	Number of cells used for width
gridheight	1	Number of cells used for height
fill	NONE	Resizing behavior
ipadx, ipady	0	Internal padding
insets	new Insets(0,0,0,0)	External padding
anchor	CENTER	Placement
weightx	0.0	Distribute row space
weighty	0.0	Distribute column space

As we experiment, we will come to understand the meaning of these default values.

The applet of Example 6.11, shown in Figure 6.22, adds two buttons to an applet using all default settings for the GridBagConstraints. We do not activate the buttons; our purpose is to discuss the layout.

FIGURE 6.22 Using the GridBagConstraints defaults in Example 6.11

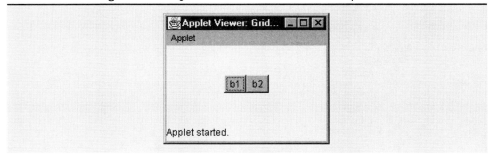

Gridx and gridy each start out at 0, indicating column 0 and row 0. The default of GridBagConstraints.RELATIVE positions the component just to the right of the previous component (or below if placing it in a column). Thus b2 is just to the right of b1. Because gridwidth and gridheight are 1, they each use one row and one column. Because weightx and weighty are each 0.0, the extra space in the applet is outside the layout of the two components, which are clustered in the center. The fill constraint has no effect as there is no extra space to fill.

EXAMPLE 6.11   **GridBag1.java**

```java
/* Uses default settings for GridBagConstraints.
 */
import java.awt.*;
import java.applet.*;
public class GridBag1 extends Applet {
    private Button b1 = new Button("b1");
    private Button b2 = new Button("b2");
    public void init() {
        GridBagLayout gbl = new GridBagLayout();
        setLayout(gbl);
        GridBagConstraints c = new GridBagConstraints();     // Note 1
        gbl.setConstraints(b1, c);                           // Note 2
        add(b1);
        gbl.setConstraints(b2, c);
        add(b2);
    }
}
```

Note 1: We only need one GridBagConstraints object, c, which we can reuse for each component. In this example, we do not change any values of its variables, using the default settings.

Note 2: Before adding a component to the applet, we call the setConstraints method to associate the current constraint settings with the component to be added.

Setting Weight and Fill

In Example 6.12, we set weightx to .7 for button b1 and weightx to .3 for b2. We see, in Figure 6.23, that b1 is centered in the left 70 percent of the applet while b2 is centered in the right 30 percent. Had we left the default fill of NONE, the buttons would have remained their normal size as in Figure 6.22. By giving b1 VERTICAL fill, it expands to fill its space vertically. Button b2, with HORIZONTAL fill, expands horizontally. Had we chosen BOTH for fill the button would have expanded to fill its entire area. Had we left weightx and weighty at the default of 0.0, no additional area would have been allocated to each component and the setting fill would have no effect. Had we left weighty at 0.0, specifying VERTICAL fill would have no effect.

FIGURE 6.23 Using weightx, weighty, and fill in Example 6.12

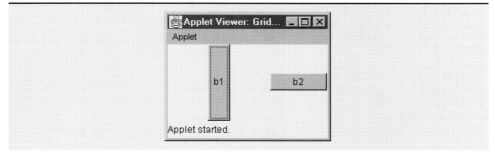

EXAMPLE 6.12	**GridBag2.java**

```java
/* Sets weightx, weighty, and fill.
 */

import java.awt.*;
import java.applet.*;
public class GridBag2 extends Applet {
    private Button b1 = new Button("b1");
    private Button b2 = new Button("b2");
    public void init() {
        GridBagLayout gbl = new GridBagLayout();
        setLayout(gbl);
        GridBagConstraints c = new GridBagConstraints();
        c.weightx = .7;                                    // Note 1
        c.fill = GridBagConstraints.VERTICAL;
        gbl.setConstraints(b1, c);
        add(b1);
        c.weightx = .3;
        c.weighty = 1.0;                                   // Note 2
        c.fill = GridBagConstraints.HORIZONTAL;
        gbl.setConstraints(b2, c);
        add(b2);
    }
}
```

Note 1: The `weightx` setting applies to the whole column of components. In this case we have only b1 in the first column, but with more components the layout manager takes `weightx` for the column to be the maximum of `weightx` for each component.

Note 2: The `weighty` setting applies to the whole row of components. The layout manager takes `weighty` for the row to be the maximum of `weighty` for each component. Thus b1 expands vertically even though its `weighty` setting is 0.0, because the `weighty` setting for the entire row is 1.0.

Anchoring and Internal Padding

Example 6.13 illustrates anchoring and internal padding. We use the anchor field to position a component that is smaller than its display area. Its possible values are CENTER, NORTH, NORTHEAST, EAST, SOUTHEAST, SOUTH, SOUTHWEST, WEST, and NORTHWEST. Internal padding adds to the dimensions of the component. Setting `ipadx` to 25 causes the buttons in Figure 6.24 to have 25 pixels of internal padding on both the left and the right. Similarly, setting `ipady` to 25 causes the buttons to have 25 pixels of top and bottom padding. We anchor b1 to the WEST of its area and b2 to the NORTH. In this example, we return to the default NONE for `fill`.

FIGURE 6.24 Anchoring and internal padding in Example 6.13

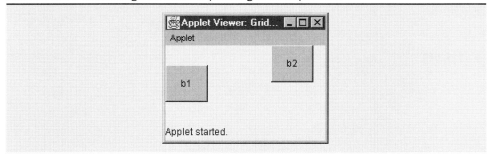

EXAMPLE 6.13   **GridBag3.java**

```
/*Illustrates anchoring and internal padding.
 */

import java.awt.*;
import java.applet.*;
public class GridBag3 extends Applet {
    private Button b1 = new Button("b1");
    private Button b2 = new Button("b2");
    public void init() {
        GridBagLayout gbl = new GridBagLayout();
        setLayout(gbl);
        GridBagConstraints c = new GridBagConstraints();
        c.weightx = .7;
        c.anchor = GridBagConstraints.WEST;
        c.ipadx = 25;
        c.ipady = 25;
```

```
                    gbl.setConstraints(b1, c);
                    add(b1);
                    c.weightx = .3;
                    c.weighty = 1.0;
                    c.anchor = GridBagConstraints.NORTH;
                    gbl.setConstraints(b2, c);
                    add(b2);
                }
            }
```

Insets

In Example 6.14, we set `fill` to BOTH so the buttons expand to fill their display areas. However, we set the external padding for b2 to provide a border of 10 pixels around it. The type of the `insets` variable is `Insets`, a class used to specify the four values for the border in each direction.

FIGURE 6.25 Insets and `fill.BOTH` in Example 6.14

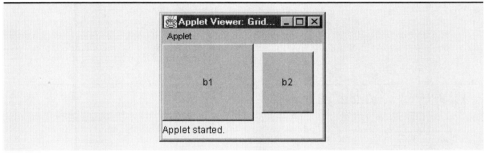

EXAMPLE 6.14 **GridBag4.java**

```java
/* Illustrates Insets and fill BOTH.
 */

import java.awt.*;
import java.applet.*;
public class GridBag4 extends Applet {
    private Button b1 = new Button("b1");
    private Button b2 = new Button("b2");
    public void init() {
        GridBagLayout gbl = new GridBagLayout();
        setLayout(gbl);
        GridBagConstraints c = new GridBagConstraints();
        c.weightx = .7;
        c.fill = GridBagConstraints.BOTH;
        gbl.setConstraints(b1, c);
        add(b1);
        c.weightx = .3;
        c.weighty = 1.0;
        c.insets = new Insets(10,10,10,10);
        gbl.setConstraints(b2, c);
        add(b2);
    }
}
```

Positioning Constraints

By setting gridx and gridy we can locate a component at a specific row and column, while setting gridwidth and gridheight defines the number of cells for a component. The constant GridBagConstraints.REMAINDER specifies the remaining space in the row or column, while GridBagConstraints.RELATIVE indicates a position next to the preceding component.

In Example 6.15, shown in Figure 6.26, we use six buttons to illustrate the use of positioning constraints.

FIGURE 6.26 The applet of Example 6.15

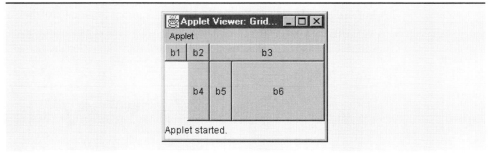

EXAMPLE 6.15   **GridBag5.java**

```java
/* Illustrates positioning constraints.
 */
import java.awt.*;
import java.applet.*;
public class GridBag5 extends Applet {
    private void makebutton(String name,                              // Note 1
            GridBagLayout gridbag, GridBagConstraints c) {
        Button button = new Button(name);
        gridbag.setConstraints(button, c);
        add(button);
    }
    public void init() {
        GridBagLayout gbl = new GridBagLayout();
        setLayout(gbl);
        GridBagConstraints c = new GridBagConstraints();
        c.fill=GridBagConstraints.BOTH;
        makebutton("b1",gbl,c);
        makebutton("b2",gbl,c);
        c.gridwidth = GridBagConstraints.REMAINDER;              // Note 2
        makebutton("b3",gbl,c);
        c.gridwidth = 1;                                         // Note 3
        c.gridx = 1;                                             // Note 4
        makebutton("b4",gbl,c);
        c.gridx = GridBagConstraints.RELATIVE;                   // Note 5
        makebutton("b5",gbl,c);
        c.weightx = 1.0;                                         // Note 6
        c.weighty = 1.0;
        makebutton("b6",gbl,c);
    }
}
```

Note 1: Because we are using six buttons, we use a method to create each button, set its constraints, and add it to the applet.

Note 2: Using REMAINDER for `gridwidth` specifies that it is the last component in its row.

Note 3: We set `gridwidth` back to its default value of 1. Had we not done that, buttons b5 and b6 would appear under b4, each filling an entire row.

Note 4: Setting `gridx` to 1 positions b4 in the second column, because a value of 0 represents the first column.

Note 5: Returning to the default RELATIVE for `gridx` keeps b5 and b6 in the same row as b4; otherwise they would each be positioned with `gridx` equal to 1, underneath b4.

Note 6: `c.weightx = 1.0;`

The other buttons in the second row, b4 and b5, have `weightx` equal to 0.0, so b6, with `weightx` equal to 1.0, gets all the extra space. Because b3 occupies the remainder of the first row, it gets stretched along with b6.

The Form Applet with a Text Area

Finally, Example 6.16 contains the code for the `Form` applet shown in Figure 6.21. Typically, the user would be filling out the form to buy a product or register for access to a restricted site. The information would go to a web site server which would return a response. In this example, since we do not really submit the data anywhere, we just display the information the user enters in a text area.

A text area provides a rectangular area in which to display information. We can specify initial text and the number of rows and columns of the display using the constructor

```
public TextArea(String initialText, int rows, int cols);
```

or use the default constructor

```
public TextArea();
```

to get a blank text area with a default size. The text area comes with scroll bars to see text not visible.

We want this form to look good when the applet is resized, so we keep the `fill` HORIZONTAL for all text fields except those with a small fixed size, which along with the buttons have `fill` NONE. The text area for messages expands in both directions.

EXAMPLE 6.16 **Form.java**

```
/* Illustrates creating a form
 * with a GridBagLayout.
 */

import java.awt.*;
import java.applet.*;
public class Form extends Applet implements ActionListener {
    private TextField first = new TextField("first name",12);
    private TextField middle = new TextField("I",1);
    private TextField last = new TextField("last name",15);
    private TextField address = new TextField("address",25);
```

continued

```java
private TextField city = new TextField("city",20);
private TextField state = new TextField("state",2);
private TextField zip = new TextField("zip",5);
private Button submit = new Button("Submit");
private Button reset = new Button("Reset");
private TextArea message = new TextArea();

public void init() {
    GridBagLayout gbl = new GridBagLayout();
    setLayout(gbl);
    GridBagConstraints c = new GridBagConstraints();
    c.anchor = GridBagConstraints.WEST;
    c.weightx=1.0;
    c.weighty=1.0;
    c.fill=GridBagConstraints.HORIZONTAL;
    c.insets = new Insets(5,5,5,5);
    gbl.setConstraints(first,c);              add(first);
    c.fill=GridBagConstraints.NONE;
    gbl.setConstraints(middle,c);             add(middle);
    c.fill=GridBagConstraints.HORIZONTAL;
    gbl.setConstraints(last,c);               add(last);
    c.gridy=1;
    gbl.setConstraints(address,c);            add(address);
    c.gridy=2;
    gbl.setConstraints(city,c);               add(city);
    c.fill=GridBagConstraints.NONE;
    gbl.setConstraints(state,c);              add(state);
    c.gridy=3;
    gbl.setConstraints(zip,c);                add(zip);
    c.gridy=4;
    c.anchor = GridBagConstraints.CENTER;
    gbl.setConstraints(submit,c);             add(submit);
    gbl.setConstraints(reset,c);              add(reset);
    c.gridy=5;
    c.gridwidth=3;
    c.fill=GridBagConstraints.BOTH;
    gbl.setConstraints(message,c);            add(message);
    submit.addActionListener(this);
    reset.addActionListener(this);
}
public void actionPerformed(ActionEvent event) {
    Object source = event.getSource();
    if (source == submit){
        String initial = middle.getText();
        if (initial != null)
            initial += ". ";
        message.setText(first.getText() + ' ' + initial + last.getText());
        message.append('\n' + address.getText());
        message.append('\n' + city.getText() + ", " + state.getText());
        message.append(' ' + zip.getText());
    }
    else if (source == reset) {
        message.setText("");
        first.setText("");
        middle.setText("");
        last.setText("");
        address.setText("");
        city.setText("");
```

```
                state.setText("");
                zip.setText("");
            }
        }
    }
```

> **The Big Picture**
> ....
> The gridbag layout gives us the most flexibility in arranging components. The various
> gridbag constraints allow us to specify the starting position of a component, the num-
> ber of rows and columns it takes, internal padding, anchoring, external padding, the
> direction(s) in which it will fill up its allocated cell, and its weighting relative to other
> components which determines how much screen space it gets relative to them. The
> gridbag layout manager uses the largest weight in a row or column to determine how
> to allocate space to that row or column. The fill for a component determines how it will
> appear within its allocated space. Experimenting, as we did in Examples 6.11–6.15,
> helps to understand the effect of the different constraints.

TEST YOUR UNDERSTANDING

TRY IT YOURSELF   13. Modify Example 6.12 so b1 has a `weighty` of 1.0 and b2 has a `weighty` of 0.0.
Explain the result.

TRY IT YOURSELF   14. Modify Example 6.15 to omit the line `c.gridwidth = 1`. Explain the result.

TRY IT YOURSELF   15. Modify Example 6.15 to omit the line

```
c.gridx = GridBagConstraints.RELATIVE;
```

Explain the result.

TRY IT YOURSELF   16. Modify Example 6.15 to omit the line

```
c.gridwidth = GridBagConstraints.REMAINDER;
```

Explain the result.

ENHANCEMENT

6.6 ▪ A GUI for Insertion Sorting

With user interface components, we can enhance our console applications, make them more
user friendly and interactive.

Defining the Problem

In this section we develop a graphical user interface to sort by insertion. In Example 4.6, we
sorted data inputted by the user in a console window, displaying the sorted array in the same
way, with no graphics. With the graphics of Chapter 5, and the user interface components of

this chapter, we can provide a user-friendly interface that lets us see the insertion sort proceed graphically, step by step.

Designing a Solution

The Exploratory Process

Sometimes problems come to us fully formulated, with our job being to develop a good solution. Other times we have a general goal, but have not yet settled on the specific requirements. For example, we have the goal of providing a GUI for insertion sort, but have not decided on a specific design. A good approach in this situation is to do a little exploratory programming, trying out some ideas on a small scale to determine what might work nicely.

We will want to display the data in a bar chart, so let us start the exploratory process by trying to display data in a chart. We can use a text field to enter the data, as we did in Example 6.5, and a canvas to display it, as we have done in several examples.

Our applet can register with the text field to be notified when the user hits the ENTER key, after entering the next value. The applet implements the `actionPerformed` method, (which the text field will call when the user hits the ENTER key), to get the data from the text field and ask the canvas to repaint, displaying the data entered so far in a bar chart. Our program will be event-driven, getting the data and displaying when the user hits the ENTER key.

Designing a Solution

Making a Chart

We need to figure out how to draw the chart. We might have very large or very small values, positive or negative values. The best approach is to simplify as much as possible, adding refinements later when we master the simpler cases. For now, let us use a 100x100 canvas, and integer data between 0 and 99, which will eliminate the problem of figuring out the vertical scale; we represent a value of 59 with a bar of height 59 pixels. To find the width of each bar, we divide the width of the canvas, 100, by the size of the data, dividing the canvas into equal parts for each bar.

To use the `fillRect` method to draw our bars, we need the upper-left corner for each bar and its height and width. Figure 6.27 shows the canvas with a few bars (unfilled).

Because the coordinate origin is in the upper-left corner of the canvas, the upper-left corner of the bar representing the array element `item[i]` is (`i*width,100-item[i]`) where `width` is the common width of each bar, given by `100/count`, and `count` is the number of elements entered thus far.

In the `paint` method, we draw a bar chart showing the values entered so far. Figure 6.28 shows the chart after the user entered the values 53, 22, and 75. Example 6.17 gives the code for this exploration.

FIGURE 6.27 Part of a bar chart

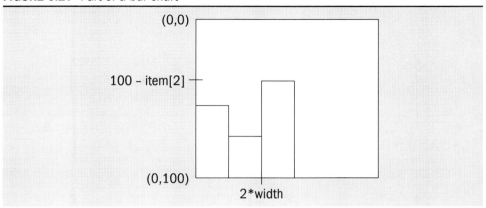

FIGURE 6.28 The applet of Example 6.17

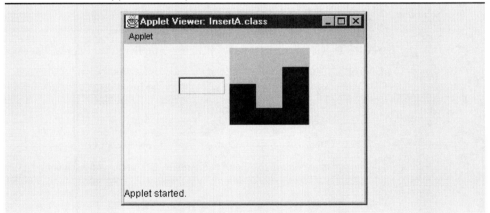

EXAMPLE 6.17   **InsertA.java**

```
/* Uses a text field to enter data and a canvas
 * to display the chart.  This is an exploratory
 * program to develop a GUI for insertion sort.
 */

import java.awt.*;
import java.awt.event.*;
import java.applet.Applet;

public class InsertA extends Applet implements ActionListener{
    public static int ITEM_SIZE = 10;
    public static int CHART_SIZE = 100;
    private TextField number = new TextField(5);
    private DrawOn canvas = new DrawOn();
    private int [] item = new int[ITEM_SIZE];            // Note 1
    private int count = 0;     // number of item entered
```

// **Note 1**

continued

```
public void init() {
    add(number);
    canvas.setSize(CHART_SIZE,CHART_SIZE);
    canvas.setBackground(Color.pink);
    add(canvas);
    number.addActionListener(this);
}
public void actionPerformed(ActionEvent event) {
    item[count++] = new Integer(number.getText()).intValue();      // Note 2
    number.setText("");
    canvas.repaint();                                              // Note 3
}
class DrawOn extends Canvas {
    public void paint(Graphics g) {
        if (count > 0){                                           // Note 4
            int width = CHART_SIZE/count;                         // Note 5
            for (int i=0; i<count; i++)
                g.fillRect(i*width, CHART_SIZE-item[i],width,item[i]); // Note 6
        }
    }
}
}
```

Note 1: We use an array of size 10, for simplicity, but later will allow different sizes of data.

Note 2: The only action listener is the applet who registers with the text field which generates an action event when the user hits the $\boxed{\text{ENTER}}$ key after entering a value. Here the applet gets the value from the text field, converts it to an integer using the expression

```
new Integer(number.getText()).intValue()
```

saves it in the next position in the array, and increments the count of the items entered so far.

Note 3: Because the user just entered another value, the applet asks the canvas to repaint the chart including this new value.

Note 4: When the applet starts, Java will call this `paint` method to draw the canvas once before the user has a chance to enter any values. While the count is still zero, we do not want to draw a chart.

Note 5: To get the width of each bar, we divide the width of the canvas, CHART_SIZE, by the number of data items, `size`.

Note 6: `g.fillRect(i*width,CHART_SIZE-item[i],width,item[i]);`

Figure 6.27 shows the canvas' coordinates and a few bars. We can see from this figure how to get the coordinates of the upper-left corner of each bar. The data value itself gives the height of the bar, because we do not need to scale the data.

Designing a Solution

Sorting

We can improve the chart later, but at least Example 6.17 shows that we are on the right track. For our next step, let us try to sort the data. Each time the user enters another value, we can insert it in the order of its predecessors. To do this we can just copy the code from Example 4.6 into an `insertNext` method to which we pass the array of data and the number of elements so far. The array holds ten elements but if the user just entered the third element, for example, the `insertNext` method should insert that value in the correct position with respect to its two predecessors, ignoring the array values beyond the third.

Figure 6.29 shows the chart just after the user entered the values 53, 22, and 75; as we can see, the values are sorted. Example 6.18 shows the additions to Example 6.17 needed to do the sorting.

FIGURE 6.29 The applet of Example 6.18

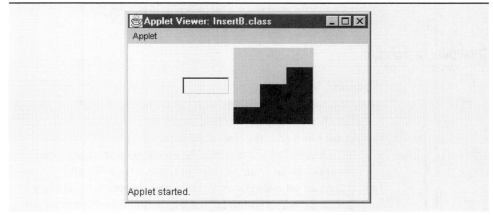

EXAMPLE 6.18 **InsertB.java**

```java
/* Add an insertNext method to insert
 * each value in order as soon as the
 * user enters it
 */

public class InsertB extends Applet implements ActionListener {
    // The rest is the same as Example 6.17

    public void actionPerformed(ActionEvent event) {
        item[count] = new Integer(number.getText()).intValue();
        insertNext(item,count++);                            // Note 1
        canvas.repaint();
    }
    public void insertNext(int [] data, int size) {          // Note 2
        int current = data[size];
        int j = 0;
        while (current > data[j]) j++;
        for (int k=size; k>j; k--)
            data[k] = data[k-1];
        data[j] = current;
    }
}
```

Note 1: The applet inserts the next value in the correct position as soon as the user enters it in the text field and hits the ENTER key to generate the action event.

Note 2: We copied the code for inserting the next element from Example 4.6.

We are making progress, but the chart does not look very nice with each bar the same color. We can greatly improve the appearance of the chart by drawing adjacent bars in different colors. To do this we add a simple statement

```
if (i%2==0) g.setColor(Color.green);
else g.setColor(Color.blue);
```

to the `paint` method to set the color to green if the index is even and to blue otherwise. We will see this improvement in our next example.

Designing a Solution

The User Interface

Now that we are comfortable with the sorting and drawing, we can think about the user interface. We will keep the text field for the user to enter data values, and add another text field for the user to specify the size of the array.

As in Example 4.6, it is nice to be able to select random input generated by the computer, or manual input from the user in the text field. We add two checkboxes in a checkbox group to make these selections. The user must select a method of input and specify the amount of input, so we add a button for the user to indicate the desired selections have been made.

Although we started our exploration using the default flow layout, the idea of putting a row of components in the North region of a border layout and the canvas with the chart in the Center appeals to us. We could create a panel and add four components to it, which are:

- A panel containing checkboxes for random or manual input, which uses a 2x1 grid layout so one checkbox will appear above the other.

- A panel containing a text field to input the size of the data and a label for the text field, which uses a 2x1 grid layout so that the label will appear above the text field.

FIGURE 6.30 The applet of Example 6.19

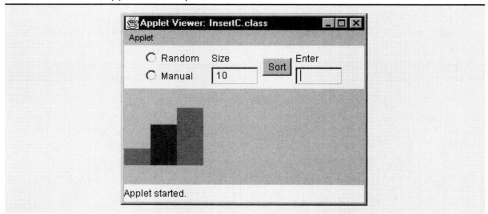

- A button for the user to start the sorting (used when random input is selected).

- A panel containing a text field for the user to enter data and a label for the text field, which uses a 2x1 grid layout so the label will appear above the text field (used when the user enters data manually).

Let us see how the GUI looks before we try to make it work. In Example 6.19 we add the new components using the border layout, but do not respond to any events other than the text entry of Example 6.18. Figure 6.30 shows our new GUI.

EXAMPLE 6.19   **InsertC.java**

```java
/* Adds components to the North region
 * of a border layout, and a canvas in
 * the Center.  Handles the same event as
 * Example 6.18, but no other events.
 */

public class InsertC extends Applet implements ActionListener {
    public static int ITEM_SIZE = 10;
    public static int CHART_SIZE      = 100;
    private TextField number = new TextField(5);
    private DrawOn canvas = new DrawOn();
    private int [] item = new int[10];
    private int count = 0;
    private CheckboxGroup acquire = new CheckboxGroup();
    private Checkbox random = new Checkbox("Random",false,acquire);
    private Checkbox manual = new Checkbox("Manual",false,acquire);
    private Label size = new Label("Size");
    private TextField getSize = new TextField("10",5);
    private Button sort = new Button("Sort");
    private Label enter = new Label("Enter");

    public void init() {
        setLayout(new BorderLayout());
        Panel p = new Panel();                           // Note 1
        Panel p1 = new Panel();                          // Note 2
        p1.setLayout(new GridLayout(2,1));
        p1.add(random);
        p1.add(manual);
        p.add(p1);
        Panel p2 = new Panel();
        p2.setLayout(new GridLayout(2,1));
        p2.add(size);
        p2.add(getSize);
        p.add(p2);
        p.add(sort);
        Panel p4 = new Panel();
        p4.setLayout(new GridLayout(2,1));
        p4.add(enter);
        p4.add(number);
        p.add(p4);                                       continued
```

```
            add(p,"North");
            canvas.setBackground(Color.pink);
            add(canvas, "Center");
            number.addActionListener(this);
        }

        // The rest is the same as Example 6.18
}
```

Note 1: Panel p is the panel we will add to the North region. We will add four components to it using its default flow layout.

Note 2: Panel p1 is the first of the four components that we add to p.

Completing the Java Code

Making the User Interface Work

Our attempt at a GUI, shown in Figure 6.30, looks reasonable enough to attempt to continue the implementation. Before getting into the details of the implementation we need to look at a few techniques that we will need.

We can disable a component c using the statement

```
c.setEnabled(false);
```

and enable it with

```
c.setEnabled(true);
```

We should disable components when they should not be used. For example, when the user selects random data, we should disable the text field that allows the user to input data manually. Until we are ready to handle repeated sorting with different data, we should disable most of the components after we complete the first sort.

In addition to getting data from a text field, or reading the label of a button, we can also set these strings to desired values. We can clear the text field after each data entry, so that the user will not have to delete the previous value to enter a new one. After the user presses the Sort button to start the sorting, we can change that button's label to Next so the user can command that we insert the next value.

Returning to our GUI, we will start with both the Sort button and the data entry text field disabled. If the user selects random input we enable the Sort button, while if the user selects manual input we enable the data entry field. No matter what type of input the user selects, we disable both checkboxes after the first selection because we will not yet handle repeated sorting, or changes in the input method. We handle the item events generated by the checkboxes in the `itemStateChanged` method.

When the user enters the first number, the applet, registered with that text field as an action listener, will be notified, and will allocate an array whose size it gets from the other text field in which the user specified the array size. We initialize this field with a default size of 10. We only accept a data entry until the user has entered the specified number of elements. After processing the latest entry we set the text field to the empty string to make it easier for the user to enter the next element.

The applet also listens for button presses. On the first press, when the button has its original label, Sort, the applet gets the desired size of the data, creates the array, and fills it with random numbers from 0 to 99. The applet then asks the canvas to display the data, and changes the button's label to Next so the user can start inserting the items one by one.

Each time the user presses the Next button, the applet inserts the next element in order with respect to its predecessors and asks the canvas to display the data. Thus we see the sorting process step by step. The applet disables the button when all the elements have been inserted and the data is completely sorted.

In Figure 6.30 the canvas fills the whole center region, but the chart, using bar widths based on the 100x100 canvas size used in the flow layout, is over on the far left side of the canvas. To use the whole canvas, we get the size of the applet and compute the width of each bar as `(applet width)/count`, where `count` gives the number of data elements.

Figure 6.31 shows the sorting partially completed. Example 6.20 shows the changes from Example 6.19 needed to make the applet respond to the events generated by the user.

FIGURE 6.31 The applet of Example 6.20

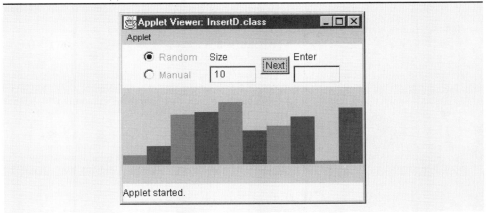

Applet started.

EXAMPLE 6.20   InsertD.java

```
/* Provides a user interface for insertion sorting.
 * The user can choose random data or input values
 * manually.  The user specifies the size of the data.
 * As the user inserts each item, the chart is redisplayed.
 */

public class InsertD extends Applet implements ActionListener, ItemListener {
     // The rest is the same as Example 6.19

    int nextCount = 1;     // next random number to insert
    int itemSize = 0;      // size of the data to sort
    Dimension d;           // size of the applet

    public void init() {
       ...
```

continued

```
            d = getSize();
            number.setEnabled(false);                                   // Note 1
            sort.setEnabled(false);
            sort.addActionListener(this);                               // Note 2
            random.addItemListener(this);                               // Note 3
            manual.addItemListener(this);
        }
        public void itemStateChanged(ItemEvent event) {                // Note 4
            String label = (String)event.getItem();
            if (label.equals("Random"))
                sort.setEnabled(true);                                  // Note 5
            else if (label.equals("Manual"))
                number.setEnabled(true);                                // Note 6
            random.setEnabled(false);                                   // Note 7
            manual.setEnabled(false);
        }
        public void actionPerformed(ActionEvent event) {
            Object source = event.getSource();
            String name = event.getActionCommand();
            if (source == number) {
                if (count == 0) {                                       // Note 8
                    itemSize = new Integer(getSize.getText()).intValue();
                    item = new int[itemSize];
                }
                if (count < itemSize) {
                    item[count] = new Integer(number.getText()).intValue();
                    insertNext(item,count++);
                    canvas.repaint();
                    number.setText("");                                 // Note 9
                }
                if (count == itemSize)  number.setEnabled(false);       // Note 10
            }
            else if (name.equals("Sort")) {
                itemSize = new Integer(getSize.getText()).intValue();
                item = new int[itemSize];
                count = itemSize;
                for (int i=0; i<itemSize; i++)
                    item[i] = (int)(100*Math.random());
                canvas.repaint();
                if (count == 1)                                         // Note 11
                    sort.setEnabled(false);
                else
                    sort.setLabel("Next");                              // Note 12
            }
            else if (name.equals("Next")) {
                insertNext(item,nextCount++);                           // Note 13
                canvas.repaint();
                if (nextCount == count) sort.setEnabled(false);         // Note 14
            }
        }

    // Also change width in paint to d.width/size.
}
```

Note 1: Until the user selects either the random or the manual data entry, we disable both the Sort button and the data entry text field.

Note 2: The applet registers to listen for button presses.

Note 3: The applet registers to listen for checkbox selections.

Note 4: The checkbox calls this method when the user makes a selection.

Note 5: When the user selects random input, we enable the Sort button.

Note 6: `if (label.equals("Manual"))number.setEnabled(true);`

When the user selects manual input, we enable the data entry text field.

Note 7: `random.setEnabled(false); manual.setEnabled(false);`

We disable both checkboxes, so that the user cannot change the selection before the sorting is finished, and cannot repeat with new data.

Note 8: `if (count == 0) {`

When the user enters the first data element, we get the size of the data and allocate an array to hold the values.

Note 9: `number.setText("");`

We erase the current entry in the text field to make it easier for the user to enter the next value.

Note 10: `if (count == itemSize) number.setEnabled(false);`

When the user has entered all the values, we disable the text field.

Note 11: `if (count == 1)`

When count is 1 we have nothing more to do so we disable the Sort button. We really do not expect the user to choose to sort only one item, but our program should be correct if that unlikely event does occur.

Note 12: `sort.setLabel("Next");`

We change the button's label to Next to allow the user to insert each random number in the correct position with its predecessors.

Note 13: `insertNext(item,nextCount++);`

When the user presses the Next button, we insert the next item in order and display all the values, so the user can see the sorting step by step.

Note 14: `if (nextCount == count) sort.setEnabled(false);`

We disable the button when the sorting is complete.

Testing the Code

By developing our program in stages, we are also able to test it in stages. We tested the making of the chart in InsertA, the sorting in InsertB, and the look of the user interface in InsertC. Finally we test the implementation of the user interface in InsertD.

We should carefully test the two main use cases of this system, sorting with random input and sorting manually. When we select random input we check that the Sort button becomes enabled and the text field remains disabled. First we check the sorting using the default size of 10 and then check with size 1. With size 10, pressing the Sort button does display the chart properly and change the button label to Next. Pressing the Next button the first time inserts the second element in the correct place with respect to the first element. Pressing the Next button nine times results in a completely sorted array, at which time the Next button becomes disabled. Using size 1 causes the Sort button to be immediately disabled after we press it, which is what we want because there is no need to do anything further to sort one element.

When we select manual input of 10 items, we see that the Sort button is disabled and the Enter text field is enabled so we can enter the data to sort. Entering each value causes it to be placed in the correct order with respect to the data previously entered. After entering 10 items, the Enter text field becomes disabled.

In industrial strength systems much effort is spent in validating the input to ensure the program does not crash if the user enters incorrect data.[7]

The Big Picture

The development of a GUI for insertion sorting illustrates the exploratory development process. First we make the chart, then we add the sorting algorithm. Next we design the user interface, and finally we make the user interface work. At each stage we have a complete program we can test.

Chapter 6 Summary

In this chapter we created graphical user interfaces. The Component class of the AWT is the superclass of the various components we can add to containers such as applets to make a user interface. The Container class describes components that are themselves containers. We use the Panel and its Applet subclass inside another window such as a browser, while we can use the window and its Frame subclass in standalone applications.[8]

The Java event model allows objects that want to be notified about an event to register with the event source as a listener. The listener promises to implement an interface consisting of a method (or methods) to handle the event(s). The event source will call this method for each registered listener when the event occurs. This event model is very much like the rate change notification of Example 4.9.

Both buttons and text fields generate action events. An object that wants to be notified of these events implements the ActionListener interface, requiring that it implement the actionPerformed method stating how it wants to handle an action event. The object registers with the button or text field requesting to be notified when the user presses the button or when the user hits the (ENTER) key after entering a string in the text field. When one of these events occurs, the event source, button or text field, calls the object's actionPerformed method, passing it an action event describing the button press or text entry event.

---

[7] We will cover the Java exception handling facility, which helps us check for and recover from errors, in Section 8.1.

[8] We cover standalone graphics applications in Section 7.4.

We use the `Canvas` class to get an object to draw on, so we do not overwrite other components by drawing directly in the container. We use the `Label` class to label text fields, and can use it for a simple text display. Neither the canvas nor the label generates any events.

Both checkboxes and choice boxes generate item events when the user makes a selection. An object that wants to be notified of these events implements the `ItemListener` interface, requiring that it implement the `itemStateChanged` method stating how it wants to handle an item event. The object registers with the checkbox or choice box requesting to be notified when the user makes a choice or checks a box. When one of these events occurs, the event source, the checkbox or the choice box, calls the object's `itemStateChanged` method, passing it an `ItemEvent` describing that selection.

Java uses layout managers to position components in a container. The layout manager adjusts the positions of the components when necessary as a consequence of resizing a component or the whole container. The default for an applet or a panel is the flow layout, which adds components from left to right until no more will fit in a row, centering any remaining components in the last row. The flow layout uses the preferred sizes of its components in laying them out.

The border layout uses five regions, North, South, East, West, and Center. We add a component to a border layout specifying the region in which to place it. It stretches components horizontally in the North and South, vertically in the East and West, and in both directions in the Center. The grid layout uses a grid of m rows and n columns, adding components from left to right across each row before moving to the next lower row. The grid layout expands each component to fill its grid cell. We can use panels to nest components, to give us more flexibility in designing user interfaces. The gridbag layout provides the most flexibility attained by appropriately setting the `gridx`, `gridy`, `gridwidth`, `gridheight`, `fill`, `ipadx`, `ipady`, `insets`, `anchor`, `weightx`, and `weighty` constraint variables.

We can use exploratory programming to develop an interface step by step. At each stage we have a working program which provides some of the functionality we want, and can focus all our attention on the next enhancement. We used this technique to provide an interface for insertion sorting which lets us see each step of the sorting process.

In this chapter we introduced the `Button`, `Checkbox`, `TextField`, `Choice`, `Label`, `TextArea`, and `Canvas` components. The AWT also contains other useful components including `List`, which we cover in Chapter 11, `Scrollbar`, and others.

Build Your Own Glossary

Find the uses, in this chapter, of the terms below. Enter a definition of each in the `glossary.txt` file on the disk that comes with this text.

action event	flow layout	item event
border layout	grid layout	Java event model
checkbox	gridbag layout	layout manager
checkbox group	GUI	panel
choice box	inner class	wrapper class

Skill Builder Exercises

1. Fill in the blanks below to show the correct listener type for the indicated object.

    a. button.add_____Listener where `button` is a button.

    b. choice.add_____Listener where `choice` is a choice box.

    c. check.add_____Listener where `check` is a checkbox.

    d. text.add_____Listener where `text` is a text field.

2. Match a layout type on the left with a description on the right of how it uses the preferred size of its components.

    a. Grid layout   i.   always respects the preferred sizes

    b. Flow layout   ii.   never respects the preferred sizes

    c. Border layout   iii.   partially respects preferred sizes

3. Fill in the blanks in the following:

    To be notified about a button press, an object must be instantiated from a class that _____ the _____ interface which means that it implements the _____ method. To be notified about a checkbox selection, an object must be instantiated from a class that _____ the _____ interface which means that it implements the _____ method.

Critical Thinking Exercises

4. Suppose we add a button, a text field, a choice box, and a checkbox to an applet using the default flow layout. Choose the best description below of how these components will be positioned in the applet.

    a. Centered in a row across the top of the applet

    b. The number of rows used depends upon the size of the applet

    c. In the top row if the applet is wide enough, and in two rows otherwise

    d. At the positions where we add them

    e. None of the above

5. For an `ItemEvent e`, the method call `e.getItem()` returns

    a. the object that was the source of the event.

    b. the object that is listening for the event.

    c. the object that represents the item selected.

    d. none of the above.

6. For checkboxes in a `CheckboxGroup`

    a. at most one may be selected.

    b. exactly one must be selected.

    c. more than one may be selected.

    d. more than one may be selected only if we set a flag in each checkbox to **true** when we construct it.

7. The border layout uses five regions in which to arrange components. To use more than five components we can

    a. add more than one component directly into each region.

b. add more than one component directly to the Center region, but not to the others.
c. add some components into a component that is a type of container and then add the container to one of the regions.
d. none of the above.

Debugging Exercise

8. The following applet attempts to draw the figure selected in a checkbox, with the name of the shape displayed inside it, when the user presses the Draw button. Find and fix any errors.

```java
import java.awt.*;
import java.awt.event.*;
import java.applet.Applet;
public class NameIt extends Applet {
    Button draw = new Button("Draw");
    DrawOn canvas = new DrawOn();
    CheckboxGroup shapes = new CheckboxGroup();
    Checkbox square = new Checkbox("Square",false,shapes);
    Checkbox circle = new Checkbox("Circle",true,shapes);
    public void init() {
        add(draw);
        add(square);
        add(circle);
        add(canvas);
        draw.addActionListener(canvas);
    }
    class DrawOn extends Canvas implements ActionListener {
        public void actionPerformed(ActionEvent event) {
            repaint();
        }
        public void paint(Graphics g) {
            if (circle.getState()){
                g.fillOval(20,20,100,100);
                g.drawString("Circle",40,40);
            }
            else {
                g.fillRect(20,20,100,100);
                g.drawString("Square",40,40);
            }
        }
    }
}
```

Program Modification Exercises

9. Modify Example 5.3 to draw on a `Canvas` object rather than drawing directly on the applet.
10. Modify Example 5.4 to draw on a `Canvas` object rather than drawing directly on the applet.
11. Modify Example 5.5 to draw on a `Canvas` object rather than drawing directly on the applet.

A LITTLE EXTRA

12. Modify Example 6.2 to override the `getPreferredSize` method to set the size of the canvas, rather than calling the `setSize` method.

13. Modify Example 6.20, the GUI for insertion sort, to indicate the number below each bar in the chart.

PUTTING IT ALL
TOGETHER

14. a. Example 5.4 displayed text in various fonts. Develop a GUI to allow the user to select a font from a choice box, to select a style from checkboxes, and to enter a point size in a text field. Display a message, using the selected font, in a canvas.

    b. Check that the message will fit on one line. If it does not fit in the point size that the user requested, reduce the point size so it does fit.

    c. Allow the user to select a color, and display the text in that color.

PUTTING IT ALL
TOGETHER

15. a. Example 5.3 displayed various shapes. Develop a GUI to allow the user to select a shape from a choice box. Use checkboxes for the user to select whether to draw the shape filled or unfilled. If the user selects a line, then disable the checkboxes. Display the selected shape in a canvas.

    b. Allow the user to select a color, and display the shapes in that color.

PUTTING IT ALL
TOGETHER

16. Example 4.1 computed the distribution of the sum of the values of two dice. Allow the user to enter the number of tosses in a text field, and display a bar chart showing the percentage of outcomes for each possibility, two through twelve.

17. Develop a GUI to reverse an array. Allow the user to choose random input, or to manually input the array. Display the array in a bar chart on the left of a canvas. Provide a button, and when the user presses the button, do one more interchange of elements, displaying the partially reversed array on the right of the canvas. When the reversal is complete, disable the button.

PUTTING IT ALL
TOGETHER

18. In Example 3.3 we defined `Shape`, `Line`, and `Circle` classes, but did not actually draw any figures. Modify the `draw` methods to draw the line and the circle. Display a circle and a line and allow the user to choose which figures to move. Add text fields to allow the user to input the amounts to move in the x and y directions, and a button for the user to request the move. Move the selected figure when the user presses the button.

PUTTING IT ALL
TOGETHER

19. Modify Exercise 24 of Chapter 5 to add a text field for the user to input the sales values for each day.

20. Modify Example 6.8 to center the figure in the canvas.

21. Modify Example 6.9 to center the figure in the canvas.

Program Design Exercises

22. a. Develop a GUI for a four-function (+,-,*,/) calculator. Use a text field for the display and a grid of buttons for the input. Just provide the interface. Do not try to implement the calculator functions.

    b. Implement the calculator functions, +,-,*, and /.

23. Write a program which lets the user input, in a text field, the number of fair coins to toss. Using random numbers to simulate the toss of a coin, repeat this experiment 100 times, and display a bar chart showing the distribution of the outcomes for each possible number of heads obtained. For example, tossing six coins 100 times might

give no heads twice, one head 10 times, two heads 20 times, three heads 30 times, four heads 24 times, five heads 13 times, and six heads once.

24. a.   A market sells eggs at $1.90 per dozen, milk at $1.47 per quart, and bread and $2.12 per loaf. Use a choice box to allow the user to select an item, and a text field for the user to input the quantity desired. Include an Order button for the user to order the specified quantity of the selected item. When the user selects an item, the price should appear in a label. When the user presses the Order button, a description and the total cost of the order should appear in a canvas. Design the applet using a layout of your choice.

   b.   Allow the user to order more than one type of food. Each time the user presses the Order button, describe the purchase. Add a Total button and when the user presses this button display the total prices of all items ordered.

25. Every brokerage firm has its own formulas for calculating commissions when stocks are purchased or sold. Many of these formulas are based on the number of round lots (groups of 100) purchased and the number of stocks in any odd lot (less than 100) purchased as well as the price of the stock. To keep things simple, our company will only charge according to the number of stocks purchases as follows:

$30 per round lot
$.50 per stock in any odd lot

For example, if the purchaser buys 110 shares of stock, he or she is charged $35 in commissions. Include three text fields for the user to enter the name of a stock, the quantity desired, and the cost of one share. Include a Buy button and when the user presses this button display the total cost of that stock, including the commission. Design the applet using a layout of your choice.

26. Write a Java applet which prints the message listing the capital of a country using the interface shown in Figure 6.32 where the style can be bold or italic, and the size, large or small, refers to the point size of the font.

FIGURE 6.32 User Interface for Exercise 26

7

Mouse, Key, and Window Events

When we use the mouse to press buttons, and to select checkbox and choice items, we do not handle the mouse events directly. The button responds to a mouse click, generating an action event describing the button press. The checkbox and choice box respond to the mouse click by generating an item event. We refer to these action and item events as high-level or semantic events because they incorporate the meaning of the mouse click as a button press or item selection.

In this chapter we learn to handle mouse events directly, and in the process introduce the concept of an adapter class, a useful tool for implementing interfaces containing multiple methods. By responding directly to low-level mouse events, we will be able to use the mouse to drag shapes, change colors, and play games. Figure 7.1 shows a simple screen for playing tick tack toe with the computer. We leave the implementation of this and a Nim game for the exercises.

When we use the keyboard to enter text in a text field, we do not handle the key events directly. When we hit the ENTER key, the text field generates an action event describing the text entry. In this chapter we learn to handle key events. We will use keys to rotate shapes in solving the tangram puzzle.

Thus far we have not needed to handle window events. Our graphics programs have been applets run by the browser or applet viewer which opens and closes the windows for us. To create standalone graphics applications we will need to handle window events, and can then

FIGURE 7.1 Tick tack toe

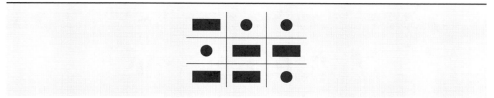

use the same graphics techniques and user interface components that we used in our applet examples. We will show the steps needed to convert an applet to a standalone application or to convert a standalone application to an applet.

OBJECTIVES

◆ Handle mouse events.
◆ Handle key events.
◆ Handle window events.
◆ Use adapter classes for a default implementations of interfaces.
◆ Convert applets to standalone applications.
◆ Convert standalone applications to applets.
◆ Use the mouse and the keyboard to solve tangram puzzles.

OPTIONS

◆ Omit 7.1: Using the Mouse (needed in 7.5, 12.3-12.4)
◆ Omit 7.2: Using the Keyboard (needed in 7.5)
◆ Omit 7.5: Tangrams with the Mouse and Keys

7.1 ▪ Using the Mouse

Our three examples illustrate three aspects of using the mouse. The first concerns the mouse in a fixed location, while the second concerns a moving mouse. Finally, for mouse events, we utilize the general technique of providing a trivial implementation of an interface that we can override to implement those methods in which we are interested.

Mouse Events

In the Java event model, a source object generates the event and a handler object handles it. Pressing a button generates a high-level action event. An event handler implementing the `ActionListener` interface has only to implement the `actionPerformed` method. With the mouse, the user can generate seven types of **low-level events** in a component source:

MOUSE_PRESSED	user pressed the mouse in a component
MOUSE_RELEASED	user released the mouse in a component
MOUSE_CLICKED	user clicked the mouse in a component
MOUSE_ENTERED	the mouse entered a component
MOUSE_EXITED	the mouse exited a component
MOUSE_MOVED	the mouse moved (no button down) in a component
MOUSE_DRAGGED	the mouse moved (button down) in a component

These are seven types of MouseEvent, which is a class in the java.awt.event package. The MOUSE_CLICKED type refers to a mouse click which consists of a mouse press followed by a mouse release, with no intervening mouse drag.

MouseListener and MouseMotionListener

Java uses two types of listener interfaces to handle mouse events. An object that wants to be notified of any of the first five mouse events, mouse pressed, released, clicked, entered, or exited, needs to register, using the addMouseListener method, as a MouseListener with the component that is the event source and must implement each of the five methods of the MouseListener interface:

```
public void mousePressed(MouseEvent e);
public void mouseReleased(MouseEvent e);
public void mouseClicked(MouseEvent e);
public void mouseEntered(MouseEvent e);
public void mouseExited(MouseEvent e);
```

An object that wants to be notified of any of the last two mouse events, mouse moved or dragged, needs to register as a MouseMotionListener, using the addMouseMotionListener method. It registers with the component that is the event source and must implement each of the two methods of the MouseMotionListener interface:

```
public void mouseMoved(MouseEvent e);
public void mouseDragged(MouseEvent e);
```

We can choose which mouse events we want to handle in our applications. If we only care about the mouse click, then we can ignore the mouse pressed, mouse released, mouse entered, and mouse exited events. To ignore a low-level mouse event we implement its handler method with an empty body. For example, implementing the mousePressed method as

```
public void mousePressed(MouseEvent e) {  }
```

with an empty body will ensure nothing will happen in response to a mouse pressed event.

As an example, let us suppose when we press the mouse inside a polygon we want to change its color to red. A scenario describing the interactions of the objects is:

The applet registers as a mouse listener with the polygon.

The user presses the mouse inside the polygon.

The applet (the component in which the user pressed the mouse) passes a mouse event describing the mouse press to the applet's mousePressed method.

The applet's mousePressed method changes the foreground color to red and asks Java to redraw it.

Java calls the applet's paint method which fills the polygon in red.

To illustrate the handling of mouse events, our first example program of this chapter will fill a triangle in red if the user presses the mouse inside it, and fill it in blue if the user releases the mouse inside it. If the user presses the mouse inside the triangle, but then drags the mouse outside of it, the triangle will remain red, as it only turns blue if the user releases the mouse inside it. Nothing happens if the user presses the mouse outside the triangle first, but if the user then drags the mouse inside the polygon and releases it, the polygon will turn blue. We also draw the strings "Got the mouse" when the mouse enters the applet, and "Lost the mouse" when the mouse leaves the applet.

Figure 7.2 shows the applet after the user pressed the mouse inside the triangle and released it outside, moving the mouse outside the applet. In color, the string and triangle are red.

FIGURE 7.2 The applet of Example 7.1

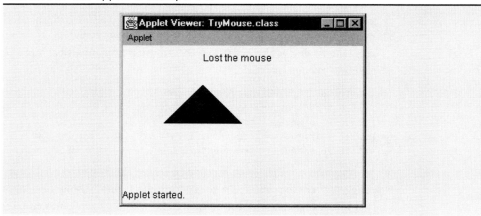

To implement these effects, the applet defines the mouse event-handling methods to change the settings and repaint so that the `paint` method can draw using the new settings. The `mousePressed` method sets the foreground color to red, while the `mouseReleased` method sets it to blue. The `mouseEntered` method sets the display string to "Got the mouse", while the `mouseExited` method sets that string to "Lost the mouse". In this example, we implement the `mouseClicked` method with an empty body, meaning we do not care about that type of event.

We can get the position of any mouse event using the `getX` and `getY` methods, as in

```
int x = event.getX();
```

where `event` is one of the seven types of mouse events. We can use the `contains` method of the `Polygon` class to see if the user pressed the mouse inside a polygon, as in

```
boolean inside = p.contains(event.getX(),event.getY());
```

where p is a polygon, and `event` is a mouse event describing a mouse press.

EXAMPLE 7.1 **TryMouse.java**

```
/* Fills a red triangle when the user presses the mouse
 * inside it.  Fills the triangle in blue when the user
 * releases the mouse inside it.  Draws "Got the mouse"
 * when the mouse enters the applet, and draws "Lost
 * the mouse" when the mouse exits the applet.  Provides
 * an empty body for mouseClicked.
 */

import java.awt.*;
import java.awt.event.*;
import java.applet.Applet;
```

```
public class TryMouse extends Applet implements MouseListener {        // Note 1
    private int [ ] x = {50,100,150};
    private int [ ] y = {100,50,100};
    private Polygon p = new Polygon(x,y,3);                            // Note 2
    private String mouse = "";                                        // Note 3

    public void init() {
        addMouseListener(this);                                       // Note 4
    }
    public void paint(Graphics g) {
        g.drawString(mouse,20,20);
        g.fillPolygon(p);
    }
    public void mousePressed(MouseEvent event) {
        if (p.contains(event.getX(), event.getY()))  {                // Note 5
          setForeground(Color.red);
          repaint();
        }
    }
    public void mouseReleased(MouseEvent event) {
        if (p.contains(event.getX(), event.getY()))  {
          setForeground(Color.blue);
          repaint();
        }
    }
    public void mouseClicked(MouseEvent event) {  }                   // Note 6
    public void mouseEntered(MouseEvent event) {
      mouse = "Got the mouse";                                        // Note 7
      repaint();
    }
    public void mouseExited(MouseEvent event) {
      mouse = "Lost the mouse";
      repaint();
    }
}
```

Note 1: A class that implements the MouseListener interface must implement the five methods mousePressed, mouseReleased, mouseClicked, mouseEntered, and mouseExited to handle each of these events.

Note 2: We construct a triangle from arrays of its x- and y-coordinates.

Note 3: The mouseEntered and mouseExited methods will set this string. We initialize it to the empty string so the string will not be **null** when the applet is initialized.

Note 4: The applet registers to listen to its own mouse events. We could have created another class to listen to these events. With this approach, we use

```
addActionListener(new MouseHandler())
```

adding the MouseHandler class given by

```
class MouseHandler implements MouseMotionListener {
    /* Code for mousePressed, mouseReleased, mouseClicked,
     * mouseEntered, and mouseExited goes here.
    }
```

In a small illustrative example we find it simpler to let the applet handle events itself, but in general, code quality improves when we separate GUI event handling from the code needed for other processing.

Note 5: The `getX()` and `getY()` methods return the position at which the user pressed the mouse. If that position is contained inside the polygon, then we set the foreground color to red, so the triangle will be redrawn in red.

Note 6: `public void mouseClicked(MouseEvent event) {  }`

A mouse listener must implement all five mouse event handling methods. We do not care about the mouse clicked event, so we just provide an empty body for the `mouseClicked` method.

Note 7: `mouse = "Got the mouse";`

When the mouse enters the applet, we set the display string to "Got the mouse" and then repaint so that the `paint` method will draw this string.

Moving the Mouse

Java separates the mouse moved and mouse dragged events from the other mouse events. Mouse motion events occur in great numbers and users not interested in them should not have to be bothered by them. As noted above, the `MouseMotionListener` interface has two methods, `mouseMoved` and `mouseDragged`, that need to be implemented by mouse motion listeners interested in these events. A mouse moved event occurs when the user moves the mouse with no buttons pressed, while a mouse dragged event occurs when the user moves the mouse while pressing a button.

A sample scenario in which the user drags a polygon to a new position is:

The applet registers as a mouse listener.

The applet registers as a mouse motion listener.

The user presses the mouse inside a polygon.

The applet sends a mouse event describing the mouse press to the applet's `mousePressed` method.

The applet's `mousePressed` method saves the position of the mouse.

The user drags the mouse to a new position.

The applet sends a mouse event describing the mouse drag to the applet's `mouseDragged` method.

The applet's `mouseDragged` method translates the polygon to the new mouse position, saves that position in case the user drags the polygon again, and asks Java to redraw the applet.

Java calls the applet's `paint` method to redraw the applet with the polygon in its new position.

Example 7.2 enables the user to drag a triangle with the mouse. Figure 7.3 shows the triangle when the applet has just started, and then after the user dragged it to the lower left.

FIGURE 7.3 The applet of Example 7.2

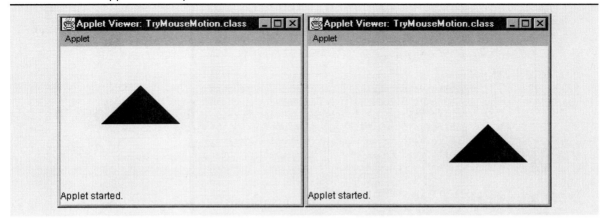

This applet implements both the MouseListener and the MouseMotionListener interfaces. When the user presses the mouse inside the polygon, the mousePressed method saves
the x- and y-coordinates of the mouse press for use if the user drags the mouse to a new
position.

When the user does drag the mouse to a new position, the mouseDragged method translates the polygon by the amount (x-oldx,y-oldy) where (x,y) is the new location of the mouse,
and (oldx,oldy) is its previous location. The mouseDragged method then updates the oldx and
oldy values to the current position, (x,y), in case the user continues to drag the mouse.

EXAMPLE 7.2 **TryMouseMotion.java**

```java
/*  Drags a triangle to a new location
 *  using the position of the mouse to
 *  determine how to move the polygon.
 */

import java.awt.*;
import java.awt.event.*;
import java.applet.Applet;

public class TryMouseMotion extends Applet
      implements MouseListener, MouseMotionListener {      // Note 1
    private int [ ] x = {50,100,150};
    private int [ ] y = {100,50,100};
    private Polygon p = new Polygon(x,y,3);
    private int oldx;                          // saves previous polygon position
    private int oldy;

    public void init() {
        addMouseListener(this);
        addMouseMotionListener(this);                      // Note 2
    }
    public void paint(Graphics g) {
        g.fillPolygon(p);
    }
```

continued

```
        public void mousePressed(MouseEvent event) {
            int x = event.getX();
            int y = event.getY();
            if (p.contains(x,y)){                              // Note 3
              oldx = x;
              oldy = y;
            }
        }
        public void mouseReleased(MouseEvent event) {  }
        public void mouseClicked(MouseEvent event) {  }
        public void mouseEntered(MouseEvent event) {  }
        public void mouseExited(MouseEvent event) {  }
        public void mouseMoved(MouseEvent event) { }
        public void mouseDragged(MouseEvent event) {
            int x = event.getX();
            int y = event.getY();
            if (p.contains(x,y)){
                p.translate(x - oldx, y - oldy);              // Note 4
                oldx = x;
                oldy = y;
                repaint();
            }
        }
    }
```

Note 1: The applet implements the MouseListener interface to respond to mouse pressed events and the MouseMotionListener interface to respond to mouse dragged events. We could have used another class to implement these listener interfaces.

Note 2: The applet registers with itself to listen for mouse events. The applet will be both the source component in which the user creates mouse events, and the listener who responds to these events.

Note 3: When the mouse is pressed, we check to see if it is inside the triangle, and if so we save the mouse position to use if the user drags the mouse to a new position.

Note 4: When the mouse is dragged to a new position, we translate the polygon to that position.

Adapter Classes

An **adapter class** connects the source of an event with its target, implementing the methods of a listener interface. In Example 7.2 the applet implements all five methods of the mouse MouseListener interface, four of them with empty bodies because we are not interested in those events. We could have used a separate class to implement the mouse listener interface, registering it as a mouse motion listener rather than the applet. Using such an adapter class would separate the event-handling code from the application code.

To simplify the handling of low-level events, Java provides adapter classes with default implementations of each mothod of the appropriate interface. The MouseAdapter class implements each of the five methods of the MouseListener interface with empty bodies. We can define a subclass of MouseAdapter to override just those methods which handle the events we are interested in.

For example, if we just want to handle the mouse pressed event, as was the case in Example 7.2, we can define a class MousePressListener which extends the MouseAdapter class and just overrides the mousePressed method. Example 7.3 revises Example 7.2 to make this change.

☞ **TIP**

For easier handling of mouse events we use the MouseAdapter class from the AWT, but this technique is generally applicable. Whenever we define an interface, say

```java
public interface Chores {
    public void washCar(Car aBigCar);
    public void feedDog(Dog iggy);
    public void makeCoffee(Coffee brew);
}
```

we should define an adapter class which implements the interface with empty bodies

```java
public abstract class ChoresAdapter implements Chores {
    public void washCar(Car aBigCar) {  };
    public void feedDog(Dog iggy) {  };
    public void makeCoffee(Coffee brew) {  };
}
```

A class that implements the Chores interface can override those methods of the ChoresAdapter class which handle events of interest to it. For example, a class that just wants to feed the dog could override the ChoresAdapter class to provide an implementation for the feedDog method, inheriting the do nothing implementations for the washCar and makeCoffee methods.

```java
public class DogFeeder extends ChoresAdapter {
    public void feedDog(Dog iggy) {
        // routine to feed the dog goes here
    };
}
```

EXAMPLE 7.3   **TryMouseAdapter.java**

```java
/*  Revises TryMouseMotion.java using
 *  an inner class to extend MouseAdapter,
 *  overriding the mousePressed method.
 */

import java.awt.*;
import java.awt.event.*;
import java.applet.Applet;

public class TryMouseAdapter extends Applet
                implements MouseMotionListener {          // Note 1
    private int [ ] x = {50,100,150};
    private int [ ] y = {100,50,100};
    private Polygon p = new Polygon(x,y,3);
    private int oldx;      // Saves previous polygon position
    private int oldy;
    public void init() {
        addMouseListener(new MousePressListener());        // Note 2
        addMouseMotionListener(this);
    }
    public void paint(Graphics g) {
        g.fillPolygon(p);
    }
```

                                                            continued

```
        public void mouseMoved(MouseEvent event) { }
        public void mouseDragged(MouseEvent event) {
            int x = event.getX();
            int y = event.getY();
            if (p.contains(x,y)){
                p.translate(x - oldx, y - oldy);
                oldx = x;
                oldy = y;
                repaint();
            }
        }

        class MousePressListener extends MouseAdapter {                    // Note 3
            public void mousePressed(MouseEvent event) {
                int x = event.getX();
                int y = event.getY();
                if (p.contains(x,y)){
                    oldx = x;
                    oldy = y;
                }
            }
        }
    }
}
```

Note 1: The applet implements the MouseMotionListener interface, but uses a MousePressListener class, which extends the MouseAdapter class, to handle mouse events. We could have defined another class, extending the MouseMotionAdapter class, to handle mouse motion events, instead of having the applet implement the mouse motion listener interface, but we will leave that to the exercises.

Note 2: We create a new MousePressListener class to listen for mouse press events. MousePressListener extends MouseAdapter so it only has to implement the methods for the events it wants to handle.

Note 3: MousePressListener is an inner class, defined inside the applet. It implements the mousePressed method to handle mouse pressed events. The MouseAdapter class provides default implementations of all five mouse listeners methods; we only have to override the handlers for the events in which we are interested.

> **The Big Picture**
>
> Unlike the ActionEvent and ItemEvent classes that represent higher-level events, MouseEvent deals with basic mouse operations. In contrast to the ActionListener interface which needs only one method, the MouseListener interface has five methods and the MouseMotionListener interface has two. The MouseAdapter and MouseMotionAdapter classes provide trivial implementations of these interfaces, in which the methods do nothing. A class can either implement all five methods of the MouseListener interface or extend MouseAdapter and override only those methods needed to handle the mouse events of interest. The choice is similar for events involving mouse motion.

1. Write a scenario showing the interactions of the objects in Example 7.1 if the user presses the mouse outside the triangle and releases the mouse inside it.

TRY IT YOURSELF

2. Add `print` statements to the `mousePressed`, `mouseReleased`, and `mouseClicked` methods in Example 7.1 to see in what order the mouse pressed, released, and clicked events are generated when the user clicks the mouse. In addition, try to press the mouse, drag it, and release it so a mouse clicked event is not generated.

TRY IT YOURSELF

3. Add a `print` statement to the `mouseDragged` method of Example 7.3 to see how frequently mouse dragged events are generated. The statement should print the value of a counter which `mouseDragged` increments each time it handles an event.

7.2 ▪ Using the Keyboard

When using the keyboard we have to distinguish between the physical keys pressed and the characters they might represent. For example, we use two keys to represent uppercase letters. With the `KeyEvent` class and the `KeyListener` interface Java lets us respond to user generated keyboard events.

Focus

We press the mouse at a specific point, in a specific component. We saw in Chapter 6 that an applet may contain several panels and other components. When we press the mouse, the component in which the mouse was pressed receives the mouse event. By contrast, pressing a key has no association with any specific component in the user interface. Java sends key events to the currently selected component, which is said to have the **focus**. A component that wants to receive key events must execute the `requestFocus` method to get the focus.

Key Events

Java defines three key events, KEY_PRESSED, KEY_RELEASED, and KEY_TYPED. Java generates the key pressed or key released events for each physical key that is pressed or released, and generates the key typed event when a Unicode character is typed. The KEY_TYPED event allows Java to give meaning to sequences of key presses used to represent a single Unicode character. We press two keys, the [SHIFT] key and a letter key, to represent an uppercase letter. This facility is very useful in adapting keyboards to input characters from diverse languages.

To represent the physical keys, Java uses key codes, (named integer constants), starting with the letters VK_ (for virtual key). Some of these key codes are listed below:

Key Code	Physical Key
VK_A, ... , VK_Z	The keys A to Z
VK_0, ... , VK_9	The keys 0 to 9
VK_SHIFT	The [SHIFT] key
VK_CONTROL	The [CTRL] key
VK_DOWN, VK_UP	The [↓] and [↑] keys
VK_RIGHT, VK_LEFT	The [→] and [←] keys

Pressing the Ⓖ key will generate a key pressed event and a key released event; it will also generate a key typed event which will indicate a G if the user pressed the [SHIFT] key while pressing the Ⓖ key , or indicate a g otherwise. The key events that occur when the user types the uppercase letter G are:

Event	Description
KEY_PRESSED	Press the VK_SHIFT key
KEY_PRESSED	Press the VK_G key
KEY_TYPED	G was typed
KEY_RELEASED	Release the VK_G key
KEY_RELEASED	Release the VK_SHIFT key

KeyListener

An object that wishes to respond to key events must register as a key listener, using the addKeyListener method, with the source of these key events. The KeyListener interface has three methods,

```
public void keyPressed(KeyEvent e);
public void keyReleased(KeyEvent e);
public void keyTyped(KeyEvent e);
```

that a class wanting to listen to key events must implement. The source of a key event passes a KeyEvent object to the appropriate method. For example, pressing the Ⓖ key will cause the component that has the focus to pass a KeyEvent object to the keyPressed method of any objects registered as key listeners with that component.

To see how we handle key events, consider a simple applet which displays the letter a user types, moving it to the left when the user presses the → key, and moving it to the right when the use presses the ← key. If we move the letter using the arrow key alone, we change the position of the letter by two pixels, but if we hold down the [CTRL] key while we move the letter using an arrow key, then we change the position of the letter by ten pixels. To implement this, the keyPressed method changes the increment to ten pixels when the user presses the [CTRL] key, and the keyReleased method changes the increment back to two pixels when the user releases the [CTRL] key.

Figure 7.4 shows, on the left, the applet of Example 7.4 initially displaying an A, and, on the right, displaying a G that the user typed and moved to the right of the applet.

To find the physical key that was pressed or released we can use the getKeyCode method, as in:

```
event.getKeyCode();
```

where event describes the key event that occurred. The getKeyCode method returns a value of type **int** which gives the virtual key code for the key that was pressed or released. For example, it returns VK_G when the user presses or releases the Ⓖ key, and VK_LEFT when the user presses or releases the ← key.

To find the character that was typed we can use the getKeyChar method, as in:

```
event.getKeyChar();
```

FIGURE 7.4 The applet of Example 7.4

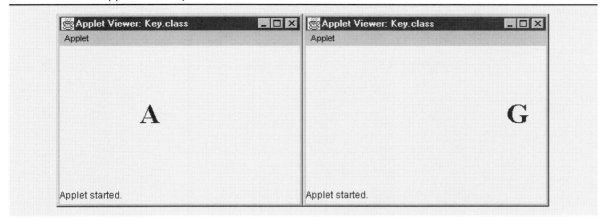

where event describes the key event that occurred. The getKeyChar method returns a value of type **char** which gives the character that was typed. For example, it will return g if the user types the Ⓖ key alone, but will return G if the user presses and releases the Ⓖ key while pressing the ⌊SHIFT⌋ key.

EXAMPLE 7.4   **Key.java**

```java
/*  Displays a key pressed by the user.  Moves the
 *  character to the right if the user presses the ->
 *  key and to the left if the user presses the <-
 *  key.  Moves ten pixels if the user holds down the
 *  Ctrl key and two pixels otherwise.
 */

import java.awt.*;
import java.awt.event.*;
import java.applet.Applet;

public class Key extends Applet implements KeyListener {          // Note 1
    public static int SLOW = 2;     // pixel change using arrow keys
    public static int FAST = 10;    // pixel change using arrow and ctrl keys
    private int x = 100,y = 100;    // position of the character displayed
    private char theKey = 'A';
    private Font f = new Font("Serif",Font.BOLD,36);
    private int deltaX = SLOW;
    public void init() {
        setFont(f);
        addKeyListener(this);                                     // Note 2
        requestFocus();                                          // Note 3
    }
    public void paint(Graphics g) {
        g.drawString(String.valueOf(theKey),x,y);               // Note 4
    }
```

```
public void keyPressed(KeyEvent event){
    int code = event.getKeyCode();                        // Note 5
    if (code == KeyEvent.VK_CONTROL) {                     // Note 6
            deltaX = FAST;
    }
    else if (code == KeyEvent.VK_RIGHT){                   // Note 7
            x += deltaX;
            repaint();
    }
    else if (code == KeyEvent.VK_LEFT) {
        x -= deltaX;
        repaint();
    }
}
public void keyReleased(KeyEvent event) {
    if (event.getKeyCode() == KeyEvent.VK_CONTROL)
        deltaX = SLOW;                                     // Note 8
}
public void keyTyped(KeyEvent event) {
    theKey = event.getKeyChar();                           // Note 9
    repaint();
}
}
```

Note 1: To handle key events a class can implement the `KeyListener` interface which consists of the methods `keyPressed`, `keyReleased`, and `keyTyped`.

Note 2: The applet registers with itself as a key listener, meaning it wants to be notified when key events occur. We could have defined another class to handle the key events.

Note 3: The applet must request the focus to receive key events. Unlike clicking the mouse, pressing a key does not associate that key press with any specific location in the applet. Any component in the applet could request the focus, but in this simple example, there are no components other than the applet itself.

Note 4: We convert the character that the user pressed (or A initially) to a string to display at the current (x,y) position.

Note 5: The `getKeyCode` method returns the number of the actual key that the user pressed. For example, pressing the G key would return VK_G from `getKeyCode`, and pressing the CTRL key would return VK_CONTROL.

Note 6: `if (code == KeyEvent.VK_CONTROL) {`

When the user presses the CTRL key, we set the increment to 10 pixels, so that if the user presses a ← or → key the letter will move by ten pixels as long as the user does not release the CTRL key.

Note 7: `else if (code == KeyEvent.VK_RIGHT){`

If the user presses the → key, we add the increment to the current value of x and ask that the applet be repainted. When Java calls the `paint` method the character will be drawn `deltaX` pixels to the right of its current position.

Note 8: `deltaX = SLOW;`

> If the user releases the [CTRL] key, we set the increment back to two pixels, so that if the user presses a [←] or [→] key the letter will move by two pixels as long as the user does not press the [CTRL] key.

Note 9: `theKey = event.getKeyChar();`

> If the user types a Unicode character then the applet will generate a `keyTyped` event and the `getKeyChar` method will return the character typed (not the key pressed). If the user presses the [G] key without pressing the [SHIFT] key, then `getKeyChar` will return g.

Just as there is a `MouseAdapter` class which provides implementations of the five methods of the `MouseListener` interface, there is a `KeyAdapter` class that provides implementations, with empty bodies, of the three methods of the `KeyListener` interface. We did not use the `KeyAdapter` class in Example 7.4.

The Big Picture
....

A component that wishes to receive key events can use the `requestFocus` method to get the focus. To handle key events, a class either implements each of the three methods of the `KeyListener` interface or extends `KeyAdapter`, overriding those methods of interest. Two of the key events, KEY_PRESSED and KEY_RELEASED, represent physical key presses, while the third, KEY_TYPED represents a Unicode character typed. The `getKeyCode` method returns the physical key code, while the `getKeyChar` method returns the character typed.

TEST YOUR UNDERSTANDING

4. List the key events that occur if the user presses the [G] key without holding down the [SHIFT] key.

TRY IT YOURSELF

5. Add `print` statements to the `keyPressed`, `keyReleased`, and `keyTyped` methods of Example 7.4 to see in what order these events are generated when the user presses the [R] key. Try this with and without holding down the [SHIFT] key.

TRY IT YOURSELF

6. Remove the `requestFocus()` from the `init` method of Example 7.4 and note what happens when you run the modified program.

7.3 ▪ Closing Windows

Thus far in our GUI examples we have not had to close any windows. Our applets are panels which are not standalone windows; they are displayed by a browser or applet viewer which is responsible for closing the window containing the applet. Preparing to create standalone GUI applications, we present the `WindowEvent` class and `WindowListener` interface that allow us to respond to window events in a **top-level window**.

Frames

Recall from our introduction to components in Section 6.1 that the `Frame` class is a container we can use as a standalone window. Frames are top-level windows, not nested inside other containers. We, not the browser or applet viewer, are responsible for creating frames and closing them when we are done using them. In this section we popup a top-level window from an applet, while in the next we will see how to write standalone applications using frames.

To create a frame, we pass a title to the `Frame` constructor

```
Frame f = new Frame("MyFrame");
```

The title will appear in the frame that comprises the window's border. To specify the size of the frame, we use the `setSize` method

```
f.setSize(150,150);
```

To make the frame visible we can call

```
f.setVisible(true);
```

while calling

```
f.setVisible(false);
```

will hide the frame. The `show` method will make a frame visible and if the frame is already visible it will bring it to the top.

Window Events and WindowListener

To listen for window events, an object must be registered as a window listener, using the `addWindowListener` method. The seven methods of the `WindowListener` interface are:

windowActivated	The window gets the focus.
windowDeactivated	The window loses the focus.
windowOpened	The window is opened.
windowClosed	The window is closed.
windowClosing	The user asks to close the window.
windowIconified	The window is minimized as an icon.
windowDeiconified	The window is restored from an icon.

In Example 7.5 we create a frame in an applet, letting the applet implement the window listener interface simply by printing a message indicating a particular window event occurred. The purpose of this example is just to observe when window events occur and to construct our first standalone frame.

Even though the applet and the frame are part of the GUI, we can still use `System.out.println` to display messages in a console window. We use `println` statements in this example to record each window event as it occurs. Using `System.out.println` is also a handy debugging technique, allowing us to print the values of variables or determine whether or not a particular line of the program is reached during execution.

Running Example 7.5 involves three windows; the first contains the applet, the second is the frame, and the third is the console window. Figure 7.5 shows the three windows after the fourth of six user actions listed in Figure 7.6.

FIGURE 7.5 The applet, frame, and console windows

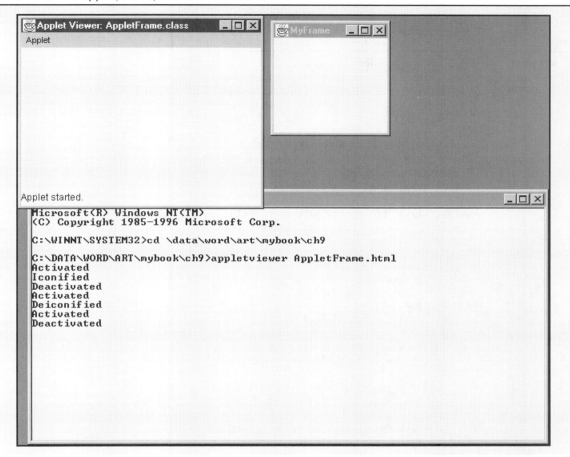

FIGURE 7.6 Six user actions and the events each generates

Event	User Action that generates the event(s).
Activated	The user started the applet from Applet Viewer. (The frame starts activated.)
Iconified Deactivated	The user clicked on the minimize button. (The frame becomes an icon and is deactivated. The applet becomes activated, but we do not listen for window events affecting the applet.)
Activated Deiconified Activated	The user clicked on the icon for MyFrame. (Clicking on the icon for the frame causes the icon to become activated and the applet to become deactivated. The icon then becomes deiconified and activated as a window.)
Deactivated	The user clicked in the applet to give it the focus.
Activated	The user clicked in MyFrame to give it the focus.
Closing Deactivated Closed	The user clicked the close button in MyFrame.

In Example 7.5, six of the seven window handling methods just print messages to show when they are invoked. The windowClosing method hides the window, and calls the dispose method to free up the system resources used for that window.

EXAMPLE 7.5 **AppletFrame.java**

```java
/*  Shows the use of a Frame as a top-level
 *  window.  Each window event causes the
 *  applet to display a message in the console.
 */

import java.awt.*;
import java.awt.event.*;
import java.applet.Applet;

public class AppletFrame extends Applet
                implements WindowListener {          // Note 1
    private Frame f;

    public void init() {
        f = new Frame("MyFrame");                    // Note 2
        f.setSize(150,150);
        f.show();
        f.addWindowListener(this);                   // Note 3
    }
    public void windowActivated(WindowEvent evt) {   // Note 4
        System.out.println("Activated");
    }
    public void windowDeactivated(WindowEvent evt) {
        System.out.println("Deactivated");
    }
    public void windowClosed(WindowEvent event) {
        System.out.println("Closed");
    }
    public void windowDeiconified(WindowEvent event) {
        System.out.println("Deiconified");
    }
    public void windowIconified(WindowEvent event) {
        System.out.println("Iconified");
    }
    public void windowOpened(WindowEvent event) {
        System.out.println("Opened");
    }
    public void windowClosing(WindowEvent event) {
        System.out.println("Closing");
        f.setVisible(false);
        f.dispose();                                 // Note 5
    }
}
```

Note 1: The applet implements the WindowListener interface, which requires that it implement the seven window event-handling methods. We will see in Example 7.6 how to use the WindowAdapter class when we do not want to handle all seven window events.

Note 2: We have to create the frame, set its size, and show it. In this example we did not use the frame, but in the next we will show we can create a subclass of Frame on which we can draw and to which we can add components.

Note 3: The applet registers itself with the frame as a listener of window events.

Note 4: Each of the seven window event-handling methods displays a message in the console, so we can see when that event occurs.

Note 5: When the user asks to close the window, we hide it to remove the display and then dispose of it, freeing up the system resources that were used for it.

WindowAdapter

In Example 7.5 we implemented all seven window event-handling methods of the WindowListener interface by printing messages to see when each event occurred. In many window applications we may only wish to respond to the user's request to close the window. Implementing the window listener interface directly in this situation requires us to provide empty bodies for the six methods which handle the six events to which we do not want to respond.

As with mouse events and key events, Java provides a WindowAdapter class containing default implementations of the seven event-handling functions that have empty bodies. We can define a subclass of WindowAdapter to handle window events and only override those methods for events we explicitly wish to handle.

In Example 7.6 we define a subclass of Frame and add a text field to it, showing that we can add components to standalone windows just as we have added them to applets and panels. When the user enters text in the text field we display it in our frame using a larger point size and a bold style. We define a subclass of WindowAdapter which overrides the windowClosing method to allow the user to close the frame but accepts the default implementations of the other six window event-handling methods inherited from the WindowAdapter class. Figure 7.7 shows the applet and the frame just after the user entered "hi there" in the text field.

FIGURE 7.7 The applet of Example 7.6

Panels and applets, which are a type of panel, use the flow layout as the default, but frames use the border layout as the default. In Example 7.6, we add the text field to our subclass of Frame in the North region.

EXAMPLE 7.6 **FrameAdapter.java**

```java
/*  Defines a subclass of WindowAdapter, overriding
 *  the windowClosing method but inheriting the default
 *  implementations of the other six window event-handling
 *  methods. Defines a subclass of Frame, adding a text
 *  field, and displaying the string the user enters.
 */

import java.awt.*;
import java.awt.event.*;
import java.applet.Applet;

public class FrameAdapter extends Applet {
    private MyFrame f;

    public void init() {
        f = new MyFrame("MyFrame");                          // Note 1
        f.setSize(150,150);
        f.show();
        f.addWindowListener(new CloseWindow());              // Note 2
    }

    class CloseWindow extends WindowAdapter {                // Note 3
        public void windowClosing(WindowEvent event) {
            f.setVisible(false);
            f.dispose();
        }
    }
}

class MyFrame extends Frame implements ActionListener{
    Font font = new Font("Serif",Font.BOLD,24);
    TextField text = new TextField(10);

    public MyFrame(String title) {                           // Note 4
        super(title);                                        // Note 5
        add(text,"North");                                   // Note 6
        text.addActionListener(this);
    }
    public void actionPerformed(ActionEvent event) {         // Note 7
        repaint();
    }
    public void paint(Graphics g) {
        g.setFont(font);
        g.drawString(text.getText(),20,100);
    }
}
```

Note 1: We need to define a subclass, MyFrame, of Frame, to override the paint method, and to add components to it.

Note 2: We create an instance of the inner CloseWindow class to handle the window events generated by our frame.

Note 3: We define the CloseWindow subclass of WindowAdapter which has default implementations of all seven window event-handling methods of the WindowListener interface. We only override the windowClosing method.

Note 4: Top-level frames add components in their constructors, in contrast to applets which add components in the init method called by the browser or applet viewer before starting the applet.

Note 5: We pass the title to the constructor of the Frame superclass.

Note 6: add(text,"North");

Frames use the border layout by default.

Note 7: public void actionPerformed(ActionEvent event)
              { repaint();}

When the user presses the (ENTER) key after entering text in the text field, we repaint the frame drawing the text entered in a larger font.

The Big Picture
....
When using a top-level window such as a Frame, we need to close it ourselves. The WindowEvent class represents seven window events. A class that handles these events can either implement the seven methods of the WindowListener interface or extend WindowAdapter, overriding the methods which handle the events of interest. To close a window, we need only override the windowClosing method.

TEST YOUR UNDERSTANDING

TRY IT YOURSELF   7. Omit the windowIconified method from Example 7.5. Compile the modified program and see what happens.

TRY IT YOURSELF   8. a. Omit the statement f.show() from Example 7.6. Run the modified program and see what happens.

  b. Change the statement f.show() in Example 7.6 to f.setVisible(true). Run the modified program and see what happens.

....

7.4 ■ Applets <-- --> Standalone Applications
• • • • • • • • • • • • • • • •

Having learned how to create frames and close them in the last section we can now write standalone applications with graphical user interfaces. We will find we can use the text and shapes we drew in applets in Chapter 5, and the components we added to applets in Chapter 6 in standalone applications. In fact, a graphical standalone application is not very much different from a graphical applet. We can easily convert an applet to a standalone application and vice versa.

Applets to Applications

To illustrate, we convert the applet, ButtonPress, of Example 7.3 to the standalone application ButtonPressFrame, which will have a main method that we call using the Java interpreter. Because we are writing event-driven programs, the main method is very simple; it just creates our subclass of Frame, sets its size, and shows it. When the user presses one of the buttons, our program responds; pressing the Print button causes the string "hi there" to be displayed, while pressing the Clear button causes it to be erased. We define the windowClosing method so clicking the button in the upper-right corner of the frame causes the program to terminate.

Figure 7.8 shows our top-level window for Example 7.7, which looks like the right window of Figure 6.8 except that this top-level window, titled MyFrame, is the one we created, whereas the applet viewer created the window of Figure 6.8.

FIGURE 7.8 The frame of Example 7.7

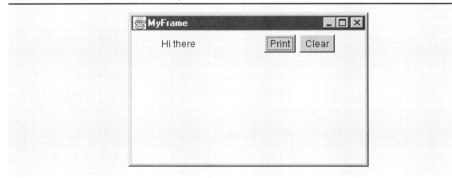

The changes from Example 6.3 to Example 7.7 in the order they appear are:

- Omit "import java.applet.Applet" because we are not using an applet.
- Extend Frame, rather than Applet, because we want a top-level window.
- Use a constructor rather than the init method of the applet.
- Add a string argument to the constructor to pass the title to the Frame. (Applets, appearing inside other windows, do not have titles.)
- Set the layout to flow layout to preserve the default layout of the applet.
- Add a window listener. (We have to close our top-level window, whereas the applet viewer or browser closed the applet.)
- Add a main method to create our subclass of Frame, set its size, and show it. (The applet viewer or browser did that for applets.)
- Add a CloseWindow subclass of WindowAdapter to handle window events.

Otherwise the code in Example 7.7 is the same as the code in Example 6.3.

EXAMPLE 7.7   **ButtonPressFrame.java**

```
/* Converts Example 6.3, ButtonPress.java
 * from an applet to a standalone
 * application.
 */
```

```
import java.awt.*;
import java.awt.event.*;

public class ButtonPressFrame extends Frame                          // Note 1
                    implements ActionListener {
    private Button print = new Button("Print");
    private Button clear = new Button("Clear");
    private Label message = new Label("Message goes here");

    public ButtonPressFrame(String title) {                          // Note 2
        super(title);
        setLayout(new FlowLayout());                                 // Note 3
        add(message);
        add(print);
        add(clear);
        print.addActionListener(this);
        clear.addActionListener(this);
        addWindowListener(new CloseWindow());                        // Note 4
    }
    public void actionPerformed(ActionEvent event) {
        Object source = event.getSource();
        if (source == print)
            message.setText("Hi there");
        else if (source == clear)
        message.setText("");
    }
    public static void main(String [ ] args) {                       // Note 5
        ButtonPressFrame f = new ButtonPressFrame("MyFrame");
        f.setSize(300,200);
        f.show();
    }

    class CloseWindow extends WindowAdapter {                        // Note 6
        public void windowClosing(WindowEvent event) {
            System.exit(0);                                          // Note 7
        }
    }
}
```

Note 1: To make a top-level window to which we can add components, we extend the Frame class.

Note 2: We put the code in the applet's init method in our frame's constructor. In addition we pass a title to the constructor of the superclass, Frame.

Note 3: The applet had a flow layout by default. To keep that same flow layout we need to set it explicitly because frames have a default border layout.

Note 4: We have to handle the window closing that the applet viewer or the browser handled for the applet.

Note 5: The main method creates our frame, sets its size, and shows it. User events such as button presses and window closing drive our program after the main method has set up the frame.

Note 6: `class CloseWindow extends WindowAdapter {`

The inner class `CloseWindow` overrides the `windowClosing` method of the `WindowAdapter` class so the user can close the window.

Note 7: `windowClosing(WindowEvent event) {`

When the user closes the window we terminate the program, returning control to the operating system.

Console Application to GUI Application

We can convert a standalone application to an applet even more easily than we converted the applet of Example 6.3 to the standalone application of Example 7.7. However we have not yet written any standalone GUI programs that we can convert to applets, so we first convert Example 1.7, `Growth.java`, to a GUI application, Example 7.8, `GrowthFrame.java`.

In Example 1.7 we input the interest rate, the beginning account balance, and the term of the account, then output the amount of the balance at the end of the term. We make a GUI similar to that of Example 6.19 in which we put the user interface components in the North region and drew on a canvas in the center. In the `Growth` program we input three values; to input three values graphically we can use three text fields, with each text field having a label.

Six components, the three labels and the three text fields, would be too wide to fit in one row side by side, so we put each text field and its label into a panel with a 2x1 grid layout, so the text field will appear below its label. We add the three panels to a containing panel which we add to the North region. We do not want to compute the resulting balance until the user has entered the desired values in all three text fields, so rather than handling the action events generated by the text fields, we add a button in the South region that the user can press when all the data has been entered.

Figure 7.9 shows the standalone graphical application of Example 7.8.

FIGURE 7.9 The standalone application of Example 7.8

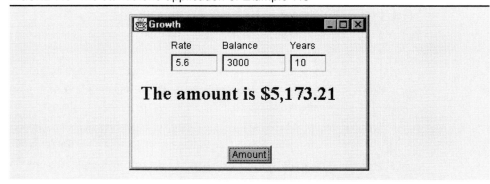

When the user presses the button, we use the `getText` method to get the text from each of the three text fields. The text in a text field has type `String`, so we need to convert the entries to type **double** for the interest rate and the balance, and to type **int** for the years. As in Section 6.2, we use the expression

```
double rate = new Double(getRate.getText()).doubleValue();
```

to convert the rate to a **double**. To convert the years from String to **int** we use the analogous expression

```
int years = new Integer(getYears.getText()).intValue();
```

In Example 7.8, the user can change one or more values and press the button again to calculate the amount that will result if the specified balance earns interest at the specified rate for the specified number of years. In addition to notes, we annotate the code with change items which identify the simple changes needed to convert this standalone application to an applet.

EXAMPLE 7.8   **GrowthFrame.java**

```
/* Revises Example 1.7, Growth.java, to make it a graphical
 * rather than a console program.
 */

import java.awt.*;
import java.awt.event.*;
import java.text.NumberFormat;
                                                                // Change 1
public class GrowthFrame extends Frame                          // Change 2
                implements ActionListener {
    private TextField getRate = new TextField(5);
    private TextField getBalance = new TextField(8);
    private TextField getYears = new TextField(3);
    private Label rate = new Label("Rate");
    private Label balance = new Label("Balance");
    private Label years = new Label("Years");
    private MyCanvas canvas = new MyCanvas();
    private Button button = new Button("Amount");
    private String amount = "Press button";                     // Note 1
    private NumberFormat nf;

    public GrowthFrame(String title) {                          // Change 3
        super(title);                                           // Change 4
        Panel p1 = new Panel();
        p1.setLayout(new GridLayout(2,1));
        p1.add(rate);
        p1.add(getRate);

        Panel p2 = new Panel();
        p2.setLayout(new GridLayout(2,1));
        p2.add(balance);
        p2.add(getBalance);

        Panel p3 = new Panel();
        p3.setLayout(new GridLayout(2,1));
        p3.add(years);
        p3.add(getYears);

        Panel p = new Panel();
        p.add(p1);
        p.add(p2);
        p.add(p3);
```

continued

```
            Panel p4 = new Panel();
            p4.add(button);
            add(p,"North");
            add(canvas,"Center");
            add(p4,"South");
            addWindowListener(new CloseWindow());            // Change 5
            button.addActionListener(this);                  // Note 2
            nf = NumberFormat.getCurrencyInstance();
        }

    public String computeGrowth() {                          // Note 3
        double rate = new Double(getRate.getText()).doubleValue();
        double balance = new Double(getBalance.getText()).doubleValue();
        int years = new Integer(getYears.getText()).intValue();
        for (int i = 1; i <= years; i++)
            balance += balance * rate / 100;
        return nf.format(balance);                           // Note 4
    }

    public void actionPerformed(ActionEvent event) {         // Note 5
        amount = "The amount is " + computeGrowth();
        canvas.repaint();
    }

    public static void main(String [ ] args) {               // Change 6
        GrowthFrame f = new GrowthFrame("Growth");
        f.setSize(300,200);
        f.show();
    }

    class CloseWindow extends WindowAdapter {                // Change 7
        public void windowClosing(WindowEvent event) {
           System.exit(0);
        }
    }

    class MyCanvas extends Canvas {
        public MyCanvas() {
            Font f = new Font("Serif",Font.BOLD,24);
            setFont(f);
        }
        public void paint(Graphics g) {
            g.drawString(amount,10,30);
        }
    }
}
```

Note 1: The applet will display the string "Press button" when it starts running.

Note 2: We only listen for action events generated by button presses, not text fields. The user presses the button after entering the desired values in each of the three text fields.

Note 3: The computeGrowth method gets the text from the text fields, converting the rate and balance to type **double** and the years to type **int**, and then computing the amount of the balance at the end of the term.

Note 4: We format balance as currency.

Note 5:   When the user presses the button, we call the `computeGrowth` method to do the computation, and ask to repaint the canvas which will draw a string showing the resulting amount.

Standalone Application to Applet

We can easily convert the standalone application of Example 7.8 to an applet, mostly by omitting lines. We show the necessary changes, referring to the change items annotating Example 7.8.

Change 1:   Add the line `import java.applet.Applet` here.

Change 2:   Replace with `public class GrowthApplet extends Applet`

Change 3:   Replace with `public void init() {`

Change 4:   Replace with `setLayout(new BorderLayout());` to preserve the border layout which is the default for the frame of the standalone application.

Change 5:   Omit this line, as the browser or applet viewer handles the windows.

Change 6:   Omit the `main` method, because an applet is called from a browser or applet viewer.

Change 7:   Omit the `CloseWindow` class because we do not need to handle window events.

Because there are so few changes to Example 7.8 needed to convert it to an applet we do not print the full code here, but provide the `GrowthApplet.java` file on the disk that accompanies this text. Running the applet gives a user interface that looks just like Figure 7.9.

The Big Picture

We started graphical user interfaces using applets, but the same drawing and user interface components can be used in standalone applications. A standalone application extends `Frame` rather than `Applet`, uses a constructor rather than an `init` method, and must close the window itself. It is easy to convert a standalone application to an applet or an applet to a standalone application.

TEST YOUR UNDERSTANDING

TRY IT YOURSELF   9. Modify Example 7.7 to omit adding a window listener and the `CloseWindow` class. Run the modified program to see what happens. (To abort the program, hold the `CTRL` key and hit the `C` key in the console window.)

TRY IT YOURSELF   10. Another way to convert a numerical string s to a double d uses the `valueOf` method, as in

```
d = Double.valueOf(s).doubleValue();
```

Modify Example 7.8 to use this method to convert the rate and balance strings entered in the text fields to type **double**. Run the modified example to see that it still works properly.

ENHANCEMENT

7.5 ▪ Tangrams with the Mouse and Keys

We return to the tangram puzzle we set up in Example 5.7. To solve tangram puzzles on the computer we need to drag polygons to new locations and rotate them. We use the mouse to drag polygons and key presses to rotate them.

Defining the Problem

In this section we will use the mouse to translate a polygon, as we did in Example 7.3, use the Ⓑ key (for back) to rotate a polygon counterclockwise, and use the Ⓕ key (for forward) to rotate the polygon clockwise. With these translation and rotation operations we can move the puzzle pieces to form other designs. We leave to the exercises the implementation of an operation to reflect the parallelogram, in effect flipping it over, to allow the user to form even more shapes.

The ability to rotate a polygon is an important piece of the solution.

Designing a Solution

Solving a Subproblem—Rotating a Polygon

To **rotate** a polygon we **translate** the polygon to center it at the origin. We then rotate each of the vertices (x,y) of the polygon by an angle z using the formulas

```
x = (cos z)x - (sin z)y
y = (sin z)x + (cos z)y
```

where sin z represents the sine of the angle z and cos z represents its cosine. Using a positive angle will cause the polygon to rotate clockwise, while using a negative angle will cause it to rotate counterclockwise. Finally we translate the rotated polygon back to its original position.

While the formulas define the correct values mathematically for the rotated points, they cause a problem for our implementation. The cosine and sine methods return **double** values so the x and y values of the exact rotated point will have **double** values. For example the triangle we use has points (50,100), (100,50), and (150,100). Translating the triangle to be centered about the origin changes its coordinates to (−50,25), (0,−25), and (50,25). Rotating counterclockwise by .05 radians (just less than three degrees) changes these points to (−48.68, 27.46), (−1.24, −24.96), and (51.18,22.46) which do not have integer values for the coordinates. We must either truncate or round to get integer pixel values that we can plot.

If we truncate the rotated points to (−48,27), (−1,−24), and (51,22), then the triangle gets slightly smaller, and as we repeat a few dozen times to rotate by a larger angle, the triangle will get smaller and smaller, eventually disappearing. We can round, rather than truncating, which keeps the triangle about the same size, but after dozens of repetitions, its shape will be a bit distorted. In the above example rounding would give the points (−49,27), (−1,−25), and (51,22).

We need to be able to rotate by a small angle to form our polygons into various shapes, but only rotating by a small angle will sometimes take too many rotations, and the roundoff error will cause the polygons to become distorted. We need to rotate by a larger angle until we get close to the final angle, and then rotate by a smaller angle to fit the polygon into the new shape. By using a larger angle first, we will reduce the number of rotations, and thus reduce the distortion to an acceptable level.

In Example 7.9, pressing the uppercase B or F will rotate the polygon by pi/6 radians (30 degrees) while pressing the lowercase b or f will rotate the polygon by pi/60 radians (3 degrees). We use the 30-degree rotation to get close to the desired total rotation angle, and the 3-degree rotation to finish. By not using too many rotations, we will not cause too much distortion.

Figure 7.10 shows the triangle when the applet starts, and then after the user has rotated it counterclockwise a few times.

FIGURE 7.10 The applet of Example 7.9

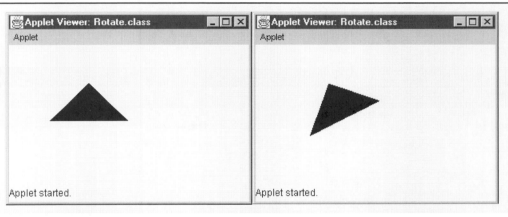

We initialize a point (rotx, roty) to the center of the triangle; this will be the point which will move to the origin when we translate the triangle before rotating it about the origin and translating it back. A polygon object has an npoints field which gives the number of points in a polygon, and xpoints and ypoints fields which are arrays that give the x- and y-coordinates of the polygon's points.

EXAMPLE 7.9   **Rotate.java**

```
/* Displays a triangle which the user can
 * rotate counterclockwise by pressing the
 * B key, and clockwise by pressing the F key.
 */

import java.awt.*;
import java.awt.event.*;
import java.applet.Applet;

public class Rotate extends Applet implements KeyListener {

    private int [] xcoord = {50,100,150};
    private int [] ycoord = {100,50,100};
    private Polygon p = new Polygon(xcoord,ycoord,3);
```

continued

```
    public static final double PCOS = Math.cos(Math.PI/60);        // Note 1
    public static final double NCOS = Math.cos(-Math.PI/60);
    public static final double PSIN = Math.sin(.Math.PI/60);
    public static final double NSIN = Math.sin(-.Math.PI/60);
    public static final double BIGPCOS = Math.cos(Math.PI/6);
    public static final double BIGNCOS = Math.cos(-Math.PI/6);
    public static final double BIGPSIN = Math.sin(Math.PI/6);
    public static final double BIGNSIN = Math.sin(-Math.PI/6);
    private int rotx = 100;        // x-coordinate of the center
    private int roty = 75;         // y-coordinate of the center

    public void init() {
        addKeyListener(this);
        requestFocus();                                            // Note 2
    }
    public void paint(Graphics g) {
        g.fillPolygon(p);
    }
    public void keyTyped(KeyEvent event) {
        char key = event.getKeyChar();
        double C;
        double S;
        switch(key) {                                              // Note 3
            case 'b' : C = NCOS; S = NSIN; break;
            case 'f' : C = PCOS; S = PSIN; break;
            case 'B' : C = BIGNCOS; S = BIGNSIN; break;
            case 'F' : C = BIGPCOS; S = BIGPSIN; break;
            default  : return;
        }
        p.translate(-rotx,-roty);                                  // Note 4
        for (int i=0; i<p.npoints; i++) {                          // Note 5
            int x = p.xpoints[i];
            int y = p.ypoints[i];
            p.xpoints[i] = (int)Math.round(x*C - y*S);             // Note 6
            p.ypoints[i] = (int)Math.round(x*S + y*C);
        }
        p.translate(rotx,roty);                                    // Note 7
        repaint();
    }
    public void keyReleased(KeyEvent event) { }
    public void keyPressed(KeyEvent event) { }
}
```

Note 1: It is generally more efficient to compute the sine and cosine values once rather than each time we press a key to rotate the triangle. We rotate by an angle of Math.PI/60 radians which is three degrees. (There are pi radians in 180 degrees.)

Note 2: To receive key events a component must request the focus.

Note 3: If the user types b we use the negative angle, -Math.PI/60, while if the user types f we use the positive angle, Math.PI/60. Typing F causes a rotation by Math.PI/6, while typing B causes a rotation by -Math.PI/6. If the user presses any other key we return immediately. By choosing the angle here we can use the same rotation formula for both forward and backward rotations.

Note 4: We translate the triangle so its center (rotx,roty) moves to the origin.

Note 5: This loop rotates each point of the polygon by the given angle.

Note 6: `p.xpoints[i] = (int)Math.round(x*C - y*S);`

The sine and cosine values have type **double**, so the rotation formulas compute **double** values. We use integer valued pixels so we need to cast the result to an **int**. If we did not round first we would keep underestimating the size of the rotated vector, so the triangle would gradually get smaller. (See Test Your Understanding question 12.) Rounding sometimes rounds up, overestimating the pixel values of the rotated point, and sometimes rounds down, underestimating the pixel values. On the average these estimations balance and the triangle stays the same size, but may be slightly distorted.

Note 7: `p.translate(rotx,roty);`

We translate the rotated triangle back to its original position.

Completing the Java Code

We combine the setup of the polygons forming the square from Example 5.7, the dragging a polygon with the mouse from Example 7.3, and the rotating a polygon from Example 7.9 to write a program to transform the polygons into other shapes.

As in Example 7.3 we store the current location of the polygon when the user presses the mouse in it, but here we have to loop through the seven polygons to find in which one, if any, the user pressed the mouse.

If the user releases the mouse inside one of the polygons, we save the point where the user releases the mouse as the new center (approximately) of the polygon in case the user decides to rotate that polygon. We also save the index of that polygon which tells us which polygon to rotate if the user types b, B, f, or F. Usually the user will rotate the polygon just after dragging it, but the user can rotate another polygon by clicking the mouse in it which will generate a mouse released event.

EXAMPLE 7.10   **TangramSolver.java**

```
/* Start with the seven polygons forming
 * a square. Drag them with the mouse, and rotate
 * them with the F and B keys to form other shapes.
 */

import java.awt.*;
import java.awt.event.*;
import java.applet.Applet;

public class TangramSolver extends Applet
                     implements MouseMotionListener, KeyListener {
    public static final double PCOS = Math.cos(Math.PI/60);
    public static final double NCOS = Math.cos(-Math.PI/60);
    public static final double PSIN = Math.sin(Math.PI/60);
    public static final double NSIN = Math.sin(-Math.PI/60);
    public static final double BIGPCOS = Math.cos(Math.PI/6);
```
continued

```
public static final double BIGNCOS = Math.cos(-Math.PI/6);
public static final double BIGPSIN = Math.sin(Math.PI/6);
public static final double BIGNSIN = Math.sin(-Math.PI/6);
int [ ][ ] x = {{0,100,200},{100,200,200},{0,100,0},{0,50,0},
                {50,150,100},{0,50,100,50},{100,200,150,50}};
int [ ][ ] y = {{200,100,200},{100,0,200},{0,0,100},{100,150,200},
                {50,50,100},{100,50,100,150},{0,0,50,50}};
Color [ ] colors = {Color.red,Color.blue,Color.yellow,Color.magenta,
                       Color.cyan,Color.pink,Color.orange};
Polygon [ ] polygons = new Polygon[7];
int [ ] oldx = new int[7];          // Saves previous polygon position
int [ ] oldy = new int[7];
int rotate;                         // Index of polygon to rotate
int rotx;                           // x-coordinate of a point inside
int roty;                           // y-coordinate of a point inside

public void init() {
    for (int i = 0; i < x.length; i++) {
        polygons[i] = new Polygon(x[i], y[i], x[i].length);
    }
    addMouseListener(new MousePressListener());
    addMouseMotionListener(this);
    addKeyListener(this);
    requestFocus();
}
public void mouseMoved(MouseEvent event) { }
public void mouseDragged(MouseEvent event) {
    int x = event.getX();
    int y = event.getY();
    for (int i = 0; i < 7; i++){
        if (polygons[i].contains(x,y)){                              // Note 1
            polygons[i].translate(x - oldx[i], y - oldy[i]);
            oldx[i] = x;
            oldy[i] = y;
            repaint();
            break;                                                   // Note 2
        }
    }
}

class MousePressListener extends MouseAdapter {
    public void mousePressed(MouseEvent event) {
        int x = event.getX();
        int y = event.getY();
        for (int i = 0; i < 7; i++)
            if (polygons[i].contains(x,y)){                          // Note 3
                oldx[i] = x;
                oldy[i] = y;
                break;
            }
    }
    public void mouseReleased(MouseEvent event) {
        int x = event.getX();
        int y = event.getY();
        for (int i = 0; i < 7; i++)
            if (polygons[i].contains(x,y)){                          // Note 4
```

```
                    rotate = i;
                    rotx = x;
                    roty = y;
                    break;
                }
            }
        }
    }

    public void showPolygon                                    // Note 5
                (Polygon p, Color c, Graphics g, boolean fill) {
        Color oldColor = g.getColor();
        g.setColor(c);
        if (fill)
            g.fillPolygon(p);
        else
            g.drawPolygon(p);
        g.setColor(oldColor);
    }
    public void paint(Graphics g) {
        for (int i = 0; i < polygons.length; i++)
            showPolygon(polygons[i], colors[i], g, true);
    }
    public void keyPressed(KeyEvent event) {
        char key = event.getKeyChar();
        double C = PCOS;
        double S = NCOS;
        switch(key) {
            case 'b' : C = NCOS; S = NSIN; break;
            case 'f' :  C = PCOS; S = PSIN; break;
            case 'B' : C = BIGNCOS; S = BIGNSIN; break;
            case 'F' : C = BIGPCOS; S = BIGPSIN; break;
            default  : return;
        }
        polygons[rotate].translate(-rotx,-roty);              // Note 6
        for (int j = 0; j < polygons[rotate].npoints; j++) {
            int x = polygons[rotate].xpoints[j];
            int y = polygons[rotate].ypoints[j];
            polygons[rotate].xpoints[j] = (int)Math.round(x*C - y*S);
            polygons[rotate].ypoints[j] = (int)Math.round(x*S + y*C);
        }
        polygons[rotate].translate(rotx,roty);
        repaint();
    }
    public void keyReleased(KeyEvent event) { }
    public void keyTyped(KeyEvent event) { }
}
```

Note 1: We perform the same translation steps as in Example 7.3, but first we check to see which polygon the mouse is dragging. The point (x,y) is the point to which the user dragged the mouse. It does not take much motion to generate a drag event (See Test Your Understanding question 3) so (x,y) will still be inside the same polygon in which the user pressed the mouse to start dragging it.

Note 2: Once we find the polygon, we break out of the loop as there is no need to check any of the other polygons.

Note 3: Again as in Example 7.3, we save the position at which the user presses the mouse, if that position is contained in one of the seven polygons.

Note 4: When the user releases the mouse inside a polygon, we save the index of that polygon and the coordinates of that mouse release. We will rotate this polygon if the user types b, f, B, or F.

Note 5: We include this method from Example 5.7 which allows us to draw the polygons with different colors, and to choose whether or not to fill them. We might prefer to draw the polygons unfilled to reduce the flicker that comes from repainting the polygons on a white background. (See Test Your Understanding question 11.)

Note 6: `polygons[rotate].translate(-rotx,-roty);`

We rotate as in Example 7.9, using the saved index, `rotate`, and the position inside polygon i, (rotx,roty), that we saved when the user released the mouse.

Testing the Code

Figure 7.11 shows the polygons dragged and rotated into the shape of a cat. Test Your Understanding questions 13 and 14 provide other tests of the tangram solver.

FIGURE 7.11 The applet of Example 7.10

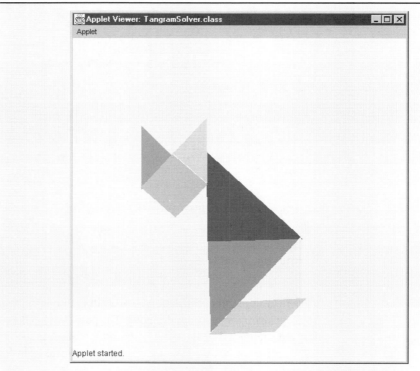

The Big Picture

We use the mouse to drag polygons and the keyboard to rotate them. Rotating distorts the polygon slightly due to the limitations of a fixed number of pixels. By using both small and large rotations we can minimize this distortion, dragging and rotating polygons to solve tangram puzzles.

TEST YOUR UNDERSTANDING

TRY IT YOURSELF

11. Modify Example 7.10 by changing the last argument in the call to showPolygons in the paint method from **true** to **false** so we draw the polygons without filling them. Also change the second argument to Color.black. Run the modified program to see these changes greatly reduce the amount of flicker.

TRY IT YOURSELF

12. Modify Example 7.9 to remove the two calls to Math.round from the mousePressed method computing the rotated point as

```
polygons[rotate].xpoints[j] = (int)(x*C - y*S);
polygons[rotate].ypoints[j] = (int)(x*S + y*C);
```

Rotate the triangle many times and see what happens.

TRY IT YOURSELF

13. In Example 7.10, in the mouseReleased method, save the coordinates (x,y) in rotx and roty before checking if (x,y) is inside a polygon. Run the modified program to see what happens. Test rotations carefully.

TRY IT YOURSELF

14. Run the tangram solver of Example 7.10 to transform the seven polygons into the boat shown in Figure 7.12.

FIGURE 7.12 Form the seven polygons into this boat

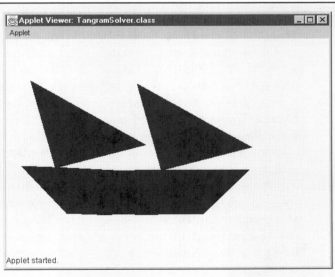

TRY IT YOURSELF

15. Run the tangram solver of Example 7.10 to transform the seven polygons into the duck shown in Figure 7.13.

FIGURE 7.13 Form the seven polygons into this duck

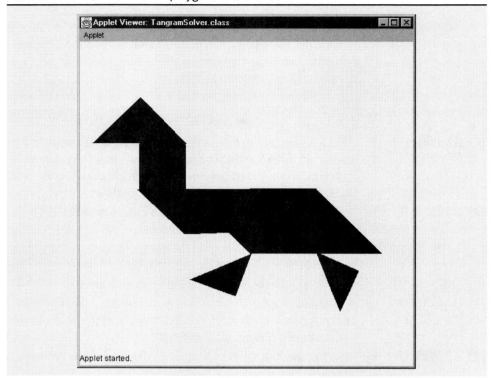

Chapter 7 Summary

The ActionListener and ItemListener interfaces of Chapter 6 each have only one method, but the interfaces for the low-level mouse, key, and window events of this chapter each have several methods. The MouseListener interface has the mousePressed, mouseReleased, mouseClicked, mouseEntered, and mouseExited methods. A class that implements the MouseListener interface must implement each of these methods, but may use an empty body for those methods which handle events in which it is not interested. Another approach to handling mouse events is to extend the MouseAdapter class which provides default implementations for the five events. By extending the MouseAdapter class, we only have to implement the methods for those events we wish to handle.

The MouseMotionListener interface specifies two methods, mouseDragged and mouseMoved. A class can implement this interface or extend the MouseMotionAdapter class. Java generates the mouse moved event when the user moves the mouse without holding down a button, while the mouse dragged event represents the user moving the mouse with a button held down. A mouse click occurs when the user presses and releases the mouse without dragging it in between.

An implementor of the KeyListener interface defines the keyPressed, keyReleased, and keyTyped methods. We refer to physical keys using integer codes such as VK_A or VK_SHIFT. Pressing the F key will generate a keyPressed event; the keyCode method will return VK_F for this event. If the user also holds down the [SHIFT] key, then the keyTyped method would return the F character, but if the user does not hold down the [SHIFT] key, then the keyTyped event would return the character f. A class that wants to handle key events can implement the three methods of the keyListener interface or subclass the KeyAdapter class, overriding only those methods for events that it wants to handle.

Frames are essential to making standalone applications. A frame is a standalone window, not a panel in the applet viewer or the browser. If we use frames, we are responsible for showing them, defining their size, and closing them properly. The WindowListener interface has seven methods, windowActivated, windowDeactivated, windowIconified, windowDeiconified, windowOpened, windowClosed, and windowClosing. If we just want to close the window, we can extend the WindowAdapter class, overriding only the windowClosing method.

Once we know how to close a frame, we can write standalone graphical applications. Applets and applications use the same graphical components, but a standalone application has to make its window visible and handle window events. We easily converted an applet to an application. After converting a standalone program to have a graphical user interface, we then converted this standalone graphical program to an applet.

In our Enhancement section we used the mouse to drag and keystrokes to rotate the polygons of the tangram puzzle into various shapes.

Build Your Own Glossary

Find the uses, in this chapter, of the terms below. Enter a definition of each in the glossary.txt file on the disk that comes with this text.

adapter class	mouseDragged	windowActivated
focus	mouseEntered	windowClosed
keyPressed	mouseExited	windowClosing
keyReleased	mouseMoved	windowDeactivated
keyTyped	mousePressed	windowDeiconified
low-level event	mouseReleased	windowInconified
mouseClicked	top-level window	windowOpened

Skill Builder Exercises

1. The following, showing the key events generated when the user types an uppercase T, does not list these events in they order in which they occur. Rearrange the list to reflect the order in which they are generated.

   ```
   KEY_TYPED, KEY_PRESSED, KEY_RELEASED, KEY_PRESSED, KEY_RELEASED
   ```

2. Fill in the blanks in the code below to create an applet that draws a red circle of radius 25 centered where the user clicks the mouse.

   ```
   import java.awt._____;
   import java.awt._____;
   import java.applet._____;
   ```

continued

```
public class MouseClickRed _____  {
    int x = 25, y = 25;
    public _____() {
        setForeground(_____);
        addMouseListener(new _____);
    }
    public void paint(Graphics g) {
        g.fillOval(x - 25,y - 25,_____ ,_____ );
    }
    class MouseHandler extends MouseAdapter {
        public void _____ (_____) {
            x = _____;
            y = _____;
            _____
        }
    }
}
```

3. Fill in the blanks in the code below to create a standalone application that draws a red
    circle of radius 25 centered where the user clicks the mouse.

```
import java._____;
import java._____;
public class MouseClickRedAlone _____ F_____ {
    int x = 25, y = 25;
    public _____() {
        setForeground(_____);
        add_____Listener(new _____());
        add_____Listener(new _____());
    }
    public void paint(Graphics g) {
        g.fillOval(x - 25,y - 25,_____,_____);
    }
    class MouseHandler extends MouseAdapter {
        public void _____(_____) {
            x = _____;
            y = _____;
            _____;
        }
    }
    class WindowClose extends WindowAdapter {
        public void _____(_____) {
            _____;
        }
    }
    public _____ void main(_____) {
        _____ m = new _____();
        m._____;
        m._____;
    }
}
```

Critical Thinking Exercises

4. The MouseAdapter class
   a. implements the five methods of the MouseListener interface.
   b. allows the user to override any of the MouseListener or MouseMotionListener methods.
   c. can be used by applets, but not by standalone applications.
   d. none of the above.
   e. all of the above.

5. Invoking event.getKeyCode(), where event is a KeyEvent will return
   a. the character that the user typed, even if it required pressing two keys, for example typing an uppercase G.
   b. an integer representing the character typed, so a lowercase g will be distinguished from an uppercase G.
   c. an integer representing the physical key generating the event.
   d. none of the above.

6. Which of the following are steps we need to perform when converting an applet to a standalone application? More than one choice may be correct.
   a. Remove "extends Applet"
   b. Add a call to the addWindowListener method
   c. Add a main method
   d. All of the above

7. To handle a window closing event, a class could
   a. implement the WindowListener interface, overriding the windowClosing method.
   b. extend the WindowAdapter class, overriding the windowClosing method.
   c. directly implement the windowClosing method, without implementing the WindowListener interface or extending the WindowAdapter class.
   d. none of the above.
   e. all of the above.

Debugging Exercise

8. The following standalone application attempts to increase the radius of a circle by three pixels when the user presses the Ⓑ key and decrease it by three pixels when the user presses the Ⓢ key. Find and correct any errors.

```
import java.awt.*;
import java.awt.event.*;
public class CircleSize extends Frame{
    int width = 50, height = 50;
    int x = 140, y = 90;
    public CircleSize()  {
        addKeyListener(this);
        addWindowListener(this);
    }
```
continued

```
public void paint(Graphics g) {
    g.drawOval(x,y,width,height);
}
public void keyPressed(KeyEvent e) {
    int i = e.getKeyCode();
    if (i == KeyEvent.VK_B) {width += 6; height += 6; x -= 3; y -= 3;}
    else if (i == KeyEvent.VK_S) {width -= 6; height -= 6; x += 3; y += 3;}
    repaint();
}
public void keyTyped(KeyEvent e) {}
    public static void main(String[] args) {
        CircleSize cs = new CircleSize();
        cs.setSize(300,200);
        cs.show();
    }
class WindowHandler extends WindowAdapter {
    public void widowClosing(WindowEvent e) {
        System.exit(0);
    }
}
}
}
```

Program Modification Exercises

9. Modify Example 7.3 to define a subclass of MouseMotionAdapter to handle mouse motion events rather than having the applet implement the MouseMotionListener interface.

10. Modify Example 7.3 so the user cannot drag the triangle out of the visible region of the applet.

11. Modify Example 7.4 to move the character using the ⬆ and ⬇ keys in addition to moving it with the ⬅ and ➡ keys.

12. Modify Example 7.4 to ensure that the character stays within the visible region of the applet.

13. Modify Example 7.6 to draw on a canvas, rather than directly in the frame itself.

14. In Example 7.8 we create three panels, each time setting the layout to the grid layout, and adding a label and a text field. Modify Example 7.8, writing a method to create and initialize these panels, and then calling this method three times when building the user interface.

PUTTING IT ALL TOGETHER
15. Convert the applet of Example 7.5 to a standalone application.

PUTTING IT ALL TOGETHER
16. Convert the insertion sort applet of Example 6.20 to a standalone application.

17. Convert the tangram solver of Example 7.10 to a standalone program.

18. Modify the tangram solver of Example 7.10 so the parallelogram flips over when the user presses the ⓡ key. Figure 7.14 shows the positions of the parallelogram before and after the flip.

FIGURE 7.14 The parallelogram positions

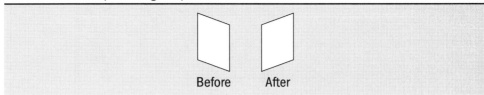

Before   After

Program Design Exercises

19. Write an applet that displays a red oval which turns green when the user moves the mouse over it and turns back to red when the user moves the mouse outside of the oval.

20. Write an applet that causes a blue rectangle to appear where the user clicks the mouse, and to disappear when the user double-clicks the mouse inside of it. (Use the getClickCount() method of the MouseEvent class to get the click count. A double-click is two mouse clicks in quick succession.)

21. Write an applet that prints the user's name wherever the user clicks the mouse.

22. Write an applet that displays a circle which changes its color to red when the user presses the Ⓡ key, to yellow when the user presses the Ⓨ key, to blue when the user presses the Ⓑ key, and to green when the user presses the Ⓖ key.

23. Write an applet that displays a string, changing it to bold if the user presses the Ⓑ key, to italics if the user presses the Ⓘ key, to all uppercase if the user presses the SHIFT key, and to all lowercase if the user presses the CTRL key.

PUTTING IT ALL TOGETHER

24. Design a graphical user interface for the QuickFood application of Example 2.7. Let the user order by clicking the mouse on icons for a burger, fries, and a drink. Write this program as a standalone application.

25. Write a Java applet to play the game of Nim with the computer. To play Nim, start with a certain number of tokens. Each player takes from one up to a maximum number of tokens at each turn, and the player taking the last token loses. Provide a GUI to input the total number of tokens, and the maximum number each player can take at each move. Randomly choose whether the user or the computer moves first. At startup, display all the tokens. Let the user select a token with a mouse click which should cause that token to be erased. When finished selecting tokens, the user should press the Ⓝ key. The computer should print a message stating how many tokens it chooses, and these tokens should be erased. When the game is over, display a message announcing the winner. Use colors to enhance the display.

26. Write a Java applet to play a game of Tick Tack Toe with the computer. You may use the interface of Figure 7.1. Let the player move by clicking the mouse in an available square.

   a. For an easier version, let the computer make its move in any available square.
   b. For a more challenging program, have the computer find its best move.

8

Using Data

In this chapter we delve into techniques for handling data in Java. Before we can use data files in Java, we must explore exceptions which signal data errors. Exception handling enables us to manage serious errors that might occur, such as trying to read from a file that does not exist. It is essential for implementing input and output operations and networking where errors beyond the control of the programmer may easily occur.

Thus far we have hidden the details of exception handling needed to use files inside the methods of our Io class found on the disk included with this text. After we discuss exception handling, we show how to read from and write to external files. Files persist after our program is finished, keeping data for later use. Business applications, and many others, are heavily dependent on good access to external data. Because handling data carefully is important, it is not easy to do it well. We left it until this chapter when we have developed sufficient background to tackle these concepts.

Inside our programs we have passed data as arguments to individual methods but not to the main method called at the start of execution or to applets. We will show how to use program arguments to pass data to the main method of a standalone application, and to use a tag in the HTML file to pass data to an applet.

In this chapter we introduce recursion, another approach to repetition. In recursion, instead of spelling out each step of the repetition the way loops do, we do one step and call the recursive method again to complete the remaining steps. We illustrate recursion with two important data processing applications, searching and sorting.

Data structures allow us to organize data for efficient processing. In this chapter we introduce two of the most important data structures, linked lists and stacks. In contrast to an array which stores its elements together, a linked list uses a link to refer to the location of its next item, making it easier to add and remove elements but harder to search for them. Choosing the right data structure is an engineering decision based on the requirements of the problem being solved.

Programmers familiar with C or C++ might be familiar with the use of pointers to create linked lists. Java does not need explicit pointers. Each object variable contains a reference to an object, so we can easily create linked lists in Java. Our linked list class includes operations to insert and remove items from a list, and other useful operations.

The stack in computer science is like a stack of books; putting data on a stack has the last in, first out (LIFO) property that the last item placed on it is the first removed. We implement our stack using an array, but could make a more flexible stack using a linked list to hold its elements.

OBJECTIVES

- ◆ Handle exceptions.
- ◆ Read and write text files using reader and writer classes.
- ◆ Input and output binary files.
- ◆ Read and write objects to and from files.
- ◆ Pass arguments to the `main` method of a standalone application.
- ◆ Pass arguments to an applet.
- ◆ Use recursive methods.
- ◆ Be familiar with binary search and merge sort algorithms.
- ◆ Use Java for the linked list and stack data structures.

OPTIONS

- ◆ Omit 8.5: Recursion: Searching and Sorting
- ◆ Omit 8.6: Data Structures: Linked Lists
- ◆ Omit 8.7: Data Structures: Stacks

8.1 ▪ Exception Handling

We do not want our programs to crash or to produce erroneous results. When inputting a test score we can check that the value entered was between 0 and 100. We can easily include this check on the value of the score in our program, but sometimes we have no control over circumstances that might affect our program. For example, if our program tries to read from a file that does not exist, it may abort. Someone else may have deleted that file without our knowledge, or the disk drive may have failed.

Java provides an exception handling facility to allow the programmer to insert code to handle such unexpected errors and allow the program to recover and continue executing, or to terminate gracefully, whichever is appropriate. An **exception** signals that a condition such as an error has occurred. We **throw** an exception as a signal, and **catch** it to handle it and take appropriate action. We will not cover all of the features of exception handling in this text, but will show how to handle exceptions that Java generates.

Exception Classes

In Java, exceptions are instances of a class derived from `Throwable`. Figure 8.1 shows the exception classes that we discuss in this chapter (the ... indicates a class that we do not use is omitted from the display.)

FIGURE 8.1 Classes of exceptions

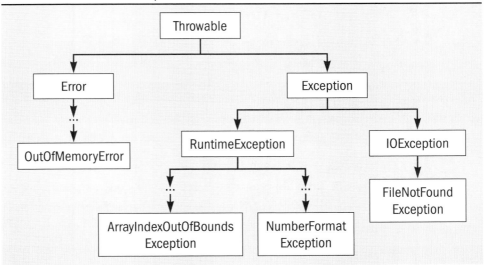

In this section we consider array index out of bounds and number format exceptions, leaving the IO exceptions to the next section. Java allows us to handle array index out of bounds and number format exceptions, but does not require us to handle them. Java requires that we handle IO errors, usually beyond our control, which would otherwise cause our program to abort.

The Array Index Out of Bounds Exception

The array

```
int[] a = {4,5,6};
```

has three elements which we can access using the indices 0, 1, and 2. If we try to use an index other than these three, as for example in the expression `i=a[3]`, Java will throw an array index out of bounds exception. Each exception is an instance of a class. We can write our own classes to define new types of exceptions, which we do in Section 8.7, but in this text we mostly use the Java exception classes shown in Figure 8.1.

Throwing an exception interrupts the program, transferring control to a user-defined catch clause, if any, which specifies how to handle the exception, or aborting if no catch clause is found. Example 8.1 shows the latter case when no catch clause is found. Java **aborts** execution with a message when it reaches the statement that uses an index that is out of bounds, and will not return to execute the rest of the program.

EXAMPLE 8.1 **Abort.java**

```java
/* Shows that Java aborts when it encounters
 * an out of bounds array index.
 */

public class Abort{
    public static void main(String [] args) {
        int [ ] anArray = {5,6,7};
        int badIndex = 5;                                  // Note 1
        int causesError;

        causesError = anArray[badIndex];                   // Note 2
        System.out.println("This statement never gets executed.");   // Note 3
    }
}
```

Output
```
java.lang.ArrayIndexOutOfBoundsException: 5
        at Abort.main(Abort.java:10)
```

Note 1: Only 0, 1, and 2 are valid indices for `anArray`.

Note 2: Java throws an array index out of bounds exception when it encounters this use of the bad index 5; it prints a message naming the type of exception, and showing the bad value. Java outputs the class, method name, and line number causing the exception.[1] Because the program does not handle this exception, Java aborts the program.

Note 3: Java aborted the program before reaching this line, thus it never gets executed.

We can easily fix Example 8.1 so it does not abort. Example 8.2 asks the user to input an index, but confirms that the index is between zero and two so that the program does not abort. If the index is out of bounds, then we display an error message.

EXAMPLE 8.2 **ValidateInput.java**

```java
/* Validates the index that the user inputs
 * so the program does not abort.
 */

import iopack.Io;
public class ValidateInput{
    public static void main(String [] args) {
        int [ ] anArray = {5,6,7};
        int index = Io.readInt("Enter an index from 0 to 2");
        int value;
```

---

[1] We need to disable the Just in Time compiler to view the line numbers in a stack trace. (See footnote 3 of Chapter 4). Some JDK version 1.1 compilers omit the second line of the error message in this example.

```
          if (index >= 0 && index <= 2) {                        // Note 1
              value = anArray[index];
              System.out.println("Execution does not get here if index is bad");
          }
          else
              System.out.println("Stick with 0, 1, or 2");
      }
  }
```

Output—First Run

```
Enter an index from 0 to 2:  1
Execution does not get here if index is bad
```

Output—Second Run

```
Enter an index from 0 to 2:  34
Stick with 0, 1, or 2
```

Note 1: We validate the user's input, accepting indices in the correct range, and displaying an error message otherwise.

☛ **TIP**

If there is a good way to validate the input, do so yourself. Much code in many applications is devoted to validating the data. Unsophisticated users may be unsure of the proper way to enter data, and even professionals make occasional errors.

....

Java does not require the programmer to handle the array index out of bounds exception, but it allows the programmer to do so. To handle an exception, we put the code that could cause that exception to occur in a **try block** followed by a catch clause to handle the exception, as in:

```
try {
     // some code that might generate an out of bounds exception
} catch(ArrayIndexOutOfBoundsException e) {
     // some code to execute when that exception occurs
}
```

where ArrayIndexOutOfBoundsException is the type of exception we are trying to catch. Java passes an instance, e, of this exception, which contains information about the array index out of bounds exception that occurred, to the catch clause.

If an exception occurs in the try block then Java looks for a catch clause that handles that exception. If it finds such a catch clause, it jumps immediately to execute that code, never returning to any code in the try block after the code which caused the exception. If Java does not find such a catch clause, it will abort the program with an error message, as happened in Example 8.1.

In Example 8.3 we put our use of the array index into a try block, and when Java throws the array index out of bounds exception, we catch it and display an error message. Our program does not abort, and execution continues after the catch clause. With a little more effort, we could use a loop to give the user another chance to input a correct value after making an error. We leave this enhancement to the exercises.

EXAMPLE 8.3 **TryException.java**

```java
/* Puts the array code in a try block and
 * catches the array index out of bounds exception
 * if it occurs.
 */

import iopack.Io;
public class TryException{
    public static void main(String [] args) {
        int value;
        try {
            int [ ] anArray = {5,6,7};
            int index = Io.readInt("Enter an index from 0 to 2");
            value = anArray[index];                              // Note 1
            System.out.println("Execution does not get here if index is bad");
        }catch (ArrayIndexOutOfBoundsException e) {              // Note 2
            System.out.println("Stick with 0, 1, or 2");
        }
        System.out.println("This is the end of the program");    // Note 3
    }
}
```

Output—First Run
```
Enter an index from 0 to 2:  2
Execution does not get here if index is bad
This is the end of the program
```

Output—Second Run
```
Enter an index from 0 to 2:  89
Stick with 0, 1, or 2
This is the end of the program
```

Note 1: Java will throw an array index out of bounds exception if index < 0 or if index >2.

Note 2: After throwing an array index out of bounds exception, Java jumps here to the handler, skipping any code remaining in the try block.

Note 3: After executing the code in the catch clause, Java continues executing here. Handling the array index out of bounds exception allows the program to continue executing whether or not Java throws an array index out of bounds exception.

Control Flow for Exceptions

If there is no catch clause for an exception immediately following the try block, then, when that exception occurs, Java looks for the catch clause in the caller of the method in which the try block is contained. In Example 8.4, we use an out of bounds array index in the getAndSetValue method, but do not include a catch clause in that method. When Java encounters the invalid index during execution it throws an array index out of bounds exception and looks for a catch clause which handles that exception. Not finding one in getAndSetValue, Java looks in the caller of the getAndSetValue method, the main method, which does have a clause that catches the exception. Java jumps to the catch clause in the main method in which we call e.printStackTrace(), where e is the array index out of bounds exception which Java passes to the catch clause.

We will implement a Stack class in Section 8.7. For now, think of a stack in the sense we use it in English as a stack of books or a stack of dishes. An important use of stacks is in implementing method calls in programming languages. We stack up arguments passed to the method and local variables. When starting to execute the main method, Java puts the data that main needs onto the stack. When the main method calls the getAndSetValue method, Java pushes the data that getAndSetValue needs onto the top of the previous data on the stack. In Example 8.4, we have only two methods, but in larger examples we could have half a dozen or more methods using the stack, many of them from the Java library packages.

The data for the method that Java is currently executing is on the top of the stack. When Java throws an array index out of bounds exception it passes an object of that exception type to the catch clause for that exception. This object contains a list of all the methods whose data is on the stack when the exception occurred. To see this list, we call the printStackTrace() method, which outputs

```
java.lang.ArrayIndexOutOfBoundsException: 98
    at TryExceptionTrace.getAndSetValue(TryExceptionTrace.java:12)
    at TryExceptionTrace.main(TryExceptionTrace.java:17)
```

The first line names the exception that occurred and displays the invalid value, 98. The next two lines are the stack entries. TryExceptionTrace is the class name for Example 8.4. The data for getAndSetValue is on the top of the stack and the exception occurred at line 12 of the program. Thus by printing the **stack trace** we can find the line at which the exception occurred. The second stack entry says that the bottom of the stack contains the data for the main method, and that the main method called the getAndSetValue method at line 17 of the program.

By reading the stack trace, we can follow the sequence of method calls that culminated in the throwing of the exception. After executing the code in the catch clause, Java continues execution with the code following the catch clause. Had we omitted the catch clause, Java would have aborted the program with an error message, as in Example 8.1. Handling the exception allows us to recover from the error and continue executing the remainder of the program.

EXAMPLE 8.4   **TryExceptionTrace.java**

```
/* Shows the use of the printStackTrace method
 * to obtain the sequence of method calls that
 * culminated in the throwing of the array index
 * out of bounds exception.
 */

import iopack.Io;
public class TryExceptionTrace{
    public static int getAndSetValue() {
        int [ ] anArray = {5,6,7};
        int index = Io.readInt("Enter an index from 0 to 2");
        return anArray[index];                          // Note 1
    }
```

continued

```
public static void main(String [ ] args) {
    int value;
    try {
        value = getAndSetValue();                            // Note 2
        System.out.println("Execution does not get here if the index is bad");
    }catch (ArrayIndexOutOfBoundsException e) {
        e.printStackTrace();
    }
    System.out.println("This is the end of the program");
}
}
```

Output

```
Enter an index from 0 to 2:   98
java.lang.ArrayIndexOutOfBoundsException: 98
    at TryExceptionTrace.getAndSetValue(TryExceptionTrace.java:12)
    at TryExceptionTrace.main(TryExceptionTrace.java:17)
This is the end of the program
```

Note 1: This is line 12 which causes the exception.[2]

Note 2: This is line 17, where the main method calls the getAndSetValue method that produces the exception.

The Number Format Exception

Java allows us to construct an Integer object from a string, as in:

```
Integer i = new Integer("375");
```

If we provide a string that is not a valid integer constant, as in:

```
Integer j = new Integer("3.75");
```

then Java will throw a number format exception. If we do not handle the exception in a catch clause, Java will abort the program with an error message.

Example 8.5 shows both valid and invalid attempts to construct Integer and Double objects from strings. As in Example 8.4, we call the printStackTrace method to determine the exception and where it occurred. In this example the stack of method calls shows three entries, the bottom from our main method and the top two from the Integer class of the Java library. Java uses <init> to denote a constructor.

EXAMPLE 8.5 StringToNumber.java

```
/* Illustrates wrapper classes used to convert
 * a string to an int or a double, and the number format
 * exception when the string has an invalid format.
 */
```

---

[2] Some JDK 1.1 compilers report this exception on line 11 because Java checks that the index is between 0 and 2 before executing line 12.

```java
public class StringToNumber {
    public static void main(String [] args) {
        try {
            int i = new Integer("435").intValue();          // Note 1
            System.out.println("i = " + i);
            int j = new Integer("45.2").intValue();          // Note 2
            System.out.println("j = " + j);
        }catch(NumberFormatException e) {
            e.printStackTrace();                             // Note 3
        }
        double d = new Double("3.14").doubleValue();         // Note 4
        System.out.println("d = " + d);
    }
}
```

Output
```
i = 435
java.lang.NumberFormatException: 45.2
    at java.lang.Integer.parseInt(Integer.java:238)
    at java.lang.Integer.<init>(Integer.java:342)
    at StringToNumber.main(StringToNumber.java:11)
d = 3.14
```

Note 1: We construct an `Integer` object from the string "435" which represents an integer literal. The `intValue` method returns a value, 435, of type **int**.

Note 2: Passing the string "45.2" to the `Integer` constructor causes Java to throw a number format exception, as 45.2 is not a valid integer literal.

Note 3: The stack trace shows two methods from the `Integer` class. The bottom line of the trace shows that line 11 of our program caused Java to throw the exception.

Note 4: After handling the exception, we wrap a valid **double** value as a `Double`.

☞ **STYLE**   Even though the catch clause in Example 8.5 has just one statement, `e.printStackTrace()`, we still need to enclose it in curly braces. If a try block contains just one statement, that statement needs to be enclosed in curly braces.

> **The Big Picture**
>
> Enclosing code that can throw an exception in a try block allows us to handle that exception in a catch clause. When Java throws an exception it will jump to a catch clause for that exception in the same method if there is one, and continue searching for a catch clause in the calling method, if there is not one. Printing the stack trace shows the methods that were in progress when the exception occurred.

TRY IT YOURSELF 1. Run Examples 8.3 and 8.4, entering a negative value for the index to see that Java throws an exception in this case.

TRY IT YOURSELF 2. Revise Example 8.5 to remove the `try` statement and the catch clause for the `NumberFormatException`. Rerun the revised code and note what happens and what code gets executed.

TRY IT YOURSELF 3. Write a small test program to create an array with elements of type `Object`. Show that you can include `Integer` and `Double` objects in the array. What happens if you try to include an **int** value?

4. Which of the following will cause Java to throw a `NumberFormatException`?
   a. `Integer i = new Integer("-7200");`
   b. `Double d = new Double("PI");`
   c. `String s = new String("PI");`
   d. `String s = new String("64000");`
   e. `Double d = new Double(".123");`

8.2 ▪ Text File Input and Output

So far we have done our character mode input from the keyboard, and our character mode output to the screen. We used the Io class to hide some of the messy details of console IO. In this section we will see how to read from and write to external files. Also we will get behind the scenes to explain how we implemented some of the methods of the Io class.

Reading from a File

The `FileReader` class allows us to read from an external file stored on a disk. We pass the name of the file to the constructor, as in:

```
FileReader input = new FileReader("myFile.data");
```

Many errors can occur during the input and output process. For example if the file `myFile.data` does not exist, (it may have been accidentally deleted, or never created), then Java will throw a `FileNotFoundException`. In our file IO programs in this section we catch `IOException` which is a superclass of all the IO exceptions so we will be notified if any IO exception occurs such as an unexpected end of file or a file not found.

We have no way of knowing that a file is missing from the disk, so we cannot validate the file name the same way we validated the array index in Example 8.2. If Java raises an IO exception and we do not handle it, the program will abort. To prevent this, Java requires that we handle the IO exception, and will not compile a call to any method which can generate an IO exception unless it is in a try block which handles that exception.

Example 8.6 shows an attempt to construct a `FileReader` which compiles because it is in a try block with a catch clause which catches the IO exception. This program generates a `FileNotFoundException` at runtime because there is no external file named `zxcvb.data`. We print the stack trace and let the program terminate. We leave it to the exercises to improve the program by allowing the user to enter another file name if the current name causes an exception.

EXAMPLE 8.6   **IoError.java**

```
/* Throws an IO exception because the
 * file zxcvb.data does not exist.
 */

import java.io.*;                                              // Note 1
public class IoError {
    public static void main(String [] args) {
        try {
            FileReader f = new FileReader("zxcvb.data");       // Note 2
        }catch(IOException e) {
            e.printStackTrace();                               // Note 3
        }
    }
}
```

Output
```
java.io.FileNotFoundException: zxcvb.data
        at java.io.FileInputStream.<init>(FileInputStream.java:64)
        at java.io.FileReader.<init>(FileReader.java:43)
        at IoError.main(IoError.java:9)
```

Note 1: We tell the compiler to look for the classes used for input or output, such as FileReader, in the package java.io.

Note 2: This line will not compile unless it is included in a try block with a catch clause for the IO exception. (See the A Little Extra section for an amendment to this rule.) If the file does not exist at runtime, Java will throw a file not found exception. The file, zxcvb.data, named in this example does not exist.

Note 3: The stack shows three methods, main from our IoError class, the FileReader constructor, and the constructor from the FileInputStream class in the Java library that the FileReader class uses.

☛ **A LITTLE EXTRA**   **Checked Exceptions and the Throws Clause**

Java divides exceptions into two categories:

unchecked	subclasses of Error or RuntimeException
checked	subclasses of Exception (but not of RuntimeException)

As we have seen, we have a choice whether or not to handle an **unchecked exception** such as ArrayIndexOutOfBoundsException or NumberFormatException. If a method does not want to handle a **checked exception** it can use a throws clause to declare that it may pass along that exception to its caller. For example, in

```
public FileReader createFile(String s) throws IOException {
    return new FileReader(s);
}
```

the createFile method does not put the FileReader in a try block, and does not catch the IOException; it declares that it may throw the IOException to whomever calls it. The main method could call the createFile method, putting that call in a try block and adding a catch clause for the IOException that createFile may throw because it does not catch the error caused when the file zxcvb.data is not found.

The code for main would look like Example 8.6, except that FileReader("zxcvb.data") would change to createFile("zxcvb.data"). A method might use the throws clause when it does not know what to do when the exception occurs. The caller of the method might have more information to help handle the error appropriately.

Reading Lines and Fields

Java provides classes for both binary and text input and output. Binary IO uses the internal representations of the data without converting them to the character representations using digits and letters that humans can more easily read. Such binary IO can be very useful for transferring data from one file to another. We will concentrate on text IO in this section and binary IO in the next.

Disk storage units have mechanical parts so we access external data much more slowly than that residing in the computer's memory. Reading one character at a time from a disk would be very inefficient. A better plan is to read a whole block of data from the disk, say 1024 characters, storing these characters in an area of the computer memory called a **buffer**. We then read the characters, when we need them, from the buffer. Because the buffer is in memory rather than on the disk, we can access the data much faster. When we have read all the characters in the buffer, we again read another block of data from the disk.

The process works in reverse for output. We write each character into a buffer, and when the buffer is full we write the entire buffer to the external disk. Java provides a BufferedReader class to allow us to read blocks of data into a buffer. We pass a FileReader to the BufferedReader constructor

```
BufferedReader f = new BufferedReader (new FileReader("messages.data"));
```

To read from the buffer, we use the readLine method which reads a line of text, returning it as a string.

Example 8.7 reads the strings from the messages.data file which we created by typing some lines of our choosing using a simple text editor (that is, not using a fancy editor with hidden formatting characters). The **while** loop terminates when the readLine method returns **null**, indicating it is at the end of the file. After reading the strings we close the file, releasing any resources used back to the operating system.

EXAMPLE 8.7 **FileReadStrings.java**

```
/* Reads strings from a file created in
 * a text editor.
 */

import java.io.*;
public class FileReadStrings {
```

```
public static void main(String [ ] args) {
    String line;
    try {
        BufferedReader f = new BufferedReader
                        (new FileReader("messages.data"));      // Note 1
        while((line = f.readLine()) != null)                    // Note 2
        System.out.println(line);
        f.close();                                              // Note 3
    }catch(IOException e) {
        e.printStackTrace();
    }
}
}
```

Output

```
Java is fun.
The three did feed the deer.
An apple a day keeps the doctor away.
```

Note 1: We pass the name of the external file, messages.data, to the FileReader constructor, and pass the FileReader that we construct to the BufferedReader constructor, so that we can read blocks of data into a buffer rather than reading one character at a time from the external file.

Note 2: We read a line from the file, assigning it to the variable line. We terminate the loop when line is **null** which indicates we are at the end of the file.

Note 3: We close the file to release the resources it used back to the operating system.

☞ A LITTLE EXTRA **The Finally Clause**

Java provides a finally clause to cleanup after it completes the execution of the code in the try clause, either normally or by throwing an exception. We could use the finally clause in Example 8.7 to close the file f. The close() method at the end of the try clause in Example 8.7 will not be invoked if the readLine method throws an exception. Putting the statement f.close() in the finally clause will make sure that it gets executed whether or not an exception occurs. The revised code fragment using the finally clause is:

```
BufferedReader f = null;
try {
    f = new BufferedReader(new FileReader("messages.data"));
    while((line = f.readLine()) != null)
        System.out.println(line);
}catch(IOException e) {
    e.printStackTrace();
}finally {
    try{
        f.close();
    }catch(IOException e) {
        e.printStackTrace();
    }
}
```

Because the close() method can generate an IOException, we put it in a try block.

In Example 8.7 we read one string from each line. Typically, data files contain records with several fields on each line. We will separate fields by a delimiter which we can choose. For example, if we use the vertical bar to separate the fields, we can write the name, product number, color, and price of an item as

```
shirt|12345|blue|15.99
```

We can use the `StringTokenizer` class to read several fields from a single line. Each field on the line is called a **token**. We first read the line using the `readLine` method. Suppose we read the line of data above for the shirt storing it in a variable named `line`, and then pass `line` to the `StringTokenizer` constructor, as in:

```
StringTokenizer strings = new StringTokenizer(line,"|");
```

where the second argument to the constructor specifies the vertical bar, |, as the delimiter, dividing the line into separate string tokens. We use the `nextToken()` method to get each token. Calling `strings.nextToken()` will return the string, shirt, the characters up to the vertical bar delimiter. Calling `nextToken()` three more times will return the strings 12345, blue, and 15.99, in that order.

In Example 8.8 we will enter the four fields, street, city, state, and zip, of the `Address` class of Example 2.8 on a single line separating them with the vertical bar, as in:

```
77 Sunset Strip|Hollywood|CA|90048
```

To read the next string we use the `nextToken` method, as in:

```
strings.nextToken();
```

which for the above line will return 77 Sunset Strip the first time it is called, Hollywood the second time, and so on. We use the `countTokens` method to check that the line has exactly four strings, skipping it if it does not.

EXAMPLE 8.8   **FileReadAddresses.java**
....

```java
/* Reads the four fields of an Address, separated by vertical bars,
 * from a single line.  Uses a StringTokenizer to get each string.
 */

import java.io.*;
import personData.*;
import java.util.StringTokenizer;                              // Note 1
public class FileReadAddresses {
    public static void main(String [ ] args) {
        String line;
        String street, city, state, zip;
        StringTokenizer strings;
        Address address;

        try {
            BufferedReader f = new BufferedReader(new
                            FileReader("addresses.data"));
            while ((line = f.readLine()) != null){
                strings = new StringTokenizer(line,"|");
                if (strings.countTokens() == 4) {
```

```
                                street = strings.nextToken();
                                city = strings.nextToken();
                                state = strings.nextToken();
                                zip = strings.nextToken();
                                address = new Address(street,city,state,zip);     // Note 2
                                System.out.println(address);                       // Note 3
                                System.out.println();
                        }
                }
                f.close();
        }catch(IOException e) {
                e.printStackTrace();
        }
    }
}
```

Output
```
77 Sunset Strip
Hollywood, CA 90048

222 Bridge Road
Grand Palabra, ND 58585
```

Note 1: We tell the Java compiler to find the `StringTokenizer` class in the `java.util` package.

Note 2: We create an `Address` object as defined in Example 2.8.

Note 3: The `println` method uses the `toString` method of the `Address` class to print the `address` object on the screen. Notice that the `address` object rearranges the fields input from the file to look like an address on two lines.

In Example 8.8 we read lines containing four strings. Often our lines may contain other types of data such as **int** or **double**. For example we might have a line with three fields, an item which is a string, an **int** quantity and a price of type **double,** as in:

```
Milk    3    2.10
```

where for variety we separate the fields using blank spaces. Using the blank spaces to delimit the fields prevents us from including an item such as ice cream which has an internal blank.

When we read each token, using the `StringTokenizer` object, we get values of type `String`. We use the wrapper classes introduced in Section 6.2 to convert the string representing the quantity to an **int**, and the string representing the price to a **double**. Java allows us not to handle the number format exception that would be generated if, for example, our file had a value of 3.5 in the field for the quantity. In that case our program would abort, so it is better to include a catch clause to handle the number format exception. We leave this improvement for the exercises.

Writing to a File

Java makes it easy to write values of different types to an external file. We first create a `FileWriter`, passing it the name of the external file on which we want to write our data, and then pass the `FileWriter` object to a `PrintWriter` constructor, as in:

```
PrintWriter p = new PrintWriter (new FileWriter("totalCost.data"), true);
```

The second argument, **true**, to the PrintWriter constructor indicates that we want Java to flush the buffer whenever it executes a println statement, instead of waiting until the buffer fills up before writing its contents to the file. Using p.print and p.println statements we will write to the external file totalCost.data overwriting its previous contents.

☞ **TIP**

To append to the end of totalCost.data instead of overwriting its contents use the constructor

```
new FileWriter("totalCost.data", true);
```

where the second argument is an append flag set to **true** to append to the file and **false** to overwrite it.

We can format the output using the NumberFormat class introduced in Section 6.2.

EXAMPLE 8.9 **Prices.java**

```
/* Reads records from a file, each containing an item, a
 * quantity, and a price.  Computes the total cost of each item,
 * uses NumberFormat objects to write a double to a file, and
 * to write in a currency format to the screen.
 */

import java.io.*;
import java.text.*;
import java.util.StringTokenizer;
public class Prices {
    public static void main(String [ ] args) {
        String line;
        String item;
        int quantity;
        double price;
        double cost;
        StringTokenizer strings;
        NumberFormat decimal = NumberFormat.getInstance();
        decimal.setMaximumFractionDigits(2);
        NumberFormat currency = NumberFormat.getCurrencyInstance();
        try {
          BufferedReader f =
              new BufferedReader(new FileReader("prices.data"));
          PrintWriter p = new PrintWriter(new FileWriter("totalCost.data"),true);
          while ((line = f.readLine()) != null){
              strings = new StringTokenizer(line);                    // Note 1
              if (strings.countTokens() == 3) {
                  item = strings.nextToken();
                  quantity = new Integer(strings.nextToken()).intValue(); // Note 2
                  price = new Double(strings.nextToken()).doubleValue();  // Note 3
                  cost = price*quantity;
                  System.out.println("Total cost of "+ item + " is " +
                                                    currency.format(cost));
                  p.print(item + " ");
                  p.println(decimal.format(cost));
              }
          }
```

```
            f.close();
            p.close();
        }catch(IOException e) {
            e.printStackTrace();
        }
    }
}
```

Output
```
Total cost of Milk is $6.30
Total cost of Coffee is $6.78
Total cost of Bread is $5.67

The file TotalCost.data

Milk 6.30
Coffee 6.78
Bread 5.67
```

Note 1: We use the StringTokenizer constructor with one argument, the string whose parts we wish to read. We use the default delimiters, space, tab, and newline, often called **whitespace** characters, so that we do not need to specify the delimiters in the second argument.

Note 2: The nextToken method returns a string which we have to convert to an **int** for the quantity field.

Note 3: The nextToken method returns a string which we have to convert to a **double** for the price field.

Inside Our Io Class

In Example 8.9 we used methods of the java.text package to do the formatting of decimal numbers we had done previously using methods of the Io class we provided with this text. Now that we have introduced exceptions and input we can see how the input methods of the Io class work.

Example 8.10 shows the readInt method, part of the Io class, that we have used many times in earlier chapters. We use a print, rather than a println, statement to print the prompt, so the user can input the integer on the same line as the prompt. The print statement adds the prompt to the buffer but does not automatically output the buffer to the screen. To output the prompt immediately, we use the flush method to flush the buffer. A System.out.println statement would flush the buffer, but we do not want to go to the next line.

When reading from a file, as in Example 8.9, we pass the file name to a FileReader constructor, and then pass a new FileReader object to a BufferedReader constructor. For input from the keyboard we pass the standard input stream, System.in, to an InputStreamReader constructor, which converts the bytes to characters, and then pass a new InputStreamReader to a BufferedReader constructor to buffer the input.

We expect the user to enter a string representing a valid integer literal which we convert to a value of type **int** using the static parseInt method of the Integer class. We could have done the conversion, as we did in Example 8.9, by constructing an Integer object from the input string, and using the intValue method to get the **int** value wrapped inside it. The conversion code using that approach would be

```
i = new Integer(s).intValue();
```

We handle exceptions inside a **while** loop so that if an error occurs we can recover and continue with the program. We initialize the integer variable i to zero. If an IO exception occurs, we exit the loop and return i which will have the default value of zero. If the user inputs a string that does not represent an **int**, such as 34.5 or cat, then Java will throw a number format exception which we catch inside the loop, printing a message so, hopefully, the user will enter a correct value at the next iteration of the loop.

EXAMPLE 8.10   PartOfIo.java

```
/* Shows the readInt method from the
 * Io class.
 */

import java.io.*;
public class PartOfIo {
    public static int readInt(String prompt) {
        boolean done = false;
        String s;
        int i = 0;
        while (!done) {
            System.out.print(prompt+ ":  ");
            System.out.flush();
            try {
                BufferedReader in = new BufferedReader
                                    (new InputStreamReader(System.in));
                s = in.readLine();
                i = Integer.parseInt(s);
                done = true;
            }catch (IOException e){                        // Note 1
                    done = true;
            }catch (NumberFormatException e1){             // Note 2
                System.out.println("Error -- input an integer -- Try again");
            }
        }
        return i;
    }
    public static void main(String [ ] args) {
        System.out.println(readInt("Enter an integer"));
    }
}
```

Output

```
Enter an integer:  57
57
```

Note 1: If an IO error occurs, we exit the loop and return the default value of zero.

Note 2: We can catch more than one type of exception that might be thrown in a try block. Here we have two catch clauses, one for an IO exception, and another for a number format exception.

The Big Picture
....

Code that might throw an IOException must be enclosed in a try block with a catch clause to handle that exception, or in a method which declares that it may throw an IOException to its caller.

    The FileReader lets us read characters from a file. To gain efficiency by buffering, we pass it to a BufferedReader which has a readLine method we use to read from the file. We can use a StringTokenizer to retrieve fields from a line. Similarly a FileWriter lets us write characters to a file. The PrintWriter, automatically buffered, lets us use the familiar print and println methods.

TEST YOUR UNDERSTANDING

TRY IT YOURSELF   5.   Create a prices.data file, in which each row has a string naming an item, an integer representing the quantity desired of that item, and a **double** representing the unit price for that item, which contains data in an invalid format, such as a value of 3.5 for the quantity. Run Example 8.9, explaining the result.

TRY IT YOURSELF   6.   Write your own messages.data file with one string on each line, and run the code of Example 8.7, checking that the program does list the strings from your file.

TRY IT YOURSELF   7   Change the file prices.data, used in Example 8.9, to include ice cream as an item. Run the program and explain what happens.

8.3 ▪ Binary and Object Input and Output
··················

We use binary input and output for data in 8-bit byte form rather than the character form needed for 16-bit Unicode characters. Such binary data can be stored in files, but is not meant to be read by humans.

    When reading and writing objects, we must be very careful in dealing with shared objects. Fortunately the Java object serialization facilities handle these details automatically, allowing us to easily store and retrieve objects.

The File Class

The File class has several methods which return properties of the file. Example 8.11 uses some of these methods, whose names nicely signify their functions.

EXAMPLE 8.11   **FileProperties.java**
....

```
/* Creates a file and returns some
 * of its properties.
 */

import iopack.Io;
import java.io.*;
```
continued

```
public class FileProperties {
    public static void main(String [ ] args) {
        String filename = Io.readString("Enter a file name");
        File f = new File(filename);
        System.out.println("Name: "+f.getName());
        System.out.println("Path: "+f.getPath());
        System.out.println("Can write: "+f.canWrite());
        System.out.println("Is directory: "+f.isDirectory());
        System.out.println("Length: "+f.length());
        System.out.println("Parent directory: "+f.getParent());
    }
}
```

Output
```
Enter a file name:  FileProperties.java
Name: FileProperties.java
Path: c:\book2\ch8\FileProperties.java
Can write: true
Is directory: false
Length: 714
Parent directory: c:\book2\ch8
```

Reading and Writing Bytes

The FileInputStream class lets us read from a file. Using the read statement we can read the entire file, and copy it to a new location with the write method of FileOutputStream. We can copy either text files or binary files this way. To illustrate we run Example 8.12 to copy the class file of Example 8.11 to a new directory. To show that Example 8.12 properly copied FileProperties.class, we run it from its new location.

The basic read method

```
public native int read( ) throws IOException;
```

where the modifier native indicates that Java implements this method in a platform-dependent manner, reads a single byte, but we use the version

```
public int read(byte[ ] b) throws IOException;
```

that reads an entire byte array. There is also a three-argument version of read,

```
public int read(byte[] b, int off, int len) throws IOException;
```

that reads into an array of bytes, with the second argument specifying the starting offset in the file, and the third giving the number of bytes to read. The various read methods return the number of bytes read, or -1 if at the end of the file.

The FileOutputStream class has three analogous versions of the write method: write one byte, write an array of bytes, or write a given length of an array of bytes from a starting offset in the file.

EXAMPLE 8.12   **FileCopy.java**

```
/* Copies a file from one location to another,
 * using read and write statements.
 */

import iopack.Io;
import java.io.*;
public class FileCopy {
    public static void main(String [ ] args) {
        try {
            String source = Io.readString("Enter file to copy");
            String target = Io.readString("Enter the new file name");
            File f = new File(source);
            FileInputStream fis = new FileInputStream(f);         // Note 1
            int length = (int)f.length();                         // Note 2
            byte [] data = new byte[length];                      // Note 3
            fis.read(data);                                       // Note 4
            fis.close();
            FileOutputStream fos = new FileOutputStream(target);  // Note 5
            fos.write(data);
            fos.close();
        }catch (IOException e) {
            e.printStackTrace();
        }
    }
}
```

Output
```
Enter file to copy: FileCopy.java
Enter the new file name: NewFileCopy.java
```

Note 1: We create a File object, f, and pass it rather than the file name, because we need a File object to get the file's length.

Note 2: The length method returns a value of type **long**, which we must cast to an **int** because the **new** operator creates an array with size given by a value of type **int**.

Note 3: We create an array large enough to hold the entire file.

Note 4: We created the data array to have the size of the file, so we use the read method that fills the entire array. This is equivalent to read(data,0,length).

Note 5: We do not need a File object for the output file, so we just pass the file name directly to the constructor.

Reading and Writing Primitive Types

The DataOutputStream class has methods for writing each of the primitive types in binary form, including writeBoolean, writeChar, writeDouble, writeFloat, writeInt, and writeLong, while DataInputStream has methods for reading these types including readBoolean, readChar, readDouble, readFloat, readInt, and readLong.

To create a `DataOutputStream`, we first create a `FileOutputStream`

```
new FileOutputStream(args[0])
```

where `args[0]` is the name of the file to which we write. We pass this `FileOutputStream` to a `BufferedOutputStream`

```
new BufferedOutputStream(new FileOutputStream(args[0]))
```

so that each `write` statement does not force an expensive write to external storage, but rather writes to a buffer which, when filled, is written to the disk. Finally, we construct the `DataOutputStream` from the `BufferedOutputStream`

```
new DataOutputStream(new BufferedOutputStream(new FileOutputStream(args[0])))
```

In Example 8.13, we use the `writeDouble` method to write the numbers from 0.0 through 9.0 to a file. The binary format, used internally, is not suitable for human reading; we use the `readDouble` method to read from the newly created file, displaying the values on the screen using the `System.out.print` method to verify the file was written correctly.

EXAMPLE 8.13   **Primitive.java**

```
/* Illustrated the DataOutputStream and DataInputStream
 * classes for primitive type IO using type double.
 */

import iopack.Io;
import java.io.*;
public class Primitive {
    public static void main(String [ ] args) {
        try {
            String datafile = Io.readString("Enter a filename for the data");
            DataOutputStream dos = new DataOutputStream
                (new BufferedOutputStream (new FileOutputStream(datafile)));
            for (double d = 0.0; d < 10.0; d++) dos.writeDouble(d);    // Note 1
            dos.close();
            DataInputStream dis = new DataInputStream
                (new BufferedInputStream(new FileInputStream(datafile)));
            double din;
            for (int i = 0; i < 10; i++) {
                din = dis.readDouble();
                System.out.print(din + "  ");
            }
            dis.close();
        }catch (IOException e) {
            e.printStackTrace();
        }
    }
}
```

Output
```
Enter a filename for the data:  primitive.data
0.0  1.0  2.0  3.0  4.0  5.0  6.0  7.0  8.0  9.0
```

Note 1: Although **for** loops usually use integer indices, using a **double** index suits this example well. We illustrate the DataOutputStream and DataInputStream classes using type **double**, leaving the use of other types for the exercises.

Random Access Files

We access FileInputStream, FileOutputStream, DataInputStream, and DataOutputStream objects sequentially. We read one item after another, going forward in the file, but cannot go back to data before the current file position. Similarly, when writing data, we cannot return to an earlier position in the file. By contrast, the **random access file**, which we can use for both reading and writing, allows us to read or write at any position in the file.

A random access file implements the methods to write primitive types that a DataOutputStream does, and provides the same methods to read primitive types as found in a DataInputStream. The seek method locates a position in the file. Calling seek(20) sets the position at the twentieth byte in the file, at which position we can either read or write. After completing a read or write operation, we can use seek(4) to move the position to the location further back in the file at byte 4.

When creating a random access file, we use the second argument in its constructor to specify the access mode, "r" for read-only or "rw" for read-write access. For example,

```
new RandomAccessFile("random.dat", "rw");
```

creates a random access file with read and write capabilities on the file random.dat. The system will create random.dat if it does not exist.

EXAMPLE 8.14   **RandomAccess.java**

```
/* Seek forward and back, and writes and reads
 * in a random access file.
 */

import java.io.*;
public class RandomAccess {
    public static void main(String [ ] args) {
        try {
            RandomAccessFile raf = new RandomAccessFile("random.dat", "rw");
            for (int i = 0; i < 10; i++)
                raf.writeInt(i);
            raf.seek(20);                                              // Note 1
            int number = raf.readInt();
            System.out.println("The number starting at byte 20 is " + number);
            raf.seek(4);                                               // Note 2
            number = raf.readInt();
            System.out.println("The number starting at byte 4 is " + number);
            raf.close();
        }catch (IOException e) {
            e.printStackTrace();
        }
    }
}
```

Output

```
The number starting at byte 20 is 5
The number starting at byte  4 is 1
```

Note 1: Each integer is 32 bits or 4 bytes, so the position at byte 20 will bypass the first five integers, 0, 1, 2, 3, and 4, in the file. (The bytes are numbered 0, 1, ..., 19.) Reading at byte 20 should result in reading the integer 5.

Note 2: Going back to byte 4 will position the file after the first integer, 0, so reading an integer at this position should return the value 1.

Reading and Writing Objects

Anyone who has tried to write objects to external storage in C++ or even to understand various C++ implementations of object persistence will appreciate how nicely Java has set up **object persistence**, the ability to write objects to and read them from external files. Each class whose objects we wish to store must implement the Serializable interface, which has no methods. Implementing Serializable shows we intend to write objects of that class to disk. For security reasons, Java did not make the capability for persistence the default, but requires programmers to explicitly permit persistence by implementing the Serializable interface.

Java, transparently to the programmer, writes type information to the file, so reading an object will automatically recover its type. Shared objects could cause problems. References are memory addresses which would be meaningless when we reload the objects. Saving a copy of a shared object each time we refer to it, would cause problems trying to maintain several copies of the formerly shared object. Java solves these problems by automatically numbering objects and using these numbers to refer to shared objects which need to be saved only once. This process of coding objects so they can be written to external storage and recovered properly is called **object serialization**.

Example 8.15 illustrates the saving and restoring of objects. An Account, general, and a SavingsAccount, savings, share an account holder, fred, of type Person. We use simplified versions of the these classes to create these objects.

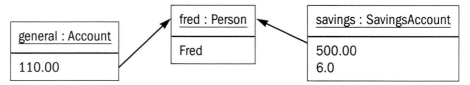

To show that Java handles types correctly, we declare both accounts as Account and create an ObjectOutputStream, calling writeObject to serialize these objects to a file. The classes, Account, BankAccount, and Person all implement Serializable.

ObjectInputStream provides the readObject method to read our objects, which we cast to their compile-time type, Account, checking that Java restored the types correctly. The aGeneral object should be just a plain Account, not an instance of the subclass SavingsAccount, while the aSavings object should be a SavingsAccount. We also check that both the aGeneral and the aSavings accounts have the identical account holder, fred, which shows that Java restored the shared object correctly. Using object serialization, Java saves all the type information and does the object numbering automatically.

EXAMPLE 8.15   **ObjectIO.java**

```java
/* Illustrates object persistence.
 */
import java.io.*;
public class ObjectIO {
    public static void main(String [] args) {
        try {
            Person fred = new Person("Fred");
            Account general = new Account(fred, 110.0);
            Account savings = new SavingsAccount(fred, 500.0, 6.0);     // Note 1
            ObjectOutputStream oos = new ObjectOutputStream(
                                new FileOutputStream("Objects.dat"));
            oos.writeObject(general);                                   // Note 2
            oos.writeObject(savings);
            oos.close();
            ObjectInputStream ois = new ObjectInputStream(
                                new FileInputStream("Objects.dat"));
            Account aGeneral = (Account)ois.readObject();
            Account aSavings = (Account)ois.readObject();              // Note 3
            if (aGeneral instanceof SavingsAccount)
              System.out.println("aGeneral account is a SavingsAccount");
            else if (aGeneral instanceof Account)
              System.out.println("aGeneral account is an Account");    // Note 4
            if (aSavings instanceof SavingsAccount)
              System.out.println("aSavings account is a SavingsAccount");
            else if (aSavings instanceof Account)
              System.out.println("aSavings account is an Account");    // Note 5
            if (aGeneral.holder == aSavings.holder)                    // Note 6
              System.out.println("The account holder, fred, is shared");
            else
              System.out.println("The account holder, fred, has been duplicated");
                ois.close();
        }catch (IOException ioe) {
            ioe.printStackTrace();
        }catch (ClassNotFoundException cnfe) {                         // Note 7
            cnfe.printStackTrace();
        }
    }
}
class Person implements Serializable {                                 // Note 8
    String name;
    Person (String name) { this.name = name; }
}
class Account implements Serializable {
    Person holder;
    double balance;
    Account(Person p, double amount) {
        holder = p;
        balance = amount;
    }
}
```

continued

```
class SavingsAccount extends Account implements Serializable {
    double rate;
    SavingsAccount(Person p, double amount, double r) {
        super(p,amount);
        rate = r;
    }
}
```

Output
```
aGeneral account is an Account
aSavings account is a SavingsAccount
The account holder, fred, is shared
```

Note 1: We declare `savings` to have type `Account`, but assign it an instance of the `SavingsAccount` subclass to check that `writeObject` saves the object's actual type correctly.

Note 2: Besides the `writeObject` method, `ObjectOutputStream` provides the primitive type output methods such as `writeDouble` and `writeInt`. The object written, `general`, is of type `Account` which implements the `Serializable` interface.

Note 3: The second object written had type `SavingsAccount`. Reading it should create an object of type `SavingsAccount`. The `readObject` method returns type `Object`, so we cast the return value to type `Account` and assign it to an `Account` reference, but will check later that its original `SavingsAccount` type has been preserved.

Note 4: The object `aGeneral` was read from the first object written, `general`, so it should be an `Account` object.

Note 5: The object `aSavings` was read from the second object written, `savings`, so it should be a `SavingsAccount`.

Note 6: `if (aGeneral.holder == aSavings.holder)`

We check that these references are equal, meaning they point to the identical object. This shows by using the object serialization facility, we preserve the structure of the objects. Objects shared before writing are still shared after being read again. In our example both accounts have the identical account holder, `fred`.

Note 7: `}catch (ClassNotFoundException cnfe) {`

The `readObject` method may throw a `ClassNotFoundException`.

Note 8: `class Person implements Serializable {`

The `Person`, `Account`, and `SavingsAccount` classes are simplified versions of these classes which we use to illustrate object serialization. Each implements the `Serializable` interface, as it must in order to be serializable.

TEST YOUR UNDERSTANDING

TRY IT YOURSELF   8. Modify Example 8.12 to add buffering using the `BufferedInputStream` and `BufferedOutputStream` classes.

TRY IT YOURSELF   9. Modify Example 8.12 to close the files in a finally clause so that the files will be closed even if an exception is thrown.

TRY IT YOURSELF   10. Modify Example 8.13 to use the `readLong` and `writeLong` methods instead of `readDouble` and `writeDouble`.

TRY IT YOURSELF   11. Modify Example 8.14 to write the same values using type **double** instead of **int**. Seek the position of 5.0, and then the position of 1.0.

The Big Picture

The `File` class allows us to get file properties. To read binary files we can use one of the three `read` methods of a `FileInputStream`. To read primitive types we construct a `DataInputStream` from the basic `FileInputStream`, and use the `readInt` and other similar methods. To read objects we construct an `ObjectInputStream` and use the `readObject` method. Such objects must implement the `Serializable` interface. Analogous classes and methods exist for writing. A random access file, used for both input and output, allows us to seek specific locations without having to process the file sequentially.

8.4 ▪ Program Arguments and Applet Parameters

In Sections 1.6 and 4.2 we showed how to pass arguments to methods of a class other than the `main` method. In this section we show how to pass arguments to the `main` method and how to pass parameters to an applet.

Program Arguments

In all our standalone application examples we have diligently declared the `main` method as

```
public static void main(String [] args)
```

but have yet to use the `String` array, `args`, that we specify as its parameter.

When we call a method, we pass it its arguments. For example, we can pass the values 4.5, 1000.0, and 7 to the `amount` method of Example 1.7, as in `amount(4.5, 1000, 7)`. However, we do not call the `main` method, the Java interpreter does. The technique for passing arguments to the `main` method depends on which environment we are using. **Program arguments**, passed to the `main` method, are often called command-line arguments because they are placed just after the program name when a **command line** is used to run the program.

In Example 8.16, we will read and display a specified number of lines from a specified file, using program arguments to specify the file from which to read, and the number of lines to read and display. For example, the first argument, `messages.data`, is the name of the file we want our program to read, while the second, 2, is the number of lines we want to read and display.[3]

---

[3] Using the JDK from Sun we would enter the command

```
java CommandLine  messages.data  2
```

See Appendix E to see how to pass program arguments using other environments.

The method to pass the arguments in to the main method depends on which environment we are using, but inside the program we use the arguments in the same way, referring to the strings passed as program arguments using the String variable, args. When executing the program, Java sets args[0] to the file messages.data, and args[1] to the number of lines, 2, passing these values to our main method. We can run our program again passing different program arguments, such as the program to read from, FileReadPrices.java, and the number of lines to read, 10. When passed these arguments, our program will read and display the first 10 lines of the program FileReadPrices.java.

If the user does not pass two program arguments we display a message and abort the program. To abort a Java program we can use the statement

```
System.exit(1);
```

which exits, returning the value of its argument to the operating system. By convention, a non-zero value indicates an error caused the program to terminate.

Each program argument is passed as a string. The second argument in our example specifies the number of lines of the file to read. Inside the program we need to convert the string args[1], representing the second argument, to an **int** to obtain the number of lines to read from the file.

EXAMPLE 8.16   **ReadFileLines.java**

```
/* Reads arg[1] lines from file args[0].
 */

import java.io.*;
public class ReadFileLines {
    public static void main(String [ ] args) {
        String line;
        int totalLines;        // number of lines to read from the file
        int count = 0;         // number of lines read so far
        if (args.length != 2){                                           // Note 1
            System.out.print("Pass the file name and number of lines");
            System.out.println(" to read as program arguments");
            System.exit(1);
        }
        try {
            totalLines = Integer.parseInt(args[1]);
        }catch(NumberFormatException e) {                                // Note 2
            totalLines = 3;
        }
        BufferedReader f;
        try {
            f = new BufferedReader(new FileReader(args[0]));
            System.out.println("Opening " + args[0]);
            while((line = f.readLine()) != null && count++ < totalLines)
                System.out.println(line);
            f.close();
        }catch(IOException e) {
            e.printStackTrace();
        }
    }
}
```

Output

(passing the arguments `messages.data` and 2)

```
Opening messages.data
Java is fun.
The three did feed the deer.
```

Note 1: We check that the user entered two arguments. If not, we print a message indicating the arguments the user needs to specify, and exit the program by calling `System.exit` with an argument of 1, indicating that an error occurred.

Note 2: If the user enters an invalid string for the number of lines to read, we catch the number format exception, setting `totalLines` to a default value of three.

Applet Parameters

An applet does not have a `main` method; a browser or an applet viewer loads an applet when it encounters the `applet` tag in an HTML file. **Applet parameters** allow us to pass information to an applet. To pass parameters to an applet, we use the `param` tag in the HTML file. For example, we can use a `param` tag to specify the name of the font in which we will draw strings. The `param` tag has an attribute, `name`, which gives the name of the parameter, and an attribute, `value`, which gives the desired value of the parameter with that name.

The `param` tag

```
<param name=fontName value="Serif">
```

describes a parameter named `fontName` whose value is `Serif`. In our code for the applet, we can obtain the value of any parameters using the `getParameter` method, as in:

```
String fontName = getParameter("fontName");
```

The argument to the `getParameter` method is the parameter name found in the `name` attribute of a `param` tag in the HTML file. In this example, the `getParameter` method finds the `param` tag with the `name` attribute and returns the string specified in the `value` attribute. For the above `param` tag, the `getParameter` method would return the string, Serif.

The `getParameter` method always returns a string. For example, we could use the `param` tag

```
<param name=size value=36>
```

to specify the point size to use for the font. In our applet, calling the `getParameter` method to get the size, as in:

```
getParameter("size");
```

will return 36 as a string which we must convert to an **int** value, as in:

```
int size = Integer.parseInt(getParameter("size"));
```

The `param` tag occurs in the HTML file between the `<applet>` and the `</applet>` tags, as in:

```
<applet code = PassToApplet.class width = 200 height = 200>
<param name = fontName value = "SansSerif">
</applet>
```

We can include several `param` tags to specify the names and values for several parameters.

Example 8.17 gets the `fontName` and `size` parameters from the HTML file, creating a bold font with the specified font name and point size.

EXAMPLE 8.17 **PassToApplet.java**

```java
/* Uses parameters set in the HTML file to
 * specify the font name and size.
 */

import java.awt.*;
import java.applet.Applet;

public class PassToApplet extends Applet {
    private Font font;
    private int size;
    private String fontName;

    public void init() {
        fontName = getParameter("fontName");
        if (fontName == null) fontName = "Serif";          // Note 1
        try {                                               // Note 2
            size = Integer.parseInt(getParameter("size"));
        }catch(NumberFormatException e) {
            size = 12;
        }
        font = new Font(fontName,Font.BOLD,size);           // Note 3
        setFont(font);
    }
    public void paint(Graphics g) {
        g.drawString("Uses applet parameters",20,70);
    }
}
```

Note 1: If the HTML file has no `param` tag with a name attribute of `fontName`, the `getParameter` method will return **null**, in which case we set the `fontName` to the default value of `Serif`.

Note 2: If the HTML file has no `param` tag with a name attribute of `size`, the `getParameter` method will return **null**, which will cause the `parseInt` method to throw a number format exception. The `parseInt` method will also throw a number format exception if the value specified does not represent a valid integer. In either case, we catch the number format exception setting the point size to a default value of 12.

Note 3: We construct the font using the `fontName` and `size` parameters obtained from the HTML file. We could also have input the style as a parameter to the applet. We leave this modification for the exercises.

Figure 8.2 shows an HTML file with two `param` tags for the font name and the point size. Figure 8.3 shows the applet of Example 8.17 run using the HTML file of Figure 8.2.

FIGURE 8.2 The file `PassToAppletA.html`

```
<applet code = PassToApplet.class width=400 height=200>
<param name=fontName value="Serif">
<param name=size value=36>
</applet>
```

FIGURE 8.3 Example 8.17 run using `PassToAppletA.html`

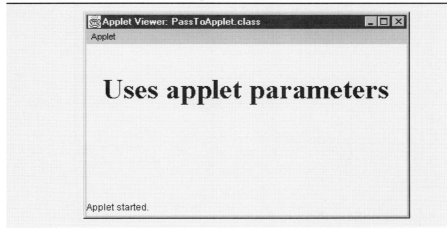

Figure 8.4 shows another HTML file, with different values for the `fontName` and `size` parameters. Figure 8.5 shows the applet of Example 8.17 run using the HTML file of Figure 8.4.

FIGURE 8.4 The file `PassToAppletB.html`

```
<applet code = PassToApplet.class width=200 height=200>
<param name=fontName value="SansSerif">
<param name=size value=14>
</applet>
```

The Big Picture

Program arguments allow us to pass data to the `main` method of a standalone program. Applet parameters, placed in the HTML file, let us pass data to an applet.

TEST YOUR UNDERSTANDING

TRY IT YOURSELF   12. Execute Example 8.16 passing it the program arguments `ReadFileLines.java` and 10. Explain the result.

TRY IT YOURSELF   13. Execute Example 8.16 passing it the program arguments `zxcvb.data` and 3. Explain the result.

TRY IT YOURSELF   14. Revise the HTML file of Figure 8.2 to omit both `param` tags. Run the applet of Example 8.17 using this revised file and explain what happens.

FIGURE 8.5 The applet viewer processing `PassToAppletB.html`

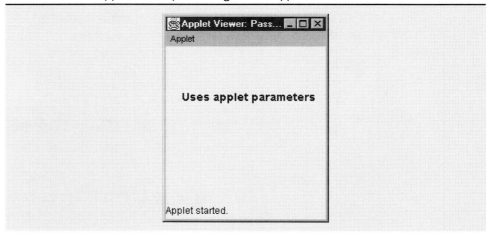

15. Explain what would happen if you removed the try and catch clauses from Example 8.17, and ran the applet using the HTML file as revised in question 14.

8.5 ■ Recursion: Searching and Sorting

Iteration and recursion allow us to repeat steps in a program. The **while**, **for**, and **do-while** loops use iteration to repeat steps where continuing the repetitions is based on the value of a test condition. Using iteration, our program shows the detailed mechanics of the repetition process. By contrast **recursion** deals with repetition at a higher level, letting Java manage the details hidden in calls to recursive methods. Some problems are much easier and more natural to solve using recursion, but in some situations using recursion can be inefficient. With experience, which comes with further study, one can judge when a recursive solution is appropriate.

An old proverb says that the journey of a thousand miles begins with the first step. This proverb captures the essence of recursion. Take one step and you start toward your goal. To achieve the goal you just have to repeat the process beginning at your new position, one step along the road. We can express this proverb in a Java method, `travel`:

```java
public void travel(int start, int finish) {
    if (start < finish) {
        takeOneStep(start);
        travel(start + 1,finish);
    }
}
```

The `travel` method is **recursive**; it recurs inside of itself. One reason recursion is an important concept is that recursive methods are easy to read. The code is like a specification of the travel process. The `travel` code says that to complete the trip from start to finish, if you are not already at the finish, then take one step and travel from that point to the finish, a method that certainly works.

Another reason for the importance of recursion is that recursive methods are usually short because we do not need to specify each detailed step of the solution. We call the method recursively, as in:

```
travel(start+1,finish);
```

letting the `travel` method fill in one more step every time it recurs. Another old proverb says that one picture is worth a thousand words, which we modernize to: One program is worth a thousand words. (The number 1000 seems to have some significance in proverbs.) Our one program is Example 8.18.

EXAMPLE 8.18 **Journey.java**

```java
/* Use a recursive method, travel, to journey from start
 * to finish, printing messages to show its progress.
 */

public class Journey {
    private static String indent = "";                            // Note 1
    public static void takeOneStep(int step) {
        System.out.println(indent + "Taking step " + step);
    }
    public static void travel(int start, int finish) {
        String oldIndent = indent;    // save indent to restore it later
        System.out.println
            (indent + "Starting travel from " + start + " to " + finish);  // Note 2
        if (start < finish) {
            takeOneStep(start);
            indent += "   ";                                      // Note 3
            travel(start + 1,finish);                             // Note 4
            indent = oldIndent;                                   // Note 5
        }
        System.out.println
            (indent + "Finishing travel from " + start + " to " + finish); // Note 6
    }
    public static void main(String [ ] args) {
        int start = Integer.parseInt(args[0]);                    // Note 7
        int finish = Integer.parseInt(args[1]);
        travel(start, finish);
    }
}
```

Output from java Journey 1 4
```
Starting travel from 1 to 4
Taking step 1
   Starting travel from 2 to 4
   Taking step 2
      Starting travel from 3 to 4
      Taking step 3
         Starting travel from 4 to 4
         Finishing travel from 4 to 4
      Finishing travel from 3 to 4
   Finishing travel from 2 to 4
Finishing travel from 1 to 4
```

Note 1: We use a static field to keep the string which specifies the amount to indent the message. Each time we call the `travel` method we increase the indent by three spaces, and each time we return we restore the previous indent three spaces to the left.

Note 2: We print the message using the current indent.

Note 3: We increase the indent before calling the `travel` method recursively.

Note 4: This recursive call starts executing the `travel` method with new arguments. This new activation of the `travel` method starts before the previous call to `travel` has returned.

Note 5: We restore the previous indent when we finish executing the `travel` method.

Note 6: `System.out.println`
```
        (indent + "Finishing travel from " + start + " to " + finish);
```
We display a message to show when this call to `travel` completes.

Note 7: `int start = Integer.parseInt(args[0]);`

We use program arguments to specify the start and finish values.

We see from the output that `travel(1,4)` takes the first step and calls `travel(2,4)` to complete the trip. `Travel(2,4)` takes the second step and calls `travel(3,4)` to complete the trip. `Travel(3,4)` takes the third step and calls `travel(4,4)` to complete the trip. `Travel(4,4)` concludes that the trip is complete and terminates. `Travel(3,4)` terminates, `travel(2,4)` terminates, and finally `travel(1,4)` terminates.

A recursive method needs an alternative that does not involve making another recursive call, or it will never terminate. The `travel` method terminates if start >= finish. At each call to `travel`, we increase start by one, getting closer to the finish, so eventually the program will terminate.

Before tackling the interesting binary search and merge sort problems we use recursion to sum an array of prices. Like the `travel` method, our recursive `sum` method adds the sum of the remaining elements of the array to the start element. We find the sum of 12.23 + 3.68 + 34.99 + 8.87 + 63.99 by adding 12.23 to the sum of 3.68 + 34.99 + 8.87 + 63.99 which we find by adding 3.68 to the sum of 34.99 + 8.87 + 63.99 and so on.

EXAMPLE 8.19   **SumPrices.java**

```java
/* The recursive sum method sums the elements
 * from start to end of the array passed to it.
 */

public class SumPrices {
    public static double sum(double[ ] p, int start, int end) {
        if (start < end)
            return p[start] + sum(p,start+1,end);              // Note 1
        else
            return 0;                                          // Note 2
    }
    public static void main(String[ ] args) {
        double[ ] prices = {12.23, 3.68, 34.99, 8.87, 63.99};
        System.out.println("The sum is $" + sum(prices,0,prices.length));
    }
}
```

Output
```
The sum is $123.76
```

Note 1: This is the recursive call where we add the start element of the array to the remaining elements.

Note 2: When start reaches end we have added all the elements and no longer need to make a recursive call. We return a zero as the sum of the remaining (none) elements.

Binary Search

A simple way to search for a value in an array compares the value with each element of the array until we find the element or reach the end of the array. In the worst case, when the element sought is not an element of the array, we must check each element.

If we keep the elements in the array in order from smallest to largest, then we can find an element using binary search, a much more efficient algorithm. We can program binary search as a recursive method or using a loop. In this section we will write a recursive method to perform a binary search.

The idea behind binary search is quite simple. Compare the element sought, called the key, with the middle element in the array. If the key equals the middle element, then we have found it and we are done.

If the key is smaller than the middle element, we know we need only search the left half of the array; because the array is ordered, a key smaller than the middle element can only be found in the left half of the array. In this case we call the binary search method recursively to search the array elements to the left of the middle.

If the key is greater than the middle element, then we need only search the right half of the array. In this case we call the binary search method recursively to search the array elements to the right of the middle.

In the journey of a thousand miles we take the first step, and then travel the rest of the way. In binary search we compare our key to the middle element of the array, and if it is not the value for which we are searching we search either the left half or the right half of the array for the key.

Figure 8.6 shows a trace of a binary search for 78 in an array, data, where data = {2,5,7,12,23,34,56,78,99,123,234,345,567}, whose leftmost element has index 0, and whose rightmost element has index 12.

FIGURE 8.6 Trace of a binary search for 78 in the data array

```
binarySearch(data,78,0,12)
                              middle = (0 + 12)/2  = 6
                              78 >  data[6]                 // data[6]==56
                                         // search index 7 to index 12.
         binarySearch(data,78,7,12)    middle = (7 + 12)/2 = 9
                              78 < data[9]                  // data[9]==123
                                         // search from index 7 to index 8.
             binarySearch(data,78,7,8)  middle = (7 + 8)/2 = 7
                              78 == data[7]                 // data[7]==78
                                   // return the index 7 to the caller.
```

Example 8.20 shows the code for binary search. The user inputs the array elements as program arguments to the main method, entering the elements in order from smallest to largest. The program prompts the user for the key to search.

EXAMPLE 8.20

BinarySearch.java

```
/* Inputs integers in order from smallest to largest on the
 * command line.  Uses a recursive method to implement binary
 * search.
 */

import iopack.*;                                                    // Note 1
public class BinarySearch {
    public static int binarySearch(int [ ] data, int key, int left, int right) {
        if (left <= right) {                                       // Note 2
            int middle = (left + right)/2;
            if (key == data[middle])
                return middle;
            else if (key < data[middle])
                return binarySearch(data,key,left,middle - 1);     // Note 3
            else                                                   // Note 4
                return binarySearch(data,key,middle + 1,right);
        }
        return -1;
    }
    public static void main(String [] args) {
        int key;                                    // the search key
        int index;                                  // the index returned
        char repeat;                                // 'Y' to repeat, 'N' to quit
        int [] data = new int[args.length];
        for (int i=0; i < data.length; i++)
            data[i] = Integer.parseInt(args[i]);                   // Note 5
        do {
            key = Io.readInt("Enter the search key");
            index = binarySearch(data,key,0,data.length-1);
            if (index == -1)
                System.out.println("Key " + key + " not found");
            else
                System.out.println("Key " + key + " found at index " + index);
            repeat = Io.readChar("Enter 'Y' to repeat, 'N' to quit");
        }while (repeat=='Y' || repeat == 'y');
    }
}
```

Output

(using program arguments 2 5 7 12 23 34 56 78 99 123 234 345 567)

```
Enter the search key:  78
Key 78 found at index 7
Enter 'Y' to repeat, 'N' to quit:  y
Enter the search key:  8
Key 8 not found
Enter 'Y' to repeat, 'N' to quit:  n
```

Note 1: For simplicity, we use the `Io` methods from package `iopack`.

Note 2: We do not use a **while** loop here, because the recursive call will start the rest of the search. We just determine one of three conditions: either we found the key, or it can only be found in the left half of the data, or it can only be found in the right half of the data. In the first case, we return the index at which we found the key. In the second and third cases, we return the result of the recursive call to the binary search method on the appropriate half of the array.

Note 3: We know that `data[middle]` is greater than the key, so we only need to search the array up to and including the element at index `middle-1`.

Note 4: In this third case, the only possibility remaining is that

```
key > data[middle]
```

so we call the binary search method to search the array from position `middle+1` to `right`.

Note 5: We convert the program arguments to **int** values, storing them in the data array that we will search.

We can also program binary search using a **while** loop, without recursion. We leave this approach for the exercises.

Merge Sort

We introduced the insertion sorting method in Section 4.4, and developed a GUI for it in Section 6.6. The **merge sort** algorithm, easily programmed recursively, is much more efficient than insertion sort especially for larger sets of data.

Merge sort uses the merge operation to sort the data. The merge operation takes two sorted arrays, merging them into a larger sorted array containing all the elements of the original two arrays. For example, the arrays {2,4,6,8} and {1,3,5,7} merge into the array {1,2,3,4,5,6,7,8}. Figure 8.7 traces a merge operation. At each step, we compare the initial elements of the first and second array, adding the smallest of these two elements to the merged array. When we have added all the elements of one array to the merged array, we simply copy the remaining elements in the other array to the merged array.

FIGURE 8.7 Merging two sorted arrays

First Array	Second Array	Merged Array
{2,5,7,8}	{3,4,9,10}	{2}
{5,7,8}	{3,4,9,10}	{2,3}
{5,7,8}	{4,9,10}	{2,3,4}
{5,7,8}	{9,10}	{2,3,4,5}
{7,8}	{9,10}	{2,3,4,5,7}
{8}	{9,10}	{2,3,4,5,7,8}
{ }	{9,10}	{2,3,4,5,7,8,9,10}

To make an analogy illustrating merge sort, we could say that a journey of a thousand miles ends with a single step. Given an array, such as {8,5,2,7,10,9,3,4}, to sort we first sort each half, {8,5,2,7} and {10,9,3,4}, obtaining the two sorted arrays {2,5,7,8} and

{3,4,9,10}, and then, as the final step, merge the two sorted halves into the final sorted array as shown in Figure 8.7.

We can write the merge sort program very simply, just as described in the last paragraph. The details of sorting the two arrays {8,5,2,7} and {10,9,3,4} are hidden in further recursive calls. To sort {8,5,2,7} we first sort each half, {8,5} and {2,7}, giving the two arrays {5,8} and {2,7}, and then merge these two sorted arrays into {2,5,7,8}. To sort {8,5} we first sort each half, {8} and {5}—the recursion stops here because single-element arrays are already sorted—and then merge these two sorted arrays into the array {5,8}. We have not traced all the steps involved in the merge sort. Tracing shows how the merge sort works, but we do not need these details to write the program.

EXAMPLE 8.21   **MergeSort.java**

```java
/* Implements the recursive merge sort
 * algorithm to sort an array that the user
 * inputs as program arguments.
 */

public class MergeSort {
    public static void mergeSort (int [ ] data, int left, int right) {
        if (left < right) {
            int middle = (left + right)/2;
            mergeSort(data,left,middle);              // sort the left half
            mergeSort(data,middle + 1,right);         // sort the right half
            merge(data,left,middle,middle + 1,right); // merge the left and right
        }
    }
    public static void merge(int[ ] data, int l1, int r1, int l2, int r2) {
        int oldPosition = l1;      // save position to copy sorted array
        int size = r2 - l1 + 1;
        int [ ] temp = new int[size];
        int i = 0;
        while (l1 <= r1 && l2 <= r2) {                          // Note 1
            if (data[l1] <= data[l2])
                temp[i++] = data[l1++];
            else
                temp[i++] = data[l2++];
        }
        if (l1 > r1)                                            // Note 2
            for (int j = l2; j <= r2; j++)
                temp[i++] = data[l2++];
        else
            for (int j = l1; j <= r1; j++)
                temp[i++] = data[l1++];
        System.arraycopy(temp,0,data,oldPosition,size);        // Note 3
    }
    public static void display(int [ ] anArray) {              // Note 4
        System.out.print("{");
        for (int i = 0; i < anArray.length; i++) {
            if (i != 0) System.out.print(",");
            System.out.print(anArray[i]);
        }
        System.out.println("}");
    }
```

```
public static void main (String [ ] args) {
    int [ ] data = new int[args.length];
    for (int i = 0; i < data.length; i++)
        data[i] = Integer.parseInt(args[i]);
    mergeSort(data,0,data.length - 1);
    display(data);
    }
}
```

Output

(using program arguments 8 5 2 7 10 9 3 4 6 11 77 1)

{1,2,3,4,5,6,7,8,9,10,11,77}

Note 1: As long as both arrays that we are merging are not empty, we copy the smallest element of both arrays into a temporary array, and increment the index, either l1 or l2, to the array containing the smallest element.

Note 2: If the first array empties first we copy the remaining elements of the second array into the temporary array, otherwise we copy the remaining elements of the first array.

Note 3: We want to copy the merged elements in the temporary array back to the original array that we are sorting. We use the `arraycopy` method of the `java.lang.System` class, which takes as its arguments

```
source array
starting index in the source array at which to
    start copying
destination array
starting index in the destination array for the
    copied elements
number of elements to copy
```

Note 4: We use the `display` method of Example 4.3.

The Big Picture

Iteration, used in **for** and **while** loops, manages each step of a repetition explicitly. Recursion takes a higher level approach to repetition, doing one step, which reduces the problem to a smaller size, and then asking to repeat that process. The system, behind the scenes, attends to the details. Binary search and merge sort, two important algorithms, illustrate recursion.

TEST YOUR UNDERSTANDING

16. As we did in Figure 8.6, trace the steps in the binary search for 8 in the array {2,5,7,12,23,34,56,78,99,123,234,345,567}.

17. Trace all the steps of the merge sort method on the array {8,5,2,7,10,9,3,4}.

TRY IT YOURSELF   18. Run the merge sort of Example 8.21 to sort the array {34,23,67,87,2,45,98,12,16,78,32}.

8.6 ▪ Data Structures: The Linked List

Data structures, ways of organizing data, are an important computer science topic because they are an important tool for programmers to develop efficient solutions to a variety of problems. Having a repertoire of data structures and the knowledge of their characteristics needed to use them effectively is essential. In this and the next section we introduce two of the most important data structures, the linked list and the stack, to add to the array that we have already introduced.

The very efficient binary search method of Example 8.20 uses an ordered array to hold the data to be searched. We can search array data efficiently, but inserting or deleting elements from an ordered array is not very efficient. For example, to insert 17 in the array {1,3,5,9,12,15,19,23,34,36,45}, we would have to move the elements greater than 17 one position to the right, assuming space is available in the array. If the array is full, then we have to allocate a new array and copy the whole array into the larger array.

The **linked list** data structure makes it easy to add or remove elements which is why it is so important, but searching for an item may be less efficient than the best array searches. In a linked list, we keep each data item in a node which also contains a reference to the next node in the list. Figure 8.8 shows a linked list containing integer data.

FIGURE 8.8 A linked list

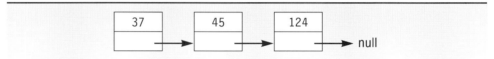

We show the reference in each node as an arrow pointing to the next node. In contrast to arrays, which store their elements contiguously, linked lists allocate each individual node as needed. No matter the size of the list, adding an element to it only requires us to change two references, which is why lists are especially useful when we have to perform many additions and deletions. For example to add 40 to the list of Figure 8.8, we need to allocate a new node, enter 40 in its data field, enter a reference to the node containing 45, and change the node containing 37 to refer to this new node. Figure 8.9 shows these changes.

FIGURE 8.9 Adding 40 to the linked list of Figure 8.8

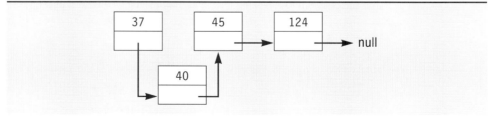

We can implement a linked list in a variety of ways. We choose to use an inner class, Node, for the individual nodes of the list. The private field, head, refers to the head node, which is the front of the list. The private field, current, refers to the current element; we can advance to the next element of a list, so at any given time we may be inspecting any of the list elements.

We need the field, previous, which refers to the list element just before the current element, when we remove the current element.

Figure 8.10 shows the list operations we include in our LinkedList data type.

FIGURE 8.10 The operations of the LinkedList class

```
public LinkedList()
    Constructs  an empty linked list, with the head, previous, and current
    nodes null.

public boolean isEmpty()
    Returns true if the list is empty and false otherwise.

public void insert(Object o)
    Creates a node containing the object o, inserting it before the current
    element.

public void remove()
    Removes the current element.

public Object getData()
    Gets the data field from the current element.  Returns null
    if the current element is null.

public boolean atEnd()
    Returns true if the current element is null,
    false otherwise.

public void advance()
    If current is not null, advances the previous and current references.

public void reset()
    Resets the current reference to refer to the head of the list.

public void display()
    Prints each element of the list on a separate line.
```

Example 8.22 shows the code for the LinkedList class. The main method tests the class. We can use LinkedList objects in other classes.

EXAMPLE 8.22 **LinkedList.java**

```java
/* Implements the LinkedList data structure.
 */

public class LinkedList {

    class Node {
        Object data;
        Node next;

        public Node(Object o, Node n){
            data = o;
            next = n;
        }
    }
```

continued

```
private Node head;
private Node previous;
private Node current;

public LinkedList() {
    head = null;
    previous = null;
    current = null;
}
public boolean isEmpty() {
    if (head == null)
        return true;
    else
        return false;
}
public void insert(Object o) {
    Node n = new Node(o,current);          // Note 1
    if (previous == null)                  // Note 2
        head = n;
    else
        previous.next = n;
    current = n;                           // Note 3
}
public void remove() {
    if (head != null){
        if (previous == null)              // Note 4
            head = head.next;
        else
            previous.next = current.next;
        current = current.next;            // Note 5
    }
}
public Object getData(){
    if (current != null)
        return current.data;
    return null;
}
public boolean atEnd() {
    return current == null;
}
public void advance(){
    if (!atEnd()){
        previous = current;
        current = current.next;
    }
}
public void reset() {
    previous = null;
    current = head;
}
public void display() {
    reset();                               // Note 6
    if (head != null)
        do {
            System.out.println("  " + getData());
            advance();
        }while (!atEnd());
}
```

```
public static void main(String [ ] args) {
  LinkedList list = new LinkedList();
  System.out.println("It is " + list.isEmpty() + " that this list in empty");
  list.insert("Happy days");
  list.insert("Pie in the sky");
  list.insert("Trouble in River City");
  System.out.println("The original list is:");
  list.display();
  list.reset();                                                    // Note 7
  list.advance();
  System.out.println("The current list element is " + list.getData());
  list.remove();
  System.out.println("The list, after removing the current element, is:");
  list.display();
  }
}
```

Output
```
It is true that this list is empty
The original list is:
  Trouble in River City
  Pie in the sky
  Happy days
The current list element is Pie in the sky
The list, after removing the current element, is:
  Trouble in River City
  Happy days
```

Note 1: We insert the new node before the current element, passing the current element to the next field of the new node.

Note 2: If we are inserting a node before the head node, we make head refer to the new node, otherwise we set the next field of the previous node to refer to the new node.

Note 3: The node being inserted becomes the current node.

Note 4: If we remove the head node, then we update the head reference, otherwise we change the next field of the previous node to refer to the node following the current node. This change unlinks the current node from the list.

Note 5: The node following the current node becomes the new current node, when we remove the current node.

Note 6: `reset();`

We reset to the beginning of the list to display the entire list.

Note 7: `list.reset();`

After displaying the list, the current node is **null**, at the end of the list. We need to reset to the beginning of the list to process the list further.

☛ A LITTLE EXTRA   **Running Out of Memory**

Whenever we use the **new** operator, as we do in creating nodes in a linked list, we are allocating memory. Java uses garbage collection to reclaim memory no longer in use by our program, but programs that use a lot of memory may cause memory to run out. If not enough memory is available, Java throws an `OutOfMemoryError` exception. As our examples have been relatively small we have not been concerned with running out of memory, and have not caught this exception which Java designates as unchecked to give us this option.

The Big Picture

In contrast to an array, a linked list does not keep its data in neighboring locations, but rather each node has a field which refers to its successor. This structure makes it easier to add and delete elements because we do not have to move as many items as we would in an array. However we can use an index to directly access an array element, but must traverse the list to reach a list element.

TEST YOUR UNDERSTANDING

TRY IT YOURSELF

19. Devise a thorough series of tests for the `LinkedList` class of Example 8.22. Run these tests and note the results.

8.7 ▪ Data Structures: Stacks

The stack is one of the most useful data structures. We have already seen it in Section 8.1 in the listing by the `printStackTrace` method of the stack of method calls in progress when an exception occurs.

Often technical terms mirror familiar terms that provide good analogies to the technical concepts. **Stack** is such a term. A stack of data is similar to a stack of books or a stack of dishes. We add a book to the top of the stack, and remove a book from the top. Sometimes we call the stack a LIFO stack, where **LIFO** stands for last in, first out. The last dish we stacked is the first that we remove, because it is on the top of the stack.

Of course computer people have to introduce some jargon, so we call the add operation **push** and the remove operation **pop**. In addition to the push and pop operations, we want an operation, isEmpty, to tell us if the stack is empty, an operation, isFull, to tell us if the stack is full, and an operation, top, to tell us what is on top of the stack without removing it.

Users of stacks just need to know the stack operations to work with stacks. As implementers of the `Stack` class, we can hide the representation of the stack by using private data fields. Let us use the elements of an array to hold the stack data, which for this example will be integers such as 15, 7, or 12. Stacks of books or dishes grow vertically, but we draw our arrays horizontally, so our stack will grow from left to right. Figure 8.11 shows a stack with space for six integers, which currently contains three integers.

FIGURE 8.11 A stack growing from left to right

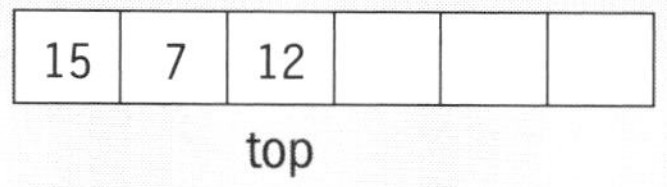

15	7	12			

top

A field, top, tells us the index of the top element on the stack. In Figure 8.11, top has the value two. The top method returns the element on the top of the stack without changing or removing it. Figure 8.12 shows an empty stack, in which case we assign top a value of −1 to indicate that nothing is on the stack yet.

FIGURE 8.12 An empty stack

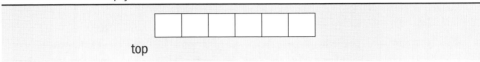

top

When the stack is full, as in Figure 8.13, then top has the value size-1, where the size field gives the number of elements allocated for the array.

FIGURE 8.13 A full stack

15	7	12	4	5	24

top

Using our array, we implement the push operation with the pseudocode:

```
if (stack full)
    throw an exception;
else  {
    Increment top;
    Add item to the array at index top;
}
```

Figure 8.14 shows the result of pushing 4 onto the stack of Figure 8.11.

FIGURE 8.14 Pushing 4 onto the stack of Figure 8.11

15	7	12	4		

top

The pop method returns and removes the element on the top of the stack, which presents a problem when the stack is empty, in which case we throw an exception. The pseudocode for the pop operation is:

```
if (stack empty) {
    throw an exception;
}
Return the top of the stack
    and decrement top;
```

Figure 8.15 shows the stack resulting from popping the stack of Figure 8.14. The integer 4 is still at index three of the array, but top is now two, so we ignore it.

When the push method encounters a full stack, it can throw a standard RuntimeException, passing it a message describing the nature of the error. Similarly the pop

FIGURE 8.15 Popping the stack of Figure 8.14

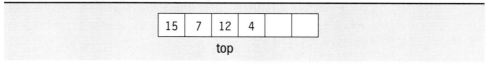

| 15 | 7 | 12 | 4 | | |

top

method can throw a RuntimeException when it tries to remove an element or return the top from an empty stack. We can also define our own exception classes with names specific to the error identified. For example, to handle a stack empty error, we can define a StackEmptyException class, which has a constructor with a string parameter which we pass to the superclass which saves it. When Java throws a StackEmptyException the printStackTrace method will display this error message, and a stack trace.

Example 8.23 shows the code for the stack data type implemented using an array for the stack data, a size field giving the number of elements allocated, and a top field which holds the index of the top element of the stack. We provide two constructors, one that allocates an array with a default size of 10, and one that allocates an array with a size passed in as an argument. Note that using an array to hold the stack values limits the size of the stack to the number of elements allocated for that array. We leave to the exercises the use of a linked list to create a stack which would remove this limitation.

EXAMPLE 8.23 | **Stack.java**

```java
/* Implements the stack data type using
 * an array and fields top and size.
 */

public class Stack {
    private int [] data;        // holds the stack data
    private int size;           //  holds the size allocated
    private int top = -1;       //  holds the index of the top       // Note 1
                                //  element, or -1 if none
    public Stack() {                                                  // Note 2
        size = 10;
        data = new int[size];
    }
    public Stack(int size) {
        this.size = size;                                             // Note 3
        data = new int[size];
    }
    public boolean isEmpty() {                                        // Note 4
        return top == -1;
    }
    public boolean isFull() {                                         // Note 5
        return top == size - 1;
    }

    public void push(int i) {
        if (isFull())
            throw new RuntimeException("Stack full -- cannot push");  // Note 6
```

```
        else
            data[++top] = i;                                            // Note 7
    }
    public int pop() {
        if (isEmpty())
            throw new StackEmptyException("Stack empty  -- cannot pop");
        else
            return data[top--];                                         // Note 8
    }
    public int top() {
        if (isEmpty())
            throw new StackEmptyException("Stack empty  -- top undefined");
        else
            return data[top];
    }
    public static void main(String [ ] args) {
        try {
            Stack stack1 = new Stack();
            Stack stack2 = new Stack(3);
            stack2.push(4);
            stack2.push(5);
            System.out.println("The top is now " + stack2.top());
            stack2.push(6);
            System.out.println("Popping stack 2 returns " + stack2.pop());
            System.out.println("Stack 1 has size " + stack1.size);
            System.out.println("Stack 1 empty? " + stack1.isEmpty());
            stack1.pop();
            System.out.println("Throws exception before we get here");
        } catch(Exception e) {
            e.printStackTrace();
        }
    }
    class StackEmptyException extends RuntimeException {
        public StackEmptyException(String message) {
            super(message);
        }
    }
}
```

Output
```
The top is now 5
Popping stack 2 returns 6
Stack 1 has size 10
Stack 1 empty? true
Stack$StackEmptyException: Stack empty -- cannot pop
        at Stack.pop(Stack.java:35)
        at Stack.main(Stack.java:57)
```

Note 1: We always construct empty stacks to start, so, no matter which constructor we use, the correct initial value of top is −1.

Note 2: This constructor initializes the stack with a default size of 10.

Note 3: This constructor initializes the stack with the size passed in as an argument. The formal parameter has the same name as the field, so we refer to the field as `this.size`, where the variable `this` refers to the current object. We do not often need to refer to the current object explicitly, but here is an example where we need to use the variable `this`.

Note 4: Top holds the index of the top element on the stack. When the stack is empty, top has the value −1.

Note 5: When the stack is full, top holds the index, size-1, of the last element in the array.

Note 6: `throw new RuntimeException ("Stack full -- cannot push");`

When the array is full we cannot add any more elements. Trying to add more would cause Java to throw an array index out of bounds exception and abort. We can detect this condition and throw an exception ourselves to allow users of a stack to catch the exception and to continue processing. We could have written our own exception class but for simplicity chose to throw a runtime exception which is unchecked so that stack users can ignore it if they wish. The message we pass to the exception object we create will be displayed by the `printStackTrace` method used in the catch clause.

Note 7: `data[++top] = i;`

The expression `++top` increments top to point to the next free space in the array, and returns the new value to use as the array index. This is equivalent to the code

```
top++;
data[top] = i;
```

Note 8: `return data[top--];`

When the stack is not empty, we return the integer, `data[top]`, at index top and decrement top. We do not have to remove the value from the array. The index top tells us where the top of the stack is; we ignore any array elements with index higher than top.

☞ A LITTLE EXTRA   **Using a Stack**

Stacks have many uses, including the evaluation of expressions. To evaluate an expression, we write it in **postfix** form, sometimes called reverse Polish notation, in which the operands occur first followed by the operator. The expression $5 + 6$ has the postfix form $5\ 6\ +$, while the expression $(7 + 8)*(4 + 5)$ has the postfix form $7\ 8\ +\ 4\ 5\ +\ *$. Remember the postfix form follows the pattern

left operand	right operand	operator

We will not cover methods for converting an infix expression to its postfix form. It helps to add parentheses and then follow the pattern. For example, given the expression $3 + 4 * 5$, multiplication has higher precedence, so we add parentheses to give $3 + (4*5)$. Following the pattern, the postfix is

```
3    4 5 *    +
```

or 3 4 5 * +.

Once we have a postfix expression, we can easily evaluate it using a stack. Let us assume, for simplicity, that all operands are single digits 0 to 9, and that the operators are the binary arithmetic operators, +, -, *, and /. The algorithm for evaluating a postfix expression is:

```
do {
    Read the next character;
    if (next character is a digit)
        Convert the digit to an integer
            and push the integer on  the stack;
    else if (next character is an operator) {
        Pop two operands from stack;
        Perform the operation;
        Push the result onto the stack;
    }
} while (more characters);
Display the top of the stack;
```

Figure 8.16 applies this algorithm to the expression 7 8 + 4 5 + * .

FIGURE 8.16 Evaluating the expression 7 8 + 4 5 + *

Read '7'	
Push the integer 7	7
Read '8'	
Push the integer 8	7 8
Read '+'	
Pop 8 and pop 7	empty
Add, getting 15	
Push 15	15
Read '4'	
Push the integer 4	15 4
Read '5'	
Push the integer 5	15 4 5
Read '+'	
Pop 5 and pop 4	15
Add, getting 9	
Push 9	15 9
Read '*'	
Pop 9 and pop 15	empty
Multiply, giving 135	
Push 135	135
Pop 135 and display it.	

☛ **TIP**

When evaluating subtraction we take the top entry on the stack and subtract it from the next to the top entry. For example, when evaluating 7 5 - we follow the steps:

```
push 7, push 5, pop 5 and 7, subtract 7-5, push 2
```

When evaluating division we take the top entry on the stack and divide it into the next to the top entry. For example, when evaluating 14 3 / we follow the steps:

```
push 14, push 3, pop 3 and 14, divide 14/3, push 4
```

We can input the postfix expression one character at a time. To convert from a character c, where the value of c is a digit such as 9, to an integer, use the method `Character.digit(c,10)` where the number 10 refers to base 10. We leave the writing of a program to evaluate a postfix expression using a stack to the exercises.

The Big Picture

A stack allows us to add and remove data at the top. Because we implemented a stack using an array, we throw an exception if the user tries to pop an empty stack or push onto a full stack. We illustrated by throwing a standard `RuntimeException` and creating our own `StackFullException` and `StackEmptyException` classes. Stacks have many uses; one is evaluating postfix expressions.

TEST YOUR UNDERSTANDING

20. Show the stack of Figure 8.11 after performing the operations `pop()`, `pop()`, and `push(19)`, in that order.

21. Show the stack of Figure 8.11 after performing the operation `push(pop())`. What can you conclude about the relationship between the push and pop operations?

A LITTLE EXTRA

22. Use a stack to evaluate the following postfix expressions:

     a.   2 3 4 * + 6 2 - +   b.   9 9 + 9 9 * -   c.   6 3 / 4 9 8 + - +

TRY IT YOURSELF

23. Run additional tests of the stack operations for the `Stack` class of Example 8.23.

Chapter 8 Summary

Java provides an exception handling facility with a hierarchy of predefined exception classes. Some of the methods in the Java library packages throw exceptions when error conditions occur. Java will throw an array index out of bounds exception when a user tries to find an array element having an index outside the bounds specified in the creation of the array. The program aborts when an unhandled exception occurs.

To handle an exception, we put the code that can throw that exception in a try block followed by a catch clause for that exception. Java passes an object representing that exception to the catch clause. Inside the catch block, we can put the code we want to execute after an exception has occurred. Java jumps from the line where the exception occurred to the catch clause, and continues execution from there, never returning to the code that caused the exception.

If there is no catch clause in the method where the exception occurs, Java will look for one in the caller of that method, and so on to its caller, finally aborting the program if no catch clause is found. Inside the catch block, we can call the `printStackTrace` method to show which exception occurred and to see the sequence of method calls that led to the exception.

To prevent the program from aborting, we can sometimes validate our data; for example, we can check that array indices are valid or that objects are non-null before we try to access their fields. For these types of exceptions Java gives us the option to handle them or not. By contrast, Java requires that we handle IO exceptions from which we could not otherwise recover and continue with the program.

Java has classes for binary IO, in which data is kept in an internal format not easily readable, and for text IO where we convert internal values to text output, and text input to binary internal values. We can read from an external text file by passing the file name to the `FileReader` constructor. For efficiency we pass the file reader to a buffered reader so we can minimize the number of accesses to the external disk, and do most of our reading from a buffer in memory. The `readLine` method reads a line from the file, returning **null** when at the end of the file. We must handle the `IOException` that the constructors and methods might generate. The `close` method releases resources back to the operating system.

We use the `StringTokenizer` class to read more than one field from a single line. The default `StringTokenizer` constructor assumes that whitespace separates the strings, but an optional second argument allows us to specify other delimiters, such as the vertical bar. The `nextToken` method returns the next item on the line as a string, which we convert to an **int** for integer data, and to a **double** for decimal values.

To write to a text file, we pass the file name to a `FileWriter` constructor. Passing the file writer to a `PrintWriter` constructor will allow us to use the familiar `print` and `println` methods to output values of any of the primitive types, or string representations of objects using the `toString` method. The methods of the `Io` class that comes with this text use the methods described here.

The `File` class has methods to obtain the properties of a file. We use the `read` method of the `FileInputStream` class to read bytes from a file and the write method of the `FileOutputStream` and `DataOutputStream` classes allow the input and output of primitive types in binary format. We can seek a specific position in a `RandomAccessFile`. The `writeObject` and `readObject` methods use object serialization to store and retrieve objects.

We can use program arguments to pass data to the `main` method of a standalone application. If the formal parameter to the `main` method is `String[] args`, then `args[0]` will represent the first program argument, `args[1]` the second, and so on. These values are strings which we may need to convert to type **int** or **double**.

To pass parameters to an applet, we use the `param` tag in the HTML file. The `name` attribute in that tag specifies the name of the parameter, while the `value` attribute specifies its value. In the applet's code, we use the `getParameter` method, with the `name` attribute as its argument, to get the value of the parameter. The value is of type `String`, which we may need to convert to an **int** or a **double**.

Recursion deals with repetition by describing a process whose structure recurs within itself. In binary search, we compare the key with the middle element. If equal, we have found it; if the key is greater than the middle element we have only to search the upper portion of this sorted array. Calling binary search recursively, we compare our key to the middle of the upper portion, either returning, if we found it, or starting yet another binary search of the upper or lower half (of the upper half of the original array). While the trace of a recursive algorithm can get complicated, the program is usually very simple.

Merge sort is a very efficient sorting method which has a very nice recursive implementation. We call merge sort recursively to sort the left and right halves of the array, and then merge the resulting sorted arrays into the final result.

The linked list data structure makes it easy to add and remove elements but loses some of the advantages of arrays for searches, and takes extra space to hold the node references. Each node of a linked list contains a data field and a reference to another node. When we add an item to the list, or remove an item from a list, we just have to change two references. If the Stack class of Section 8.7 used a linked list to store its data, rather than an array, it would be able to grow in size without being restricted to the size of its internal array.

The useful Stack type has operations push, pop, top, isEmpty, and isFull. Push adds an item to the top of the stack. Pop removes the item on the top of the stack, while top inspects the top item without removing it. IsEmpty and isFull tell us whether the stack is empty or full. We operate on stacks using just these operations and constructors. The representation of the stack, which is hidden from the user, uses an array to hold the data, a size field to hold the size of the array, and a top field which gives the index of the top element of the stack, or holds -1 if the stack is empty.

Build Your Own Glossary

Find the uses, in this chapter, of the terms below. Enter a definition of each in the `glossary.txt` file on the disk that comes with this text.

abort	merge sort	seek
applet parameters	object persistence	stack
buffer	object serialization	stack trace
catch	pop	throw
checked exception	postfix	token
command line	program arguments	try block
exception	push	unchecked exception
LIFO	random access file	whitespace
linked list	recursion	

Skill Builder Exercises

1. For each code fragment in the left column choose the exception from the right column that it might throw.

    a. `r.readLine()`

    b. `a[index]`

    c. `new Integer(s)`

    d. `new FileReader("abc.data")`

    i. `ArrayIndexOutOfBoundsException`

    ii. `NumberFormatException`

    iii. `IOException`

2. Fill in the blanks in the code to create a reader to read text from a file `text.data` buffering it for efficiency.

```
BufferedReader buffer
    = new _____ (new _____ (_____));
```

3. Fill in the blanks in the code to create a writer to write text to a file `text.out` using `print` and `println` methods.

```
PrintWriter writer
    = new _____ (new _____ (_____));
```

Critical Thinking Exercises

4. If we use the tag

```
<param name="font"  value="Monospaced">
```

to pass the font to an applet, which of the following expressions can be used in the applet to retrieve the font name?

   a. `getName(font);`   d. `getParameter("font");`
   b. `getParameter(font);`   e. `none of the above`
   c. `getParameter(name);`

5. If we include the statement

```
FileReader reader = new FileReader("test.dat");
```

in our program without putting it in a try block with a catch clause for the `IOException` (and without declaring that the method containing this line can throw an `IOException`)

   a. the compiler will report an error.
   b. the program will compile, but will abort when running if there is no file named `test.dat` on the user's machine.
   c. the program will always run, using a default file if `test.dat` is not available.
   d. none of the above.

6. If we include the statement

```
Integer i = new Integer("3.14");
```

in our program without putting it in a try block with a catch clause for the `NumberFormatException` (and without declaring that the method containing this line can throw a `NumberFormatException`)

   a. the compiler will report an error.
   b. the program will compile, but will abort when it reaches this statement.
   c. the program will always run, truncating the value 3.14 to 3.
   d. none of the above.

7. Given the statement `StringTokenizer s = new StringTokenizer("123|abc|456","|");` which statement below will cause the **int** variable `i` to have the value 123?

   a. `int i = s.nextToken();`   d. `int i = nextToken(s);`
   b. `int i = Integer.parseInt(s.nextToken());`   e. none of the above.
   c. `int i = Integer.parseInt(s);`

Debugging Exercise

8. The following program attempts to compute the product of the first n integers where n is passed as a program argument. For example, passing a program argument of 4 should generate a result of 24 because 4*3*2*1 = 24. Find and correct any errors in this program.

```
public class Factorial {
    public static int product(int n) {
        return n*product(n - 1);
    }
    public static void main(String[ ] args) {
        System.out.println(product(Integer.parseInt(args[0])));
    }
}
```

Program Modification Exercises

9. Modify Example 8.3 to give the user another chance to enter a correct value after Java throws an exception.

PUTTING IT ALL TOGETHER

10. Modify Example 4.4, `StudentScore`, to read the test scores from a file. The first line will contain the number of students. Put each student's scores on a separate line, with each score separated by a blank. Prompt the user to enter the file name.

11. Modify the `Stack` class of Example 8.23 to use the `LinkedList` of Example 8.22, rather than an array, to hold the stack elements.

12. Modify Example 8.9, `Prices`, to catch the number format exceptions that might be generated. Test with a file that includes some values that will cause the exception to be thrown.

13. Modify Example 8.7 to allow the user to enter another file name if an exception is thrown.

PUTTING IT ALL TOGETHER

14. Modify Example 1.7, `Growth`, to use a `NumberFormat` instance to print the balance rather than the `println$` method of the `Io` class that comes with this text.

PUTTING IT ALL TOGETHER

15. Modify Example 4.4, `StudentScore`, to use a `NumberFormat` instance to format the outputting of the average score, rather than the `Io.println` method.

16. Modify Example 8.17, `PassToApplet`, to input the font style using a `param` tag in the HTML file.

17. Modify Example 8.20, `BinarySearch`, to use the `java.io` package for IO rather than the `Io` class in the `iopack` package.

18. Modify the merge sort program of Example 8.21 to input the numbers to be sorted from a file, rather than as program arguments. Pass the file name as a program argument.

19. Modify the merge sort program of Example 8.21 to sort `String` objects. Input the strings from a file, rather than as program arguments, and write the sorted strings to a file. Pass the file name as a program argument.

20. Modify Example 8.23 by writing an exception class `StackFullException` and throwing this exception instead of `RuntimeException`.

21. Modify Example 8.22 to include an inner class which implements the Enumeration interface. Add an `elements` method to the `LinkedList` class which returns an Enumeration for the list. In the `main` method, use the `nextElement` and `hasMoreElements` methods of the Enumeration interface to display the elements of a list.

Program Design Exercises

22. Write a Java standalone application that presents data read from a file in a bar chart. Enter the file name as a program argument.

23. Write a Java program that searches a file for a string. Pass the string and the file name as program arguments.

24. Write a Java program that reads a text file, removing any extra spaces between words, and writes the output to a file. Enter the file names to read from and write to as program arguments.

25. Write a Java program to update an inventory file. Each line of the inventory file will have a product number, a product name, and a quantity separated by vertical bars. The items in the inventory file will be ordered by product number. The transaction file will contain a product number and a change amount which may be positive, for an increase, or negative, for a decrease. Assume the transaction file is also ordered by product number. Use the transaction file to update the inventory file, writing a new inventory file with the updated quantities. Assume there is at most one transaction for each item in the inventory, and that no new items occur in the transaction file.

26. Write a Java program which provides a GUI to copy Java programs to the screen or to another file. List the Java programs in a choice box. Use checkboxes in a checkbox group to indicate whether to copy the file to the console window or to another file. Use a text box to enter the name of the file receiving the copy.

27. Write an applet which draws a circle and a rectangle. Use applet parameters to specify the color of the circle and the color of the rectangle.

28. Write an applet which draws a rectangle. Use applet parameters to specify the position and size of the rectangle.

29. Write a Java program, which does not use recursion, to perform a binary search of an array input from a file.

30. Write a Java program to search for an item in a linked list. Use the `LinkedList` class of Example 8.22. Use a recursive method to do the search.

31. Write a Java program to check if a string is a palindrome (reads the same backward and forward, as, for example, "toot"). Use a recursive method to do the checking.

32. Suppose we have n disks on a peg, each of different sizes, stacked in order of size, with the largest on the bottom, and two other pegs, as shown in Figure 8.17.

What is the sequence of moves needed to transfer the rings to the second peg, in the same configuration, in order from largest to the smallest, with the largest at the

FIGURE 8.17 The Towers of Hanoi puzzle

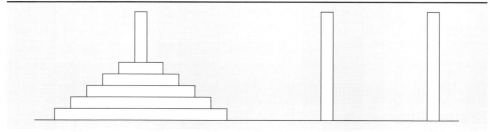

bottom, if we can move only one disk at a time, and we cannot place a larger disk on top of a smaller disk? We may use the last peg to store disks, but the same rules apply. Use a recursive method to provide the solution. To move n disks from peg 1 to peg 2, move n-1 disks to peg 3, move the bottom disk to peg 2, then move n-1 disks from peg 3 to peg 2. For this problem, print out the sequence of moves, as for example, "Move disk from peg 1 to peg 2", and so on. In Chapter 9 we will animate the solution.

A LITTLE EXTRA

33. In Section 8.7, we showed how to use a stack to evaluate a postfix expression. Write a program to allow the user to input a postfix expression, of the form described in Section 8.7, in a text field. Add a button for the user to ask the program to process the next character. Make a graphical display which shows the stack as it grows and shrinks, and display the result of the evaluation.

34. Write a Java program that uses a stack to reverse an array.

35. Write a Java program which uses a recursive method to reverse an array.

PUTTING IT ALL TOGETHER

36. Create a user interface to manipulate and display a linked list. Include buttons to insert and remove a string entered in a text field, and a button to reset the list to its head. Display the list using rectangles for the nodes.

9

Threads and Multimedia

To go on with Java we need to exploit its powerful capabilities for multimedia and networking. In this chapter we introduce threads which allow two or more pieces of a program to execute in their own threads of control, appearing to work simultaneously. Concurrent programming deals with synchronization of and communication among threads. We hope to avoid a deadlock in which all threads are stuck.

Animation provides a good illustration of thread use. Running an animation in one thread allows the program to continue other processing. We can animate graphics we draw on the screen or images we can download from remote sites. In Java, we can play sound clips, and we add these to our animations.

OBJECTIVES

- Create several threads in one program.
- Use threads by extending the Thread class.
- Use threads by implementing the Runnable interface.
- Use locks to ensure exclusive access to data.
- Communicate using the wait and notify methods.
- Understand deadlock.
- Perform animations using threads.
- Load and draw an image.
- Use a media tracker to wait for an image to load.

◆ Remove flicker with double buffering.
◆ Play sounds.

OPTIONS
◆ Omit 9.3: Animation (used in 9.4 and 9.5)
◆ Omit 9.4: Images
◆ Omit 9.5: Sound

9.1 ▪ Introduction to Threads

The term **thread** is short for thread of control. Someone who can read a book and watch television at the same time is processing two threads. For awhile she concentrates on the book, perhaps during a commercial, but then devotes her attention to a segment of the TV program. Because the TV program does not require her undivided attention, she reads a few more lines every now and then. Each thread gets some of her attention. Perhaps she can concentrate on both threads simultaneously, like musicians who are able to follow the different parts of the harmony.

If we have only one thread of control, then we have to wait whenever that thread gets delayed. For example, suppose our thread is downloading a picture from the Internet. It may have to wait while the system transfers all the pixels of the picture from some remote site. If our program can create a second thread to wait for the input from the remote site, it can go with other processing while the new thread is waiting. When some new data comes in from the remote site, the new thread can receive it while the first thread waits for awhile. The two threads share the processor. Figure 9.1 illustrates this sharing.

FIGURE 9.1 Two threads sharing the processor

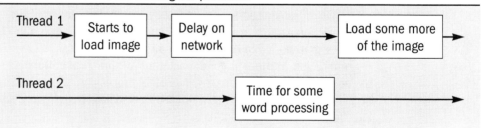

The Thread Class

Java allows us to use threads in our program. Our program can define two or more threads, and the processor will divide its attention among the active threads. Each thread that we define executes a run method when it gets the processor. We can define a thread by extending the Thread class, overriding the run method to specify what our thread will do when it has control, as in

```
public class MyThread extends Thread {
    ...
    public void run () {
        // put code here for thread to run
        // when it gets control
    }
}
```

To make a thread, we can create a new thread and call its `start` method, as in

```
MyThread t = new MyThread();
t.start();
```

This will make the thread ready, and when it gets scheduled, it will execute its `run` method. When another thread gets a turn, the thread, t, will stop executing the code in its `run` method, but will start again from where it left off, when it gets another turn.

Optionally, we can name our thread, passing the name to the constructor which in turn passes it to the `Thread` superclass constructor, as in:

```
public MyThread(String name) {
    super(name);
    ...
}
```

The `Thread` class has a static method, `sleep(int milliseconds)`, which will cause its caller to **sleep** (be blocked from using the processor) for the specified number of milliseconds. While one thread sleeps, another will get a turn. We call the `sleep` method in a try block, as in

```
try {
    Thread.sleep(1000);
}catch(InterruptedException e) { return;}
```

where we have to catch the `InterruptedException` which would occur if another thread interrupted this one. We will not consider interruption or other more advanced thread concepts in this text.

In Example 9.1 we create two threads, Bonnie and Clyde, and let them write their names five times, sleeping after each writing. Bonnie will sleep for 1000 milliseconds after each writing, while Clyde will sleep for 700 milliseconds. Processors are so fast that a thread could do a large amount of output while it has its turn. We sleep here to slow the thread down to human scale, so we only have to read a few lines of output. The `main` method runs in a thread, different from the two we create, so we will actually have three threads sharing the processor, writing their names when they get their turns. We let `main` sleep for 1100 milliseconds after it writes.

EXAMPLE 9.1   **NameThread.java**

```
/* Creates two threads that write their names and
 * sleep. The main thread also writes its name
 * and sleeps.
 */

import java.io.*;

public class NameThread extends Thread {
    int time;              // time in milliseconds to sleep

    public NameThread(String n, int t) {
        super(n);
        time = t;
    }
```

continued

```
        public void run() {
          for (int i = 1;i <= 5;i++) {                          // Note 1
            System.out.println(getName() + " " + i);            // Note 2
            try {
              Thread.sleep(time);
            } catch (InterruptedException e) {return;}
          }
        }

        public static void main(String argv[]) {                // Note 3
          NameThread bonnie = new NameThread("Bonnie",1000);
          bonnie.start();
          NameThread clyde =  new NameThread("Clyde",700);
          clyde.start();
          for (int i = 1; i <= 5; i++) {                        // Note 4
            System.out.println(Thread.currentThread().getName() + " " + i);
            try {
              Thread.sleep(1100);
            } catch (InterruptedException e) {return;}
          }
        }
      }
```

Output
```
main 1
Bonnie 1
Clyde 1
Clyde 2
Bonnie 2
main 2
Clyde 3
Bonnie 3
Clyde 4
main 3
Clyde 5
Bonnie 4
main 4
Bonnie 5
main 5
```

Note 1: Each thread will print its name five times and sleep after each time, returning where it left off when it gets the processor again. We can see in the output that after Bonnie prints her name the first time, Clyde gets a turn and manages to print his name twice before Bonnie returns printing her name the second time. Main started first because it had the processor first, at the start of the program.

Note 2: The getName method of the Thread class returns the thread's name.

Note 3: We put the main method in the NameThread class for simplicity. We could have created another class, say TryNameThread, with a main method to create the threads.

Note 4: The main method also writes its name five times. Because we did not create this thread in our program we get it using the static currentThread method of the Thread class.

Figure 9.2 helps us to understand the order in which the three threads execute in Example 9.1. Each thread spends most of its time sleeping; printing its name takes a mere fraction of the time it sleeps. By graphing the sleep times in Figure 9.2 we can get a good idea of when each thread will be ready to run.

FIGURE 9.2 Threads of Example 9.1 sleeping and waking up

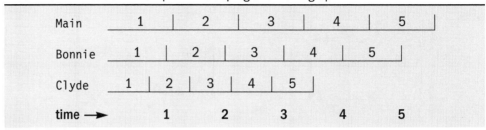

Main prints its name first and then sleeps for 1.1 seconds. Bonnie starts next, printing her name and sleeping for 1 second. Clyde prints his name and sleeps for .7 second. When he wakes up the other two threads are still sleeping so he prints his name again (#2). Because Bonnie woke up before main, she prints her name (#2) first, followed by main (#2). (Picking the thread that becomes ready first is a choice made by the thread scheduler.) When main finishes both Bonnie and Clyde are awake, but Clyde woke up first and executes first (#3). We leave it to the reader to continue following the diagram in Figure 9.2 to explain the results of Example 9.1.

The Runnable Interface

In Example 9.1 we extended the Thread class, creating a NameThread subclass which overrode the run method to provide the code for a NameThread object to execute in its thread of control. The Runnable interface provides an alternate method to use threads. In this approach we use an interface to perform a callback as we have been doing with the various listener interfaces used in event handling.[1]

The Runnable interface contains just the one run method.

```java
public interface Runnable  {
    public void run();
}
```

A concrete class that implements the Runnable interface must implement the run method. An object of this class must pass itself to a thread so that when that thread gets the processor it will execute the object's run method. In Example 9.2 we rewrite Example 9.1, creating a class that implements the Runnable interface rather than extending Thread.

EXAMPLE 9.2   **NameUsingThread.java**

```java
/* Revises Example 9.1 to implement the Runnable
 * interface rather than extending Thread.
 */

import java.io.*;
```
continued

---

[1] See Section 4.6 for an introduction to the use of interfaces to perform callbacks.

```java
public class NameUsingThread implements Runnable {
    private int time;
    private Thread thread;  // the thread to execute the run method

    public NameUsingThread(String n, int t) {
        time = t;
        thread = new Thread(this,n);                              // Note 1
        thread.start();
    }
    public void run() {                                           // Note 2
        for (int i = 1;i <= 5;i++) {
            System.out.println(thread.getName() + " " + i);
            try {
                Thread.sleep(time);
            } catch (InterruptedException e) {return;}
        }
    }

    public static void main(String argv[ ]) {
        NameUsingThread bonnie = new NameUsingThread("Bonnie",1000);
        NameUsingThread clyde =  new NameUsingThread("Clyde",700);
        for (int i = 1;i <= 5; i++) {
            System.out.println(Thread.currentThread().getName() + " " + i);
            try {
                Thread.sleep(1100);
            } catch (InterruptedException e) {return;}
        }
    }
}
```

Output

Output is the same as that from Example 9.1.

***

Note 1: We create a new thread, passing it the current object of type NameUsingThread, which implements the run method that the thread will run when it gets the processor, and the name of the thread. The next line starts the thread, making it ready to run when it gets its turn.

Note 2: The NameUsingThread class implements the run method which the thread will call when it gets the processor.

Either Example 9.1 or 9.2 works fine; there is no reason to prefer one approach. When a class already extends a class, it cannot extend the Thread class, therefore only the approach of Example 9.2, implementing the Runnable interface, would work.

***

The Big Picture

Threads appear to execute simultaneously. When a thread gets the processor it executes the code in a run method. A thread which extends the Thread class has its own run method. A thread may instead execute the run method of a class which implements the Runnable interface. In either case, the start method makes the thread ready to run. Threads that are ready to run share the processor in a manner determined by a scheduler. We use the sleep method to pause a thread for a specified period of time.

TRY IT YOURSELF   1.   In Example 9.1, change the sleep amounts for threads Bonnie and Clyde to 300 and 200 milliseconds respectively. How does the output change when you rerun the example?

TRY IT YOURSELF   2.   In Example 9.2, change the sleep times for the `main` thread to 200 milliseconds. How does the output change when you rerun the example?

TRY IT YOURSELF   3.   What do you predict the output would be if you omit all the `sleep` statements from Example 9.1? Rerun the program with these changes, and see if your supposition is correct.

9.2 ▪ Concurrent Programming

Having seen threads running independently in the last section, we take up the interesting and difficult problem of threads which share data and need to cooperate with each other to operate correctly. To illustrate the problem, suppose that two threads are depositing to an account, and that a deposit involves two steps:

1.   computing the new balance
2.   recording the change in a log

We assume each thread computes the balance separately, but shares the log to enter the result.

A thread runs for a certain time period, and then another thread gets its turn. If each thread completes both steps when it has its turn, the balance and the log will be consistent, but perhaps thread1 loses its turn after completing step 1.

FIGURE 9.3 A problem with threads

thread1	thread2
balance = $100	
	balance = $200
	enter $200 in log
enter $100 in log	

The execution sequence of Figure 9.3 shows that after thread1 computes a balance of $100 it loses its turn to thread2 which computes a balance of $200 and records the new balance in the log. When thread1 gets its turn again it finishes where it left off, entering $100 in the log, which is now incorrect.

To create a simple program to illustrate this phenomenon, we use a buffer that contains an integer, number, which two threads share. The `increment` method adds 1 to number and reports its new value, while the `decrement` method subtracts 1 and reports the new value. When the threads operate correctly, successive reports will display values that differ by one.

For example, starting with `number` at 5, two increments give

```
thread1 increments to 6
thread1 reports 6
                        thread2 increments to 7
                        thread2 reports 7
```

while an increment followed by a decrement results in

```
thread1 increments to 6
thread1 reports 6
                        thread2 decrements to 5
                        thread2 reports 5
```

Only when the threads operate incorrectly do we get the same value on successive reports. For example, again starting at 5,

```
thread1 increments to 6
                        thread2 decrements to 5
                        thread2 reports 5
thread1 reports 5
```

Example 9.3 shows this behavior, but not very often. The `increment` and `decrement` methods are so short they are rarely interrupted before completion. For larger methods, this error would occur more often.

EXAMPLE 9.3　**TallyWrong.java**

```
/* Two threads occasionally err in reporting
 * values because they get interrupted before
 * finishing to execute a method.
 */

public class TallyWrong {
    class Buffer {
        int number = 0;        // the number to increase or decrease
        int previous = 0;
        int total = 0;         // total number of operations performed
        int errors = 0;        // number of errors

        public void increment() {
            number++;
            report(number);
        }
        public void decrement() {
            number--;
            report(number);
        }
        public void report(int n) {
            total++;
            if (n == previous)                                    // Note 1
                System.out.println(++errors + "\t" + total );
            previous = n;
        }
    }
}
```

```
class Plus extends Thread {
   Buffer buf;
   Plus(Buffer b) {
     buf = b;
   }
   public void run() {
       while (true)                          // Note 2
           buf.increment();
   }
}
class Minus extends Thread {
   Buffer buf;
   Minus(Buffer b) {
       buf = b;
   }
   public void run() {
       while (true)
           buf.decrement();
   }
}
public static void main(String[ ] argv) {
   TallyWrong tw = new TallyWrong();
   Buffer b = tw.new Buffer();               // Note 3
   Plus p = tw.new Plus(b);
   Minus m = tw.new Minus(b);
   p.start();
   m.start();
}
}
```

Output

1	456250
2	4649634
3	7245105
4	12916833
5	14452346
6	23241347
7	24257947
8	29420248
9	30436644
10	36110307

Note 1: When the current and previous values of number agree, an error has occurred, and we output the number of errors and the total number of operations performed. We illustrated the error when a thread loses its turn after changing number but before reporting. It could also lose its turn while executing the report method. The point is that errors can occur due to the interleaving of execution of various threads.

Note 2: This thread keeps incrementing number in an unending loop. We need to abort execution manually.

Note 3: We used inner classes to make the class names local to the TallyWrong class. To access these classes we use an instance of TallyWrong.

Synchronization

To correct the problem exhibited by Example 9.3, we need to enable the `increment` and `decrement` methods to execute completely once they have begun. Java provides the **synchronized** keyword to enforce this behavior. We declare the `increment` method as

```
public synchronized void increment() {
    number++;
    report(number);
}
```

When thread1, say, calls this method of the `buf` object,

```
buf.increment();
```

if no other thread is executing any method of `buf`, thread1 gets a **lock** for the object. Therefore no other thread can use `buf` until it has finished executing the `increment` method. If another thread is executing a method of `buf`, then thread1 must wait until that operation completes. If several threads are waiting to get a lock on an object, the thread scheduler determines who will get it when it becomes available.

FIGURE 9.4 Thread1 locks `buf` while thread2 waits

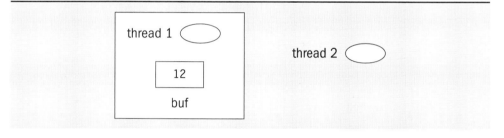

We only need make the `increment` and `decrement` methods in Example 9.3 synchronized to ensure that they work correctly when accessed by multiple threads. With synchronized methods, behavior like that shown in Figure 9.3 cannot occur because once a thread starts executing a synchronized method it is allowed to finish.

EXAMPLE 9.4 **TallyRight.java**

```
/* Uses synchronized methods to correct the
 * error in Example 9.3.
 */

public class TallyRight {

    // The rest of the code is the same as Example 9.3.

    public synchronized void increment() {
        number++;
        report(number);
    }
```

```
public synchronized void decrement() {
    number--;
    report(number);
}
```
}

Output
There is none, because no errors occur.

☛ **A LITTLE EXTRA**   Synchronizing methods exacts a performance penalty. If only a portion of the code of a large method is critical, we can improve performance by only synchronizing that portion of the code. Schematically,

```
public returnType someMethod(someParameters) {
        // some code not synchronized

    synchronized(this) {
        // synchronize the critical code here
    }

        // more non-synchronized code
}
```

When a thread calls this method it will only lock the object while executing the synchronized block.

....

Communication

Synchronization allows threads to complete portions of code without interruption. Sometimes threads also need to communicate with one another to signal the occurrence of a condition that may affect their ability to proceed. For our example, we look at the classic **producer-consumer problem**, in which both producer threads and consumer threads access a data buffer. Producers add data to the buffer, while consumers remove it.

Assuming a fixed-size buffer, a producer cannot add more than the buffer can hold, while a consumer cannot retrieve data from an empty buffer. Each Java object has wait and notify methods which are useful in this situation.

When a producer has a lock on the buffer, and cannot add data because the buffer is full, it executes the wait method causing it to release the lock and wait to be notified the state of the buffer has changed. When a consumer removes an item from a full buffer, it executes the notify method to notify a waiting thread that the buffer is no longer full.

Similarly, when a consumer has a lock on the buffer, and cannot remove data because the buffer is empty, it executes the wait method causing it to release the lock and wait to be notified that the state of the buffer has changed. When a producer puts an item into an empty buffer, it executes the notify method to notify a waiting thread that the buffer is no longer empty.

Example 9.5 solves this producer-consumer problem. We input sleep times for each thread to see how the behavior varies depending on which thread has more time. We use one producer and one consumer thread.

EXAMPLE 9.5   PutGet.java

```java
/* Uses wait and notify to enable
 * producer and consumer threads to
 * cooperate in using a buffer.
 */
public class PutGet {
    public static final int size = 3;
    class Buffer {
        int[ ] buffer = new int [size];          // the data
        int putpos = 0;                          // next position to put a value
        int getpos = 0;                          // next position to get a value
        int number = 0;                          // number of items in the buffer

        public synchronized void put(int value)
                                throws InterruptedException {       // Note 1
            if (number == size) {
                System.out.println("Cannot put -- Buffer full");
                wait();                                            // Note 2
            }
            number++;
            buffer[putpos] = value;
            System.out.println("Put "+value);
            putpos = (putpos + 1) % size;                          // Note 3
            if (number == 1) notify();                             // Note 4
        }
        public synchronized int get() throws InterruptedException {
            if (number == 0) {
                System.out.println("Cannot get -- Buffer empty");
                wait();
            }
            number--;
            int n = buffer[getpos];
            System.out.println("Get "+n);
            getpos = (getpos + 1) % size;
            if (number == size - 1) notify();                      // Note 5
            return n;
        }
    }
    class Producer extends Thread {
        Buffer buf;
        int time;
        Producer(Buffer b, int t) {
            buf = b;
            time = t;
        }
        public void run() {
            for(int i = 1; i <= 10; i++)                           // Note 6
                try {
                    buf.put(i);
                    sleep(time);                                   // Note 7
                }catch (InterruptedException e){
                    e.printStackTrace();
                }
        }
    }
```

```
class Consumer extends Thread {
    Buffer buf;
    int time;
    Consumer(Buffer b, int t) {
        buf = b;
        time = t;
    }
    public void run() {
        for (int i = 1; i <= 10;i++)
            try {
                buf.get();
                sleep(time);
            }catch (InterruptedException e) {
                e.printStackTrace();
            }
    }
}
public static void main(String[ ] args) {
    PutGet pg = new PutGet();
    Buffer b = pg.new Buffer();
    Producer p = pg.new Producer(b,Integer.parseInt(args[0]));
    Consumer c = pg.new Consumer(b,Integer.parseInt(args[1]));
    p.start();
    c.start();
}
}
```

Output—from java PutGet 300 500

```
Put 1
Get 1
Put 2
Get 2
Put 3
Put 4
Get 3
Put 5
Put 6
Get 4
Put 7
Get 5
Put 8
Cannot put -- Buffer full
Get 6
Put 9
Cannot put -- Buffer full
Get 7
Put 10
Get 8
Get 9
Get 10
```

Note 1: The `wait` method may throw an `InterrupedException`, which we declare here to pass it on to the caller of the `put` method who will handle it.

Note 2: The `wait` method is a member of the `Object` class, so any object, such as a buffer in this example, may invoke it. `Wait` must be invoked in a synchronized method so the thread invoking it has a lock on the object.

Note 3: We use a circular buffer. Visualizing the array as a circle shows that after filling position 2, we move to position 0 again. This formula computes indices in this way: 0, 1, 2, 0, 1, 2, ... and so on.

Note 4: After putting an item into an empty buffer, we call `notify`. The scheduler will notify one thread, making it ready to run. In our example, we have at most one thread waiting, so if it is waiting it will be notified that the buffer is non-empty. When more than one thread may be waiting, calling `notifyAll` will wake them all up.

Note 5: By removing an item from a full buffer, the consumer has just made it possible for a waiting producer to add an item. Calling `notify` will wake up a waiting producer.

Note 6: `for(int i = 1; i <= 10; i++)`

We let the producer produce 10 numbers, and the consumer consume these 10. We could have run these threads in an unending loop, and aborted the program manually.

Note 7: `sleep(time);`

Typically there will be some extended computation to produce the value to put in the buffer. For simplicity we sleep to simulate some computational time.

Deadlock

When threads wait for locks to be freed that cannot be freed we have **deadlock**. We can easily modify Example 9.5 to produce deadlock. If we change the condition for the producer to put only when the buffer is empty and the consumer to get only when the buffer is full we reach a deadlocked state almost immediately. The buffer starts out empty so the producer can put one item into it, but no more until the consumer removes that item, making the buffer empty again. The consumer cannot remove the one item in the buffer until the producer adds two more items to fill the buffer. Both the producer and consumer are stuck, each waiting an action by the other that can never occur.

Good programming is the only prevention for deadlock. In more complicated situations it can be very difficult to determine if deadlock can occur.

EXAMPLE 9.6 **Deadlock.java**

```
/* Modifies Example 9.5 to illustrate deadlock.
 */

public class Deadlock {

   // the rest of the code is the same as Example 9.5

   public synchronized void put(int value) throws InterruptedException {
      if (number != 0) {
```

```
                    System.out.println("Cannot put -- Buffer not empty");
                    wait();
                }
            ...
        }
        public synchronized int get() throws InterruptedException {
            if (number != size) {
                System.out.println("Cannot get -- Buffer not full");
                wait();
            }
            ...
        }
    }
```

Output
```
Put 1
Cannot get -- Buffer not full
Cannot put -- Buffer not empty
```

(At this point the program hangs up because neither the producer nor the consumer can proceed.)

The Big Picture

Concurrent programming coordinates multiple threads. When threads share data we can synchronize access so that a thread using the data will be able to complete its operation before another thread gets access to that data. A thread gets a lock on the object which contains the data until it finishes the synchronized method or block.

Threads can wait on a condition, to be notified by other threads when changes occur that may make the condition satisfied. Deadlock occurs when threads wait for locks that can never be freed and no thread can proceed.

TEST YOUR UNDERSTANDING

TRY IT YOURSELF   4.   Vary the sleep times when running Example 9.5, and determine how that affects the results.

TRY IT YOURSELF   5.   Modify Example 9.5 to start two producers and two consumers, and explain the resulting behavior.

9.3 ▪ Animation

Animation, creating the illusion of motion by displaying a sequence of frames, provides an interesting application for threads. We will use a thread to show a ball moving across the screen. This thread computes the new position of the ball, asks the system to repaint the screen, showing the ball in the new position, and sleeps for a fraction of a second, so we can see the ball moving at human speed.

In downloading a file from a remote site, using only one thread slows the system down preventing other processing. In animation, we need a thread to drive the processing. Our standalone programs so far have had a single thread of control. The Java interpreter starts executing the statements of the main method, perhaps calling other methods, but always remaining in control. Our applets so far have been quite passive, responding to events generated by button presses, entries in text fields, or other user actions. The Java interpreter and the browser or applet viewer keep control of execution.

Pressing a Button to Move a Ball

Suppose we want to write an applet to move a ball across the screen. We can add a button, and whenever the user presses the button we can change the position of the ball, asking the system to repaint the screen. Figure 9.5 shows the actionPerformed method that handles the button press.

FIGURE 9.5 Handling a button press to move a ball

```
public void actionPerformed(ActionEvent event) {
    x += 9;
    y += 9;
    repaint();
}
```

Each time we press the button, the browser or applet viewer executes the code of Figure 9.5. The call to repaint asks Java to paint the screen which, because x and y have increased, will show the ball in its new position. When the browser releases control, the system paints the screen. We can press the button several times, each time moving the ball farther.

Suppose we try to have the user initiate several moves with one button press. Figure 9.6 shows the actionPerformed method with a loop to repeat the move.

FIGURE 9.6 An attempt to move the ball 10 times with one button press

```
public void actionPerformed(ActionEvent event) {
    for (int i = 0; i < 10; i++) {
        x += 9;
        y += 9;
        repaint();
    }
}
```

When the user presses the button, the browser or applet viewer executes the actionPerformed method. Every time around the loop, it requests the system to paint the ball in its new position, but it does not release control until the loop is finished. Java combines all these repaint requests that it could not get to while the browser was executing the actionPerformed method, and just draws the ball in its final position.

Example 9.7 shows the complete code using the actionPerformed method of Figure 9.5, which moves the ball once with each button press. We leave it as an exercise to modify this example to use the actionPerformed method of Figure 9.6 and see what happens (or does not happen).

FIGURE 9.7 The applet of Example 9.7

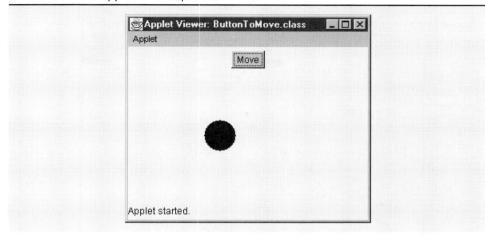

Figure 9.7 shows the ball which the user can move by pressing the button.

EXAMPLE 9.7 **ButtonToMove.java**

```java
/* Moves a ball when the user presses a
 * button. Modifying the actionPerformed method
 * to put the code in a loop will cause the applet
 * to display only the ball's final position.
 */

import java.awt.*;
import java.applet.Applet;
import java.awt.event.*;

public class ButtonToMove extends Applet
                                implements ActionListener {
    private int x = 50, y = 50;
    private Button move;

    public void init() {
        move = new Button("Move");
        add(move);
        move.addActionListener(this);
    }
    public void actionPerformed(ActionEvent event) {
        x += 9;
        y += 9;
        repaint();
    }
    public void paint(Graphics g) {
        g.fillOval(x,y,40,40);
    }
}
```

Animating a Ball

An applet needs its own thread of control to do animation. By using a separate thread, other system functions can proceed in parallel, sharing the processor, so the animation will not monopolize the system's resources. As in Example 9.2, the applet implements the `Runnable` interface providing the `run` method, shown in Figure 9.8, which the new thread it creates will execute when it gets its turn.

FIGURE 9.8 The `run` method of a thread for an animation

```java
public void run() {
    for(int i = 0; i < 10; i++) {
        x += 9;
        y += 9;
        repaint();
        try {
            Thread.sleep(300);
        }catch(InterruptedException e) {
            e.printStackTrace();
        }
    }
}
```

Our thread changes the position of the ball, calls the `repaint` method to show the ball in its new location, and sleeps for 300 milliseconds to do the animation on a human scale and allow the system to paint the screen. When a thread sleeps, other threads that are waiting to run, if any, get their turn to use the processor.

When the user presses the button, the applet creates an instance of `Thread`, passing itself as an argument, as in

```java
Thread t = new Thread(this);
```

so the thread will know whose `run` method to execute. To set up the thread and make it ready to run, the applet calls the thread's `start` method, as in

```java
t.start();
```

As soon as it gets its turn, the thread will start executing the applet's `run` method, terminating when it has finished; while executing it takes turns with the system which paints the ball in the new position. Example 9.8 moves the ball ten times when the user presses the button. Each time the user presses the button, we return the ball to its start position and repeat the ten moves. Try it out; a static figure will not show how it works.

EXAMPLE 9.8 **MoveBall.java**

```java
/* Uses a thread to move a ball ten times
 * when the user presses a button.
 */

import java.awt.*;
import java.applet.Applet;
import java.awt.event.*;
```

```
public class MoveBall extends Applet
                              implements ActionListener, Runnable {
    private int x = 50, y = 50;
    private Button move;

    public void init() {
        move = new Button("Move");
        add(move);
        move.addActionListener(this);
    }
    public void actionPerformed(ActionEvent event) {
        x = 50; y= 50;
        Thread t = new Thread(this);
        t.start();
    }
    public void paint(Graphics g) {
        g.fillOval(x, y, 40, 40);
    }
    public void run() {
        // See Figure 9.8
    }
  }
}
```

The animation occurs in a thread and does not interfere with other processing, so we can modify Example 9.8 to keep it going indefinitely, as long as the user remains viewing the page containing the applet. If the user moves to another web page, we want to terminate the thread. If the user returns to the page containing this applet, we start a new thread. The browser or applet viewer calls the start and stop methods of the Applet class to start the applet when we reenter the web page it is on, and to stop it when we leave that page. We override the start method to create a new thread and start it when the user initially views the page containing our applet, or returns to it. We override the applet's stop method to set a **boolean** flag when we leave the web page containing the applet. The run method periodically checks this flag, terminating when it becomes **true**.

☛ **TIP**

The Thread class has methods suspend, resume, and stop which were used to control a thread when the user leaves a web page or returns to it. In Java 2(1.2) these methods are **deprecated**, meaning they are still available so old programs will still run, but should not be used in new code.[2]

....

To move the ball continuously, we use a nonterminating loop in the run method, moving the ball ten times down and to the right, followed by ten moves up and to the left. While the ball

---

[2] Using the JDK, the Java compiler will give you a warning if you use any deprecated methods. To find out which methods caused the warning, compile the program using the deprecation option, as in:

`javac -deprecation progname.java`

Look up the deprecated methods in the JDK documentation to find suggested alternatives. For example, this documentation explains the reasons why suspend, resume, and stop were deprecated and suggests the approach of setting a flag that we use in Example 9.9.

moves, notice we can edit text in another window, or run another program. As we will see in Section 9.5, we can perform other tasks in the same applet, such as whistling a tune, while the animation is continuing. The animation shares the processor rather than monopolizing it. In Example 9.9 the applet uses a thread to move the ball back and forth until the user aborts it.

EXAMPLE 9.9   **AnimateBall.java**
....

```java
/* Animates a ball, moving it until the applet
 * quits.  Uses a thread to allow the system to
 * continue other processing.
 */

import java.awt.*;
import java.applet.Applet;

public class AnimateBall extends Applet implements Runnable {
    private int x, y;
    private boolean done;

    public void start() {
        x = 50; y = 50;                                    // Note 1
        done = false;
        Thread t = new Thread(this);
        t.start();
    }
    public void stop() {                                   // Note 2
        done = true;
    }
    public void paint(Graphics g) {
        g.fillOval(x,y,40,40);
    }
    public void run() {
        int dx = 9, dy = 9;
        while (true) {                                     // Note 3
            for(int i = 0; i < 10; i++) {
                if (done) return;                          // Note 4
                x += dx;
                y += dy;
                repaint();
                try {
                    Thread.sleep(300);
                }catch(InterruptedException e) {
                    e.printStackTrace();
                }
            }
            dx = -dx; dy = -dy;                            // Note 5
        }
    }
}
```

Note 1: The browser or applet viewer calls the applet's start method when the user enters or returns to the web page containing the applet. We start the ball at (50,50) and set the done flag, which the thread checks while running, to **false**.

Note 2: When the user leaves the web page containing the applet, the browser or applet viewer calls the applet's stop method, which we override to set the done flag to

true. The thread checks this flag; it will terminate when the flag is **true**. When the user leaves the page containing the applet, we want the thread to finish and release the resources it uses.

Note 3: Often threads run in a nonterminating loop. Because they share the processor with other threads, they can keep running until the user terminates the applet. By contrast, a nonterminating loop in a program without threads would monopolize the processor, not allowing the program to perform other tasks.

Note 4: The thread checks the done flag at every iteration of the loop. When **true**, it returns, terminating itself.

Note 5: To get the applet to move back and forth, we change the direction of the increment after every ten moves.

☛ **TIP**   The browser or applet viewer may call the start method of the applet many times, whenever the user returns to the web page containing the applet. We call the start method of a thread only once to make it ready to run. We may call the applet's stop method many times, whenever the user leaves the web page containing it.

Animating a Face

We used a thread to animate a ball; we can use the same technique to do fancier animations, for example, to draw a sequence of figures that would change the appearance of each frame of the animation. To make a modest start in this direction, we animate a circular face with circles for eyes, a line for a nose, and an arc for a mouth. We use three different frames, and continually draw one after the other, sleeping for 300 milliseconds between each repainting with the next frame. Example 9.10 performs this animation. Figure 9.9 shows two of the three different faces we use in the animation.

FIGURE 9.9 Two faces from the applet of Example 9.10

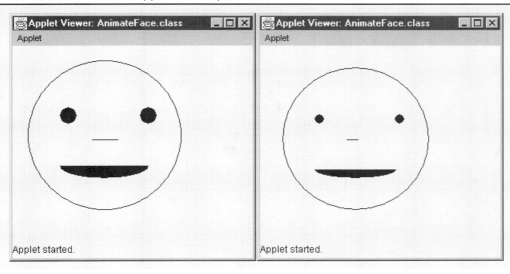

EXAMPLE 9.10 AnimateFace.java

```java
/* Animates a face, drawing it in three different sizes
 * with an interval of 300 milliseconds between each
 * repainting.
 */

import java.awt.*;
import java.applet.Applet;

public class AnimateFace extends Applet implements Runnable {
    private int x, y, width, height;                      // circle for face
    private int leftEyeX, leftEyeY, lwidth, lheight;      // circle for left eye
    private int rightEyeX, rightEyeY, rwidth, rheight;    // circle for right eye
    private int leftNoseX, leftNoseY, rightNoseX, rightNoseY; // line for nose
    private int mouthX, mouthY, mouthWidth, mouthHeight;  // arc for mouth
    private boolean done;

    public void start() {
        x = 10; y = 10; width = 200; height = 200;
        leftEyeX = 50; leftEyeY = 70; lwidth = 30; lheight = 30;
        rightEyeX = 150; rightEyeY = 70; rwidth = 30; rheight = 30;
        leftNoseX = 90; leftNoseY = 120; rightNoseX = 140; rightNoseY = 120;
        mouthX = 50; mouthY = 130; mouthWidth = 120; mouthHeight = 40;
        done = false;
        Thread t = new Thread(this);
        t.start();
    }
    public void stop() {
        done = true;
    }
    public void paint(Graphics g) {
        g.drawOval(x, y, width, height);
        g.fillOval(leftEyeX, leftEyeY, lwidth, lheight);
        g.fillOval(rightEyeX, rightEyeY, rwidth, rheight);
        g.drawLine(leftNoseX, leftNoseY, rightNoseX, rightNoseY);
        g.fillArc(mouthX, mouthY, mouthWidth, mouthHeight, 180, 180);
    }
    public void run() {
        int dx = 9, dy = 9;
        while (true) {
            for (int i = 0; i < 3; i++) {
                if (done) return;
                x += dx; y += dy; width -= dx; height -= dy;        // Note 1
                leftEyeX += dx; leftEyeY += dy; lwidth -= dx; lheight -= dy;
                rightEyeX += dx; rightEyeY += dy; rwidth -= dx; rheight -= dy;
                leftNoseX += dx; rightNoseX -= dx;
                mouthX += dx; mouthY += dy;
                mouthWidth -= dx; mouthHeight -= dy;
                repaint();
                try {
                    Thread.sleep(300);
                }catch(InterruptedException e) {
                    e.printStackTrace();
                }
            }
            dx = -dx; dy = -dy;
        }
    }
}
```

Note 1: We add a small amount to the position of each component of the face to move it slightly to the right three times, returning it to its original position in the next three moves. We decrease the size of the face, and its eyes, nose, and mouth as we move it to the right, increasing it to the original size as we move back to the original position.

The Big Picture

Trying to move the ball inside the actionPerformed method did not work because the event thread executing that method had to finish handling the button press before it could take on another task. Using a separate thread for the animation lets it proceed concurrently. In an animation, we draw different frames successively. In our examples, we changed the positions of the objects and redrew them. We could have designed separate frames not obtained from one another by these changes in coordinates.

TEST YOUR UNDERSTANDING

TRY IT YOURSELF
6. Change the actionPerformed method of Example 9.7 to the one of Figure 9.6. Run the code and observe the result. Does it draw the ball 10 times?

TRY IT YOURSELF
7. Start the animation of Example 9.9 and leave it running. Can you edit another program while the animation is running? Can you open another console window and run another Java program while the animation is running?

TRY IT YOURSELF
8. In Example 9.10, try different values for the sleep time and decide which you think makes the animation look best. How might we make it easier to try the applet with these different sleep times?

9.4 ▪ Images

An applet or an application can download an image from a remote site. Any component can create an image on which we can draw. An applet uses the getImage method to download an image.

Locating an Image

We used URLs (Uniform Resource Locator) in Section 5.1 to specify resources available on computers at remote sites. We entered the URL in the browser and the browser connected to the remote site, displaying the resource for us in its window. In this section we use a URL to specify the location of an image. We can specify the complete URL, creating a URL object, as in:

```
URL url = new URL("http://www.engr.csulb.edu/~artg/gittleman.gif");
```

and pass that object to the getImage method, as in:

```
Image pic = getImage(url);
```

More commonly, we specify the location of the image relative to another URL, as in:

```
Image pic = getImage(getDocumentBase(),"gittleman.gif");
```

where we use the `getDocumentBase` method which returns the URL of the HTML document in which the applet is embedded, and the string `gittleman.gif` represents a path to a file relative to the location of the HTML file. In this case the image is in the same directory as the HTML file. We could also use the relative URL

```
Image pic = getImage(getCodeBase(),"gittleman.gif");
```

where `getCodeBase` returns the URL of the applet's `.class` file.

Using MediaTracker

Downloading an image, especially from a remote site, may be time consuming, so the `getImage` method returns immediately, letting a separate thread load the image. If we try to draw the image before it is fully loaded we may get a blank screen or a partial image. We can use the `MediaTracker` class to keep track of the image. We create a media tracker, passing it the component on which we will draw our images, as in:

```
tracker = new MediaTracker(this);
```

We add each image to the tracker with a priority, as in:

```
tracker.addImage(pic, 0);
```

where the image `pic` will have priority 0. We can add more than one image with the same priority. Images with lower priority values will be loaded first.

Before we try to draw an image, we wait until it has been loaded, calling the `waitForID` method, to which we pass the ID of the image for which we are waiting, as in:

```
tracker.waitForID(0);
```

When Java has loaded the image we can draw it using the `drawImage` method of the `Graphics` class, as in:

```
g.drawImage(pic,10,10,null);
```

where the first argument is the image, and the next two arguments are the (x,y) coordinates of the position to display the image. The last argument is an `ImageObserver` which we could use to track the loading of the image. Because we use a media tracker to wait for the image to load completely, we do not use this argument, passing **null**.

The four-argument version of the `drawImage` method draws the image in its normal size. A six-argument version of `drawImage` lets us scale the image, as in:

```
g.drawImage(pic,10,10,100,100,this);
```

which scales the image, drawing it in a 100x100 region.

We can use the `getWidth` and `getHeight` methods to get the width of an image, as in:

```
imageWidth = pic.getWidth(null);
imageHeight = pic.getHeight(null);
```

where the **null** argument indicates we do not use an image observer because we call these methods after the image has been fully loaded.

Animating an Image

Our first program to animate images follows the format of our animation programs of Section 9.3. In those examples we used a loop in the `run` method to change the location and/or the dimensions of the shapes to be drawn, to repaint, and to sleep. In Example 9.11 we load an image of the author from his home page, using a media tracker to wait until the image is completely loaded before trying to draw it. In the `run` method, we change the desired size of the image, repaint, and sleep for 300 milliseconds. The image expands, filling up more of the applet, and then contracts, repeating this behavior over and over again. It does flicker; we will eliminate the flicker in the Example 9.12.

☞ **TIP**

Browsers have severe security restrictions and may not allow an applet to load files, including images, from the local machine. When we deploy our applet on a web server, we can put any image and sound files it uses in the same directory as the HTML file. Applets are able to load files from the same site from which they were loaded. To illustrate the deployment of an applet we have placed Examples 9.11 and 9.12 on our web page. Use the URL

```
http://www.engr.csulb.edu/~artg/AnimateImage.html
```

in your browser to view the applet of Example 9.11. Those not connected to the Internet may use an applet viewer to view Examples 9.11 and 9.12 locally. Using an applet viewer allows us to test our applets before deploying them on our web site.

Figure 9.10 shows the distorted image of the author which occurs as part of the animations of Examples 9.11 and 9.12.

FIGURE 9.10 A step in the animation of Example 9.12

EXAMPLE 9.11   AnimateImage.java

```java
/* Animates an image, loading it from a URL.  Uses a
 * thread to allow the system to continue other processing.
 */

import java.awt.*;
import java.applet.Applet;
import java.net.URL;                                    // Note 1

public class AnimateImage extends Applet implements Runnable {
    private boolean done;                   // should the thread stop?
    private Image pic;                      // the image
    private int imageWidth, imageHeight;
    private MediaTracker tracker;           // tracks image loading
    private boolean ready = false;          // image is loaded

    public void init() {
        pic = getImage(getDocumentBase(),"gittleman.gif");    // Note 2
        tracker = new MediaTracker(this);
        tracker.addImage(pic, 0);                             // Note 3
    }
    public void start() {
        done = false;
        Thread t = new Thread(this);
        t.start();
    }
    public void stop() {
        done = true;
    }
    public void paint(Graphics g) {
        if (ready)                                            // Note 4
            g.drawImage(pic,10,10,imageWidth,imageHeight,null);
        else
            g.drawString("Loading image",30,30);
    }
    public void run() {
        try {
            tracker.waitForID(0);                             // Note 5
        }catch (InterruptedException e) {}
        imageWidth = pic.getWidth(null);                      // Note 6
        imageHeight = pic.getHeight(null);
        ready = true;              // image loaded
        int dx = 20, dy = 5;
        while (true) {
            for(int i = 0; i < 10; i++) {
                if (done) return;
                imageWidth += dx;                             // Note 7
                imageHeight += dy;
                repaint();
                try {
                  Thread.sleep(300);
                }catch(InterruptedException e) {
                    e.printStackTrace();
                }
            }
            dx = -dx; dy = -dy;
        }
    }
}
```

Note 1: We need to tell the compiler where to find the URL class.

Note 2: The image `gittleman.gif` will be loaded from the directory containing the file `AnimateImage.html`. When connected to the Internet, open the URL

http://www.engr.csulb.edu/~artg/AnimateImage.html

in a browser or applet viewer. For the convenience of those not connected to the Internet, we placed the image on the disk included with this text. Use an applet viewer that comes with your development environment to run this applet locally, if your browser will not allow an applet to get an image from the local machine. See the A Little Extra section below to use a standalone application to show an image.

Note 3: We add the image to the media tracker so we can wait until it is loaded before trying to draw it.

Note 4: If the image is not loaded yet, we draw a string, otherwise we draw the image scaled to fit a rectangle whose width is `imageWidth` and whose height is `imageHeight`. We do not use the final argument, and just pass in **null**.

Note 5: We wait for the image with ID zero to be fully loaded.

Note 6: `imageWidth = pic.getWidth(null);`

We get the width and height after the image is fully loaded, so we will get the correct values. We do not use the image observer argument, and pass a **null** value for it.

Note 7: `imageWidth += dx;`

At each iteration of the loop we change the size of our scaled image. We loop ten times increasing the size of the display, alternating with ten iterations decreasing the size back to its original value.

☛ **A LITTLE EXTRA**   **Images in Standalone Applications**

For applets, the browser or applet viewer handles the platform-dependent details of getting images. Standalone applications need to use the `Toolkit` class of the `java.awt` package. First we get a `Toolkit` object using the static `getDefaultToolkit` method, as in:

Toolkit t = Toolkit.getDefaultToolkit();

We use the `Toolkit` object to get an image, as in:

Image pic = t.getImage(url);

where, for example, `url` is the URL for the famous Mona Lisa

URL url = new URL("http://www.paris.org/
        Musees/Louvre/Treasures/gifs/MonaLisaa.gif/");

When the image file resides on the local machine we can use a `String` argument to the `getImage` method, as in:

Image author = t.getImage("gittleman.gif");

Double Buffering to Remove Flicker

Running Example 9.11 animates the image in different sizes, but the animation flickers a lot, making it somewhat unpleasant to view. The flickering occurs because of the way the system repaints the image. Our call to the `repaint` method does not directly cause a call to the `paint` method, but rather causes a call to the `update` method which first clears the screen and then calls the `paint` method. Each time we repaint, we see a blank screen followed by the image. This alternation of blank screen and image appears as a distracting **flicker**.

To eliminate the flicker, we use **double buffering**. We do our drawing, whether of graphical shapes, text, or images, on an offscreen buffer. Then when the drawing is done do we copy it to the visible screen. We use the `createImage` method to get an offscreen image on which to draw, passing it the size of our applet for its size, as in:

```
Image buffer = createImage(getSize().width,getSize().height);
```

Having created the offscreen image, we need to get a graphics context to use to draw on it. We get the `Graphics` object using the `getGraphics` method, as in:

```
bufferGraphics = buffer.getGraphics();
```

We do all our drawing in the buffer, only at the end copying the completed buffer to the screen. First we clear the buffer of the previous image. We then draw the scaled image, and a rectangular frame around it, to show we can draw shapes as well as images in the offscreen buffer. The code is:

```
bufferGraphics.clearRect(0, 0, getSize().width, getSize().height);
bufferGraphics.drawRect(8, 8, imageWidth + 3, imageHeight + 3);
bufferGraphics.drawImage(pic, 10, 10, imageWidth, imageHeight, null);
g.drawImage(buffer, 10, 10, null);
```

The last change we need to make is to override the `update` method to just call `paint`, not to clear the screen every time. We do the clearing in the offscreen buffer, where it does not cause flicker. Example 9.12 revises Example 9.11 to eliminate flickering. We only show the changes here, but the complete example is on the disk included with this text.

EXAMPLE 9.12 **AnimateImageNoFlicker.java**

```
/* Modifies Example 9.11 to use double buffering to eliminate the flicker.
 */

public class AnimateImageNoFlicker extends Applet {
                    //  the rest is the same as Example 9.11
   Image buffer;
   Graphics bufferGraphics;

   public void init() {
      ...
      buffer = createImage(getSize().width, getSize().height);
      bufferGraphics = buffer.getGraphics();
   }
   public void update(Graphics g) {                                      // Note 1
      paint(g);
   }
   public void paint(Graphics g) {
      if (ready) {
```

```
                    bufferGraphics.clearRect(0, 0, getSize().width, getSize().height);
                                                                        // Note 2
                    bufferGraphics.drawRect(8,8,imageWidth + 3, imageHeight + 3); // Note 3
                    bufferGraphics.drawImage(pic,10,10, imageWidth, imageHeight, null);
                    g.drawImage(buffer, 10, 10, null);                  // Note 4
                }
                else
                    g.drawString("Loading Image", 30, 30);
        }
}
```

Note 1: We override the `update` method to paint without clearing the screen first.

Note 2: We clear the offscreen buffer, which does not cause flicker. We clear a rectangle which has the size of the image. All the drawing on the offscreen image uses the `bufferGraphics` object that provides the graphics context for this image.

Note 3: To show we can draw shapes on an offscreen image, as well as images, we draw a rectangle to frame the picture, making it a few pixels larger than the image.

Note 4: Having completed our drawing in the offscreen buffer, we copy it to the screen using the graphics context, g, for the applet.

The Big Picture

....

Java loads images in a separate thread. The `getImage` method returns immediately, not waiting for the image to load. We use a media tracker to wait for the image to load before trying to draw it. We can animate an image by changing its shape to make the next frame, or can also use different images for the frames. Every time we repaint we clear the screen, which alternating with the image, causes flicker in an animation. Using double buffering, drawing on an offscreen image and copying the result to the screen, eliminates the flicker.

TEST YOUR UNDERSTANDING

TRY IT YOURSELF　　9. Remove the call to `waitForID` from the run method in Example 9.11. What happens when you run the revised example? What values do `getWidth` and `getHeight` return for `imageWidth` and `imageHeight`?

TRY IT YOURSELF　　10. Vary the sleep time in Example 9.11, running the example each time to see how the change in the sleep time affects the amount of flicker.

TRY IT YOURSELF　　11. What happens in Example 9.12 when you omit the call to `clearRect` in the `paint` method?

....

9.5 ▪ Sound
....

Java lets us play sound clips in our applets. Java versions 1.0 and 1.1 support the AU format for audio files, while version 2 (1.2) adds support for MIDI (type 0 and type 1), RMF, WAVE, and AIFF files. Playing audio requires a machine with a sound card.

We can play an audio clip using the `play` method of the applet class, either with an absolute URL

```
public void play(URL url);
```

or a relative URL

```
public void play(URL url, String name);
```

The `java.Applet` package contains the `AudioClip` interface with methods `play`, `loop`, and `stop`. The `play` method plays an audio clip; the `loop` method plays an audio clip repeatedly until the `stop` method is invoked. The `Applet` class has methods

```
public AudioClip getAudioClip(URL url);
```

and

```
public AudioClip getAudioClip(URL url, String name);
```

enabling us to get an audio clip.

In Example 9.13 we add sounds to the `AnimateBall` applet of Example 9.9. Each time the ball moves we play a spoken word "boom". While the ball moves, we use another thread to loop a whistling rendition of the first few bars of a familiar tune. When the applet stops, we stop the animation and the whistling (mercifully).

We can use a browser or applet viewer to open the URL

```
http://www.engr.csulb.edu/~artg/PlayBall.html
```

If not connected to the Internet, we can test the applet on our own machine.

Example 9.13   **PlayBall.java**

```
/* Add sounds to the AnimateBall applet
 * of Example 9.9.
 */

import java.awt.*;
import java.applet.*;                                          // Note 1
import java.net.*;

public class PlayBall extends Applet
                          implements Runnable {
    private int x, y;
    private boolean done;
    private URL boomURL;
    AudioClip whistle;

    public void init() {
        try {
            URL url1 = new URL(getDocumentBase(),"whistle.au");  // Note 2
            boomURL = new URL(getCodeBase(),"boom.au");          // Note 3
            whistle = getAudioClip(url1);                        // Note 4
        }catch (MalformedURLException e) {
            e.printStackTrace();
        }
    }
```

```
    public void start() {                                        // Note 5
        x = 50; y = 50;
        done = false;
        Thread t = new Thread(this);
        t.start();
        Sound sound = new Sound(whistle);
        sound.start();
    }
    public void stop() {
        done = true;
        whistle.stop();                                          // Note 6
    }
    public void paint(Graphics g) {
        g.fillOval(x,y,40,40);
    }
    public void run() {
        int dx = 9, dy = 9;
        while (true) {
            for(int i = 0; i < 10; i++) {
                if (done) return;
                x += dx;
                y += dy;
                play(boomURL);                                   // Note 7
                repaint();
                try {
                    Thread.sleep(1000);                          // Note 8
                }catch(InterruptedException e) {
                    e.printStackTrace();
                }
            }
            dx = -dx; dy = -dy;
        }
    }
}
class Sound extends Thread {                                     // Note 9
    AudioClip clip;
    public Sound(AudioClip a) {
        clip = a;
    }
    public void run() {
        clip.loop();                                            // Note 10
    }
}
}
```

Note 1: We import java.applet.* because we use both the Applet class and the AudioClip interface from the java.applet package.

Note 2: The whistle.au file is in the same directory as the document containing the applet, PlayBall.html. All versions of Java play the AU format used in this example (the 8000 Hz frequency can be played in all Java versions).

Note 3: The boom.au file is in the same directory as the applet code file, PlayBall.class.

Note 4: We use an audio clip because we want to loop this sound. The applet itself only includes a `play` method.

Note 5: When the applet starts we start the threads to do the animation and play the whistling.

Note 6: `whistle.stop();`

The `stop` method of the `AudioClip` class stops the looping of that clip.

Note 7: `play(boomURL);`

We use the `play` method of the applet to speak the word "boom" every time we move the ball.

Note 8: `Thread.sleep(1000);`

The thread sleeps for a full second, longer than in Example 9.9, to give the sound time to play.

Note 9: `class Sound extends Thread {`

We use a thread for the looping sound because we want it to continue concurrently with the animation and the other sound.

Note 10: `clip.loop();`

The `loop` method repeats the clip until the `stop` method is called.

The Big Picture

The `play` method lets us play a sound clip in an applet. The `getAudioClip` method returns an `AudioClip` class which has `play`, `loop`, and `stop` methods. The `loop` method plays a sound clip until we invoke the `stop` method. We locate a sound clip using a URL. Java 1.1 supports the AU format, while Java 2 supports additional formats.

TEST YOUR UNDERSTANDING

TRY IT YOURSELF
12. Run Example 9.13 with `PlayBall.html` in a different directory from the one containing `PlayBall.java`. In which of these directories should you place `whistle.au`? Which for `boom.au`?

TRY IT YOURSELF
13. What happens if you omit the call to `whistle.stop` from Example 9.13?

TRY IT YOURSELF
14. Modify Example 9.13 to loop the boom sound in the Sound thread rather than play it in thread t.

Chapter 9 Summary

Java allows several threads of control to proceed simultaneously, sharing the processor. Each thread executes the code in a `run` method when it gets the processor. We can extend the `Thread` class to override the default `run` method, or another class can declare that it implements the Runnable interface. Such a class must implement the run method and pass itself to a thread which will execute its run method.

Calling the start method of a thread makes it ready to run. When it gets the processor it will execute either its run method, if it is an extension of the Thread class, or the run method that a Runnable object passes to it otherwise. An applet often starts a thread in its start method, which is called by the browser or applet viewer whenever the user returns to the web page containing the applet. The best way to stop the thread is to set a flag and have the thread check the flag periodically during the execution of its run method. If the flag becomes **true** then the thread returns from the run method, terminating itself.

A thread can sleep for a specified number of milliseconds. A call to the static sleep method occurs in a try block with a catch clause to handle the InterruptedException that might be generated. We can construct a thread with a name which we can get later with the getName method. To get a thread that we did not create ourselves, such as the thread that runs the main method, we can use the static currentThread method of the Thread class.

When threads share data we must be careful to ensure correct access. Using the keyword synchronized locks an object either for the duration of a method or block of code. The thread holding the lock may complete the method or block of code without interruption.

The wait and notify methods help threads communicate. The wait method signals a condition is not satisfied, so the thread executing it must wait. The notify method wakes up one thread, signaling a condition has been satisfied. Concurrent programming requires great care to avoid problems such as deadlock where threads are unable to proceed, halting the system.

Animations run in threads, sharing the processor with the main method, or the browser, or applet viewer. We can run an animation in a nonterminating loop because other activities can proceed in other threads concurrently. In a simple example we display the same figure at different positions, allowing the thread to sleep for a fraction of a second between repaints of the screen. To make the animation more interesting, we change the figure before repainting it.

We can use images and sounds in our Java programs. Applets use the getImage method to obtain an image given its URL. We can specify either an absolute URL or a relative URL. Using getDocumentBase(), we specify the path relative to the location of the document containing the applet, while using getCodeBase(), we specify the path relative to the location of the applet's code.

Image loading, especially from remote sites, can be slow, so the getImage method returns immediately. If we do not wish to try to draw the image before it is fully loaded, we can register it with a MediaTracker object. Using the waitForId method will enable us to wait for the image to load. We can then get its width and height using the getWidth and getHeight methods.

The drawImage method has several versions. We use the simplest, with four arguments: the image, the (x,y) coordinates of the position at which to draw the image, and an ImageObserver which we do not use because we wait for the image to load fully. This version of drawImage draws the image in its normal size. We also use the six-argument version of drawImage which includes two arguments representing the scaled size we prefer for the image.

The images we animate will flicker a lot because the repaint method calls the update method which clears the screen before calling the paint method to repaint the screen. The alternation of clear background and image causes the flicker. We can eliminate the flicker by the technique of double buffering. We create an offscreen image whose size is that of the applet. We override `update` to paint only, and not to clear the screen first. In the `paint` method we do all the drawing of graphics and images on the offscreen buffer. As the last step of `paint`, we draw the offscreen buffer on the screen. We can clear the offscreen image between repaints; we do not see this background, so this does not cause flicker.

Java versions 1.0 and 1.1 support the AU audio format, while Java 2 supports additional formats, including Windows WAVE files. The applet class has a `play` method which can play a sound file specified by a URL. The applet can also get an audio clip. The `AudioClip` interface has methods to play a sound, to loop it, playing it until the `stop` method is invoked to terminate it. We can add threads to play sounds in the background while an animation is in progress. We can also add sounds to the thread in which the animation is running.

Build Your Own Glossary

Find the uses, in this chapter, of the terms below. Enter a definition of each in the `glossary.txt` file on the disk included with this text.

animation	double buffering	sleep
circular buffer	flicker	synchronized
deadlock	lock	thread
deprecated	producer-consumer problem	

Skill Builder Exercises

1. For each method on the left, choose a property on the right which applies to it.

    a. `start()` for a `Thread`   i.   should not be used
    b. `start()` for an `Applet`   ii. may be executed many times
    c. `stop()` for a `Thread`   iii. executed once only
    d. `stop()` for an `Applet`

2. For each method on the left, choose the exception on the right that it might throw.

    a. `Thread.sleep(100)`   i.   none
    b. `new URL("pic.gif")`   ii. `InterruptedException`
    c. `tracker.waitForID(0)`   iii. `MalformedURLException`
    d. `clip.loop()`   iv. `ImageNotFound`

3. Match each method on the left with the class on the right of which it is a member.

    a. `showDocument`   i.   `MediaTracker`
    b. `loop`   ii. `AppletContext`
    c. `addImage`   iii. `Thread`
    d. `Sleep`   iv. `AudioClip`

Critical Thinking Exercises

4. To run itself in a separate thread an applet can

    a. extend the `Thread` class.

b. implement the `Runnable` interface.
c. either a or b.
d. none of the above.

5. Adding the line `sleep(150.5)` to a Java program would cause an error because

    a. the argument must be an integer.
    b. the prefix `Thread.` should be used.
    c. exception handling has not been provided.
    d. none of the above.
    e. all of the above.

6. The `getDocumentBase()` method

    a. finds the location on our local file system containing the .class file for the applet.
    b. finds the location on our local file system containing the .html file for the applet.
    c. finds the location on the host containing the .class file for the applet.
    d. finds the location on the host containing the .html file for the applet.

7. To eliminate flicker from an animation we can use

    a. double buffering.
    b. a media tracker.
    c. a faster animation rate.
    d. none of the above.

Debugging Exercise

8. The following applet attempts to draw a rectangle in colors chosen at random, changing the color every two seconds. Find and correct any errors.

```java
public class ColoredBoxes extends Applet {
    Color c;
    boolean done;
    public void start() {
        done = false;
        Thread t = new Thread(this);
        t.start();
    }
    public void stop() {
        done = true;
    }
    public void paint(Graphics g) {
        g.drawRect(100,100,30,30);
    }
    public void run(){
        while(! done) {
            c = new Color((int)(255*Math.random()),
                    (int)(255*Math.random()), (int)(255*Math.random()));
            setForeground(c);
            Thread.sleep(200);
        }
    }
}
```

Program Modification Exercises

9. Modify Example 9.10 to specify the sleep time as an applet parameter in the HTML file.

10. Modify Example 9.11 to specify the sleep time as an applet parameter in the HTML file.

11. Modify Example 9.9 to use double buffering to remove the flicker.

12. Modify Example 9.10 to use double buffering to remove the flicker.

13. Modify Example 9.9 to draw the ball in a color of your choice, other than black.

14. Modify Example 9.9 to draw the ball in a different color for each of the ten iterations of the loop.

15. Modify Example 9.10 to draw the eyes in one color and the mouth in another, both different from black.

16. Modify Example 9.9 so instead of arbitrarily moving the ball ten times in each direction, we use the size of the applet, the size of the ball and the size of the increment to determine how many times we can move the ball before it goes off the screen.

PUTTING IT ALL TOGETHER

17. Modify the BankAccount class of Example 2.3 to allow multiple threads to make deposits and withdrawals from an account.

Program Design Exercises

18. Write a Java applet which uses two threads, one to animate a ball and one to animate a face. Make sure the ball and the face do not collide.

PUTTING IT ALL TOGETHER

19. Example 6.20 provides a user interface for insertion sorting. In that example, the user inserts one element at a time, pressing a button to perform the next insertion. Write a Java program which will display the bar chart as the sorting progresses, without requiring the user to press a button to insert each item. Use a thread and let the sorting proceed, sleeping between each insertion to make it easier to view.

20. Write a Java program that displays a digital clock which shows the correct time. To get the current time, use the getInstance() method of the Calendar class, in the java.util package, to get a Calendar object, c. Then use the Calendar get method to get the hours, minutes, and seconds, as in:

```
int hour = c.get(Calendar.HOUR);
int minute = c.get(Calendar.MINUTE);
int second = c.get(Clandar.SECOND);
```

Use a thread to allow the clock to keep the correct time.

21. a. Write a Java program that randomly moves a ball around the screen. If you have access to sound, make a sound at each move. If the user clicks the mouse on the ball, increment a score showing in the corner of the screen. If the user reaches a score of 5, make the ball move faster so it is more difficult to catch. If you are including sound, do so in a separate thread so it does not slow the ball down. If the user reaches a score of 10 make the ball smaller.

    b. In part a, use an image of your choice instead of a ball.

22. Write an applet that provides a choice box naming several web sites. When the user makes a choice, have the browser show the document the user has chosen.

PUTTING IT ALL TOGETHER— TOWERS OF HANOI

23. Suppose we have n disks on a peg, each of different sizes, stacked in order of size, with the largest on the bottom, and two other pegs, as shown in Figure 9.11.

FIGURE 9.11 The Towers of Hanoi puzzle

What is the sequence of moves needed to transfer the rings to the second peg, in the same configuration, in order from largest to the smallest, with the largest at the bottom, if we can move only one disk at a time, and we cannot place a larger disk on top of a smaller disk? We may use the last peg to store disks, but the same rules apply. Use a recursive method to provide the solution. To move n disks from peg 1 to peg 2, move n-1 disks to peg 3, move the bottom disk to peg 2, then move n-1 disks from peg 3 to peg 2. Use a thread to animate the solution, using various colors for the disks.

10 Networking

J ava makes it easy to connect to other computers, using classes from the java.net package. We first write client programs that connect to a server, which is a program that performs a task useful to its clients, and then write our own servers. A web server provides files such as web pages, Java applets, and images.

To write client programs, we first use Java classes that hide the details of the interaction between client and server, then try classes that let us customize the connection, and present classes that give us full control of the communication, but require us then to know the details of the request and response commands.

The simplest clients connect to a server using a URL object, or for more flexibility, a URLConnection object. The URLConnection talks to a server using a specific protocol. We introduce HTTP, the Hypertext Transfer Protocol used to connect to a web server. Understanding HTTP allows us to customize a connection using URLConnection methods, and to write our own simple web client and web server using Socket and ServerSocket objects, which moreover enable us to develop and use our own protocols for clients and servers to communicate.

On a higher level, Remote Method Invocation (RMI) allows us to invoke objects on remote machines, introducing powerful distributed computing using Java.

OBJECTIVES

◆ Use a URL to connect to a remote site from an applet or an application.

◆ Introduce HTTP to understand how web clients and servers communicate.

◆ Use `URLConnection` to customize a connection.
◆ Use `Socket` and `ServerSocket` objects to have full control over the communication.
◆ Write a very simple browser and a very simple web server.
◆ Use threads to enable a server to handle multiple clients.
◆ Introduce distributed computing using RMI.

OPTIONS

◆ The entire chapter is optional. Section 10.5 could be omitted.

10.1 ▪ Using a URL to Connect

Computers use **protocols** to communicate. A **client** sends requests using the commands provided by the protocol in the order specified in the protocol, and the **server** responds similarly. The URL class encapsulates several popular protocols, handling their details thereby making it easier for Java programmers to make network **connections** to display a page, retrieve a file, or get mail, for example.

We describe a URL more fully than we did in Chapter 5, then show how to use an applet to display a resource downloaded from a remote site, and finally we use a standalone application to make a connection using a URL.

The Uniform Resource Locator (URL)

A **URL** has four parts (we only used three in Chapter 5): the protocol name, the host address, the port, and the path to the resource file. The **port** number specifies a specific communication link between computers. For example, the full URL for Sun's Java home page is

```
http://java.sun.com:80/index.html
```

where HTTP is the protocol, `java.sun.com` is the host address, 80 is the port, and `/index.html` is the path to the resource. Because 80 is the default port for the HTTP service, and `index.html` is the default file name, we can write the same URL more concisely as

```
http://java.sun.com/
```

By connecting using the URL for Sun's Java home page, our Java client connects our machine to the web server on Sun's host machine. This server program must understand HTTP, which we will introduce in the next section. Using a URL, in this section, and a `URLConnection` in the next, to write our client programs, we let Java handle the details of the messages specified by the HTTP protocol to communicate with a **web server**.

Connecting from an Applet

We can use the URL class in a Java applet to have the browser get and display a resource for us. We pass a string specifying a URL to the URL constructor, as in:

```
URL url = new URL("http://java.sun.com/");
```

for Sun's Java home page, or

```
URL myURL = new URL("http://www.engr.csulb.edu/~artg/");
```

for the author's home page.

Each applet has a getAppletContext() method which returns an object that implements the AppletContext interface. The **applet context** is the **browser** or applet viewer that started the applet. Each object that implements the AppletContext interface has a showDocument method which will display the document specified by the URL object, if it is able to do so. We can use the applet context to display the resource using the showDocument method, as in:

```
getAppletContext().showDocument(url);
```

☞ TIP

If the applet context is a browser, then the showDocument method will cause the browser to display the page requested in the URL. However, if the applet context is an applet viewer, then the showDocument method will only display the applets contained in the document, and not the rest of the page, as the applet viewer just runs applets and does not interpret HTML.

Example 10.1 shows how an applet can get the browser to display a document. We input the URL as a parameter in the HTML file to make it easier to change. The HTML file we use to run the applet of Example 10.1 is:

```
<applet code = ShowURL.class width=300 height=400>
<param name = url value = http://www.engr.csulb.edu/~artg/>
</applet>
```

Naturally, when requesting a resource from a remote site, we must be connected to the Internet.

EXAMPLE 10.1   ShowURL.java

```
/* Running this applet in a browser will cause
 * the browser to display the resource specified
 * in the URL parameter in the HTML file used to run
 * the applet.  The applet viewer cannot
 * show a document.
 */

import java.net.*;
import java.applet.Applet;

public class ShowURL extends Applet {
    public void init() {                                   // Note 1
        try {
            URL url = new URL(getParameter("url"));
            getAppletContext().showDocument(url);
        }catch(MalformedURLException e) {
            e.printStackTrace();
        }
    }
}
```

Note 1: Java will throw a MalformedURLException if the argument to the constructor does not have the correct form for a URL.

FIGURE 10.1 Example 10.1 displaying a web site

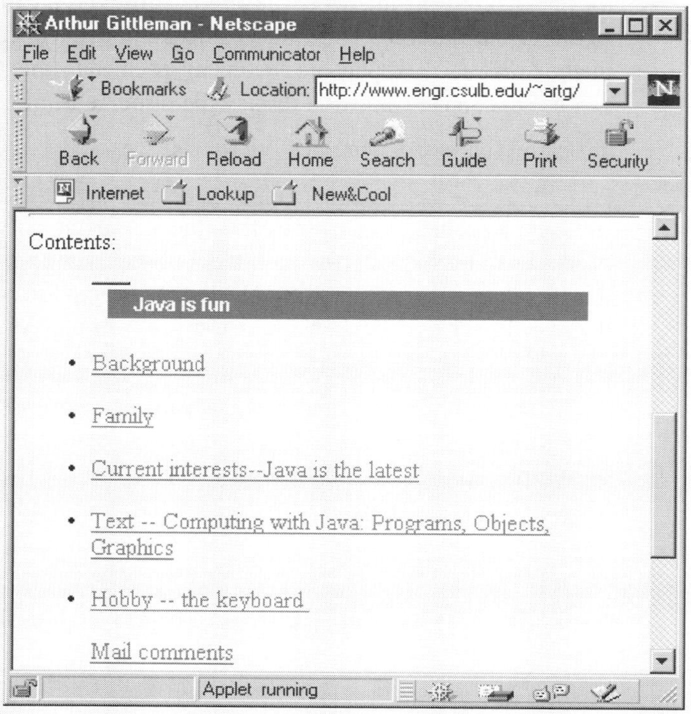

Connecting from a Standalone Application

We can use a URL object in a standalone program, reading characters from the resource specified by the URL. The URL class has a method, openStream, which opens a connection to the server, and allows us to read its responses to our client's request. The openStream method returns an InputStream which we pass to an InputStreamReader to convert the bytes to characters, and then pass the InputStreamReader to a BufferedReader

```
BufferedReader input = new BufferedReader
                (new InputStreamReader(url.openStream()));
```

to buffer the input so we can read one line at a time.

Example 10.2 reads one line at a time from the URL specified on the command line, writing each line to the screen. Our Java program is not a browser, so when we read HTML files we get output that looks like Example 5.1 with the embedded HTML tags. Passing the program argument

```
http://java.sun.com/
```

will list the HTML file for Sun's Java home page.[1] Most files that we access on web servers are HTML files, but we can get other types of files too.

---

[1] Using the JDK we enter the URL on the command line,

```
java TryURL http://java.sun.com/
```

See Appendix E for the method of passing program arguments when using other environments.

Passing the program argument

```
http://www.engr.csulb.edu/~artg/TryURL.java/
```

will connect to the web server at the California State University Long Beach College of Engineering, retrieving the source code file, TryURL.java. We could read a file from the local machine passing a file URL as a program argument, as in:

```
file:///java/TryURL.java/
```

where the file TryURL.java is in the directory c:\java on a Windows system.[2] Leaving the host name empty defaults to **localhost**, which is the user's machine rather than a remote host. Including localhost, as in,

```
file://localhost/java/TryURL.java/
```

gives the same URL.

☞ TIP   Use forward slashes in writing URLs, even on Windows machines for which the default sep-
             arator is the backslash.

EXAMPLE 10.2   **TryURL.java**

```
/* Displays the resource specified by the URL
 * entered on the command line.
 */

import java.net.*;
import java.io.*;

public class TryURL {
    public static void main(String[] args) {
        BufferedReader input;
        try {
            URL url = new URL(args[0]);
            input = new BufferedReader
                (new InputStreamReader(url.openStream()));
            String s;
            while ((s = input.readLine()) != null)
                System.out.println(s);
            input.close();
        }catch(Exception e) {                              // Note 1
            e.printStackTrace();
        }
    }
}
```

Output
(The output will display the contents of any file we pass as the program argument. Passing

```
http://www.engr.csulb.edu/~artg/TryURL.java/
```

will display the code for this example.)

---

[2] On Windows systems, we could include the drive letter, as for example

```
file:///c:/java/TryURL.java/
```

Note 1: The URL constructor may throw a `MalformedURLException`, and the `readLine` method may throw an `IOException`. We could write a catch clause for each, but we are not taking the trouble to do anything special for these exceptions so we just catch the superclass `Exception` which is the parent of both of these exceptions. If Java throws either a `MalformedURLException` or an `IOException`, control will jump here to print the stack trace and the message indicating which exception occurred.

The Big Picture

Network clients and servers communicate using protocols. The URL class hides the details of some popular protocols, letting us make connections more easily. An applet can ask its context, the browser, to show a document. A standalone client uses streams to send data to the servers and receive its response. A URL consists of a protocol, a server address, a port, and a path to the resource.

TEST YOUR UNDERSTANDING

TRY IT YOURSELF

1. Try using an applet viewer, rather than a browser, to run Example 10.1. What is the result?

TRY IT YOURSELF

2. Use Example 10.2 to display the file `TryURL.java`. Use a file URL to get the program from the local disk. For example, using Windows, if the file is in the directory `c:\java`, use the URL

   `file:///c:/java/TryURL.java`

TRY IT YOURSELF

3. Use Example 10.2 to connect to the author's home page, `http://www.engr.csulb.edu/~artg/`. Explain the result.

10.2 ■ Protocols with a URLConnection

The `URLConnection` class makes a connection using a URL, but it adds methods for us to customize the connection and get its properties. It still hides the details of the protocol, while giving the programmer more control.

The Hypertext Transfer Protocol (HTTP)[3]

We used the **HTTP** protocol for our URL in Example 10.1, and suggested the HTTP or file protocols for the URL in Example 10.2. Java supports other protocols, including FTP (File Transfer Protocol) and mailto, when connecting using a URL, but we concentrate on HTTP. Each protocol allows a formal exchange of messages using well-specified formats. Before going further with networking we describe HTTP, which we will use when customizing a `URLConnection`, and when writing our own HTTP client and server.

An HTTP client sends a **request** to the server in which the first line has the form

| Method used | Identifier for the resource | Protocol version |

---

[3] See `http://www.w3.org/` for the complete HTTP specification.

The following lines of the request are various **request headers** which provide information about the capabilities of the client. After the request headers comes the data (if any) to be sent to the server. Figure 10.2 shows the request sent by the Java client of Example 10.2 when we pass it the argument

```
http://www.engr.csulb.edu/~artg/TryURL.java
```

FIGURE 10.2 The HTTP client request from Example 10.2

```
GET  /~artg/TryURL.java  HTTP/1.0
User-Agent: Java1.1.6
Host: www.engr.csulb.edu:80
Accept: text/html, image/gif, image/jpeg, *; q=.2, */*; q=.2
Connection: keep-alive
```

In the first line of Figure 10.2, GET is the method used; we are asking the server to get us a file. The path to that file is the second part of that first line, while the protocol, HTTP/1.0, is the third.

The next four lines of Figure 10.2 are request headers of the form

```
name:   value
```

The User-Agent field indicates that Java 1.1.6 is running our client. The Host field identifies the server. The Accept field specifies the types of files that the client is prepared to accept. Each type has a preference associated with it given by the value of q (for quality) which ranges from a low of 0 to the default of 1. The three types[4] text/html, image/gif, and image/jpeg have the highest preference (the default of q=1 is not shown). If the server cannot send these types, then the client will accept any type, denoted by *, or any subtype of any type, denoted by */*. These latter generic types have preferences of q=.2. The Connection field value of keep-alive expresses the client's wish to keep the connection alive for multiple requests.

Our Java client has sent a request followed by four request headers selected from various header types available. We will see in the next section how to determine the client's request, and will see that different clients such as Netscape and Internet Explorer send different request headers to the server.

An HTTP server responds to a request with a **status line** followed by various **response headers**. The status line has the form

```
HTTP Version    Status Code    Reason
```

We will use Example 10.3 to find the server's response to a request. Figure 10.3 shows the server's response to the client request of Figure 10.2.

The server sends a status line showing the HTTP version, 1.0 in this example, a code, 200, and the reason for the code, Document follows. The Date response header gives the day and Greenwich Mean Time. The Server header names the web server used. The Content-type is text/plain because the server is sending a Java program. Content-length is the number of bytes in the file. As we shall see in Example 10.3, other servers use different response headers.

---

[4] The type names are MIME (Multipurpose Internet Mail Extensions) types.

FIGURE 10.3 The HTTP server response

Status line:

```
HTTP/1.0 200    Document follows
```

Response headers:

```
Date: Mon, 07 Dec 1998 21:12:05 GMT
Server: NCSA/1.4.2
Content-type: text/plain
Last-modified: Wed, 11 Feb 1998 19:19:01 GMT
Content-length: 439
```

☛ **A LITTLE EXTRA**   Status codes have five types, distinguished by their first digit. Some are:

Informational	(Not used with HTTP/1.0 clients.)	
Success	200	OK
Redirection	301	Moved Permanently
Client Error	400	Bad Request
	404	Not Found
	406	Not Acceptable
Server Error	501	Not Implemented

Using a URLConnection

Using a URL we can download a file. For more flexibility, we can use a URLConnection to set some capabilities of a connection, including the client request header fields, and to retrieve the server response status and headers.

We implicitly used a URLConnection in Example 10.2 because the URL openStream method is supplied by Java as a convenient shorthand for

```
openConnection().getInputStream()
```

where openConnection is a URL method that returns a URLConnection whose getInputStream method returns an InputStream to the caller. By using URLConnection explicitly, we can use the other URLConnection methods.

To get the names of the response header fields that the server sends, we use the getHeaderFieldKey method, and to get the field values we use getHeaderField. The first field returned is the status line.

EXAMPLE 10.3   **GetResponses.java**

```
/* Uses a URLConnection to find the response status
 * and headers sent by the server.
 */

import java.net.*;
import java.io.*;
```

```java
public class GetResponses {
    public static void main(String[] args) {
        BufferedReader input;
        try {
            URL url = new URL(args[0]);
            URLConnection c = url.openConnection();                      // Note 1
            System.out.println("Status line: ");
            System.out.println('\t' + c.getHeaderField(0));              // Note 2
            System.out.println("Response headers:");
            String value = "";
            int n = 1;
            while (true){                                                 // Note 3
                value = c.getHeaderField(n);
                if (value == null) break;
                System.out.println('\t' + c.getHeaderFieldKey(n++) + ": " + value);
            }
        }catch(Exception e) {
            e.printStackTrace();
        }
    }
}
```

Output

Connecting to http://www.engr.csulb.edu/~artg/TryURL.java:

See Figure 10.3

Output

Connecting to http://java.sun.com/:

Status line:
```
HTTP/1.1 200 OK
```
Response headers:
```
Date: Mon, 07 Dec 1998 21:14:55 GMT
Server: Apache/1.3.3 (Unix)
Connection: close
Content-Type: text/html
```

Output

Connecting to http://ibm.com/:

Status line:
```
HTTP/1.0 200 IBM-Planetwide Document OK
```
Response headers:
```
MIME-Version: 1.0
Server: Domino-Go-Webserver/4.6
Title: IBM Corporation
Date: Mon, 07 Dec 1998 21:17:45 GMT
Last-modified: Mon, 07 Dec 1998 21:17:45 GMT
Connection: keep-alive
Expires: Tue, 08 Dec 1998 01:17:45 GMT
Window-target: _top
Vary: User-agent
Reply-to: webmaste@us.ibm.com
Content-type: text/html
Content-Language: en-us
Content-Length: 9919
```

Note 1: In contrast to Examples 10.1 and 10.2, we explicitly open a URLConnection. With its various methods we can customize a connection as we shall see in Example 10.4.

Note 2: We use the getHeaderField method that takes an integer argument which is the number of the header in the order sent by the server. Given the argument 0, getHeaderField returns the status line sent by the server.

Note 3: The loop continues indefinitely until getHeaderField returns a **null** value, which it does when there is no header with number n.

In the output from the second run of Example 10.3 we see that the Sun server sends a Connection: close response header meaning that it closes the connection after each response. It sends Content-type: text/html because we have requested its Java home page which is an HTML file.

The IBM server, in the third run, includes an Expires response with an expiration date. This is to aid the client in using a **cache** to store the response. Suppose we connect again to IBM's home page. We can save the time and effort of downloading that HTML file again by using the file that we saved in the cache the last time we browsed IBM's site. However when reading the cache, the client should connect with IBM's web server again if the date is later than the expiration date. By including an Expires response header, the server advises the client when it might be necessary to download a fresh copy of the file.

The Vary response header states the file the server returns may vary based on the fields indicated, in this case User-agent. IBM might have versions customized for particular browsers which do not always display HTML files in the same way.

The Content-Language header describes the language for the intended audience for the response, using first a language abbreviation such as en for English, and then a country code such as us for United States, to represent a dialect of the language.

☞ **A LITTLE EXTRA**   The language abbreviations are registered with the International Standards Organization (ISO-639). Some abbreviations are:

Chinese	zh
French	fr
German	de
Greek	el
Spanish	sp

The country codes follow ISO-3166. Some country codes are:

Canada	CA
China	CN
Germany	DE
Great Britain	GB
Greece	GR
Spain	ES
Switzerland	CH
Taiwan	TW

In Example 10.3, we used the generic `getHeaderField` method to list all the response headers sent by the server. We can also get response headers using their names, as in

```
getHeaderField("Content-length")
```

The `URLConnection` class has separate methods for the most common header requests, so we could also get the content length using

```
getContentLength()
```

which returns −1 if the server does not send a Content-length response.

We can use the `setRequestProperty` method to customize the request headers sent by the client. For example,

```
setRequestProperty("Accept", "text/plain");
```

would indicate a client preference for a plain text file. The server should send a 406 (not acceptable) status code if it cannot supply an entity of that type for the request, but it is not required to do so.

Example 10.4 modifies Example 10.2 to use a `URLConnection` explicitly, using some of the URL methods to customize the connection and get information about the response.

☞ **TIP**

When requesting a large file, the response will scroll out of the command window. In many operating systems we can redirect the output to a file.

For example, the command

```
java TryURLConnect http://java.sun.com/   text/html
```

causes the output of the response to scroll out of the window, but

```
java TryURLConnect http://java.sun.com/   text/html   >out
```

writes the output to a file named `out` which we can read using a text editor.

EXAMPLE 10.4   **TryURLConnect.java**

```
/* Displays the resource specified by the URL passed as the
 * first program argument, with the MIME types acceptable for
 * the response passed as the second program argument.
 * Uses URLConnection methods.
 */

import java.net.*;
import java.io.*;
import java.util.*;

public class TryURLConnect {
    public static void main(String[] args) {
        BufferedReader input;
        try {
            URL url = new URL(args[0]);
            URLConnection c = url.openConnection();
```

continued

```
                  c.setRequestProperty("Accept", args[1]);                // Note 1
                  input = new BufferedReader
                     (new InputStreamReader(c.getInputStream()));
                  String s;
                  while ((s = input.readLine()) != null)
                    System.out.println(s);
                  input.close();
                  System.out.println();
                  System.out.println("Content type: " + c.getContentType());
                  System.out.println("Content length: " + c.getContentLength()); // Note 2
                  System.out.println("Length using getHeaderField: "
                                     + c.getHeaderField("Content-length"));    // Note 3
             }catch(Exception e) {
               e.printStackTrace();
             }
           }
         }
```

Output

Arguments `http://www.engr.csulb.edu/~artg/TryURL.java text/plain`

```
(The file of Example 10.2, not shown)
Content type: text/plain
Content length: 439
Length using getHeaderField: 439
```

Arguments `http://www.engr.csulb.edu/~artg/ShowURL.java text/plain`

```
(The file of Example 10.1, not shown)
Content type: text/plain
Content length: 305
Length using getHeaderField: 305
```

Arguments `http://java.sun.com/ text/html`

```
(Sun's HTML file for its Java home page, not shown)
Content type: text/html
Content length: -1
Length using getHeaderField: null
```

Arguments `http://ibm.com/ text/html`

```
(IBM's HTML file for its home page, not shown)
Content type: text/html
Content length: 9924
Length using getHeaderField: 9924
```

Note 1: Passing text/plain as a program argument will change the client's default Accept request header to request text/plain. We could use the setRequestProperty method to set any of the client's request headers.

Note 2: In Example 10.3 we listed the response headers actually sent by the server. Here when we ask for the content length, the server may not have sent it, in which case the method returns −1.

Note 3: This version of the `getHeaderField` returns a response header for the field name passed as its argument. Its return value has type `String` so if the server has not sent any header for that field, it returns **null**.

The Big Picture

Using HTTP to communicate with a web server, the client sends a request followed by various request headers giving information about the client. The server responds with a status line and various response headers describing the server and the response. The `URLConnection` class still hides the details of the protocol, but allows us to customize the request and inquire about the response.

TEST YOUR UNDERSTANDING

4. For the Accept request header given by

   ```
   Accept: text/plain; q=0.5, text/html,
           application/zip; q=0.8, image/gif
   ```

   which two file types are most preferred, which is next, and which is least preferred?

TRY IT YOURSELF 5. Rewrite Example 10.2 to explicitly use `URLConnection` rather than implicitly using it via the `openStream` method.

TRY IT YOURSELF 6. Use Example 10.3 to connect to five different web sites, in addition to those tried in the text.

TRY IT YOURSELF 7. In Example 10.4 use the `getHeaderField` method to get the content type instead of `getContentType`.

10.3 ▪ Clients and Servers Using Sockets

The `URL` and `URLConnection` classes hide the details of a few common protocols, most importantly HTTP, so we can easily write programs to connect to, say, a web server. With the `Socket` and `ServerSocket` classes, we can write clients and servers using existing protocols, and develop our own protocols for communicating between client and server. After introducing ports, through which we connect, we use our own protocol, writing both a server and a client to illustrate the use of **sockets**. Finally, we use the HTTP protocol to write a server which echoes the requests sent by the client.

Server Ports

Each server listens on a numbered port. The system servers use port numbers below 1024; we can use higher numbered ports for our servers. The familiar services use standard port numbers. For example web servers usually use port 80, **SMTP** servers (Simple Mail Transfer Protocol) for sending mail use port 25, and **POP3** servers (Post Office Protocol-version3) for receiving mail use port 110.

We could use Java to write a client to connect to a system server. For example we could get our email by writing a client for a POP3 server. In writing such a client we would have to follow the Post Office Protocol-version 3 which specifies the form of the communication between the client and the server. Figure 10.4 shows sample interaction between a client and a POP3 server.

FIGURE 10.4 Interacting with the POP3 server

```
Client:   USER username              // client sends user's name
Server:   +OK                        // server responds OK
Client:   PASS password              // client sends the password
Server:   +OK 23 messages 3040 octets  // server sends message info
Client:   RETR 23                    // asks for message 23
Server:   text of message 23, ending
          with a '.' alone on a line
Client:   QUIT
```

A Client-Server Example

If we write our own server we can use our own protocol for communicating with a client. We write a very simple server which reverses the text that the client sends it. Figure 10.5 shows the client window and the server window.

FIGURE 10.5 ReverseClient and ReverseServer

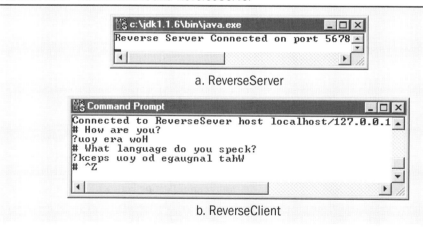

a. ReverseServer

b. ReverseClient

Java provides a `Socket` class for the client to connect to a server on a specific port, and a `ServerSocket` class for the server to listen for clients who wish to make a connection. Once the client connects with the server, they use the `reader` and `writer` classes to send and receive data to and from one another. Example 10.5 shows the code for a server that reverses whatever the client sends it.

We choose an arbitrary port number, 5678, on which our server will listen. The `accept` method waits for a client to make a connection, just like the `Io.readInt` method waits for the user to enter an integer. When a client connects, the `accept` method returns a client socket and our server prints a message announcing the connection. The client socket has a `getInputStream` method which the server uses to create a `BufferedReader` to read from the client. The server uses the client's getOuputStream method to create a PrintWriter to write to the client. The server reads one line at a time from the client, reversing it and sending it back.

EXAMPLE 10.5 **ReverseServer.java**

```
/* Listens on port 5678.  When a client connects, the server
 * reverses whatever the client sends, and sends it back.
 */

import java.net.*;
import java.io.*;

public class ReverseServer {
    public static void main(String [] args) {
        String s;  // the string to reverse
        int size;  // the length of the string
        char[] c;  // the reversed characters
        try {
            ServerSocket server = new ServerSocket(5678);         // Note 1
            Socket client = server.accept();                      // Note 2
            System.out.println("Reverse Server Connected on port 5678");
            BufferedReader input = new BufferedReader
                        (new InputStreamReader(client.getInputStream()));
            PrintWriter output = new PrintWriter
                        (client.getOutputStream(),true);          // Note 3
            while ((s = input.readLine()) != null){
                size = s.length();
                c = new char[size];                               // Note 4
                for (int i = 0; i < size; i++)
                    c[i] = s.charAt(size - 1 - i);                // Note 5
                output.println(c);                               // Note 6
            }
            input.close();
            output.close();
            client.close();
        }catch(Exception e) {
            e.printStackTrace();
        }
    }
}
```

Note 1: We create a server to listen for connections on port 5678.

Note 2: The accept statement blocks any further progress in the program until a client connects; it then returns the client socket.

Note 3: We set the second argument to the PrintWriter constructor to **true** so println statements will flush the output, rather than waiting until the buffer fills up.

Note 4: We cannot change a String object, so we create an array of characters to hold the reverse of the line the client inputs. We could also have used a StringBuffer object which is like a String but allows changes.

Note 5: We use the charAt method to get the character which is i positions from the right end of the string and copy it into element i of the char array.

Note 6: output.println(c);

The array c contains the reversed characters of the string sent by the client. We send these reversed characters back to the client.

We can run the server on the same machine as the client or on a different machine.[5] The server does not terminate until we abort it, so we should run it in the background in its own thread, and we can do other things while it is waiting for clients to connect.[6] Figure 10.5 shows the client and server running on the same machine. The client connects using the address of the host which we pass as a program argument. The name localhost denotes the local machine so the client connects to the server on the same machine with the command

```
java ReverseClient localhost
```

Using two machines, we would have started the server on one machine and let the client connect to it using the server's name or its **IP address**.[7] We usually refer to machines by their names, as for example, www.engr.csulb.edu, but underlying each name is a four byte IP (Internet Protocol) address, as for example 134.139.67.68. (The local machine, named localhost, has the IP address 127.0.0.1.) In connecting in a small lab, whose computers are linked to the Internet, we may just use these basic IP addresses. If we start ReverseServer on machine 134.139.67.68, we would connect to the server from another machine using the command

```
java ReverseClient 134.139.67.68.
```

The client creates a socket using the port, 5678, on which the server is listening. The client uses the getInetAddress method of the socket to display the address of the host to which it is connected. As in the server of Example 10.5, the client and the server use readers and writers to communicate with each other. The client uses the getInputStream method of the socket to create a BufferedReader to read from the server, and uses the getOutputStream method of the socket to create a PrintWriter to write to the server. The client also creates a BufferedReader to get the input from the user.

The client enters a loop printing a prompt, getting a line from the user, sending it to the server, getting the reversed line from the server, and displaying it on the screen, exiting when the user signals the end of input (Control + Z in Windows). Example 10.6 shows the client program which connects to a server that reverses its input.

---

[5] When running a server on the same machine as the client in Windows, the machine does not need to be connected to the Internet, but the TCP/IP protocol should be installed. (Click on the Start button, Settings, Control Panel, Network icons, then Protocol tab to check the installed protocols.)

[6] Using the JDK, in Windows systems, use the start command to run the server in the background: start java ReverseServer. On Unix systems, run in the background using the command java ReverseServer &. If using an integrated development environment, use separate projects for the server and client, running the server first.

[7] The IP address is actually associated with a network interface card.

EXAMPLE 10.6   **ReverseClient.java**

```java
/* Connects to a server which reverses whatever
 * the user inputs.  Specifies the host of the
 * server on the command line.
 */

import java.net.*;
import java.io.*;

public class ReverseClient {
    public static void main(String[] args) {
        String s; // the string to reverse
        if (args.length != 1){                                    // Note 1
            System.out.println("Pass the server's address");
            System.exit(1);
        }
        try {
            Socket server = new Socket(args[0],5678);             // Note 2
            System.out.println("Connected to ReverseSever host "
                            + server.getInetAddress());
            BufferedReader fromServer = new BufferedReader
                    (new InputStreamReader(server.getInputStream()));
            PrintWriter toServer = new PrintWriter
                                (server.getOutputStream(),true);
            BufferedReader input = new BufferedReader(
                                new InputStreamReader(System.in));
            while (true) {
                System.out.print("# ");
                System.out.flush();
                if ((s=input.readLine()) == null)
                    break;
                toServer.println(s);
                System.out.println(fromServer.readLine());
            }
            fromServer.close();
            toServer.close();
            input.close();
            server.close();
        }catch(Exception e) {
            e.printStackTrace();
        }
    }
}
```

Note 1: We check that the user passed the address of the server's host machine as a program argument. If not, we abort the program with a message indicating the omission.

Note 2: The client creates a socket connection to the server on port 5678, the port on which the server is listening.

A Request Header Server

While ReverseServer (Example 10.5) works fine when connected to by ReverseClient (Example 10.6), we will see it does not respond properly to an HTTP client such as our TryURL program of Example 10.2. ReverseServer does not respond properly to TryURL because it does not follow HTTP. We modify it, producing a server, HeaderServer, that echoes the lines sent by the client.

We used ReverseClient to connect to ReverseServer, but we can use other clients to connect to ReverseServer. Using our TryURL client using the command

```
java TryURL  http://localhost:5678/~artg/TryURL.java
```

gives the result shown in Figure 10.6.

FIGURE 10.6 Connecting to ReverseServer using TryURL

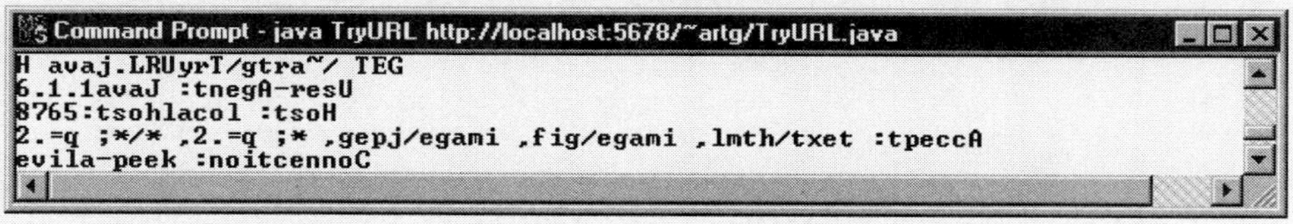

We wrote the TryURL program to connect to a server and request it to send us a file. TryURL does not send any user data to the server; it only sends its request headers. ReverseServer reverses those headers and sends them back to the client which displays them as if they were the sought after file. Thus Figure 10.6 looks almost like Figure 10.2, except in reverse, of course, because ReverseServer dutifully reverses everything sent to it before sending it back.

One difference is minor. Line 3 (reversed so that we can read it) shows the host as localhost:5678 because we connected to our ReverseServer on the local machine rather than to a web server at a remote site. The important difference is in the first line which is shortened in Figure 10.6 to

```
H avaj.LRUyrT/gtra~/
```

which reversed is GET /~artg/TryURL.java H.

The explanation for this mysterious shortening is that the client, TryURL, is using HTTP, and expects the server to follow that protocol. A web server should send the status line, any response headers, and then a blank line to signal the beginning of file to the client. The browser, or other web client, such as TryURL, uses the response headers internally but does not display them. It waits for the blank line, using the header information it received to appropriately display what follows the blank line.

ReverseServer, not using HTTP, does not send a blank line. This confuses TryURL which discards the first seven characters of the first line of data sent by ReverseServer, which just reverses and returns what is sent to it. The first line that TryURL sent was the request, and we see in the result that the first seven characters, 0.1/PPT, of the reversed request line are missing from Figure 10.6.

By modifying ReverseServer, we can make a very simple server to show us the request and headers that an HTTP client sends. We send a status line

```
HTTP/1.0  200  OK
```

and the single response header

```
Content-type: text/plain
```

then send a blank line before sending any data to the client. We no longer want to reverse what the client sends, so we can remove the code from ReverseServer that does the reversing. Rather than trying to satisfy the client's request, our server will return to the client whatever it sent. An HTTP client sends a request and perhaps some request headers, so this is what the server will return.

FIGURE 10.7 A Netscape client connected to HeaderServer[8]

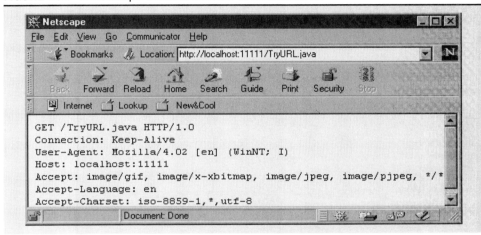

Because the server indicates that every file has type text/plain, InternetExplorer, when connecting to HeaderServer, pops up a Notepad editor window, shown in Figure 10.8, to display the text.

FIGURE 10.8 An Internet Explorer client connected to HeaderServer

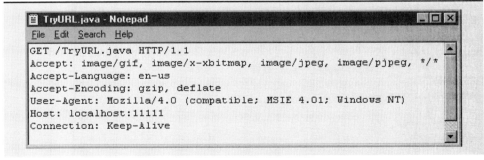

We see that each client has its own variants of the request headers that it sends to the server.

---

[8] Use the command `start java HeaderServer` 11111, where 11111 is the port number, to start header server running in the background in Windows. This server will terminate after one client connects.

EXAMPLE 10.7   HeaderServer.java

```java
/* When an HTTP client connects,  the server sends
 * the client's request and its request headers back to it.
 */

import java.net.*;
import java.io.*;

public class HeaderServer {
    public static void main(String[] args) {
        String s;
        try {
            ServerSocket server = new ServerSocket(Integer.parseInt(args[0]));
            Socket client = server.accept();
            BufferedReader fromClient = new BufferedReader
                    (new InputStreamReader(client.getInputStream()));
            PrintWriter toClient = new PrintWriter
                    (client.getOutputStream(), true);
            toClient.println("HTTP/1.0 200 OK");              // Note 1
            toClient.println("Content-type: text/plain");     // Note 2
            toClient.println();
            while ((s = fromClient.readLine()) != null){
                if(s.equals("")) break;                       // Note 3
                toClient.println(s);
            }
            fromClient.close();
            toClient.close();
            client.close();
        }catch(Exception e) {
            e.printStackTrace();
        }
    }
}
```

Output

From `java TryURL http://localhost:11111/TryURL.java`. (See also Figures 10.7 and 10.8)

```
GET /TryURL.java HTTP/1.0
User-Agent: Java1.1.6
Host: localhost:11111
Accept: text/html, image/gif, image/jpeg, *; q=.2, */*; q=.2
Connection: keep-alive
```

Note 1: A web server must send a status line to the client which reflects the status of the request. For simplicity, we always send the code 200 meaning OK even though we make no attempt to serve the file requested. We leave improvements to the server to the exercises. Were we to omit sending the status line, HTTP clients would drop some characters from the response.

Note 2: If we do not send this header to the client, then both Netscape and Internet Explorer run all the headers together in a long line rather than displaying each header on a separate line as we see in Figures 10.7 and 10.8. This shows that HTTP clients use the response headers sent by the server to display the requested resource.

Note 3: Connecting to ReverseServer, which does not have this test for an empty line, with the TryURL client causes it to become unresponsive, requiring the user to abort. The readLine method returns **null** when it detects the end-of-file character. The ReverseClient user sends an end-of-file which terminates ReverseServer. TryURL just sends a request and then request headers terminated by an empty line, but not an end-of-file. Without this line, the server reads the empty line, which is not **null**, and returns to the readLine statement in the **while** condition which waits for the client to send more data which never happens. Having the server break out of the loop when it receives an empty line solves this problem, causing the server to terminate.

The Big Picture

Using a Socket to connect gives the most flexibility, but both client and server must follow the appropriate protocol. When writing our own server, we can define the protocol by which client and server communicate. The server creates a ServerSocket and uses an accept statement to wait for a client to connect.

TEST YOUR UNDERSTANDING

TRY IT YOURSELF   8. Start the ReverseServer of Example 10.5. Connect to it with a client. After sending some strings for the server to reverse, send an end-of-file to the server. What happens to the client and the server programs? In the exercises we will suggest modifications to the ReverseServer to change this behavior.

TRY IT YOURSELF   9. Revise Example 10.7 to omit sending the status line to the client. What error occurs in the result?

TRY IT YOURSELF   10. Revise Example 10.7 to omit the Content-type response header. How does the output change when using either Netscape or Internet Explorer as the client?

10.4 ■ Browsers and Web Servers

A browser is an HTTP client, and in addition may use other protocols, while a web server is an HTTP server. In this section we write a very simple browser and a very simple web server, leaving to the exercises many improvements to make them more functional. We conclude with a threaded web server, which can handle multiple clients connected simultaneously.

A Very Simple Browser

An HTTP client sends a request to the server followed by request headers and a blank line. It then reads the status line, response headers, and the requested file from the server. A browser typically can handle several types of files, the most important being HTML files which define web pages. The browser has to interpret the HTML tags to guide it in displaying the page. With so many file types to handle, and such intricate processing necessary for web pages, a useful browser is not a small or simple undertaking. Our very simple browser just handles plain text files.

Figure 10.9 shows VerySimpleBrowser connecting to the author's web site to download a file using the command

```
java VerySimpleBrowser  www.engr.csulb.edu 80
          /~artg/TryURL.java
```

where 80 is the standard HTTP port on which the server is running. An alternative approach would pass a URL, as in the command

```
java VerySimpleBrowser
          http://www.engr.csulb.edu/~artg/TryURL.java
```

and use the URL methods `getFile`, `getHost`, and `getProtocol`, which each return a String, and `getPort`, which returns an **int**, to break the URL into the parts needed in Example 10.8.

FIGURE 10.9 VerySimpleBrowser downloads a file

```
C:\book2\ch10>java VerySimpleBrowser   www.engr.csulb.edu   80 /~artg/TryURL.java

Connected to host www.engr.csulb.edu/134.139.147.4
import java.net.*;
import java.io.*;

public class TryURL {
   public static void main(String [] args) {
     BufferedReader input;
     try {
       URL url = new URL(args[0]);
       input = new BufferedReader(new InputStreamReader(url.openStream()));
       String s;
       while ((s=input.readLine())!= null)
         System.out.println(s);
     }catch(Exception e) {
         e.printStackTrace();
     }
   }
}
```

VerySimpleBrowser always sends a GET request, and a Host request header. It ignores the status line and response headers sent by the server, rather than trying to use them to get information that would help it to display the requested resource.

EXAMPLE 10.8 VerySimpleBrowser

```
/* Connects to a web server to download a text file.
 * Exercises suggest extensions to handle other file types.
 */

import java.net.*;
import java.io.*;
```

```
public class VerySimpleBrowser {
    public static void main(String[] args) {
        String s;
        if (args.length != 3){
            System.out.println("Usage: java VerySimpleBrowser host port file");
            System.exit(1);
        }
        try {
            int port = Integer.parseInt(args[1]);                    // Note 1
            Socket server = new Socket(args[0],port);
            System.out.println("Connected to host "
                               + server.getInetAddress());
            BufferedReader fromServer = new BufferedReader
                    (new InputStreamReader(server.getInputStream()));
            PrintWriter toServer = new PrintWriter
                    (socket.getOutputStream(),true);
            toServer.println("GET " + args[2] + " HTTP/1.0");        // Note 2
            toServer.println("Host: " + args[0]+ ':' + args[1]);
            toServer.println();
            while ((s = fromServer.readLine()) != null)             // Note 3
                if(s.equals("")) break;
            while ((s = fromServer.readLine()) != null)             // Note 4
                System.out.println(s);
            fromServer.close();
            toServer.close();
            server.close();
        }catch(Exception e) {
            e.printStackTrace();
        }
    }
}
```

Note 1: The standard HTTP port is 80, but some servers use 8080. We run our simple web server on port 11111. The user should pass the server part number as the second program argument.

Note 2: We use the path to the resource, `/~artg/TryURL.java` in Figure 10.9, sending the host address and port in a separate Host header. Alternatively, we could have sent the GET command

```
GET www.engr.csulb.edu:80/~artg/TryURL.java HTTP/1.0
```

Note 3: We read and ignore the status line and headers sent by the server, looking for the blank line that signals the end of the headers and the start of the file we requested. We leave it to the exercises to improve the browser to make use of this information.

Note 4: This loop reads the file we requested from the server, displaying it in the command window. Extending this very simple browser to display HTML would use graphics extensively.

A Very Simple Web Server

An HTTP server reads the request from the client, any headers, and in some cases additional data. It sends a status line followed by headers and the requested resource, if any. Web servers often transmit data from the client to other programs for processing before returning results to the client. Our very simple web server only responds to GET requests, and only serves text files. We leave it to the exercises to add features to make this server more functional.

To start VerySimpleWebServer, we use the command

```
start java VerySimpleWebServer 11111
```

which, on Windows systems, starts the server in a new window. Figure 10.10 shows Netscape connecting to VerySimpleWebServer to download a file. We could also have used VerySimpleBrowser as the client.

FIGURE 10.10 A Netscape client connecting to VerySimpleWebServer

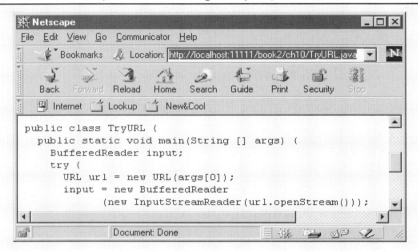

EXAMPLE 10.9 **VerySimpleWebServer.java**

```
/* Serves a text file to an HTTP client submitting a GET
 * request.  Exercises suggest extensions to make the
 * server more functional.
 */

import java.net.*;
import java.io.*;
import java.util.StringTokenizer;

public class VerySimpleWebServer {
    public static void main(String[] args) {
        String s;
        try {
            ServerSocket server = new ServerSocket(Integer.parseInt(args[0]));
            Socket client = server.accept();
            System.out.println
                    ("VerySimpleWebServer Connected on port " + args[0]);
            BufferedReader fromClient = new BufferedReader
                    (new InputStreamReader(client.getInputStream()));
            PrintWriter toClient = new PrintWriter
                    (client.getOutputStream(), true);
            s = fromClient.readLine();                          // Note 1
```

```
            StringTokenizer tokens = new StringTokenizer(s);        // Note 2
            if (!(tokens.nextToken()).equals("GET")) {              // Note 3
               toClient.println("HTTP/1.0 501 Not Implemented");
               toClient.println();
            }
            else {
               String filename = tokens.nextToken();                // Note 4
               while ((s = fromClient.readLine()) != null)          // Note 5
                  if(s.equals("")) break;
               BufferedReader file =
                     new BufferedReader(new FileReader(filename));   // Note 6
               toClient.println("HTTP/1.0 200 OK");                 // Note 7
               toClient.println("Content-type: text/plain");
               toClient.println();
               while ((s = file.readLine()) != null)                // Note 8
                     toClient.println(s);
               file.close();
            }
            fromClient.close();
            toClient.close();
            client.close();
         }catch(Exception e) {
            e.printStackTrace();
         }
      }
}
```

Note 1: We read the first line from the client to find the method and the identifier for the resource.

Note 2: Blanks separate each item of the request. We use a `StringTokenizer` to get the method and identifier parts of the request.

Note 3: If the request method is anything other than GET, the server sends a status line with a code of 501 to indicate the method is not implemented.

Note 4: The file name comes after GET, separated by a blank, in the request from the client. We save it here before we read the next line from the client which will overwrite the string s.

Note 5: We read and ignore any request headers sent by the client, looking for the blank line that separates the request and headers from any data the client might send.

Note 6: `BufferedReader file = new BufferedReader(new FileReader(filename));`

If file cannot be found or another error occurs, the exception thrown will cause control to jump to the catch clause and the server to terminate. We leave to the exercises the improvement of the server to handle the error and send an error message to the client.

Note 7: `toClient.println("HTTP/1.0 200 OK");`

Having created the file to send, the server sends a status line with code 200 meaning OK, and follows with one header describing the content type to help the client to display it.

Note 8: `while ((s = file.readLine()) != null)`

This loop sends the file to the client.

A Threaded Web Server

Our VerySimpleWebServer has the very unusual behavior for a server in that it serves one request and terminates. We can easily modify Example 10.9 to put the server code in a loop. After it responds to one request, it can respond to another, and keep serving clients one at a time. Each client has to wait until the server finishes with the preceding client before being served.

Web servers may get requests from many clients at many dispersed locations. Using threads would allow the server to serve many clients simultaneously. The server interacts with one client while others are preparing their requests or displaying responses. It divides its attention among all connected clients so that they share the server. Large web sites may have a number of servers sharing the load of serving many, many clients.

Our ThreadedWebServer runs in an unending loop. Each time a client connects, the server creates a thread to handle its processing with that client. The Client thread creates the files needed to communicate with the server in its constructor and starts itself running. Its `run` method contains the code from Example 10.9 in which the server responds to the client.

A good test for a threaded server would check how it handles simultaneous requests. We can make a step in that direction by starting two VerySimpleBrowser clients, each requesting a large file so the server will give each request some of its attention. We will see both browser windows scrolling the text of the file. The server will alternate, sending some of the first file

FIGURE 10.11 Two browsers receiving text from ThreadedWebServer

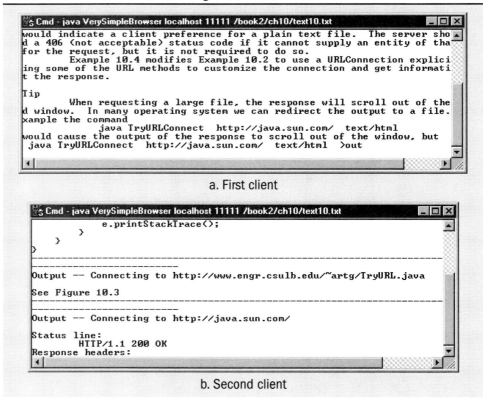

a. First client

b. Second client

to the first browser, then some of the second file to the second browser, then returning to the first, and so on until it has satisfied both requests.

We start ThreadedWebServer using the command

```
start java ThreadedWebServer
```

on Windows systems, and

```
java ThreadedWebServer &
```

on Unix systems. Once the server is running, we start the two clients. Figure 10.11 shows both clients receiving the text of this chapter from a ThreadedWebServer.

EXAMPLE 10.10   **ThreadedWebServer.java**

```java
/* When an HTTP client connects,  the server creates a thread
 * to respond to the client's request, so that multiple clients
 * can be connected simultaneously.
 */

import java.net.*;
import java.io.*;
import java.util.StringTokenizer;

public class ThreadedWebServer {
    public static void main(String [] args) {
        try {
            ServerSocket server = new ServerSocket(Integer.parseInt(args[0]));
            ThreadedWebServer web = new ThreadedWebServer();          // Note 1
            while(true) {                                             // Note 2
                Socket client = server.accept();
                web.new ClientThread(client);                        // Note 3
                System.out.println("ThreadedWebServer Connected to "
                                        + client.getInetAddress());
            }
        }catch(Exception e) {
            e.printStackTrace();
        }
    }
    class ClientThread extends Thread {
        Socket client;
        BufferedReader fromClient;
        PrintWriter toClient;
        public ClientThread(Socket c) {
            try {
                client = c;
                fromClient = new BufferedReader
                    (new InputStreamReader(client.getInputStream()));
                toClient = new PrintWriter
                    (client.getOutputStream(), true);
                start();                                             // Note 4
            }catch(Exception e) {
                e.printStackTrace();
            }
        }
```

continued

```
        public void run() {                                          // Note 5
            try {
                String s;
                s = fromClient.readLine();
                StringTokenizer tokens = new StringTokenizer(s);
                if (!(tokens.nextToken()).equals("GET")) {
                    toClient.println("HTTP/1.0 501 Not Implemented");
                    toClient.println();
                }
                else {
                    String filename = tokens.nextToken();
                    while ((s = fromClient.readLine()) != null)
                        if(s.equals("")) break;
                    BufferedReader file =
                            new BufferedReader(new FileReader(filename));
                    toClient.println("HTTP/1.0 200 OK");
                    toClient.println("Content-type: text/plain");
                    toClient.println();
                    while ((s=file.readLine()) != null)
                        toClient.println(s);
                    file.close();
                }
                fromClient.close();
                pw.close();
                toClient.close();
            }catch(Exception e) {
                e.printStackTrace();
            }
        }
    }
}
```

Note 1: We need an instance of ThreadedWebServer to create the Client thread to run each client. Client is an inner class and must be created using a reference to its containing class, ThreadedWebServer.

Note 2: The server runs in an unending loop, until aborted, continuing to serve clients as they connect.

Note 3: By making Client an inner class, we avoid conflicts in the global namespace. Client has the full name ThreadedWebServer$Client, so the name Client may be used in other contexts without conflict. Because Client is an inner class of ThreadedWebServer, we create a Client instance using the instance, web, of ThreadedWebServer. We leave the alternative, defining Client outside of ThreadedWebServer, as an exercise.

Note 4: This makes this thread runnable, so when it gets scheduled, Java will execute its run method.

Note 5: The run method contains the code from Example 10.9 by which the server and client communicate. If several threads are active, then the server will be communicating with several clients who are all connected simultaneously.

The Big Picture

Browsers and web servers use HTTP to communicate. Our ThreadedWebServer spawns a new thread to handle a connection from a client, so many clients may be connected to this server simultaneously. Rather than terminating after a client connects, ThreadedWebServer remains in a loop waiting for the next client.

TEST YOUR UNDERSTANDING

TRY IT YOURSELF
11. Modify Example 10.8 to use the HEAD method which just sends headers and does not ask for a resource in response. Connect with the very simple web server of Example 10.9. What happens?

TRY IT YOURSELF
12. Put the server of Example 10.9 into a loop so instead of terminating after each connection, it waits for another client to connect. Make the loop unending, so the server will have to be aborted to terminate it.

TRY IT YOURSELF
13. Test the threaded web server of Example 10.10 by connecting to it from two simple web browser clients at close to the same time. Find long text files to request so both clients will be connected to the server at the same time while downloading the requested files. Describe what you observe.

10.5 ▪ Remote Method Invocation (RMI)

Remote Method Invocation (**RMI**) takes networking to a higher level, providing distributed computing for Java programs. In distributed computing a program can be composed of parts located on more than one computer. So far we have used input and output streams to communicate between a client and server. These streams transfer data from one machine to another. Using RMI we can distribute our objects on various machines, invoking methods of objects located on remote sites.

Distributed Computing: The RMI Solution

We use a very simple example, a fortune server, to show how RMI works without introducing the extra complications of an involved example. The fortune server may be running on one machine. Clients, from remote sites, can request a fortune. In making these requests, clients will invoke a method of an object on the server. A **distributed computing** system must provide the following capabilities:

1. Clients must know what services the fortune server provides.

   RMI Solution: A Fortune interface lists the methods available to remote clients.

2. Clients must find a Fortune object on the server.

   RMI Solution: The fortune server registers an object with a special server, called the **rmiregistry**, so that clients can look it up by name.

3. Clients must be able to pass arguments to, and invoke a method of the Fortune object located on the server.

   RMI Solution: A special compiler, called **rmic**, creates a **stub** class for use on the client, and a **skeleton** class for use on the server. The stub takes the request from the client, passes the arguments to the skeleton which invokes the Fortune method and sends the return value back to the stub which passes it to the client.

Figure 10.12 shows how RMI operates.

FIGURE 10.12 The parts of RMI illustrated

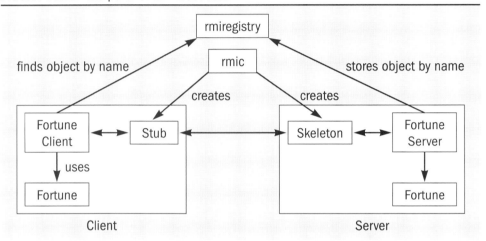

For our example which provides a fortune server that will enable clients to request fortunes (the fortune cookie kind, not the billionaire kind), we need to write the following programs:

Fortune   The interface that shows the client what remote methods it can invoke, the getFortune method in this example.

FortuneServer The implementation of the Fortune interface to provide the remote object which is served to clients who wish to use its method to get a fortune.

FortuneClient A client that gets a reference to a Fortune object and invokes its getFortune method remotely.

We run this example on one machine but use two directories, one for the client and one for the server, to simulate the use of a remote site. To compile and run our example we follow these steps:

In the server directory

1. Compile Fortune.java, FortuneServer.java, and FortuneClient.java.

2. Create the stub and skeleton classes using the command[9]

   ```
   rmic FortuneServer
   ```

---

[9] If this command generates an error, it may be necessary to set the classpath.

3. Copy `Fortune.class`, `FortuneClient.class`, and `FortuneServer_Stub.class` to the client directory.

4. Start the registry server using the command

   ```
   start rmiregistry
   ```

   on Windows systems, or

   ```
   rmiregistry &
   ```

   on Unix systems.

5. Start FortuneServer using the command[10]

   ```
   start java FortuneServer localhost
   ```

   on Windows systems, or

   ```
   java FortuneServer localhost &
   ```

   on Unix systems.

   In the client directory

6. Run FortuneClient using the command[11]

   ```
   java FortuneClient localhost
   ```

FIGURE 10.13 An RMI example

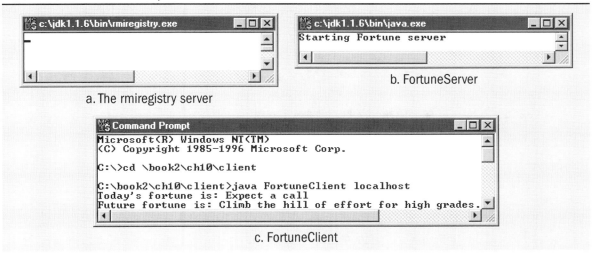

a. The rmiregistry server

b. FortuneServer

c. FortuneClient

---

[10] When using the Java™ 2 Platform (1.2), due to changes in the security model, we need to specify a security policy on the command line, so the command would be

```
start java -Djava.security.policy=d:\policy FortuneServer localhost
```

where `d:\policy` is the file containing the security policy. For testing RMI, the policy file `grant { permission java.security.AllPermission; }` will work.

[11] When using the Java™ 2 Platform (1.2), we need to make the same change as in footnote 10.

☞ TIP	To run using two machines, follow the above steps, changing localhost, in steps 5 and 6, to the address of the server.

The Interface

Let us build each part of this example RMI application. The Fortune interface specifies the getFortune method. We ensure that objects that implement the Fortune interface can be used remotely by making Fortune extend the Remote interface. Because any network communication may fail, every method used remotely must declare that it may throw a RemoteException.

EXAMPLE 10.11 **Fortune.java**

```
/* The server implements this interface and clients call
 * its method remotely.  Clients get a fortune from the server.
 */

import java.rmi.*;                                            // Note 1
public interface Fortune extends Remote {
    public static final String NOW = "Now";                  // Note 2
    public static final String LATER = "Later";
    public String getFortune(String when) throws RemoteException;  // Note 3
}
```

Note 1: The java.rmi package contains the basic classes and interfaces needed for RMI, such as Remote and RemoteException in this example.

Note 2: In addition to methods, we can declare constants in an interface. We declare two String constants here to show clients the choices they have for arguments to the getFortune method. They can either request a fortune for NOW or for LATER.

Note 3: Using RMI, Java passes arguments and return values across the network using object serialization. Objects passed must have a type, such as String, that implements the Serializable interface.

The Server

On the server, the FortuneServer class implements the Fortune interface. Figure 10.12 shows the client uses the Fortune interface, and only sees that part of the FortuneServer object declared in the interface. The client only sees the getFortune method, and cannot access any other methods of the FortuneServer object. The getFortune method, which returns a message of good fortune to the client, must declare that it throws RemoteException because it will be called from a remote site. By contrast, the find method, used only within the FortuneServer class and declared as private, does not throw RemoteException.

We declare the getFortune method as synchronized because many clients may try to access the server simultaneously. Although our example is so simple it is really not necessary to synchronize access, in many cases we want a client to have exclusive access to the server object so any changes made will be completed before another client gets access.

The java.rmi.server package contains the classes needed for implementing servers of objects accessed remotely using RMI. Our FortuneServer class directly extends UnicastRemoteObject, a remote object sent to one destination at each request.

In the main method, we set up a server for a FortuneServer object, so remote clients can call its getFortune method to get a fortune. When using RMI, we may need to load class files from a remote site and need a security manager to ensure the safety of such operations. Java provides the RMISecurityManager class for this purpose, which we install using the setSecurityManager method of the System class.

The server and the client use the URL syntax with the RMI protocol to refer to remote objects. The server gives the object a name and places it, using that name, in the registry server, using the form

```
rmi://host:port/name
```

which the client also uses to locate that object. The Naming class handles interactions with the registry. Here the server uses the rebind method which registers the object with the rmiregistry server, replacing an earlier object of that name (if any) in the registry. In our example, the server binds a FortuneServer object in the registry under the name Seer.

EXAMPLE 10.12   **FortuneServer.java**

```
/* Implements the Fortune interface. Establishes a server
 * for remote clients to use a FortuneServer object.
 */

import java.rmi.*;
import java.rmi.server.UnicastRemoteObject;
import java.util.Vector;

public class FortuneServer extends UnicastRemoteObject implements Fortune {
    public static final int SIZE = 3;
    private Vector now = new Vector(SIZE);       // Fortunes for NOW
    private Vector later = new Vector(SIZE);     // Fortunes for LATER
    public FortuneServer() throws RemoteException {
        now.addElement("A friend is near");
        now.addElement("Expect a call");
        now.addElement("Someone misses you");
        later.addElement("Wealth awaits -- if you desire it.");
        later.addElement("Climb the hill of effort for high grades.");
        later.addElement("The door to success is open to you.");
    }
    private Vector find(String when) {                            // Note 1
        if (when.equals(Fortune.NOW))
            return now;
        else return later;
    }
    public synchronized String getFortune(String when) throws RemoteException {
        int number = (int)(SIZE*Math.random());                  // Note 2
        Vector fortunes = find(when);
        return (String)fortunes.elementAt(number);               // Note 3
    }
```

continued

```
public static void main(String[] args) {
    System.setSecurityManager(new RMISecurityManager());
    try {
        Fortune fortune = new FortuneServer();
        String url = "rmi://" + args[0] + "/Seer";         // Note 4
        Naming.rebind(url,fortune);                         // Note 5
        System.out.println("Starting Fortune server");
    }catch(Exception e) {
        e.printStackTrace();
    }
}
}
```

Note 1: We only use the `find` method within `FortuneImpl` to select a vector of fortunes based on the argument passed by the client. Just in case the client passes something other than NOW or LATER, we do not put a condition in the **else** clause and return LATER if the client passes anything but NOW.

Note 2: We get a random number between 0 and SIZE-1 to choose one of the fortunes to return to the client.

Note 3: Because `elementAt` returns type `Object`, we must cast the return value to type `String`. We know the elements of the vector have type `String` because we created them that way.

Note 4: We use the name Seer for the `fortune` object. The client uses the same name when looking up a reference to this object on the rmiregistry server. We pass the host and port as a program argument so that we can run our server on different machines without having to recompile the program. By default the rmiregistry server runs on port 1099; if we do not wish to change the port, we may omit it from the URL, just passing the host name as the program argument. Note that although we use the URL syntax, we do not declare a URL object, but rather a string.

Note 5: We associate the fortune object with the Seer URL when binding to the registry, so the client can find it.

The Client

FortuneClient is the last program we need to complete our RMI example. It sets RMISecurityManager, just as the server does, and looks up a reference to the Seer object in the rmiregistry server. When the client calls the `getFortune` method of this remote object, it receives a fortune as the return value. From the client's point of view there is no difference between calling a method of a remote object and a method of a local object. RMI handles the details of the remote method call, using the `stub` and `skeleton` classes to pass arguments to the server and return values to the client.

EXAMPLE 10.13 **FortuneClient.java**

```
/* Looks up a Fortune object in the registry.
 * Invokes its getFortune method remotely to
 * get a fortune for now and for later.
 */
```

```
import java.rmi.*;

public class FortuneClient {
    public static void main(String[] args) {
        System.setSecurityManager(new RMISecurityManager());
        try {
            String url = "rmi://" + args[0] + "/Seer";          // Note 1
            Fortune fortuneTeller = (Fortune)Naming.lookup(url); // Note 2
            String fortune = fortuneTeller.getFortune(Fortune.NOW); // Note 3
            System.out.println("Today's fortune is: " + fortune);
            fortune = fortuneTeller.getFortune(Fortune.LATER);
            System.out.println("Future fortune is: " + fortune);
        }catch(Exception e) {
            e.printStackTrace();
        }
    }
}
```

Note 1: We pass the host name of the server as a program argument, omitting the port number, because we have no need to change the default port of 1099 used by the rmiregistry server.

Note 2: The `lookup` method returns a reference, obtained from the rmiregistry server, to an object of type `Remote`, meaning it implements the `Remote` interface. We need to cast this reference to type `Fortune` to invoke the `getFortune` method.

Note 3: This remote method invocation looks like a method call of a local method, but the `fortuneTeller` object is actually on the server, which may be at a remote site. The `FortuneServer_Stub` sends the argument to `FortuneServer Skel` on the host which invokes the `getFortune` method of the `fortuneTeller` object on the host. `FortuneServer Skel` gets the return value from the `getFortune` method and sends it to the `FortuneServer Stub` back on the client which returns it to the caller here to display. Fortunately, RMI handles all these details of communication using object serialization to transfer values.

The Big Picture

RMI lets us distribute our program across the network. We can bind an object in the rmiregistry server, so an object on another machine can look it up, and invoke its methods. The rmic compiler creates the skeleton and stub needed to call remote methods. Remote objects use interfaces to declare their remote methods.

TEST YOUR UNDERSTANDING

14. In Example 10.12 which classes from the `java.rmi` package are we using?

TRY IT YOURSELF
15. Run the RMI example of this section using port 3000 for the rmiregistry server. Start the registry with the command `start rmiregistry 3000`. Change the localhost program argument to `localhost:3000`.

Chapter 10 Summary

Java makes it easy to connect to other computers. The URL class encapsulates some of the common communication protocols, such as HTTP (Hypertext Transfer Protocol) and FTP (File Transfer Protocol). Its four parts are the protocol name, the host name, port, and path to the resource. In an applet, we can use the showDocument method of the applet context to display a resource specified by a URL. The openStream method allows us to download the object referred to by a URL.

The Hypertext Transfer Protocol specifies the messages by which browsers and web servers communicate. A browser or other web client sends a request, then some request headers, and a blank line to the server. The server sends a status line, followed by response headers, a blank line, and the requested resource. The examples show web clients and servers vary in the headers they choose to send.

The URLConnection class has methods which allow us to set properties of the connection to customize it, or to get information about the connection. The getHeaderField method lets us determine the response headers sent by the server. The setRequestProperty method allows us to set request headers to send to the server. Special methods such as getContentLength return the value of specific headers.

Sockets allow us to communicate using standard protocols, or to devise protocols of our own for use with clients and servers. A ServerSocket accepts connections on a numbered port. Standard services have default ports such as 80 for web servers, 25 for sending mail, and 110 for receiving it. Once we make a connection, we use input and output streams to send data back and forth between the client and the server. The ReverseClient example sends strings to the ReverseServer which sends them reversed back to the client. Modifying the ReverseServer produces HeaderServer, which just sends back to a web client the headers that it sent.

To display a web page, a browser must interpret all the HTML tags embedded in that page. Our VerySimpleBrowser uses HTTP to communicate with web servers. It is a bare outline of a browser, following HTTP but ignoring the response headers and only displaying plain text, not HTML or images. Similarly our VerySimpleWebServer ignores any headers the client sends, and puts no effort into accurately sending response headers. Nevertheless our browser can connect and download files from various web servers and our web server can respond to plain text request from browsers. The exercises suggest many improvements. Our ThreadedWebServer permits several clients to be connected at the same time, each served in its own thread.

RemoteMethodInvocation (RMI) takes networking to a higher level in which a client can invoke methods of a remote object on the server. The rmiregistry server allows the client to find a reference to a remote object. The rmic compiler creates the stub and skeleton files used to pass arguments and return values across the network. An interface, implemented on the server, specifies the remote methods available to the client.

Build Your Own Glossary

Find the uses, in this chapter, of the terms below. Enter a definition of each in the glossary.txt file on the disk that comes with this text.

applet context	localhost	rmiregistry
browser	port	server
cache	request (HTTP)	skeleton (RMI)
client	request header (HTTP)	socket
connection	response header (HTTP)	status line (HTTP)
distributed computing	RMI	stub (RMI)
IP address	rmic	web server

Skill Builder Exercises

1. Fill in the blanks in the following:

   Using RMI, the _____ compiler creates the stub and skeleton files, while the _____ server lets clients look up references to remote objects.

2. List the classes from the `java.net` package used in this chapter.

3. Match each method on the left with the class or classes on the right of which it is a member.

a.	`getContentLength`	i.	`ServerSocket`
b.	`accept`	ii.	`Socket`
c.	`getInputStream`	iii.	`URLConnection`
d.	`getHeaderFieldKey`	iv.	`URL`

Critical Thinking Exercises

4. Which of the following URLs is incorrectly formed?

   a. `http://java.sun.com:80/`
   b. `file:///ShowURL.java`
   c. `http://java.sun.com:80/TryURL.java`
   d. `file://localhost/ShowURL.java`
   e. none of the above

5. Which HTTP header is not used as a response header?

   a. Content-length
   b. Accept
   c. Last-Modified
   d. Expires
   e. none of the above

6. We can read from a URL object using the

   a. `openStream()` method.
   b. `getInputStream()` method.
   c. `openReader()` method.
   d. none of the above.
   e. all of the above.

7. We can read from a `Socket` object using the

   a. `openStream()` method.

   b. `getInputStream()` method.

   c. `openReader()` method.

   d. none of the above.

   e. all of the above.

Debugging Exercise

8. Trying to modify Example 10.9, VerySimpleWebServer, to send the content type text/html for an HTML file and text/plain otherwise, led to changing the **else** part of the **if-else** statement in Example 10.9 to

```
String filename = tokens.nextToken();
String content;
int dot = filename.indexOf('.');
if (filename.substring(dot,filename.length()) == ("html"))
    content = "Content-type: text/html";
else
    content = "Content-type: text/plain";
while ((s = br.readLine()) != null)
    if(s.equals("")) break;
BufferedReader file = new BufferedReader(new FileReader(filename));
pw.println("HTTP/1.0 200 OK");
pw.println(content);
pw.println();
while ((s = file.readLine()) != null)
    pw.println(s);
file.close();
```

Find and correct any errors in this modified program.

Program Modification Exercises

9. Modify Example 10.3 to use a `String` argument to `getHeaderField` instead of an **int**. Use `getHeaderFieldKey` to find the names of the headers.

10. Example 10.4 uses two program arguments. Add a check that the user passed two arguments and if not, print a message showing the proper usage and exit the program.

11. Modify Example 10.8 to pass a URL such as

    `http://www.engr.csulb.edu/TryURL.java`

    rather than passing the host, port, and resource path program arguments. The URL methods `getHost`, `getPort`, and `getProtocol` will be helpful.

12. Modify Example 10.9 to send a status line with code 404 and reason Not Found when the file requested is not available on the server.

13. Modify Example 10.9 so when the server has responded to one client it can accept a request from another.

14. Modify Example 10.10 to avoid using any inner classes.

15. Modify Example 10.9 to send a Content-length header giving the length of the file in bytes.

16. Modify Examples 10.5 and 10.6 to pass the port number as a program argument.

17. Modify Example 10.5 to put the `accept` statement and the code following it into a nonterminating loop, so the server can accept another client as soon as the current client finishes.

18. Modify Example 10.5 so the server can handle several clients simultaneously. After the server accepts a connection from the client, the server should create a separate thread to handle the communication with that client and loop back to the `accept` statement waiting for another client to connect.

19. Modify Example 10.13 to ask if the user wants another fortune. When testing, let two clients stay connected at the same time.

20. Modify Examples 10.11, 10.12, and 10.13 to allow the client to request a lucky number. The server will return a number from 1 to 10 at random.

Program Design Exercises

21. Write an applet which lists URLs in a choice box. When the user selects a URL, use the `showDocument` method to display the page to which it refers.

22. Write a multithreaded server which will pass whatever message line it receives from a client to all the other clients that are connected. Write a client program to connect to this server, which sends its lines and receives the lines sent by the other clients.

23. Write a mail client which will connect to a POP3 server (find the address of your server) and retrieve the first message. Specify the server address, user name, and password as program arguments. The protocol of Figure 10.4 may be helpful.

24. Write a browser that displays a plain text file in a text area, rather than in the command window as VerySimpleBrowser does.

25. Write a piece of a browser which will display HTML files. This piece will only display text within header tags, `<h1>` ... `<h6>`. Use the largest point size for text between `<h1>` and `</h1>` tags, and the smallest for text between `<h6>` and `</h6>` tags.

26. Add to the browser of Exercise 25 the capability to handle `<em>` and `<strong>` tags.

27. Write a new version of ReverseClient in which the user enters the text to be reversed in a text field.

28. Use RMI to allow clients to connect to a broker to get the price of a stock, or to buy and sell some stock. Use just three stocks, StockA, StockB, and StockC, each with a price that varies randomly within a range. Assume two accounts numbered 1111 and 2222.

    a. For simplicity, do not maintain account information, so no records are kept about buy and sell orders.

    b. Add account information, so each account keeps a record of how many of each stock it contains.

Programming Projects

29. Improve the very simple browser of Example 10.8. The browser should properly interpret HTML tags <h1>, ..., <h6>, <em>, <strong>, <ul>, <li>, <br>, <p>, <a>, and <img>.

30. Improve the very simple web server of Example 10.9. Use the status codes 200, 301, 400, 404, 406, and 501 appropriately. Send Date, Last-modified, Content-type, and Content-length response headers.

31. Implement a chess game in which the server relays moves from one player to the other. Two clients play against one another, with each showing the board and the moves as they are made. Players will use the mouse to drag a piece to its new position. (Alternatively, substitute another game for chess. For example, checkers would be simpler.)

32. Make a user interface for the mail-reading client of Exercise 23. The screen will show the message headers and allow the user to choose which message to read. For additional information on the POP3 protocol, search the Internet for RFC 1939 which contains its specification.

33. Use RMI to make the AtmScreen applet of Example 1.5 a distributed program. This will be a peer-to-peer application in which both the AtmScreen and the Teller function as server and client. Only one object needs to bind with the registry, as the other can use a method to pass itself to it.

34. Implement an SMTP client to send email. Provide a user interface to compose and send the message. Testing requires access to an SMTP server. A example session is:

```
Server:  220 charlotte.engr.csulb.edu ESMTP Sendmail 8.8.4/8.8.4;
             Thu, 11 Feb 1999 15:31:27 -0800 (PST)
Client:  HELO gordian.com                      // sent from
Server:  250 charlotte.engr.csulb.edu Hello ppool3.gordian.com
             [207.211.232.196], pleased to meet you
Client:  MAIL FROM:<artg@csulb.edu>            // email address
Server:  250 <artg@csulb.edu>... Sender ok
Client:  RCPT TO: <artg@csulb.edu>             // recipient
Server:  250 <artg@csulb.edu>... Recipient ok
Client:  DATA                                  // signals message
Server:  354 Enter mail, end with "." on a line by itself
Client:  This is                               // message
         a test.
         .                                     // signals end
Server:  250 PAA27651 Message accepted for delivery
Client:  QUIT
Server:  221 Closing connection.
```

11 Java Database Connectivity (JDBC)

For small applications, we can use files to store data, but as the amount of data that we need to save gets larger the services of a database system become invaluable. A database system allows us to model the information we need while it handles the details of inserting, removing, and retrieving data from individual files in response to our requests.

Of course each database vendor provides its own procedures for performing database operations. The Java Database Connectivity (JDBC) programming interface hides the details of different databases; our programs can work with many different databases on many different platforms. JDBC can be used as part of large scale enterprise applications. In this chapter we cover the JDBC concepts using a small example which allows many extensions, some of which we pursue in the exercises.

The example programs illustrate JDBC concepts using console applications so as not to obscure them with the details involved in building a GUI. In the last section our extended case study develops a graphical user interface to a database.

OBJECTIVES

◆ Introduce relational database tables.

◆ Introduce SQL (Structured Query Language).

◆ Register a database as an ODBC data source.

◆ Connect to a database from Java, using JDBC.
◆ Build a database using JDBC and SQL.
◆ Use Java to query a database.
◆ Use metadata to obtain the properties of a database or a result set.
◆ Introduce selected aggregate functions.
◆ Use prepared statements for efficiency.
◆ Process database transactions.
◆ Provide a GUI for the user to query a database.

OPTIONS
◆ The entire chapter is optional. Section 11.5 and/or 11.4 and 11.6 could be omitted.

11.1 ▪ Database Tables and SQL Queries

Database design is best left to other texts and courses. We introduce a few database concepts here to provide an example with which to illustrate the Java Database Connectivity techniques for working with databases using Java. Relational databases provide an implementation-independent way for users to view data. The Structured Query Language (SQL) lets us create, update, and query a database using standard commands that hide the details of any particular vendor's **database system**.

Relational Database Tables

When designing a **database** we need to identify the entities in our system. For example, a company might use a database to keep track of its sales and associated information. In our company, an order has one customer who can order several items. A salesperson may take several orders from the same customer, but each order is taken by exactly one salesperson.

Using a **relational database**, we keep our data in **tables**. In our example, we might have a Customer table with fields for the customer ID, name, address, and balance due as shown in Figure 11.1.

FIGURE 11.1 The Customer table

CustomerID	CustomerName	Address	BalanceDue
1234	Fred Flynn	22 First St.	1667.00
5678	Darnell Davis	33 Second St.	130.95
4321	Marla Martinez	44 Third St.	0
8765	Carla Kahn	55 Fourth St.	0

Each row of the table represents the information needed for one customer. We assign each customer a unique customer ID number. Customer names are not unique; moreover they may change. CustomerID is a **key** that identifies the data in the row. Knowing the CustomerID we can retrieve the other information about that customer.

☞ TIP   Do not embed spaces in field names. Use CustomerID rather than Customer ID.

Figures 11.2 and 11.3 show the Salesperson and Item tables which we define in a similar manner. A more realistic example would have additional fields, but our purpose here is only to illustrate JDBC.

FIGURE 11.2 The Salesperson table

SalespersonID	SalespersonName	Address
12	Peter Patterson	66 Fifth St.
98	Donna Dubarian	77 Sixth St.

FIGURE 11.3 The Item table

ItemNumber	Description	Quantity
222222	radio	32
333333	television	14
444444	computer	9

The SalepersonID serves as the key for the Salesperson table, while we use the ItemNumber to identify an item in the Item table. We have to be more careful in designing the Orders table, as an order can have multiple items. We use a second table, the OrderItem table, to list the items in each order. Figure 11.4 shows the Orders table with the fields OrderNumber, CustomerID, SalespersonID, and OrderDate. The OrderNumber is the key. CustomerID and SalespersonID are **foreign keys** that allow us to avoid redundancy by referring to data in other tables. For example, including the CustomerID lets us find the customer's name and address from the Customer table rather than repeating it in the Orders table.

FIGURE 11.4 The Orders table

OrderNumber	CustomerID	SalepersonID	OrderDate
1	1234	12	4/3/99
2	5678	12	3/22/99
3	8765	98	2/19/99
4	1234	12	4/5/99
5	8765	98	2/28/99

☛ **TIP**   When choosing field names, avoid names like Number, Value, Order, Name, or Date that might conflict with reserved names in the database system.

The OrderItem table uses a **compound key** consisting of both the OrderNumber and the ItemNumber to identify a specific item which is part of an order. Figure 11.5 shows that each pair (OrderNumber, ItemNumber) occurs only once, identifying a row containing the data for a specific item in a particular order. For example, the first row shows that for order number one, and item 222222, four units were ordered at a price of $27 each.

Now that we have defined our Sales database, we want to see how to get information from it, and how to make changes as needed.

FIGURE 11.5 The OrderItem table

OrderNumber	ItemNumber	Quantity	UnitPrice
1	222222	4	27.00
1	333333	2	210.50
1	444444	1	569.00
2	333333	2	230.95
3	222222	3	27.00
3	333333	1	230.95
4	444444	1	569.00
5	222222	2	27.00
5	444444	1	725.00

Structured Query Language (SQL)

The Structured Query Language (**SQL**) is a standard language with which to get information from or make changes to a database. We can execute SQL statements from within Java. The SQL statements we shall use are CREATE, SELECT, INSERT, DELETE, and UPDATE. We illustrate these statements using the Sales database defined above. The names for the data types may depend on the actual database system used. Our examples work with Microsoft Access.

We could use the CREATE statement

```
CREATE TABLE Customer (CustomerID CHAR(4), CustomerName
    VARCHAR(25), Address VARCHAR(25), BalanceDue CURRENCY)
```

to create the Customer table, the statement

```
CREATE TABLE Orders (OrderNumber VARCHAR(4), CustomerID
    CHAR(4), SalepersonID CHAR(2), OrderDate DATE)
```

to create the Orders table, and the statement

```
CREATE TABLE OrderItem (OrderNumber VARCHAR(4), ItemNumber
    CHAR(6), Quantity INTEGER, UnitPrice CURRENCY)
```

to create the OrderItem table. We use character fields for CustomerID, OrderNumber, SalepersonID, and ItemNumber, even though they use numerical characters, because we have no need to do arithmetic using these values. By contrast, we use the type INTEGER for the Quantity field because we may wish to compute with it.

Standard SQL uses various types which are not all supported in every database system. Figure 11.6 shows the SQL types we use in this text.

FIGURE 11.6 SQL data types

Type	Standard SQL	Description
CHAR(N)	Yes	Fixed size string of length N
VARCHAR(N)	Yes	Variable size string up to length N
INTEGER	Yes	32-bit integer
DATE	Yes	year, month, and day
CURRENCY	No	dollars and cents

The type DECIMAL(M,N), where M is the maximum number of digits and N is the maximum number of digits after the decimal point, is standard SQL, but is not supported in Access.

To insert the first row in the Customer table, we could use the INSERT statement

```
INSERT INTO Customer VALUES (1234,'Fred Flynn','22 First St.',1667.00)
```

☛ **TIP**   Use the single quote, ', to enclose strings within an SQL statement.

The statement

```
INSERT INTO Orders VALUES (1,1234,12,'Apr 3, 1999')
```

inserts the first row into the Order table. We write dates in the form

```
Month Day, Year
```

to avoid confusion among date formats used in various locales and to indicate the century explicitly. The database system translates this form to its internal representation, and can present dates in various formats in its tables.

The DELETE statement

```
DELETE FROM OrderItem WHERE OrderNumber = '1'
```

will delete the first three rows of the OrderItem table in Figure 11.5. These rows contain the data for the three items comprising the order with an OrderNumber of 1.

☛ **TIP**   Use the single equality sign, =, in the equality test, OrderNumber = 1, instead of the Java equality symbol, ==.

To delete just the televisions from the order in Figure 11.5 and leave the order for radios and a computer, we could use the statement

```
DELETE FROM OrderItem
WHERE OrderNumber = '1' AND ItemNumber = '333333'
```

To update an existing row we use the UPDATE statement. For example, to reduce the number of radios in order number 1 to 3, we can use the statement

```
UPDATE OrderItem SET Quantity = 3
WHERE OrderNumber = '1' AND ItemNumber = '222222'
```

When we change an order we will also want to change the balance due in the Customer table, which we can do using

```
UPDATE Customer SET BalanceDue = 1640.00
WHERE CustomerID = '1234'
```

☛ **TIP**   Because the OrderItem table uses a compound key

```
(OrderNumber, ItemNumber)
```

to identify a row, we needed to specify values for both in the WHERE clause. In updating the Customer table we only needed to specify the value of the single CustomerID key to identify a row.

The CREATE statement creates a table, and the INSERT, DELETE, and UPDATE statements make changes in a table. In many applications, we retrieve information from the database more frequently than we create a table or make changes to a table. To retrieve information we use the SELECT statement.

The simplest **query** we can make is to retrieve the entire table. For example, the statement

```
SELECT * FROM Customer
```

retrieves the entire Customer table. We use the star symbol, *, which matches every row. To retrieve the names and addresses of the customers we use the statement

```
SELECT CustomerName, Address FROM Customer
```

If we do not want data from the entire table, we can use a WHERE clause to specify a condition that the data of interest satisfy. For example, to retrieve all orders for radios we could use the statement

```
SELECT * FROM OrderItems
WHERE ItemNumber = '222222'
```

The power of database systems becomes evident when we use SQL to get information combined from several tables. For example, suppose we would like to know the names of all customers who placed orders on March 22, 1999. We can find that information using the statement

```
SELECT CustomerName FROM Customer, Orders
WHERE Customer.CustomerID = Orders.CustomerID
AND  OrderDate = {d '1999-03-22'}
```

where {d '1999-03-22'} is an escape sequence.

Date formats vary among database systems. To make programs general Java uses a generic string format yyyy-mm-dd with a four-digit year, a two-digit month, and a two-digit day. The curly braces, {}, enclose the escape sequence which tells the driver to translate it to the specific form used by the database system. The keyword, d, signifies that a date follows. The date format for the Access database we are using is #3/22/99#, which we could have used, but the escape sequence makes the code more general.

☛ **TIP**

When a field such as Address occurs in more than one table, prefix the field name with the table name, as in Customer.Address, to state precisely which Address field you desire. Similarly, use the prefixes Customer and Orders to refer to the CustomerID fields in each of these tables.

In finding the names of customers who placed orders on March 22, 1999, the database joins two tables. Customer names occur in the Customer table, while we find order dates in the Orders table, so we list both the Customer and the Orders tables in the FROM part of the query. We want to find which orders each customer placed. CustomerID, the primary key of the Customer table, is also a foreign key of the Orders table. For each CustomerID in the Customer table we only want to inspect the rows of the Orders table which have the same CustomerID, so we include the condition

```
Customer.CustomerID = Orders.CustomerID
```

in our query.

The first row of the Customer table has a CustomerID of 1234. The first and fourth rows of the Orders table have the same CustomerID of 1234 but neither of the OrderDate fields equals 3/22/99. The second row of the Customer table has CustomerID 5678 as does the second row of the Orders table and the OrderDate is 3/22/99 so the system adds Darnell Davis to the result set of customers placing orders on March 22, 1999. Continuing the search turns up no further matches. A three-line SQL statement can cause many steps to occur in the process of retrieving the requested information. The database handles all the details. We will use other interesting examples of SELECT statements when we develop our Java programs later in this chapter.

Figure 11.7 shows the general pattern for the SQL statements we have introduced so far.

FIGURE 11.7 Some patterns for SQL statements

```
CREATE TABLE tablename
    (fieldname1 TYPE1, fieldname2 TYPE2, ... , fieldnameN TYPEn)

INSERT INTO tablename
    VALUES (field1value,field2value, ..., fieldNvalue)

DELETE FROM tablename
WHERE fieldname1 = value1 ... AND fieldnameN = valueN

UPDATE tablename SET fieldnameToSet = newValue
WHERE fieldname1ToCheck = value1ToCheck

SELECT fieldname1, ..., fieldnameN FROM table1, ..., tableM
WHERE condition1 ... AND conditionN
```

The Big Picture

In a relational database, we keep our data in tables, making sure not to enter information redundantly. Using SQL, we can write statements to create a table, insert, delete, and update elements, and query the database. Generally SQL is standardized so queries do not reflect implementation details of specific database systems.

TEST YOUR UNDERSTANDING

1. Why is it a good idea to use SalespersonID as the key in the Salesperson table, rather than the salesperson's name?

2. Write an SQL statement to create the Salesperson table with the fields shown in Figure 11.2.

3. Write SQL statements to insert the data shown in Figure 11.2 into the Salesperson table.

4. Write an SQL statement to add a new salesman, Paul Sanchez, who lives at 88 Seventh St., and has an ID of 54, to the Salesperson table of Figure 11.2.

5. Write an SQL statement to delete Carla Kahn's order of a computer from the Sales database.

6. Write an SQL statement to find the names of all salespersons in the Sales database.

7. Write an SQL statement to find the order numbers of all orders taken by Peter Patterson.

11.2 ■ Connecting to a Database

A **driver** translates JDBC statements to the specific commands of a particular database system. Several different categories of drivers exist, but in this text we use the JDBC to ODBC bridge to allow JDBC to work with Microsoft Access which has an existing **ODBC** driver (for an earlier technology, Open Database Connectivity, that is still used).

Before using Java we need to register our database as an ODBC data source. To connect to our database using Java, we need only specify our JDBC driver and the URL for the database. To create and **populate** our database we pass SQL statements as arguments to JDBC methods.

Database and Application Servers

In building large systems, a **database server** may reside on one machine to which various clients connect when they need to access the stored data.

FIGURE 11.8 Client-server database access

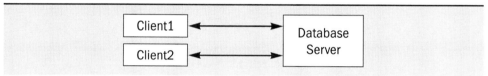

In a three-tiered design, business logic resides in a middle machine, sometimes called an **application server**, which acts as a server to various application clients. These clients provide user interfaces to the business applications on the middle machine which is itself a client of the database server.

FIGURE 11.9 A three-tiered system architecture

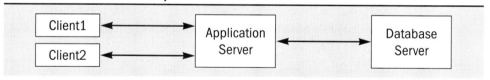

For example, a business may have an accounting department that runs a payroll client providing a user interface to the payroll application on the middle machine which itself is a client of the database server. The marketing department might have several client programs running in their sales offices enabling salespersons to get necessary information. Rather than configuring each salesperson's machine to process all the details of the application, the company just allows the sales staff to interact with the sales application on the middle machine. This sales program gets data from the database server as needed.

Java Database Connectivity (**JDBC**) allows us to write Java programs that will work no matter which database system we use. We can work entirely on one machine or use a two-tier, three-tier, or even more complex architecture for our system. What we need for any database system we wish to use is a JDBC driver. The driver provides a uniform interface to our Java programs. Many database vendors provide JDBC drivers for use with their products.

Creating an ODBC Data Source

The Microsoft Open Database Connectivity (ODBC) interface, introduced prior to the development of Java, provides an interface to many databases. Sun makes a JDBC to ODBC bridge available, so if a database has an ODBC driver, we can access it using the JDBC to ODBC bridge as our JDBC driver.

In this chapter, our examples will use Microsoft Access databases on Windows 95, Windows 98, or Windows NT. Only the driver name and the data source URL need to be changed to use another database system.

The first step is to register the database we will be creating as an ODBC data source.

1. Click on the MyComputer icon and the Control Panel icon to open the Control Panel.
2. Click on the ODBC icon in the Control Panel, which pops up the ODBC Data Source Administrator window.
3. Click Add which pops up the Create New Data Source window.
4. Select Microsoft Access Driver and click Finish which pops up the ODBC Microsoft Access Setup window.
5. Fill in the Data Source Name. We use this name to refer to this database in our Java programs. We use the name Sales for our example.
6. Fill in a short description, such as "Record sales orders" in the Description field.
7. Click Create, as we are creating a new database.
8. In the New Database window, navigate to the directory in which to place the new database, give it a name, such as Sales.mdb, and click OK.
9. If all went well, a message that the database was successfully created will appear.

☞ TIP

On a Windows 95, 98, or NT system without Microsoft Access, much of what we do in this chapter can be done with ordinary text files. The steps for using text files are:

1. Click on the MyComputer icon and the Control Panel icon to open the Control Panel.
2. Click on the ODBC icon in the Control Panel, which pops up the ODBC Data Source Administrator window.
3. Click Add which pops up the Create New Data Source window.
4. Select Microsoft Text Driver and click Finish which pops up the ODBC Text Setup window.
5. Fill in the Data Source Name. We use this name to refer to this database in our Java programs. We use the name Sales for our example.
6. Fill in a short description, such as "Record sales orders" in the Description field.
7. Deselect the UseCurrentDirectory box, click Select Directory, navigate to the desired directory for the files, and click OK in all the open windows.

Connecting from Java

We want our Sales database to contain the five tables with the data shown in Figures 11.1-11.5. We could create these tables and populate them within Access, but prefer to show how to do this using Java.

Every Java program that uses JDBC to access a database must load the driver that it will use and connect to the desired database. To load the driver we create a new driver object. Sun provides the JDBC classes in the `java.sql` package and the JdbcOdbcDriver in the `sun.jdbc.odbc` package. The core Java packages all start with the `java` prefix. Sun includes the JdbcOdbcDriver with the JDK but it is not one of the core Java classes. Sun also includes the helper file, `JdbcOdbc.dll`, with the JDK.

The statement

```
new JdbcOdbcDriver();
```

will load a new `driver` object, calling its constructor. The drawback of using the **new** operator to load JdbcOdbcDriver is that if we want to use a different driver, we have to modify the program. Java has the ability to load classes while the program is running, so we could pass a class name in as a program argument and let Java load whichever driver we decide to use. We use the `forName` method of the class `Class` in the `java.lang` package

```
Class.forName("JdbcOdbcDriver");
```

to load JdbcOdbcDriver. To make the loading dynamic we could use

```
Class.forName(arg[0]);
```

which would load the class whose name we pass as the first program argument.

Once we load the driver using one of these methods, the driver registers with the DriverManager which keeps a vector of drivers to use when making a connection to a database. We connect to a database using the static `getConnection` method of the `DriverManager` class which returns a connection representing our session with the database.

We use a URL to locate the database to which we wish to connect. We could use a database server which would require a remote connection in which case the URL would include the Internet address of the server. The URL has the form

```
jdbc:<subprotocol>:<subname>
```

where the **subprotocol** is the name of the driver or a database connectivity mechanism such as odbc which is what we will use. The **subname** identifies the database. For the case of ODBC drivers we just need the name of the database that we registered with the ODBC Data Source Administrator, which is Sales for our example. Thus the URL we will use is

```
jdbc:odbc:Sales
```

The developer of the JDBC driver defines the URL needed.

Example 11.1 will just connect to the Sales database. The code we use will occur at the beginning of all of our examples in this chapter.

EXAMPLE 11.1   **Connect.java**
....

```
/* Connects to a Microsoft Access database
 * using the JDBC-ODBC bridge.
 */

import java.sql.*;                                              // Note 1
import java.io.*;
import sun.jdbc.odbc.*;                                         // Note 2
```

```
class Connect {
    public static void main (String args[]) {
        try{
            new JdbcOdbcDriver();                              // Note 3
            String url = "jdbc:odbc:Sales";                    // Note 4
            String user = "";                                  // Note 5
            String password = "";
            Connection con =
                DriverManager.getConnection(url, user, password);    // Note 6
            System.out.println("Made the connection to the Sales database");
        }catch (Exception e) {e.printStackTrace();}           // Note 7
    }
}
```

Output

Made the connection to the Sales database

Note 1: The `java.sql` package contains the JDBC classes.

Note 2: The `sun.jdbc.odbc` package contains the JdbcOdbcDriver we use to access our ODBC data source, a Microsoft Access database.

Note 3: We create a new JdbcOdbcDriver which registers itself with the DriverManager which stores a vector of all the registered drivers. The driver hides the details of the specific database. We use JDBC generic methods which the driver translates to the specific procedures provided by the database vendor. By changing the driver, the same program can work with data on a different database system. We do not need to assign the driver to a variable, because we will not refer to it again explicitly.

Note 4: The supplier of the driver defines the URL needed. If we change the driver then we also need to change the URL. Making these changes will allow our program to work with another database system. Our programs would be more flexible if, rather than hard coding the driver and URL, we pass them as program arguments. We leave this modification to the exercises.

Note 5: For the Access database we are using we do not need a user name or a password. For other databases we may need to log in to the server. For generality we left the user and password fields in the program and set them both to empty strings.

Note 6: `Connection con = DriverManager.getConnection(url, user, password);`

This method looks through the vector of registered drivers to find a driver that can connect to this database, and throws an exception if one is not found. If it finds a suitable driver, it attempts to make the connection. In this example, we could have used the method

`DriverManager.getConnection(url);`

which omits the user and password arguments.

Note 7: `catch (Exception e) {e.printStackTrace();}`

We catch all exceptions here. We could have used the `SQLException` class to catch exceptions relating to SQL.

Building the Database

Once we have a connection to the database we can execute SQL statements. The `createState-ment` method returns a `Statement` object that we use to send SQL statements to the database.

Some SQL statements, such as those used to create tables and insert values in a table, change the database but do not return any values to the program. To execute SQL `CREATE` and `INSERT` statements we use the `executeUpdate` method. The argument to `executeUpdate` is a string, which will be sent to the database. The string argument should represent an SQL statement in a form understandable by the database system. If not, Java will throw an exception. As an example,

```
con.executeUpdate ("INSERT INTO Item VALUES (4,'CD player',10)");
```

would insert a fourth row into the Item table.

Example 11.2 uses Java to create and populate the Sales database. We create the five tables shown in Figures 11.1-11.5, using a `CREATE` statement to create each table, and `INSERT` statements to add the rows. Figure 11.10 shows the resulting Access Sales database and Figure 11.11 shows the Customer table that results from executing Example 11.2.

FIGURE 11.10 The Access Sales database created by Example 11.2

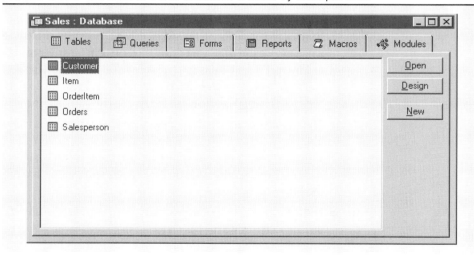

FIGURE 11.11 The Customer table created by Example 11.2

CustomerID	Name	Address	BalanceDue
1234	Fred Flynn	22 First St.	$1,667.00
5678	Darnell Davis	33 Second St.	$130.95
4321	Marla Martinez	44 Third St.	$0.00
8765	Carla Kahn	55 Fourth St.	$0.00

Record: 1 of 4

☛ TIP

After running this program, the database contains the five tables. Therefore running the program again will cause an error, unless the tables are first deleted from the database.

EXAMPLE 11.2 **Create.java**

```java
/* Creates and populates the Sales database.
 */

import java.sql.*;
import java.io.*;
import sun.jdbc.odbc.*;

public class Create {
    public static void main (String args[]) {
        try{
            new JdbcOdbcDriver();
            String url = "jdbc:odbc:Sales";
            String user = "";
            String password = "";
            Connection con = DriverManager.getConnection(url, user, password);
            Statement stmt = con.createStatement();

            stmt.executeUpdate ("CREATE TABLE Customer (CustomerID "
                + "VARCHAR(4), CustomerName VARCHAR(25), Address "
                + "VARCHAR(25),  BalanceDue CURRENCY)");            // Note 1
            stmt.executeUpdate ("INSERT INTO Customer "
                + " VALUES (1234,'Fred Flynn','22 First St.',1667.00)");  // Note 2
            stmt.executeUpdate ("INSERT INTO Customer "
                + " VALUES (5678,'Darnell Davis','33 Second St.',130.95)");
            stmt.executeUpdate ("INSERT INTO Customer"
                + " VALUES (4321,'Marla Martinez','44 Third St.',0)");
            stmt.executeUpdate ("INSERT INTO Customer "
                + " VALUES (8765,'Carla Kahn','55 Fourth St.', 0)");

            stmt.executeUpdate("CREATE TABLE Salesperson (SalepersonID "
                + " VARCHAR(2), SalespersonName VARCHAR(25), "
                + " Address VARCHAR(25))");
            stmt.executeUpdate ("INSERT INTO Salesperson "
                + " VALUES (12,'Peter Patterson','66 Fifth St.')");
            stmt.executeUpdate ("INSERT INTO Salesperson "
                + " VALUES (98,'Donna Dubarian','77 Sixth St.')");

            stmt.executeUpdate("CREATE TABLE Item (ItemNumber VARCHAR(6),"
                + "Description VARCHAR(20), Quantity INTEGER)");
            stmt.executeUpdate("INSERT INTO Item VALUES (222222,'radio',32)");
            stmt.executeUpdate("INSERT INTO Item VALUES (333333,'television',14)");
            stmt.executeUpdate("INSERT INTO Item VALUES (444444,'computer',9)");

            stmt.executeUpdate("CREATE TABLE Orders (OrderNumber VARCHAR(4),"
                + " CustomerID VARCHAR(4), SalepersonID VARCHAR(2),"
                + " OrderDate DATE)");
            stmt.executeUpdate
                ("INSERT INTO Orders VALUES (1,1234,12,'Apr 3, 1999')");
            stmt.executeUpdate
                ("INSERT INTO Orders VALUES (2,5678,12,'Mar 22, 1999')");
```

continued

```
      stmt.executeUpdate
         ("INSERT INTO Orders VALUES (3,8765,98,'Feb 19, 1999')");
      stmt.executeUpdate
         ("INSERT INTO Orders VALUES (4,1234,12,'Apr 5, 1999')");
      stmt.executeUpdate
         ("INSERT INTO Orders VALUES (5,8765,98,'Feb 28, 1999')");

      stmt.executeUpdate("CREATE TABLE OrderItem (OrderNumber CHAR(4),"
         + " ItemNumber CHAR(6), Quantity INTEGER, UnitPrice CURRENCY)");
      stmt.executeUpdate("INSERT INTO OrderItem "                      // Note 3
         + " VALUES  (1,222222,4,27.00)");
      stmt.executeUpdate("INSERT INTO OrderItem "
         + " VALUES (1,333333,2,210.50)");
      stmt.executeUpdate("INSERT INTO OrderItem "
         + " VALUES (1,444444,1,569.00)");
      stmt.executeUpdate("INSERT INTO OrderItem "
         + " VALUES (2,333333,2,230.95)");
      stmt.executeUpdate("INSERT INTO OrderItem "
         + " VALUES (3,222222,3,27.00)");
      stmt.executeUpdate("INSERT INTO OrderItem "
         + " VALUES (3,333333,1,230.95)");
      stmt.executeUpdate("INSERT INTO OrderItem "
         + " VALUES (4,444444,1,569.00)");
      stmt.executeUpdate("INSERT INTO OrderItem "
         + " VALUES (5,222222,2,27.00)");
      stmt.executeUpdate("INSERT INTO "
         + " OrderItem VALUES (5,444444,1,725.00)");

      stmt.close();                                                   // Note 4
   }catch (Exception e) {e.printStackTrace();}
  }
}
```

Note 1: Just as with any string, we need to split the SQL statement over multiple lines using the concatenation operator so that each string constant fits on one line.

Note 2: When splitting the SQL statement over multiple lines we must be sure to add spaces to separate identifiers. Without the spaces either after Customer or before VALUES, then the juxtaposition of CustomerVALUES would cause an error.

Note 3: Using nine statements to insert the nine rows into the OrderItem table is cumbersome, and would be more so if the table were larger. A better method is to read the data to enter from a file. We leave this improvement to the exercises.

Note 4: The Statement object, stmt, is closed automatically by the garbage collector and its resources freed, but it is good programming practice to close it explicitly.

The Big Picture

JDBC uses a driver to translate its platform-independent interface to work in a specific database system. We use the JDBC to ODBC bridge to connect to Access or a text file which have ODBC drivers. A URL, specific to the database system, locates the database. Once connected to the database, we can create tables and insert data into them from a Java program. Optionally we could have created the tables outside of Java.

TRY IT YOURSELF   8. Register a new Sales database as an ODBC data source.

TRY IT YOURSELF   9. Modify Example 11.1 to input the JDBC driver and the database URL as program arguments.

TRY IT YOURSELF   10. Modify Example 11.2, as described in Note 2, to omit the spaces after `Customer` and before `VALUES`. What is the effect of this change?

11.3 ▪ Retrieving Information

Now that we have created the Sales database, we can use JDBC to extract information from it. When executing an SQL statement that returns results, we use the `executeQuery` method which returns a `ResultSet` containing the rows of data that satisfy the query. Executing

```
ResultSet rs = stmt.executeQuery
    ("SELECT CustomerName, Address FROM Customer");
```

returns the rows containing the names and address of all entries in the `Customer` table.

Viewing Query Results

To view the results, the `ResultSet` has `getXXX` methods where XXX is the Java type corresponding to the SQL type of the data field we are retrieving. Because CustomerName and Address both have the `VARCHAR` SQL type, we use the `getString` method to retrieve these fields. We can retrieve fields by name or by field number. The loop

```
while(rs.next())
    System.out.println(rs.getString(1) + '\t' + rs.getString("Address"))
```

will list the rows of names and addresses from the Customer table. We retrieve the CustomerName field using its column number 1 and the Address field using its name. The `next()` method returns **true** when another row is available and **false** otherwise. Figure 11.12 shows the Java methods corresponding to the SQL types we use.

FIGURE 11.12 Java methods for SQL types

Java method	SQL type
getInt	INTEGER
getString	VARCHAR
getBigDecimal	CURRENCY
getDate	DATE

SELECT Statement Options

The `SELECT` statement has additional options. The ORDER clause allows us to display the results sorted with respect to one or more columns. The query

```
SELECT CustomerName, Address FROM Customer
ORDER BY CustomerName
```

returns the result set by name in alphabetical order. We could use

```
SELECT CustomerName, Address FROM Customer
ORDER BY 1
```

to achieve the same result using the column number in the ORDER clause.

Sometimes a query may return duplicate rows. For example, in selecting customers who ordered computers we would get the result

> Fred Flynn
> Fred Flynn
> Carla Kahn

because Fred Flynn bought computers in orders 1 and 4. We can remove duplicates by using the SELECT DISTINCT variant of the SELECT statement.

This query

```
SELECT DISTINCT CustomerName
FROM Customer, Item, Orders, OrderItem
WHERE Customer.CustomerID = Orders.CustomerID
AND Orders.OrderNumber = OrderItem.OrderNumber
AND OrderItem.ItemNumber = Item.ItemNumber
AND Description = 'computer'
```

joins rows from four tables to produce the result.

The UPDATE and DELETE statements change the database, but do not return results, so we use the executeUpdate method to execute them.

EXAMPLE 11.3   **ExtractInfo.java**

```
/* Demonstrates the use of SQL queries from
 * a Java program.
 */

import java.sql.*;
import java.io.*;
import sun.jdbc.odbc.*;

public class ExtractInfo {
  public static void main (String args[]) {
    try{
        new JdbcOdbcDriver();
        String url = "jdbc:odbc:Sales";
        String user = "";
        String password = "";
        Connection con = DriverManager.getConnection(url, user, password);
        Statement stmt = con.createStatement();

        String query = "SELECT CustomerName, Address FROM Customer "
              + "ORDER BY CustomerName";
        ResultSet rs = stmt.executeQuery(query);                          // Note 1
        System.out.println("   Names and Addresses of Customers");
        System.out.println("Name\t\tAddress");                            // Note 2
        while (rs.next())
          System.out.println(rs.getString("CustomerName") + '\t'
              + rs.getString(2));
```

```java
query = "SELECT * FROM OrderItem "
        + "WHERE ItemNumber = '222222'";
rs = stmt.executeQuery(query);
System.out.println();
System.out.println("   Order items for radios");
System.out.println("OrderNumber\tQuantity\tUnitPrice");
while (rs.next())
   System.out.println(rs.getString(1) + "\t\t"
        + rs.getInt(3) + "\t\t" + rs.getBigDecimal(4,2));     // Note 3

query =  "SELECT CustomerName FROM Customer, Orders "
        + "WHERE Customer.CustomerID = Orders.CustomerID "
        + "AND  OrderDate = {d '1999-03-22'}";
rs = stmt.executeQuery(query);
System.out.println();
System.out.println("   Customer placing orders on Mar 22, 1999");
while(rs.next())
   System.out.println(rs.getString("CustomerName"));

query = "SELECT DISTINCT CustomerName "
        + "FROM Customer, Item, Orders, OrderItem "
        + "WHERE Customer.CustomerID = Orders.CustomerID "
        + "AND Orders.OrderNumber = OrderItem.OrderNumber "
        + "AND OrderItem.ItemNumber = Item.ItemNumber "
        + "AND Description = 'computer'";
rs = stmt.executeQuery(query);
System.out.println();
System.out.println("   Customers ordering computers");
while(rs.next())
   System.out.println(rs.getString(1));                       // Note 4

query =  "SELECT OrderNumber FROM Orders "
        + "WHERE OrderDate "
        + "BETWEEN {d '1999-04-01'} AND {d '1999-04-30'}";
rs = stmt.executeQuery(query);
System.out.println();
System.out.println("   Order numbers of orders from 4/1/99 to 4/30/99");
while(rs.next())
   System.out.println(rs.getString("OrderNumber"));

String sql;
sql = "INSERT INTO Item VALUES (555555,'CD player',10)";       // Note 5
stmt.executeUpdate(sql);
sql = "UPDATE Item SET Quantity = 12 "
      + "WHERE Description = 'CD player'";
stmt.executeUpdate(sql);                                       // Note 6
System.out.println();
System.out.println("   Added and updated a new item");

System.out.println("Description");
query = "SELECT Description FROM Item";
rs = stmt.executeQuery(query);
while(rs.next())
   System.out.println(rs.getString(1));

sql = "DELETE FROM Item WHERE Description = 'CD player'";
stmt.executeUpdate(sql);
query = "SELECT Description FROM Item";
```

continued

```
                     rs = stmt.executeQuery(query);
                     System.out.println();
                     System.out.println("   Deleted the new item");
                     System.out.println("Description");
                     while(rs.next())
                         System.out.println(rs.getString(1));

                     stmt.close();
                 }catch (Exception e) {e.printStackTrace();}
             }
         }
```

Output

```
Names and Addresses of Customers
Name            Address
Carla Kahn      55 Fourth St.
Darnell Davis   33 Second St.
Fred Flynn      22 First St.
Marla Martinez  44 Third St.

    Order items for radios
OrderNumber     Quantity        UnitPrice
1               4               $27.00
3               3               $27.00
5               2               $27.00

    Customer placing orders on Mar 22, 1999
Darnell Davis

    Customers ordering computers
Carla Kahn
Fred Flynn

    Order numbers of orders from 4/1/99 to 4/30/99
1
4

    Added and updated a new item
Description
radio
television
computer
CD player

    Deleted the new item
Description
radio
television
computer
```

Note 1: The SQL SELECT statement returns the selected rows in a ResultSet. We use the executeQuery method to execute SELECT statements.

Note 2: We embed tab characters, \t, in the string to space the data horizontally.

Note 3: We omitted field 2, ItemNumber, from the display because we selected all results to have ItemNumber = 222222. We could insert a single tab character using single quotes, '\t', but inserting two characters requires the double-quoted string, "\t\t". We use the `getInt` method because field 3, Quantity, has SQL type `INTEGER`. The first argument, 4, to `getBigDecimal`, is the field number, while the second, 2, is the number of places after the decimal point. This UnitPrice field has type Currency in the database.

Note 4: We used the field number, 1, but could have used the field name, CustomerName, as the argument to `getString`. We will see in the next section how to get the number of fields and their names from the database if we do not know them.

Note 5: We add a new row to illustrate the `UPDATE` and `DELETE` statements which change the database. We update the new row, and then delete it, leaving the database unchanged when we exit the program. This is nice while learning JDBC because we can try various `SELECT` statements running the same program repeatedly without changing the data.

Note 6: `stmt.executeUpdate(sql);`

Because the `UPDATE` and `DELETE` statements do not return values, we use the `executeUpdate` method to execute them.

The Big Picture

....

When querying the database, a result set contains the selected rows. We use methods such as `getString` to display a value from a row of the result set. The SQL types have corresponding Java methods, so the Java `getInt` method retrieves `INTEGER` values, for example. We can write our SQL queries to order the results or to eliminate duplicate rows. A query may have to join several tables on common fields to obtain the desired information.

TEST YOUR UNDERSTANDING

11. Write an SQL statement to find names of salespersons and the customers that have placed orders with them. Be sure to eliminate duplicates.

TRY IT YOURSELF 12. Modify Example 11.3 to use only field names in the `getString`, `getInt`, and `getBigDecimal` methods.

TRY IT YOURSELF 13. Modify Example 11.3 to use only field numbers in the `getString`, `getInt`, and `getBigDecimal` methods.

TRY IT YOURSELF 14. Modify Example 11.3 to list CustomerID in addition to CustomerName and Address. Arrange the output rows so that the CustomerID numbers appear in numerical order.

15. Write a `SELECT` statement to find the names and addresses of customers who placed orders with Peter Patterson. Be sure to eliminate duplicates.

....

11.4 ▪ Metadata and Aggregate Functions

Java, with JDBC, allows us to get information about the database (**metadata**) with which we are working, and about any result sets we obtain. We can use SQL functions to compute with the data.

Database Metadata

The DatabaseMetaData methods return information about the database to which we are connected. To use these methods we first execute

```
DatabaseMetaData dbMetaData = con.getMetaData();
```

where con is the connection to the database. We can ask what level of SQL the database system supports by using the three methods

```
dbMetaData.supportsANSI92EntryLevelSQL();
dbMetaData.supportsANSI92IntermediateSQL();
dbMetaData.supportsANSI92FullSQL();
```

where ANSI (pronounced an'-see) stands for the American National Standards Institute. Java requires that JDBC drivers support ANSI92 entry level SQL so the first method must always return **true**. The Microsoft Access version 7.0 that we use supports ANSI89 but does not support ANSI92 intermediate or full SQL.

The method

```
dbMetaData.getIdentifierQuoteString();
```

returns the character used to delimit strings; in our database that is the single quote, '. Executing

```
ResultSet rs = dbMetaData.getTypeInfo();
```

gives us the type names used in the database itself, which may be different from the standard SQL types, or Java types. For example Microsoft Access uses the CURRENCY type and internally uses TEXT for the SQL VARCHAR type. We can list the type names from the result set using the loop

```
while(rs.next())
    System.out.println(rs.getString("TYPE_NAME"));
```

The very handy getTables method lets us obtain the names of the tables in our database. For example,

```
dbMetaData.getTables(null,null,"%",tables);
```

will return the names of the five tables in the Sales database. We pass **null** for the first two arguments because they represent the catalog and schema facilities, present in more elaborate database applications, that we do not use. The third argument to getTables is a string representing a search pattern for the tables we are seeking. In a large database with many tables, we might search for all tables starting with "Payroll" by using the string "Payroll%" where the % character matches zero or more characters. Because we want all tables, we use the string "%" which matches any string. To match a single character we could use the string "_", so "Payroll_" would match strings such as Payroll1, Payroll2, and so on.

The fourth argument to getTables uses an array of strings to specify the types of table for which to search. In addition to the tables that we created, there are various system tables in the database in which we are not interested. To limit our search we declare the fourth argument as

```
String[] tables = {"TABLE"};
```

which restricts the search to user-defined data tables.

We can also use database metadata to find the column names and types for each table. The method call

```
ResultSet rs = dbMetaData.getColumns(null,null,"Customer","%");
```

returns information about each column of the Customer table. As with the `getTables` methods we pass **null** arguments for the catalog and schema which we do not use. The third argument is a pattern for the tables to search; we pass the name Customer to get its columns. The fourth argument allows a string pattern to select the columns. We pass "%" to retrieve all columns. The details of interest about each column are its name and type which we access using

```
rs.getString(COLUMN_NAME);
```

and

```
rs.getString(TYPE_NAME);
```

As we shall see, the result set contains other information about each column.

Result Set Metadata

JDBC allows us to get information about each result set. We use

```
ResultSetMetaData rsMetaData = rs.getMetaData();
```

to get the `rsMetaData` object, and then use the methods

```
rsMetaData.getColumnCount();
```

to return the number of columns in the result set,

```
rsMetaData.getColumnLabel(i);
```

to return the name of column `i`, and

```
rsMetaData.getColumnTypeName(i);
```

to return its type.

Using the `getColumns` method, we suggested just listing the `COLUMN_NAME` and `TYPE_NAME` fields of the result set returned. In Example 11.4 we use result set metadata to list all the fields of the result set describing each column of the database. Perhaps not surprisingly, we found these result set fields differ from those described in the documentation included with the JDK, using version 1.1.6.

The `colNamesTypes` method in Example 11.4 uses the `getColumnLabel` and `getColumnTypeName` methods to return the names and types of each column of its result set argument. We can use it with any result set. For example, using it with the result set returned by

```
stmt.executeQuery("SELECT * FROM Item");
```

would list the all columns, with their types, from the Item table, because using the star, *, in the SELECT clause returns all the columns of the table.

Aggregate Functions

Aggregate functions compute values from the table data, using all the rows to produce the result.

For example, the query

```
SELECT SUM(BalanceDue),
    AVG(BalanceDue),
    MAX(BalanceDue)
FROM Customer
```

returns the sum, average, and maximum of all the balances due in the customer table. These functions operate on the BalanceDue column for all rows in the Customer table. Using a WHERE clause, as in

```
SELECT COUNT(*), MIN(Quantity) FROM OrderItem
WHERE ItemNumber = '222222'
```

will limit the computation to the rows of the OrderItem table which correspond to orders for radios. The function COUNT(*) will return the total number of rows satisfying this condition, while MIN(Quantity) returns the minimum quantity of radios ordered in one of the three rows of the OrderItem table which represent orders for radios (item number 222222).

☛ **TIP**

If using the JDK, the output will not fit in the command window. Piping it to a file allows us to open the file in an editor to see all the output. For the JDK, the command

```
java DatabaseInfo >out
```

will send the output to the file out rather than to the command window. Other environments use different names, not java, for their command line interpreters, but the same technique will work.

EXAMPLE 11.4 **DatabaseInfo.java**

```
/* Illustrate DatabaseMetaData, ResultSetMetaData
 * and SQL aggregate functions.
 */

import java.sql.*;
import java.io.*;

class DatabaseInfo {
  public static void main (String args[]) {
    try{
      ResultSet rs;
      Class.forName("sun.jdbc.odbc.JdbcOdbcDriver");          // Note 1
      String url = "jdbc:odbc:Sales";
      Connection con = DriverManager.getConnection(url);      // Note 2

      DatabaseMetaData dbMetaData = con.getMetaData();
      System.out.println("Supports entry level SQL: " +
              dbMetaData.supportsANSI92EntryLevelSQL());
      System.out.println("Supports intermediate SQL: " +
              dbMetaData.supportsANSI92IntermediateSQL());
      System.out.println("Supports full SQL: " +
              dbMetaData.supportsANSI92FullSQL());
      System.out.println("Supports stored procedures: "+
              dbMetaData.supportsStoredProcedures());
```

```
      System.out.println("Quote string: "
                  + dbMetaData.getIdentifierQuoteString());
      System.out.println("Types used in the database:");
      System.out.print('\t');

      rs = dbMetaData.getTypeInfo();
      while (rs.next())
        System.out.print(rs.getString("TYPE_NAME") + " ");      // Note 3
      System.out.println();

      String[] tables ={"TABLE"};
      rs = dbMetaData.getTables(null,null,"%",tables);
      System.out.println("Tables in the Sales database:");
      System.out.print('\t');
      while(rs.next())
        System.out.print(rs.getString("TABLE_NAME") + " ");     // Note 4
      System.out.println();

      rs = dbMetaData.getColumns(null,null,"Customer","%");
      System.out.println("Columns in the Customer table");
      while(rs.next())
        System.out.println('\t'+rs.getString("COLUMN_NAME")+"    // Note 5
                        "+rs.getString("TYPE_NAME"));
      displayStrings("Fields describing each column",colNamesTypes(rs)); // Note 6

      String query;
      query = "SELECT * FROM Item";
      Statement stmt = con.createStatement();
      rs = stmt.executeQuery(query);
      displayStrings("Item Columns",colNamesTypes(rs));         // Note 7

      query = "SELECT SUM(BalanceDue),AVG(BalanceDue), "
            + "MAX(BalanceDue) FROM Customer";
      rs = stmt.executeQuery(query);
      displayStrings("Function columns",colNamesTypes(rs));     // Note 8
      System.out.println("Sum, average, and maximum balance due");
      while(rs.next())
        System.out.println("$" + rs.getBigDecimal(1,2) + " $"
                        + rs.getBigDecimal(2,2) + " $" + rs.getBigDecimal(3,2));

      query = "SELECT COUNT(*), MIN(Quantity) FROM OrderItem "
            + "WHERE ItemNumber = '222222' ";
      rs = stmt.executeQuery(query);
      while(rs.next()){
        System.out.println("Number of radio order items: " + rs.getInt(1));
        System.out.println("Minimum quantity of radios ordered in any order item: "
                        + rs.getInt(2));
      }

      stmt.close();
    }catch (Exception e) {e.printStackTrace();}
}
public static String[] colNamesTypes(ResultSet rs) throws SQLException {
    ResultSetMetaData rsMetaData = rs.getMetaData();
    int cols = rsMetaData.getColumnCount();
    String[] s = new String[cols];                             // Note 9
    String label, tab;                                              continued
```

```
      for (int i =1; i <= cols; i++) {
        label = rsMetaData.getColumnLabel(i);
        if (label.length() < 8) tab = "\t\t"; else tab = "\t";        // Note 10
        s[i - 1] = '\t' + label + tab
                         + rsMetaData.getColumnTypeName(i);            // Note 11
      }
      return s;
    }
    public static void displayStrings(String description, String[]s) { // Note 12
      System.out.println(description);
      for(int i = 0; i < s.length; i++)
        System.out.println(s[i]);
    }
}
```

Output

```
Supports entry level SQL: true
Supports intermediate SQL: false
Supports full SQL: false
Supports stored procedures: true
Quote string: `
Types used in the database:
      BIT BYTE LONGBINARY VARBINARY BINARY LONGTEXT CHAR
CURRENCY LONG COUNTER SHORT SINGLE DOUBLE DATETIME TEXT
Tables in the Sales database:
      Customer Item OrderItem Orders Salesperson
Columns in the Customer table
      CustomerID TEXT
      CustomerName TEXT
      Address TEXT
      BalanceDue CURRENCY
Fields describing each column
      TABLE_QUALIFIER TEXT
      TABLE_OWNER     TEXT
      TABLE_NAME      TEXT
      COLUMN_NAME     TEXT
      DATA_TYPE       SHORT
      TYPE_NAME       TEXT
      PRECISION       LONG
      LENGTH          LONG
      SCALE           SHORT
      RADIX           SHORT
      NULLABLE        SHORT
      REMARKS         TEXT
      ORDINAL         LONG
Item Columns
      ItemNumber   TEXT
      Description  TEXT
      Quantity     LONG
Function Columns
      Expr1000   CURRENCY
      Expr1001   CURRENCY
      Expr1002   CURRENCY
Sum, average, and maximum balance due
$1797.95 $449.49 $1667.00
Number of radio order items: 3
Minimum quantity of radios ordered in any order item: 2
```

Note 1: To show how it works, we use the `forName` method to load the JDBC driver. The advantage of this approach is we could easily modify this program to pass the driver name as a program argument. We have no need to do that here but it might be useful in writing a general application designed to work with different databases.

Note 2: We use the form of the `getConnection` method that does not require a user name or a password because these are not needed for the Microsoft Access database system we are using.

Note 3: According to the JDK 1.1.6 documentation, the `getTypeInfo` method returns 18 columns of information for each type provided by the database system. We only list one, `TYPE_NAME`, leaving as an exercise the use of the `colNamesTypes,` in Example 11.4, to list the names and types of all columns of this result set to see if they correspond to the 18 listed in the documentation.

Note 4: According to the JDK 1.1.6 documentation, the `getTables` method returns five columns of information for each table in the database. We list only `TABLE_NAME`. It is column 3 so we could have used `getString(3)` to retrieve it, but using the column name is much more helpful.

Note 5: According to the JDK 1.1.6 documentation, the `getColumns` method returns 18 fields to describe each column. (However, see Note 6.) We list `COLUMN_NAME` and `COLUMN_TYPE`.

Note 6: `displayStrings ("Fields describing each column",colNamesTypes(rs));`

Using the `colNamesTypes` method to list the names and types of the columns in the result set returned by the `getColumns` method we see that the result set actually contains 13 columns, some of which are named differently than the columns listed in the JDK documentation.

Note 7: `displayStrings("Item Columns",colNamesTypes(rs));`

This shows that we can use the `colNamesTypes` method to display the names and types of the columns of any result set, in this case the one which selects all columns from the `Item` table.

Note 8: `displayStrings("Function columns",colNamesTypes(rs));`

The result set gives the values of the SUM, AVG, and MAX functions. These are not columns of the Customer tables, but rather they are values computed from the BalanceDue column. We use the `colNamesTypes` method to list the names and types of the columns in the result set. Because these columns have no names Java creates the names Expr1000, Expr1001, and Expr1003 for them. They have internal database types of CURRENCY which does not correspond to a SQL type. The Java `getBigDecimal` method will return the value of each function.

Note 9: `String[] s = new String[cols];`

We use a string array to hold the name and type of each column in the result set argument to this method.

Note 10: To keep the type column aligned we use two tab characters as a separator when the column name is less than eight characters in length, but one tab character otherwise.

Note 11: `s[i-1] = '\t' + label + tab + rsMetaData.getColumnTypeName(i);`

Column numbers start at 1, while array indices start at 0, so we store column i in array component i-1.

Note 12: The `displayStrings` method displays a description and a list of the elements of its `String` array argument.

The Big Picture

Database metadata tells us properties of the database such as the names of its tables, and the names and types of the columns in a table. Result set metadata lets us find properties of a result set. We can find the number of columns in the result set and the label and type of each column. We can use result set metadata on a result set from a database metadata method or on a result set from an SQL query. Aggregate functions compute values from the rows of a table.

TEST YOUR UNDERSTANDING

TRY IT YOURSELF   16. Modify Example 11.4 to pass **null** as the fourth argument to the `getTables` method, instead of the `tables` array. This will list all tables in the database, including the system tables.

TRY IT YOURSELF   17. Modify Example 11.4 to change the third argument to the `getTables` method to find the tables in the Sales database which start with `Order`.

TRY IT YOURSELF   18. Modify Example 11.4 to use the `colNamesTypes` method to list all the fields of the result set returned by the `getTypeInfo` method. Compare these fields to those listed in the JDK documentation, if available.

TRY IT YOURSELF   19. Modify Example 11.4 to use the `colNamesTypes` method to list all the fields of the result set returned by the `getTables` method. Compare these fields to those listed in the JDK documentation, if available.

11.5 ▪ Prepared Statements and Transactions[1]

A **prepared statement** lets us translate a statement to low-level database commands once, and execute it many times, thus avoiding the inefficient repetition of the translation process.

When making changes to a database we must be very careful that we complete all steps of the transaction. It would not do to withdraw funds from one account, but not have it deposited in another. **Transaction processing** allows us to explicitly control when changes become final, so that we only **commit** changes when all those desired have completed correctly.

---

[1] The ODBC text driver does not handle prepared statements or transactions.

Using Prepared Statements

Often we may wish to execute a query repeatedly using different conditions each time. The query

```
SELECT * FROM OrderItem
WHERE ItemNumber = '222222'
```

selects all order items with number 222222. To execute this query for each item, we could use a loop such as

```
String[] numbers = {"222222","333333","444444"}
ResultSet rs;
for (int i = 0; i < numbers.length, i++) {
    rs =  executeQuery("SELECT * FROM OrderItem "
        + "WHERE ItemNumber = '" + number + '\'');
    // process results
}
```

We have only three products in our database, but realistically we might have had many more. For each product, the database system must process the SQL query analyzing how to find the requested data from the database in the most efficient way possible. Our query is quite simple, but realistically it could have been much more complex. Each time we call executeQuery, we have to process the query again, spending the time over and over again to find the best way to find the results that satisfy it.

The prepared statement allows the database system to process an SQL query once, determining the best way to get the results. We can then use this prepared statement over and over again with different data but without the overhead of translating it again.

We use the question mark, ?, to denote the arguments to query that we wish to change from one execution to the next. To make a prepared statement from our previous query, we write it as

```
String query = "SELECT * FROM OrderItem "
    + "WHERE ItemNumber = ?";
```

where the question mark stands for the item number that we will pass in. Next we create a prepared statement using
```
PreparedStatement pStmt = con.prepareStatement(query);
```
where con is the connection to the database.

To pass arguments to a query we use setXXX methods where XXX stands for the type of the argument. In our example, ItemNumber has type VARCHAR which corresponds to the string type in Java, so we use the setString method, as in

```
pStmt.setString(1,"222222");
```

where we enclose the item number in double quotes because we are inside Java and not writing an SQL statement for this database system. The first argument to setString is the number of the argument to which we want to pass the value specified. We number the arguments in the order they appear in the query, with the first argument having number 1. The statement

```
rs = pStmt.executeQuery();
```

executes the prepared query with the argument 222222.

☛ **TIP**

The executeQuery method takes no arguments when used with a prepared statement because we have already passed the query to the prepareStatement method.

We process the result set as we did with simple statements. The code

```
System.out.println("OrderNumber\tQuantity\tUnitPrice");
while (rs.next())
    System.out.println(rs.getString(1) + "\t\t" + rs.getInt(3)
        + "\t\t$" + rs.getBigDecimal(4,2));
```

extracted from Example 11.3 will return the other columns of all rows in the OrderItem table having the specified item number.

After closing the result set with

```
rs.close();
```

we could pass another argument to the query and execute the query again as in

```
pStmt.setString(1,"333333");
rs = pStmt.executeQuery()
```

which would find the rows of the OrderItem table whose item number is 333333, representing a television order.

To pass multiple arguments we use additional question marks in the query. In the query

```
query = "SELECT OrderNumber FROM Orders "
    + "WHERE OrderDate BETWEEN ? AND ?";
```

the arguments represent the starting and ending dates of orders. After creating the prepared statement, we pass the arguments using the setDate method as in

```
pStst.setDate(1, Date.valueOf("1999-04-01"));
```

which replaces the first question mark with April 1, 1999, and

```
pStst.setDate(2, Date.valueOf("1999-04-30"));
```

which replaces the second question mark with April 30, 1999. The Date class, in the java.sql package, extends java.util.Date. The valueOf method translates a string representing the date to a Date which can be used in the database system.

In preparing a statement to which we pass an argument which is a currency amount, we use the setBigDecimal method to pass the currency value. For example, the query

```
SELECT CustomerName FROM Customer
WHERE BalanceDue > ?
```

has an argument for the BalanceDue value. To pass such a value to the prepared statement created from this query, we use

```
pStmt.setBigDecimal(1, new java.math.BigDecimal(0.0));
```

The BigDecimal, created from the double 0.0, will represent the amount $0.00, so our query will return the names of all customers with a non-zero balance.

Transaction Processing

Often when using a database we need to execute several statements to perform the desired transaction. For example, if a customer places a new order we will update the Order table with another order, the OrderItem table with the items ordered, and the Customer table with a new BalanceDue. We would be unhappy if an error occurred after some, but not all of these changes were made. Java allows us to manage transactions so we only commit the changes to the database when they complete without error.

The JDBC default is to commit the change as soon as we execute the update. The statement

```
con.setAutoCommit(false);
```

changes from the default behavior to require that we explicitly commit changes using

```
con.commit();
```

If we have already executed some updates and decide we do not want to commit them, we can roll back to the point when we executed the last commit, undoing these changes using

```
con.rollback();
```

For example, if we have removed the auto commit default, after executing the queries

```
INSERT INTO Item VALUES (555555,'CD player',10)
```

and

```
UPDATE Item SET Quantity = 12
WHERE Description = 'CD player'
```

we can either commit them, making the changes permanent, using the commit method, or undo them using the rollback method.

EXAMPLE 11.5   **Prepare.java**

```
/* Illustrates prepared statements
 * and transaction processing.
 */

import java.sql.*;
import java.io.*;

class Prepare {
  public static void main (String args[]) {
    try {
      ResultSet rs;
      Class.forName("sun.jdbc.odbc.JdbcOdbcDriver");
      String url = "jdbc:odbc:Sales";
      Connection con = DriverManager.getConnection(url);

      String query;
      query = "SELECT Quantity FROM Item "
            + "WHERE Description = ?";                          // Note 1
      PreparedStatement pStmt = con.prepareStatement(query);
      pStmt.setString(1, "radio");
      rs = pStmt.executeQuery();
      System.out.println
              ("   Using a prepared statement to find quantity of radios");
      while(rs.next())
         System.out.println(rs.getInt("Quantity"));
      rs.close();                                              // Note 2
```

continued

```
pStmt.setString(1, "computer");
rs = pStmt.executeQuery();
System.out.println
        ("   Using a prepared statement to find quantity of computers");
while(rs.next())
    System.out.println(rs.getInt("Quantity"));
rs.close();

query = "SELECT OrderNumber FROM Orders "
      + "WHERE OrderDate BETWEEN ? AND ?";
pStmt =con.prepareStatement(query);
pStmt.setDate(1, Date.valueOf("1999-04-01"));
pStmt.setDate(2, Date.valueOf("1999-04-30"));
rs = pStmt.executeQuery();
System.out.println
        ("   Using a prepared statement to find orders in April");
while(rs.next())
    System.out.println(rs.getInt("OrderNumber"));
rs.close();

query = "SELECT CustomerName FROM Customer "
      + "WHERE BalanceDue > ?";
pStmt = con.prepareStatement(query);
pStmt.setBigDecimal(1, new java.math.BigDecimal(0.0));
rs = pStmt.executeQuery();
System.out.println("   Using a prepared statement to find customers "
        +   "with non-zero balance");
while(rs.next())
    System.out.println(rs.getString("CustomerName"));
pStmt.close();                                               // Note 3

Statement stmt = con.createStatement();
con.setAutoCommit(false);                                    // Note 4
String sql;
sql = "INSERT INTO Item VALUES (555555,'CD player',10)";
stmt.executeUpdate(sql);

sql = "UPDATE Item SET Quantity = 12 "
    + "WHERE Description = 'CD player'";
stmt.executeUpdate(sql);
System.out.println();
System.out.println
        ("   Before commit or rollback -- table changed, but can rollback");
System.out.println("Description");
query = "SELECT Description FROM Item";
rs = stmt.executeQuery(query);
while(rs.next())
    System.out.println(rs.getString(1));                     // Note 5

con.rollback();                                              // Note 6
System.out.println();
System.out.println("   Rolled back insert and update -- table unchanged");
System.out.println("Description");
query = "SELECT Description FROM Item";
rs = stmt.executeQuery(query);
while(rs.next())
    System.out.println(rs.getString(1));
```

```
        sql = "INSERT INTO Item VALUES (555555,'CD player',10)";      // Note 7
        stmt.executeUpdate(sql);
        sql = "UPDATE Item SET Quantity = 12 "
            + "WHERE Description = 'CD player'";
        stmt.executeUpdate(sql);
        con.commit();                                                 // Note 8
        System.out.println();
        System.out.println("   Committed insert and update -- table changed");
        System.out.println("Description");
        query = "SELECT Description FROM Item";
        rs = stmt.executeQuery(query);
        while(rs.next())
            System.out.println(rs.getString(1));

        sql = "DELETE FROM Item WHERE Description = 'CD player'";      // Note 9
        stmt.executeUpdate(sql);
        con.commit();                                                 // Note 10
        query = "SELECT Description FROM Item";
        rs = stmt.executeQuery(query);
        System.out.println();
        System.out.println("   Deleted the new item");
        System.out.println("Description");
        while(rs.next())
            System.out.println(rs.getString(1));

        stmt.close();
      }catch (Exception e) {e.printStackTrace();}
    }
}
```

Output

```
Using a prepared statement to find quantity of radios
32
  Using a prepared statement to find quantity of computers
9
  Using a prepared statement to find orders in April
1
4
  Using a prepared statement to find customers with non-zero balance
Fred Flynn
Darnell Davis

  Before commit or rollback -- table changed, but can rollback
Description
radio
television
computer
CD player

  Rolled back insert and update -- table unchanged
Description
radio
television
computer

  Committed insert and update -- table changed
```

continued

```
Description
radio
television
computer
CD player

  Deleted the new item
Description
radio
television
computer
```

Note 1: The question mark indicates where we can substitute one of the descriptions, radio, television, or computer.

Note 2: With simple statements we do not need to close the result set after each query. The `stmt.close` method also closes the result set when we are done with the statment, `stmt`. For prepared statements we need to close the result set after each query.

Note 3: Closing the prepared statement, `pStmt`, automatically closes the last result set too.

Note 4: With auto commit off, we must execute the `commit` statement in order to make our updates permanent.

Note 5: Outputting the descriptions of the items in the database shows that the database system has entered the item CD player (with the updated quantity of 12 not shown). We have not yet executed the `commit` statement so we still have a chance to rollback this change.

Note 6: `con.rollback();`

After we rollback the updates we find only the original three items in the Item table.

Note 7: `sql = "INSERT INTO Item VALUES (555555,'CD player',10)";`

Now we make the same updates, this time actually committing them to the database.

Note 8: `con.commit();`

This will commit the previous updates to the database and prevent further rollbacks of them. Only updates executed after this can be rolled back.

Note 9: `sql = "DELETE FROM Item WHERE Description = 'CD player'";`

We delete the new row from the database, to leave it as we found it in this pedagogical example.

Note 10: `con.commit();`

We must commit the DELETE transaction for it to take effect.

The Big Picture

A prepared statement lets us translate a query once and substitute values for its parameters to execute it repeatedly. By deciding explicitly when to commit changes to the database, we reserve the option to rollback some changes if the entire transaction cannot be completed.

20. Write a SELECT statement to return the names of customers who ordered an item given by its description in the Item table, which we pass in as an argument, so we can create a prepared statement from the query.

TRY IT YOURSELF 21. Modify Example 11.5 to omit the first rs.close() statement. Does any error result? If so, which?

TRY IT YOURSELF 22. Modify Example 11.5 to find the order numbers of orders placed in March. Use the same prepared statement.

TRY IT YOURSELF 23. Modify Example 11.5 to omit the last commit statement. Run the modified program and check the database afterward to see that the new row has not been deleted; it must be deleted manually.

11.6 ▪ A GUI for Database Queries

Our case study builds a graphical user interface for querying our Sales database. This example illustrates the JDBC techniques covered in this chapter, adding the user interface concepts studied earlier. Even though large for an introductory example, it would need many extensions, some of which may be found as exercises, and much polishing to make it a really useful application.

The SearchSales program allows the user to create a SELECT query, and executes it, displaying the resulting rows. We use the gridbag layout to arrange the components. Figure 11.13 shows the initial screen.

FIGURE 11.13 The SearchSales initial screen

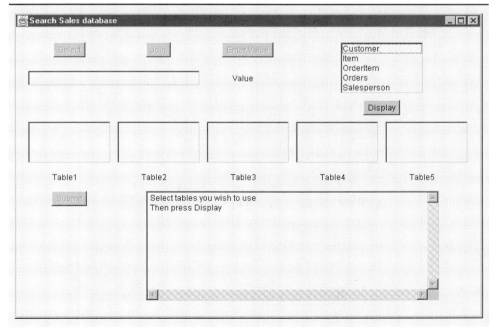

The List (we cover the List component below) at the upper-right shows the five tables of the Sales database. The user selects the tables to search. The names of these tables will appear after FROM in the query. The text area at the bottom gives instructions to the user, and displays the final results of the search. We disable all buttons, except Display, until we are ready to use them.

Figures 11.14–11.18 show the steps in the creation and execution of the query

```
SELECT CustomerName FROM Customer, Orders
WHERE Customer.CustomerID = Orders.CustomerID
AND OrderDate = {d '1999-03-22'}
```

Figure 11.14 shows the screen after the user has selected the Customer and Orders tables, and pressed the Display button. We have disabled the Display button because the user has already chosen the tables. The column names for the Customer table appear in the leftmost List, while those for the Orders table appear in the fourth List. The labels underneath now show the table names. Thus far our query is

```
SELECT ... FROM Customer, Order.
```

FIGURE 11.14 Screen to choose columns for the result

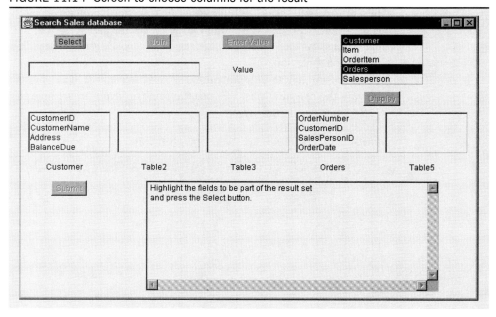

The user now selects the columns to be part of the result set, in this example choosing CustomerName and pressing the Select button. The partially constructed query is now

```
SELECT CustomerName FROM Customer, Order
```

Figure 11.15 shows the next screen in which we disabled the Select button, because we only select the fields of the result once. We deselect all fields so that the user will not have to deselect the fields before going on to the next step. At this point we enable the Join, Enter Value, and Submit buttons. The user would be ready to execute queries without conditions, such as

```
SELECT CustomerName FROM Customer,
```

so we enable the Submit button. The Join and Enter Value buttons allow us to add conditions that restrict the scope of the query.

FIGURE 11.15 After pressing the Select button

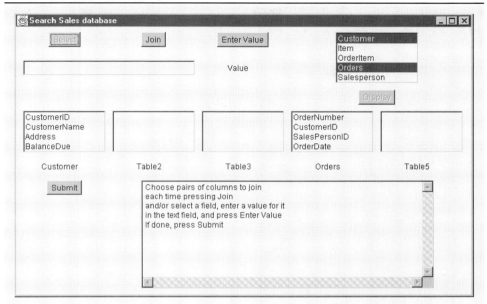

In our example query we join the Customer and the Orders tables, requiring the condition

```
Customer.CustomerID = Orders.CustomerID.
```

We impose this condition to join the information from the two tables properly, and also impose the condition

```
OrderDate = {d '1999-03-22'}
```

to select orders placed on March 22, 1999.

Figure 11.16 indicates the user has selected the CustomerID field in the Customer table and the CustomerID field in the Orders table.

After pressing the Join button the partially completed query will be

```
SELECT CustomerName FROM Customer, Order
WHERE Customer.CustomerID = Orders.CustomerID
```

The next screen, Figure 11.17 has the same options as in Figure 11.16, because we can add conditions or submit the completed query. We choose the OrderDate column from the Orders table and enter the value {d '1999-03-22} in the text field. Pressing the Enter Value button will add to our query the condition that the order date be March 22, 1999. We only use the equality relation in our conditions leaving the extension to less than and greater than to the exercises.

We could add more conditions, but this completes our query, so we press the Submit button. Figure 11.18 shows the resulting list (of only one customer, Darnell Davis) displayed in the text area. We disable all buttons, leaving for the exercises the option to continue executing additional queries.

FIGURE 11.16 Adding a Join condition

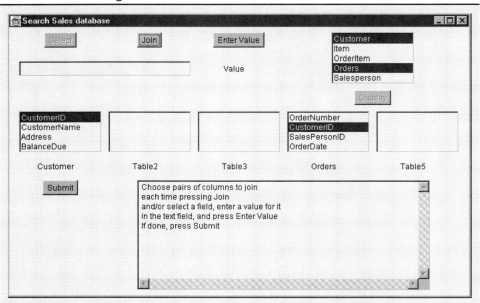

FIGURE 11.17 Entering the OrderDate condition

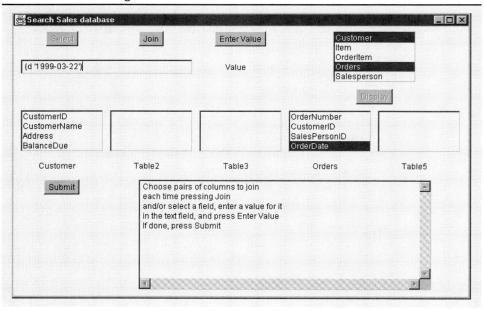

FIGURE 11.18 The query result

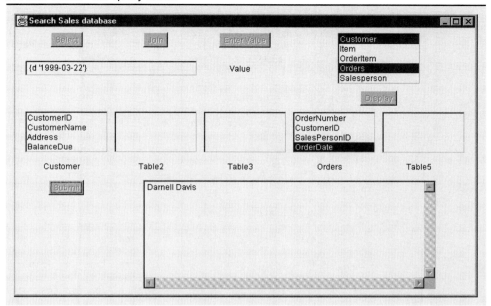

The List Component

The List component we use in Example 11.6 differs from a choice box in that we can specify how many entries to display and select multiple entries. The constructor

```
List(5,true)
```

specifies a box that will show five entries, providing a scroll bar if the list contains more than five items. A second argument of **true** permits multiple selections.

We add items to a List using the add method, as in

```
tables.add("Customer");
```

Single-clicking the mouse on a list item generates an ItemEvent, while double-clicking on an item generates an ActionEvent, but we do not handle these events in this example, preferring to wait until the user selects all the desired items and presses the appropriate button to ask us to process the selections.

EXAMPLE 11.6   **SearchSales.java**

```
/* Provides a GUI to execute an SQL query
 * on the Sales database.
 * /

import java.awt.*;
import java.awt.event.*;
import java.sql.*;
import java.net.*;
```

continued

```java
public class SearchSales extends Frame implements ActionListener{
  public static final int SIZE = 5;
  List tables = new List(SIZE,true);        // tables in Sales database
  List[] columns = new List[SIZE];          // columns in each table
  Label[] colLabel = new Label[SIZE];       // label for each table's col list
  Label value = new Label("Value");
  TextField fieldValue = new TextField(12); // enter a value in a condition

  Button submit = new Button("Submit");      // submit the query
  Button join = new Button("Join");          // choose common columns in a condition
  Button enter = new Button("Enter Value");  // enter the value for the condition
  Button select = new Button("Select");      // choose the columns for result set
  Button display = new Button("Display");    // display the selected tables' columns
  TextArea result = new TextArea();          // display prompts and final result

  Connection con;
  Statement stmt;
  DatabaseMetaData dbMetaData;
  String[] tableName = new String[SIZE];  // names of the selected Sales tables
  int[] indices = null;                   // indices of Sales tables selected
  String resultCols = "";                 // result set columns, after SELECT
  boolean firstJoin = true;               // first time for join
  String joinClauses = "";                // clauses to be joined, after WHERE or AND
  String condition = "";                  // condition clauses, after WHERE or AND
  String fromTables = "";                 // tables used, for FROM part of query
  String query = "SELECT ";               // the query to be executed
  int count = 0;                          // number of cols in result set

  public SearchSales(String title) {
    setTitle(title);
    for(int i = 0; i < SIZE; i++)                                       // Note 1
      columns[i] = new List(4,true);
    for(int i = 0; i < SIZE; i++)                                       // Note 2
      colLabel[i] = new Label("Table" + (i+1));
    GridBagLayout gbl = new GridBagLayout();
    setLayout(gbl);
    GridBagConstraints c = new GridBagConstraints();
    c.insets = new Insets(5,5,5,5);                                     // Note 3
    gbl.setConstraints(select,c);      add(select);
    gbl.setConstraints(join,c);        add(join);
    gbl.setConstraints(enter,c);       add(enter);

    c.gridwidth = GridBagConstraints.REMAINDER;                        // Note 4
    c.gridheight = 2;
    gbl.setConstraints(tables,c);      add(tables);

    c.gridx = 0;
    c.gridy = 1;
    c.gridwidth = 2;
    c.gridheight = 1;
    c.fill = GridBagConstraints.HORIZONTAL;                            // Note 5
    gbl.setConstraints(fieldValue,c); add(fieldValue);

    c.fill = GridBagConstraints.NONE;
    c.gridwidth = 1;
    c.gridx = 2;
    gbl.setConstraints(value,c);       add(value);
```

```
    c.gridx = 3;
    c.gridy = 2;
    c.gridwidth = GridBagConstraints.REMAINDER;
    gbl.setConstraints(display,c);    add(display);

    c.gridy = 3;
    c.gridwidth = 1;
    c.gridheight = 2;
    for(int i = 0; i < SIZE; i++) {                        // Note 6
      c.gridx = i;
      gbl.setConstraints(columns[i],c);
      add(columns[i]);
    }

    c.gridheight = 1;
    c.gridy = 5;
    for(int i=0; i<SIZE; i++) {
      c.gridx = i;
      gbl.setConstraints(colLabel[i],c);
      add(colLabel[i]);
    }
    c.gridx = 0;
    c.gridy = 6;
    gbl.setConstraints(submit,c);       add(submit);

    c.gridx = 1;
    c.gridheight = 2;
    c.gridwidth = GridBagConstraints.REMAINDER;
    gbl.setConstraints(result,c);       add(result);

    addWindowListener(new WindowClose());
    display.addActionListener(this);
    select.addActionListener(this);
    select.setEnabled(false);
    join.addActionListener(this);
    join.setEnabled(false);
    submit.addActionListener(this);
    submit.setEnabled(false);
    enter.addActionListener(this);
    enter.setEnabled(false);

    try {
      Class.forName("sun.jdbc.odbc.JdbcOdbcDriver");
      String url = "jdbc:odbc:Sales";
      con = DriverManager.getConnection(url);
      stmt = con.createStatement();
      result.setText("Select tables you wish to use\nThen press Display");
      dbMetaData = con.getMetaData();
      String[] tableTypes ={"TABLE"};
      ResultSet rs = dbMetaData.getTables(null,null,"%",tableTypes);
      int i = 0;
      while(rs.next())
        tables.add(tableName[i++] = rs.getString("TABLE_NAME"));   // Note 7
    }catch (Exception e) {e.printStackTrace();}
  }

public void actionPerformed(ActionEvent event) {
  Object source = event.getSource();                          continued
```

```
      if (source == display) {
        indices = tables.getSelectedIndexes();                             // Note 8
        for(int i = 0; i < indices.length; i++){
          colLabel[indices[i]].setText(tableName[indices[i]]);             // Note 9
          colLabel[indices[i]].invalidate();                               // Note 10
          fromTables += tableName[indices[i]] +',';                        // Note 11
        }

        fromTables = fromTables.substring(0,fromTables.length() - 1);      // Note 12
        display.setEnabled(false);
        result.setText("Highlight the fields to be part of the result set\n"
             + "and press the Select button.");

        for(int i = 0; i < indices.length; i++) {
          try {
            ResultSet rs =
              dbMetaData.getColumns(null,null,tableName[indices[i]],"%");// Note 13
            while(rs.next())
              columns[indices[i]].add(rs.getString("COLUMN_NAME"));        // Note 14
          }catch(SQLException e) {e.printStackTrace();}
        }
        select.setEnabled(true);
        validate();                                                        // Note 15
      }

      else if (source == select) {
        for(int i = 0; i < indices.length; i++)
          count += columns[indices[i]].getSelectedIndexes().length;       // Note 16
        resultCols = build("",',');                                       // Note 17
        resultCols = resultCols.substring(0,resultCols.length() - 1);
        result.setText("Choose pairs of columns to join\n"
                     + "each time pressing Join\n"
                     + "and/or select a field, enter a value for it\n"
                     + "in the text field, and press Enter Value\n"
                     + "If done, press Submit");
        join.setEnabled(true);
        enter.setEnabled(true);
        select.setEnabled(false);
        deselectAll(columns,indices);                                      // Note 18
        query += resultCols + " FROM " + fromTables;                       // Note 19
        submit.setEnabled(true);
      }

      else if (source == join) {
        String keyword = "";
        if (firstJoin) {                                                   // Note 20
           keyword = " WHERE ";
           firstJoin = false;
        }

        else
          keyword = " AND ";
        joinClauses = build(keyword,'=');
        joinClauses = joinClauses.substring(0,joinClauses.length()-1);
        deselectAll(columns,indices);
        query += joinClauses;
      }
```

```
    else if (source == enter){
      String keyword = "";
      if (firstJoin) {
         keyword = " WHERE ";
         firstJoin = false;
      }

      else
         keyword = " AND ";
      condition = build(keyword,'=');
      condition += fieldValue.getText();
      query += condition;
    }

    else if (source == submit) {
      try {
        ResultSet rs = stmt.executeQuery(query);
        result.setText("");
        while(rs.next()) {
           String s = "";
           for(int i = 1; i <= count; i++)
              s += rs.getString(i) + ' ';                      // Note 21
           s += '\n';
           result.append(s);
        }
      }catch(Exception e) {
         e.printStackTrace();
      }
      submit.setEnabled(false);
      join.setEnabled(false);
      enter.setEnabled(false);
    }
  }

  public String build(String start, char c ) {                 // Note 22
    String s = start;
    String[] colNames;
    for(int i = 0; i < indices.length; i++) {
      colNames = columns[indices[i]].getSelectedItems();
      for (int j = 0; j < colNames.length; j++) {
        s += tableName[indices[i]]+ '.' + colNames[j] + c;
      }
    }
    return s;
  }

  public void deselectAll(List[] columns, int[] indices) {     // Note 23
    for(int i = 0; i < indices.length; i++)
      for(int j = 0; j < columns[indices[i]].getItemCount(); j++)
        columns[indices[i]].deselect(j);
  }

  public static void main(String[] args) {
    SearchSales search = new SearchSales("Search Sales database");
    Toolkit toolkit = Toolkit.getDefaultToolkit();             // Note 24
    Dimension d = toolkit.getScreenSize();                     // Note 25
    search.setSize(d.width,d.height - 30);
    search.setLocation(0,30);                                  // Note 26
    search.setVisible(true);
  }
```

continued

```
public class WindowClose extends WindowAdapter {
  public void windowClosing(WindowEvent e) {
    System.exit(0);
  }
}
}
```

Note 1: We create a List box to hold the columns in each of the five tables in the Sales database.

Note 2: We create the Label objects to which the Label array, colLabel, refers. The constructor for colLabel initialized these references to **null**.

Note 3: Except for setting the insets to provide a border of five pixels around each component, we use the default values for the gridbag constraints.

Note 4: Setting the gridwidth to GridBagContraints.REMAINDER lets the list of tables use the remainder of the row. Because the List has a default width of 1, it is centered in the remaining two columns.

Note 5: We want the text field to fill two columns, so we set the gridwidth to 2 and the fill to GridBagConstraints.HORIZONTAL so it will expand horizontally to fill the two-column space available.

Note 6: `for(int i = 0; i < SIZE; i++) {`

This loop adds the five List boxes, one for each table of the Sales database.

Note 7: `tables.add(tableName[i++]=rs.getString("TABLE_NAME"));`

This statement concisely achieves several objectives. It gets the next table name from the result set, assigns it to the tableName array for use later, increments the index i, and finally adds the table name to the tables list.

Note 8: `indices = tables.getSelectedIndexes();`

We constructed the tables list to allow the user to select multiple items. The getSelectedIndexes method returns the array of index numbers corresponding to selected items.

Note 9: `colLabel[indices[i]].setText(tableName[indices[i]]);`

Initially, we labeled the five tables, Table1,..., Table5. We change the labels underneath the selected tables to their actual table names. We could have labeled all five tables correctly, but chose this to differentiate those tables the user selected from the unselected ones.

Note 10: `colLabel[indices[i]].invalidate();`

Because the new label may have a different length, we later invoke the validate method to get the layout manager to redo the layout. Each changed label calls invalidate here so the layout manager will know that it needs to be laid out with its new size.

Note 11: `fromTables += tableName[indices[i]] +',';`

We save the names of the selected tables in a string, separated by commas, to use after FROM when we construct the SQL SELECT query.

Note 12: `fromTables = fromTables.substring(0,fromTables.length()-1);`

This removes the last comma.

Note 13: `ResultSet rs = dbMetaData.getColumns`
`        (null,null,tableName[indices[i]],"%");`

For each table the user selected, we get the names of its columns.

Note 14: `columns[indices[i]].add(rs.getString("COLUMN_NAME"));`

We add each column name to the List box representing the selected table.

Note 15: `validate();`

The gridbag layout manager will redo the layout, so components whose size has changed will be laid out properly.

Note 16: `count += columns[indices[i]].getSelectedIndexes().length;`

We save the total number of columns in the result set for the query, obtaining it by adding up the number of columns selected in each List box. After executing the query, we use `count` to list the results.

Note 17: `resultCols = build("",',');`

The `build` method combines the selected item into a string, using the second argument as the separator. The first argument is the initial value of the string.

Note 18: `deselectAll(columns,indices);`

The `deselectAll` method deselects each of the selected items so the user does not have to manually deselect the previous choices before making selections at the next step toward building the query.

Note 19: `query += resultCols + " FROM " + fromTables;`

We continue to build the query we wish to execute, adding the pieces we have constructed so far.

Note 20: `if (firstJoin) {`

The first condition, if any, in the query follows WHERE, while the remaining conditions follow AND. We use the **boolean** variable `firstJoin` to specify whether or not this is the first condition.

Note 21: `s += rs.getString(i) + ' ';`

For simplicity, we have not dealt with the types of each table column. Knowing the column type would allow us to use a more specific method than `getString`. For example, knowing the column has type `INTEGER` would allow us to use the `getInt` method, but the `getString` method will also work for every type, although sometimes the formatting will not be as nice.

Note 22: `public String build(String start, char c ) {`

The `build` method combines the selected List items into a string. The `start` argument is the initial value of the string, while the argument c is the character used to separate the selected items.

Note 23: `public void deselectAll (List[] columns, int[] indices) {`

For simplicity, this method deselects every column, even those the user had not selected. The arguments are the array of List boxes, one for each table, and the array of indices specifying which tables the user selected.

Note 24: `Toolkit toolkit = Toolkit.getDefaultToolkit();`

The `Toolkit` class allows us to access some properties of the host platform.

Note 25: `Dimension d = toolkit.getScreenSize();`

We get the screen size of the user's machine so we can size the frame to fill the screen. Setting the size using a fixed number of pixels, such as 500 by 300, as we have done in previous examples, will cause the frame to appear smaller on a higher resolution screen, and may make the frame too large for a low resolution screen.

Note 26: `search.setLocation(0,30);`

The `setLocation` method allows us to position the frame; otherwise we get the default of (0,0) for its upper-left corner.

The Big Picture
....

A graphical user interface lets the user compose a query. At each stage the user presses a button which causes some actions to occur and instructions to appear in the text area. The user first selects the tables to be used, then the fields to be displayed. The user may add conditions by joining tables or requiring a field have a specific value. After pressing the Submit button, the user sees the results in the text area.

TEST YOUR UNDERSTANDING

TRY IT YOURSELF 24. Run Example 11.6 to execute the query which returns the customer names who placed orders on March 22, 1999, but this time add the condition that the OrderDate is March 22, 1999 before the join condition that `Customer.CustomerID = Orders.CustomerID`. This shows we can enter conditions in any order.

TRY IT YOURSELF 25. Modify Example 11.6 to remove the call to the `validate` method. Run the modified program and describe any changes from the original version.

TRY IT YOURSELF 26. Modify Example 11.6 to omit setting `gridwidth` to `REMAINDER` for the tables list. Run the modified program and describe any changes from the original version.

TRY IT YOURSELF 27. Modify Example 11.6 to omit setting the `fill` for the `fieldValue` text field to HORIZONTAL. Run the modified program and describe any changes from the original version.

Chapter 11 Summary

Java Database Connectivity (JDBC), in the `java.sql` package, allows us to create database tables, insert, update, and delete data, and query a database from a Java program. Relational databases store data in tables, and each table has a key that uniquely identifies each row. As our example, we use the `Sales` database with five tables. The Customer table has CustomerID as its key. The Orders table has OrderNumber as its key, but also includes the foreign keys CustomerID and SalepersonID which refer to entries in the `Customer` and `Saleperson` tables so the information does not have to be duplicated in the Orders table. The OrderItem table has a compound key (`OrderNumber, ItemNumber`); we need both values to identify an order item.

Structured Query Language (SQL) provides an interface to database systems from different vendors. Users can write statements that each database will translate to process the desired request. In this text, we use the `CREATE, INSERT, UPDATE, DELETE,` and `SELECT` statements. The `CREATE` statement defines data in a table. This statement may use data types that are valid in a particular database system. In this text, we use `VARCHAR(N)`, a variable size character string of maximum size `N`, `INTEGER`, and `DATE`, all of which are standard, and `CURRENCY` which is used in Microsoft Access.

To use the JDBC we need a driver to translate from the JDBC interface to the commands used by the database system, which may reside on the user's machine or at a remote site. Loading the driver, using the **new** operator, or the `forName` method, causes it to register with DriverManager. The `getConnection` method connects to the database using a URL to specify the location of the database. In this text we use the `jdbc:odbc:Sales` URL because we use the JdbcOdbcDriver to translate to the older ODBC commands, which then use the ODBC driver for the Microsoft Access database system. Sales is the name of our ODBC database. Other database system will provide the URL's needed to access them. In this text, we do not discuss the other types of JDBC drivers available.

Once connected to the database, we use the `createStatement` method to create a statement, whose `executeUpdate` method we can use to execute SQL statements to create a new table or to insert values into a table. We could also create and populate tables using the database system, outside of Java.

To retrieve information from the database, we use the `executeQuery` method, which returns a `ResultSet`, to execute SQL SELECT statements. The `ResultSet` contains the rows that satisfy the query. To get the fields in a row, we use the `getXXX` method, where `XXX` is the type of the data, so we use `getInt` for an INTEGER field and `getString` for a VARCHAR field. We pass either the column number of the field or its name, so we could use `getString(1)` or `getString(CustomerName)` if CustomerName is the first column of the result set.

The `SELECT` statement has various options, including a WHERE clause to add conditions, `SELECT DISTINCT` to remove duplicates, and `ORDER BY` to sort the result. A `SELECT` statement may refer to one table or may join information from several tables.

Metadata describes data. The `DatabaseMetaData` class provides many methods which give information about the database. We can find the data types in uses, the names of its tables, and the names and types of the columns of each table. We use the `ResultSetMetaData` class to find the number of columns in a result set and the names and types of each column.

Aggregate functions compute values using all the rows of the table. We use `SUM, MAX, MIN, AVG,` and `COUNT` in our examples. Prepared statements allow us to pass arguments to a statement to reuse it without having to repeat its translation to an efficient implementation in the database system. Transactions permit us to rollback SQL commands in the event the whole sequence did not complete successfully. The default is to commit each command as soon as it is executed, but we can change the default and use the `commit` statement to make the changes permanent only when appropriate.

Our case study builds a graphical user interface for the Sales database, allowing users to specify various parts of a SELECT statement and execute it.

Build Your Own Glossary

Find the uses, in this chapter, of the terms below. Enter a definition of each in the glossary.txt file on the disk included with this text.

aggregate functions	foreign key	relational database
application server	JDBC	result set
commit	key	SQL
compound key	metadata	subname
database server	ODBC	subprotocol
database system	populate	table
driver (JDBC)	prepared statement	transaction processing
entity	query	

Skill Builder Exercises

1. Match the method on the left with the class to which it belongs on the right.

    a. getConnection   i. DatabaseMetaData

    b. executeUpdate   ii. Statement

    c. getTypeInfo   iii. ResultSet

    d. getString   iv. DriverManager

    e. getColumnCount   v. ResultSetMetaData

2. Show the result of executing the following SQL statement using the Sales database:

    ```
    SELECT DISTINCT CustomerName
    FROM Customer, Item, Orders, OrderItem
    WHERE Customer.CustomerID = Orders.CustomerID
    AND Orders.OrderNumber = OrderItem.OrderNumber
    AND OrderItem.ItemNumber = Item.ItemNumber
    ```

3. In Example 11.4, we call the two DatabaseMetaData methods

    ```
    rs = dbMetaData.getTables(null,null,"%",tables);
    ```

    and

    ```
    rs = dbMetaData.getColumns(null,null,"Customer","%");
    ```

    Explain the choice of arguments that we pass in each of these examples.

Critical Thinking Exercises

4. Which of the following are not correctly formed SQL statements?

    a. INSERT INTO Item VALUES (5,"Tape player",7)

    b. SELECT Address FROM Customer

    c. CREATE TABLE Customer (CustomerID VARCHAR(4))

    d. SELECT * FROM Cusotmer, Orders

5. Which of the following are false?

   a. The executeQuery statement always returns a result set.
   b. The executeUpdate statement always modifies the database.
   c. The executeUpdate statement only returns a result set when executing an UPDATE statement.
   d. The executeQuery statement never modifies the database.
   e. none of the above

6. Which of the following are false?

   a. Using a prepared statement can make repeating the same query with different data more efficient.
   b. Using a prepared statement can make repeating different queries with the same data more efficient.
   c. A prepared statement may have more than one parameter.
   d. none of the above

7. Which of the following are false?

   a. Executing setAutoCommit(false) requires us to execute a commit statement to make permanent changes to the database.
   b. Executing the rollback statement reverses any changes to the database made in the current program.
   c. Executing the commit statement may not change the database.
   d. none of the above

Debugging Exercise

8. The following program attempts to list all tables of the Sales database that have a CustomerID column. Find and correct any errors.

```java
import java.net.*;
import java.sql.*;
import java.io.*;
import java.util.Vector;

public class CustomerKey {
    public static void main (String args[]) {
        try{
            ResultSet rs;
            Class.forName("sun.jdbc.odbc.JdbcOdbcDriver");
            String url = "jdbc:odbc:Sales";
            Connection con = DriverManager.getConnection(url);
            DatabaseMetaData dbMetaData = con.getMetaData();
            String[] tables ={"TABLE"};
            rs = dbMetaData.getTables(null,null,null,tables);
            Vector tableNames = new Vector();
            Vector keys = new Vector();
            while(rs.next())
                tableNames.addElement(rs.getString("TABLE_NAME"));
            String table = "";
```
continued

```
                    for(int i = 0; i < tableNames.size(); i++)
                        table = (String)tableNames.elementAt(i);
                        rs = dbMetaData.getColumns(null,null,null,null);
                        while(rs.next()){
                            String s = rs.getString("COLUMN_NAME");
                            if(s.equals("CustomerID"))
                                keys.addElement(table);
                        }
                    System.out.println(keys);
                }catch (Exception e) {e.printStackTrace();}
            }
        }
```

Program Modification Exercises

9. Modify Example 11.3 to pass the JDBC driver and the database URL as program arguments.

10. Modify Example 11.4 to pass the JDBC driver and the database URL as program arguments.

11. Modify Example 11.2 to read the data from a file to insert into the tables.

12. Modify Example 11.6 to use the most appropriate getXXX method rather than the getString method referred to in Note 21.

13. Modify Example 11.6 to allow >=, <=, >, and < operators in addition to =.

14. Modify Example 11.6 to check that exactly two columns, from different tables, have been selected when the user presses the Join button.

15. Modify Example 11.6 to add a checkbox to require that the query remove duplicates from the result.

16. Modify Example 11.6 to check that exactly one column has been selected when the user presses the Enter Value button.

17. Modify Example 11.6 to add column headings in the output.

18. Modify Example 11.6 to allow the user to keep executing queries.

19. Modify Example 11.2 to create a Sales1 database which is like Sales except it has LastName and FirstName fields, instead of CustomerName, in the Customer table.

Program Design Exercises

PUTTING IT ALL TOGETHER

20. Create a database for the Atm simulation of Example 3.6, and revise that example to use the database.

21. Write a graphical user interface for the Sales database which lists all customer names in one Choice box and all products in another. When the user selects a customer name and a product, and presses the Submit button, display a list with the customer name, product, quantity, and date of orders by customers with that name for that product. Use prepared statements wherever possible.

22. Write a graphical user interface for a salesperson using the Sales database. The salesperson should be able to enter new orders. Rollback the order, if, after part of an order has been entered, a part of the order cannot be filled because of insufficient quantity of a product.

Programming Projects

23. Develop an Account database and a Person database to use with the AtmScreen simulation of Example 1.5. Use a three-tiered architecture in which the middle tier connects to the database. Optionally, use RMI to connect to the middle tier on the same machine that hosts your applet. The three tiers are logical and do not have to reside on separate machines, although they might.

PUTTING IT ALL TOGETHER

24. Design and populate a database for the car rental system of Exercise 3.19. Implement the car rental system to use the database.

25. Design and populate a database for a record collection. Provide a screen for the collection's owner to add and remove items, to change entries, and to search.

26. Design and populate a database for sports records. Use an almanac or search the Web for sample data. Provide a screen for the user to add and remove items, to change entries, and to search.

12

Java Beans

Java Beans provide a platform-independent component technology that makes it easier to build Java programs. Hardware designers commonly integrate various components in building a new system. For example, a computer manufacturer uses available memory chips and hard drives, rather than designing them again. Unfortunately many software systems create the whole application from scratch even though many parts of it may be familiar from previously developed systems.

With Java Beans, software developers can integrate components obtained from various sources. For example, a reservation system might want to make a calendar available to the user. The developer might obtain a calendar bean which helps the user make a reservation. When the user enters a date in a text field, the calendar for the three months nearest that date appears.

We first show how to use beans to build an application, and then show how to write our own beans. Most commonly, developers use a visual tool to create applications using beans, but as we shall see they may also include beans directly in Java programs.

OBJECTIVES

◆   Use the BeanBox to build an applet from bean components.

◆   Show how to use beans with bound and constrained properties.

◆   Write a simple Java Bean.

◆   Package a Java Bean in a JAR (Java Archive) file.

◆ Write beans which have bound properties.

◆ Introduce anonymous inner classes.

◆ Use BeanInfo to make beans more user friendly.

◆ Use beans in a Java program.

OPTIONS

◆ The entire chapter is optional. Sections 12.4 and/or 12.5 could be omitted.

12.1 ▪ Building with Beans

Typically developers use a visual tool to build applets from Java Bean **components**. Our example demonstrates this process, using beans from the demo directory included with the Bean Development Kit.

Several vendors provide visual tools with which to develop applications using Java Beans. Sun offers the Beans Development Kit (BDK) as a free download from its web site, java.sun.com. The **BDK** contains a simple development environment, the **BeanBox**, which is available for learning and testing, much as the applet viewer is a simple tool for testing applets, rather than a full browser.

FIGURE 12.1 Running the BeanBox

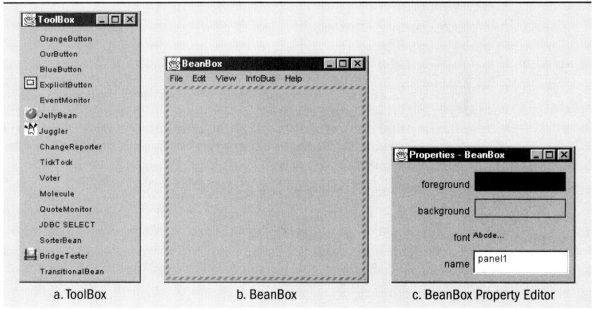

a. ToolBox   b. BeanBox   c. BeanBox Property Editor

Starting the BeanBox

Once the BDK has been installed, we can start the BeanBox by executing the run command from the command prompt in the C:\BDK\beanbox directory on Windows systems, where C:\BDK is the directory in which the BDK is installed. The three windows shown in Figure 12.1 will pop up. Figure 12.1a, the ToolBox, contains a list of beans we can drag into the BeanBox,

Figure 12.1b. The border inside the BeanBox surrounds the currently selected bean, in this case the BeanBox itself. Every bean is configurable, meaning that we can change its properties.

Figure 12.1c shows the four properties of the BeanBox that we can change, its foreground and background colors, font, and name. Clicking on the gray rectangle, showing the current light-gray background, will pop up the color editor shown in Figure 12.2, in which the rectangle on the left shows the current background, the numbers 192,192,192 in the center text field are the red, green, and blue components of light gray, and the choice box lists other colors from which to select. We can either enter new RGB values or select another color, such as yellow, from the Choice box. Doing so will immediately change the BeanBox to have the chosen background color.

FIGURE 12.2 The color editor for the BeanBox background

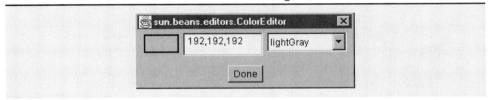

FIGURE 12.3 The Juggler and ExplicitButton beans

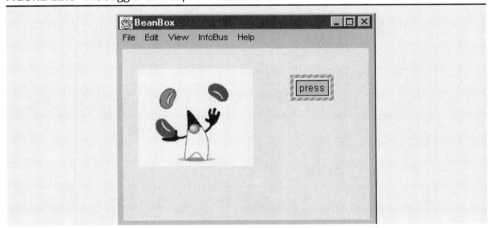

Building an Applet with Beans

To use the BeanBox to build an applet, we drag beans from the **ToolBox** to the BeanBox, customize the beans using the property editors from the **Properties window**, and connect the beans using hookup classes generated by the BeanBox.

For example, we drag the Juggler bean and the ExplicitButton bean into the BeanBox, each time clicking the mouse where we want to position the bean. Figure 12.3 shows the BeanBox, with the ExplicitButton selected. In color, the BeanBox background would appear yellow, and when run on the computer the Juggler is actually juggling the beans. Because we selected ExplicitButton, the Properties window in Figure 12.4 shows its properties that we can change. By typing Stop instead of Press in the text field for the label property and pressing the ENTER key, the button will have Stop as its label.

FIGURE 12.4 Changeable properties of an ExplicitButton

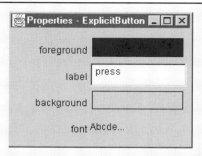

When the user presses the Stop button, Java generates an `ActionEvent`, passing it to the `actionPerformed` method of an object registered with the button as an `ActionListener`. The code inside the `actionPerformed` method implements the behavior desired when the user presses the button. We might call the class containing the `actionPerformed` method the **hookup** (or **adapter**) **class** because it connects the source of the action event with the target.

Writing our own applets in Chapter 6 we implemented all these steps to make button presses have the desired effects. Using Java Beans, the BeanBox (or other bean development tool) creates the hookup class for us, generating the `actionPerformed` method. In our example, we would like the Stop button to cause the Juggler to stop juggling. We highlight the Stop button, click on the Edit menu, then Events, then button push, then actionPerformed, as Figure 12.5 shows.

FIGURE 12.5 Connecting the Stop button

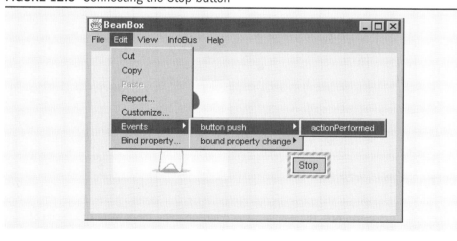

Clicking on the actionPerformed item causes a red line, shown in Figure 12.6, to emanate from the Stop button. We drag the line to the target object, Juggler, in this example, and click the mouse which brings up the EventTargetDialog, shown in Figure 12.7, which asks us to select the target method for the `actionPerformed` method to invoke when the user presses the Stop button.

FIGURE 12.6 Connecting the source to the target

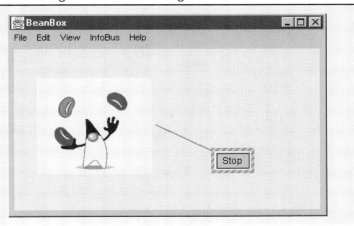

FIGURE 12.7 The target methods

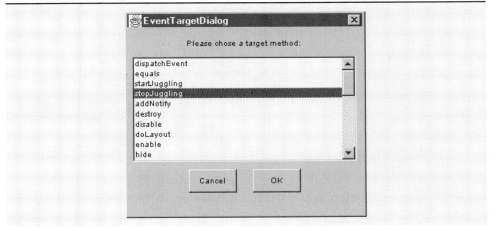

We select the stopJuggling method and click OK, receiving the message that the BeanBox is generating and compiling an adapter class to hook up the Stop button source to the Juggler target. Now, pressing the Stop button will cause the Juggler to stop juggling.

To be able to start the Juggler juggling again we add another button to the BeanBox. This time we drag an OurButton rather than an ExplicitButton to the BeanBox, changing its label to Start (see Figure 10.8).

When we select the Start button, click on Edit, and click on Events, we get a different list of events, shown in Figure 12.9, than we got for the Stop button. We will explain the reason for this difference later in the chapter.

Continuing, we connect the Start button to the Juggler, causing the BeanBox to generate the hookup class that will cause the startJuggling method of the Juggler to execute when the user presses the Start button.

FIGURE 12.8 Adding a Start button

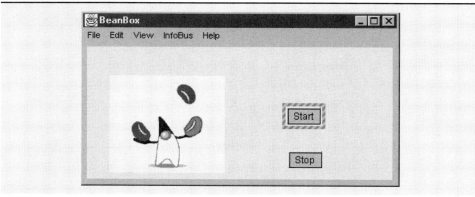

FIGURE 12.9 Connecting the Start button

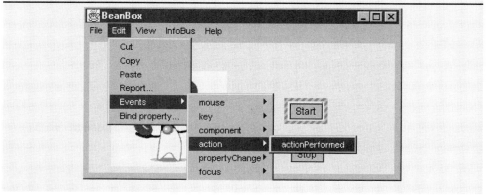

The BeanBox is a tool we can use to build an applet from beans. We have customized the BeanBox to have a yellow background and the buttons to have Start and Stop labels. We let the BeanBox create hookup classes which implement the actions for the events the buttons generate. Now we click on File, Make Applet to let the BeanBox make this creation into an applet.

FIGURE 12.10 Making an applet

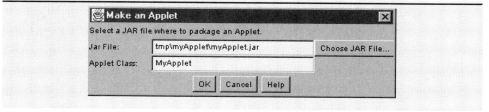

We will discuss JAR (Java Archive) files later in this chapter. In the Make an Applet window we change the default myApplet name to ButtonJuggle in both text fields and click OK. We get a sequence of messages listing the steps the BeanBox is taking to create the applet. The BeanBox puts all the files for the ButtonJuggle applet in the directory c:\BDK\beanbox

\tmp\ButtonJuggle. We can test the applet using the applet viewer with the ButtonJuggle.html file or we can run it using the Internet Explorer browser.

The Big Picture

The JavaBeans™ component technology allows us to build applets by customizing and connecting bean components, without doing any programming. The BeanBox is a simple visual tool for building with beans. Each bean has a property list providing editors to change each property. We can connect a source event in one bean to a target method in another. The BeanBox will create the hookup class which will call the target method when the source event occurs. The bean developer writes beans which the applet designer uses to make an applet run by the end user.

TEST YOUR UNDERSTANDING

TRY IT YOURSELF 1. Start the BeanBox. Change its background color to orange.

TRY IT YOURSELF 2. Following the steps in the text, use the BeanBox to create the ButtonJuggle applet, additionally changing the background color of the Start and Stop buttons to blue and the foreground color to white.

12.2 ▪ Bound and Constrained Properties

With **bound properties**, beans can use Java events to notify other components of changes in property values. **Constrained properties** allow those components to veto the changes.

In building the ButtonJuggle applet, we used the BeanBox to hook up the Start button source of an ActionEvent with the Juggler target, so the Juggler will start juggling when the user presses the Start button. In this section, we show how the BeanBox can hook up sources and targets of property changes and vetoable changes.

Simple Properties

Java objects may have various properties that we can change. For example, every component, such as a Button, has foreground and background properties that we can change using the setForeground or setBackground methods in a Java program. When using a tool such as the BeanBox, we often want to change some properties of beans before we incorporate them into an applet. In building the ButtonJuggle applet, we changed the labels on the buttons, for example.

By using the BeanBox we avoid the details of calling the setLabel method in a program, using instead the property editor for the label in the Properties window. The Properties window, shown in Figure 12.4, lets us change the foreground, label, background, and font properties of an ExplicitButton.

Bound Properties

A bound property lets other components know when it is changed. We can use the BeanBox to connect beans so changing the color, for example, of the source bean will notify the target bean to also change its color. If a bean has any bound properties there will be a Bind Property item in its Edit menu. Not all beans have any bound properties; the Juggler has no Bind Property item on its Edit menu.

For example, let us drag an ExplictButton and a JellyBean into the BeanBox. Selecting the Explicit Button and clicking Edit, Bind Property pops up the window of Figure 12.11a to let us choose the property whose changes we want to transmit to other beans. If we choose the background property and click OK a red line will appear which we can drag, clicking on the JellyBean, popping up the list of properties shown in Figure 12.11b.

FIGURE 12.11 Bound Properties

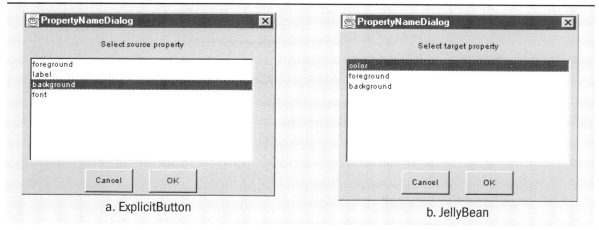

a. ExplicitButton   b. JellyBean

Selecting color and clicking OK binds the background of the ExplicitButton to the color of the JellyBean. Before binding the two beans in this way, changing the background of the ExplicitButton to red leaves the JellyBean orange (Figure 12.12a). After binding, changing the background of the ExplicitButton to red also changes the color of the JellyBean to red (Figure 12.12b).

FIGURE 12.1 Background color before and after binding

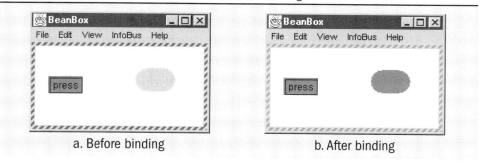

a. Before binding   b. After binding

☞ **TIP**

We must distinguish between the applet designer and the applet user. Using the BeanBox, we are functioning as the applet designer. We did not make an applet out of the bound property example above, because the applet user would have no way to change the background color of the button. If the user could change that background color, then the JellyBean's color would also change. As the applet designer we have the property editors in the Properties window to use to change the background color of the ExplicitButton, to show how bound properties work. Later we will design an applet that uses bound properties effectively.

Constrained Properties

Simple properties change without affecting properties of other beans. Changes in bound properties may change properties of other beans. Constrained properties give other beans the chance to veto the change before it takes effect. For example, the `priceInCents` property of the JellyBean is vetoable. The Voter bean, in the ToolBox, normally vetoes every change presented to it.

To see how constrained properties work, we drag the JellyBean and the Voter bean into the BeanBox. Selecting the JellyBean, we show in Figure 12.13 the Edit, Events, vetoableChange, vetoableChange menu items. The JellyBean can generate a VetoableChange event which is handled by the `vetoableChange` method. The bean at the right of Figure 12.13 displays a large No signifying that it will veto every property change presented to it.

Clicking on the vetoableChange item will cause the red line to appear, which we drag and click on the Voter bean popping up the list of Figure 12.14. Selecting vetoableChange and clicking OK will cause the BeanBox to hook up the JellyBean source to the Voter target. Highlighting the JellyBean and trying use the property editor to change its `priceInCents` to 3 will cause the message

```
WARNING: Vetoed; reason is: NO!
```

to appear on the console screen. The `priceInCents` remains 2, as the Voter bean has vetoed the change. If we change the `vetoAll` property of the Voter bean from **true** to **false**, then the Voter bean will display a large Yes instead of No, and will allow the change to `priceInCents` to proceed.

FIGURE 12.13 A vetoable change event

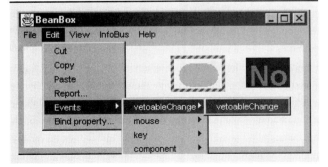

FIGURE 12.14 Choosing a target method vetoableChange

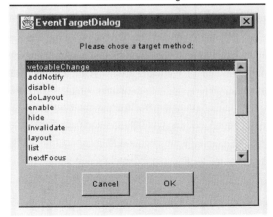

> **The Big Picture**
>
> In the BeanBox, or similar tool, we can bind a bound property of one bean to a property of the same type of another bean. When the source bean's bound property changes value so will the target's. A constrained property is like a bound property with the additional feature that a target bean may veto the change, forcing the source bean to withdraw it.

TEST YOUR UNDERSTANDING

TRY IT YOURSELF   3.   Use the BeanBox to bind the `priceInCents` of the JellyBean to the `animationRate` of the Juggler. Enter different values in the Properties window editor for the `priceInCents` and observe the change in the `animationRate`.

TRY IT YOURSELF   4.   Use the BeanBox to bind the foreground color of an ExplicitButton to the color of a JellyBean and to the background color of an OurButton. Change the foreground color of the ExplicitButton several times, using the Properties window color editor, and observe the effect.

TRY IT YOURSELF   5.   Use the BeanBox to bind the `priceInCents` of the JellyBean to the `animationRate` of the Juggler. Also hook up the `vetoableChange` event from the JellyBean source to the `vetoableChange` handler method of the Voter bean. Try to change the `priceInCents`, using the Properties window editor, and observe what happens.

12.3 ▪ Writing and Packaging a Bean

Now that we have used the BeanBox to build an applet with the demo beans provided by Sun, we will learn to write our own beans and add them to the BeanBox. We package beans in JAR (Java Archive) files that combine individual files to avoid multiple connections when downloading a bean, and compress files to save space and time. We introduce anonymous inner classes, which we could have used earlier in the text, to conveniently define event handlers.

A Simple Bean

A bean is just a Java program. Our first example, TextBean, has just a constructor and two methods, `setNumber` and `getNumber`, to set and get a single integer value. It extends `TextField` to provide a display for an integer value. Using the standard prefixes `set` and `get`, followed by the property name, as in `setNumber` and `getNumber`, allows the BeanBox to add these properties in the Properties window.

Example 12.1 has no `main` method because we will use it in the BeanBox, rather than as a standalone program. Adding a `main` method would allow it to run as a standalone application as well as to be used as a bean in the BeanBox. All programs to be used as beans should extend the `Serializable` interface, because Java uses object serialization to save the designer's customizations which can be reloaded and used in their customized state.

EXAMPLE 12.1   **TextBean.java**

```
/* A simple program to use
 * as a bean.
 */

import java.awt.*;
import java.awt.event.*;
import java.io.*;

public class TextBean extends TextField implements Serializable {
    int myNumber = 10;                          // Value to set and get
    public TextBean() {
        super(5);                                           // Note 1
        setBackground(Color.white);                         // Note 2
    }
    public void setNumber(int x) {
        myNumber = x;
        setText("" + x);                                    // Note 3
    }
    public int getNumber() {
        return myNumber;
    }
}
```

Note 1: We call the constructor of the superclass, `TextField`, to set the width of the text field to 5.

Note 2: Setting the background of the text field to white helps it to stand out from the background of the BeanBox.

Note 3: The `setText` method displays the text in the text field. Binding another bean, such as the JellyBean, to set the `number` property when its `priceInCents` is changed, will cause the `priceInCents` to be displayed.

JAR Files

JAR (Java Archive) **files** allow us to package several files—class files, images, and sounds—together as one file. We will use them here to package our beans, but another important use is to package the files needed for an applet. When we download an applet from a web site, we must make a separate connection to download each of the files used by the applet. Packaging all the files in a JAR file allows the applet to be downloaded with one connection saving much time. The JAR file may also compress the files, further reducing the download time.

We can use the jar utility program to create JAR files. For Java Beans we should always include a **manifest** file which describes the contents of the JAR file, indicating which files are beans. To package our TextBean, we first create the manifest file, `TestBean.mf`, containing just two lines.

```
Name: TextBean.class
Java-Bean: True
```

The jar command, given at the command prompt, is

```
jar cfm TextBean.jar TextBean.mf TextBean.class
```

where the options are

    c   create a new archive
    f   archive name is the first file on the list   (TextBean.jar here)
    m   manifest file is the second file on the list   (TextBean.mf here)

Executing this jar command will create the JAR file `TextBean.jar`. To put this JAR file into the BeanBox, we copy it to `c:\BDK\jars` which already contains the Sun demo beans. Starting the BeanBox shows the TextBean in the ToolBox (Figure 12.15a). Dragging the TextBean and the JellyBean into the BeanBox (Figure 12.15b) and selecting the TextBean shows the number property appears in the Properties window (Figure 12.15c).

FIGURE 12.15 The BeanBox with the TextBean

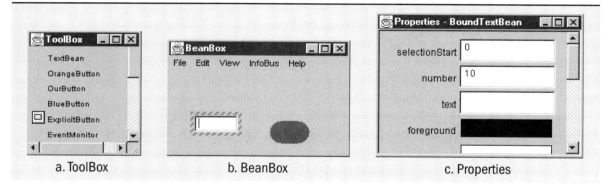

a. ToolBox   b. BeanBox   c. Properties

We will configure these beans so when the `priceInCents` of the JellyBean changes the new value will appear in the TextBean. Selecting the JellyBean and clicking on Edit, BindProperty pops up the PropertyNameDialog. Selecting `priceInCents`, and clicking OK produces the red line which we drag to the TextBean. Clicking on the TextBean pops up the PropertyNameDialog for the target property shown in Figure 12.16. Selecting the number property and clicking OK will complete the binding.

Selecting the JellyBean and entering 5 for its `priceInCents` will now cause the value 5 to be displayed in the text field of the TextBean (Figure 12.17).

A Bean with a Bound Property

Next we make the number property a bound property with the capability to notify property change listeners when `myNumber` is changed. We will use this BoundTextBean later in building an applet.

The `java.beans` package has a `PropertyChangeEvent` class. The `PropertyChangeListener` interface has one method,

```
public void propertyChange(PropertyChangeEvent e)
```

which Java calls when a property value changes. We use the `PropertyChangeSupport` methods to do most of the work implementing bound properties. Instead of managing a vector of

FIGURE 12.16 Selecting the target property in the TextBean

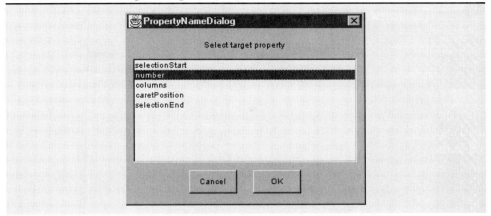

FIGURE 12.17 TextBean displaying `priceInCents`

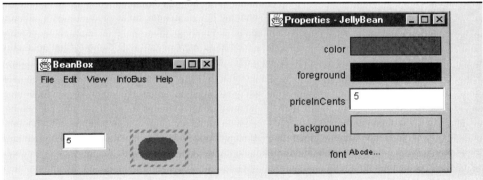

PropertyChangeListeners ourselves, we delegate that to the addPropertyChangeListener and removePropertyChangeListener methods. Instead of creating a PropertyChangeEvent and sending it to each listener we can just call the firePropertyChange method of the PropertyChangeSupport class. Example 12.2 shows these additions to Example 12.1.

EXAMPLE 12.2 **BoundTextBean.java**

```java
/* Makes the number property a bound property.
 * Notifies PropertyChangeListeners when
 * myNumber changes.
 */

import java.awt.*;
import java.awt.event.*;
import java.io.*;
import java.beans.*;

public class BoundTextBean extends TextField
                    implements Serializable, ActionListener {
```
continued

```
        private int myNumber = 10;
        private PropertyChangeSupport pChange
                       = new PropertyChangeSupport(this);        // Note 1
        public BoundTextBean(){
            super(5);
            setBackground(Color.white);
            addActionListener(this);
        }
        public void setNumber(int x) {
            Integer oldVal = new Integer(myNumber);              // Note 2
            Integer newVal = new Integer(x);
            myNumber = x;
            setText("" + x);
            pChange.firePropertyChange("number",oldVal,newVal);  // Note 3
        }
        public int getNumber() {
            return myNumber;
        }
        public void actionPerformed(ActionEvent e) {
            setNumber(Integer.parseInt(getText()));              // Note 4
        }
        public void addPropertyChangeListener(PropertyChangeListener l) { // Note 5
            pChange.addPropertyChangeListener(l);
        }
        public void removePropertyChangeListener(PropertyChangeListener l) {
            pChange.removePropertyChangeListener(l);
        }
    }
```

Note 1: We create a PropertyChangeSupport object to manage registering PropertyChangeListener objects, and sending a PropertyChangeEvent to them.

Note 2: The firePropertyChange method, which notifies property change listeners when a property changes, takes Object arguments for the old and new values of the changing property. Because myNumber is a primitive **int** value we need to wrap it in an Integer object to pass as an argument to firePropertyChange.

Note 3: Our PropertyChangeSupport object, pChange, notifies each registered PropertyChangeListener of the change to myNumber, using the firePropertyChange method. The three arguments are a string with the name of the property; number is the name we used in the setNumber method, but with lowercase, that is, "number" not "Number". The second argument is the old value of the property, wrapped as an object if necessary, while the third argument is the new value of the property, also wrapped as an object.

Note 4: We want to enter the value of myNumber in the text field. When the user presses the ENTER key, Java generates an ActionEvent which we handle in the actionPerformed method by calling the setNumber method to change myNumber to the value entered.

Note 5: Using the BeanBox, we can connect other beans who want to be notified when myNumber changes. These beans call the addPropertyChangeListener method to register to receive notification of changes. Rather than keep a vector of these listeners ourselves, we let `pChange` keep them for us and inform them of changes using the `firePropertyChange` method.

Smiley Bean

To illustrate a bean with additional features, we create Smiley bean with a happy face; pressing its nose generates an `ActionEvent`. Smiley's age is a bound property. Before discussing the code for Smiley, we use it in the BeanBox. Figure 12.18 shows the BeanBox containing our BoundTextBean from Example 12.2, an Explicit Button, Smiley, and a Juggler.

FIGURE 12.18 Using Smiley in an applet

We customized the ExplicitButton to have the label Start and the Smiley bean to have a red background color. We bound the number property of the BoundTextBean to the Smiley's age, so when we enter a value in the BoundTextBean, Smiley's age will also have that value. We bound Smiley's age to the animation rate of the Juggler, so entering a value in the BoundTextBean will actually change the Juggler's animation rate.

Using the BeanBox, we hooked up Smiley to the Juggler, so when the user clicks on Smiley's nose, generating an `ActionEvent`, the Juggler stops juggling. The Start button starts the Juggler juggling. Figure 12.19 shows Internet Explorer running the SmileApplet applet made in the BeanBox by clicking on the File, Make Applet menu item.

Now that we have used Smiley as part of `SmileApplet`, we discuss its implementation. Implementing the age property follows the same steps used for the number property in Example 12.2. Smiley generates an `ActionEvent` when the user presses its nose. We use the `mousePressed` method to detect when the mouse is over Smiley's nose. We are making Smiley's nose function like a button, and like a button we generate an `ActionEvent` and call the `actionPerformed` method of all listeners.

Smiley implements the `addActionListener` and `removeActionListener` methods to allow objects to register to be notified when the user presses Smiley's nose. A vector holds these listeners. When the user presses Smiley's nose, the `fireAction` method calls the `actionPerformed` method of all the registered listeners.

FIGURE 12.19 Running SmileApplet

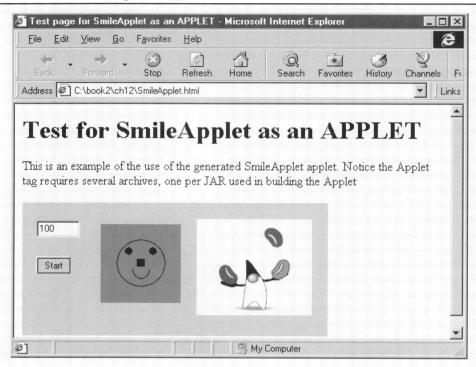

Anonymous Inner Classes

Sometimes we use an inner class just once, perhaps to define an event handler. Our Smiley example uses an inner class that overrides MouseAdapter to handle the MOUSE_PRESSED event. We could have named it, for example, as

```
class MousePress extends MouseAdapter {
    public void mousePressed(MouseEvent e) {
        //  code to handle mouse press goes here
    }
}
```

and registered it using

```
addMouseListener(new MousePress());
```

Because we only use the MousePress class once, we do not need to give it a name. Java lets us use it anonymously, as in

```
addMouseListener(new MouseAdapter() {
    public void mousePressed(MouseEvent e) {
        //  code to handle mouse press goes here
    }
} );
```

which uses an extension of the **new** operator for the syntax. Writing 'new MouseAdapter()' signifies an anonymous inner class that extends MouseAdapter. The class definition follows, as usual, between curly braces. An **anonymous inner class** is an unnamed class defined within a block of code.

While we do not name an anonymous class, Java does name it. Looking in the package smiley, after compiling Smiley.java, we find the class file, Smiley$1.class, containing the byte code for the anonymous extension to MouseAdapter. We include the Smiley$1.class file in the manifest with an entry

```
Name: smiley/Smiley$1.class
```

and add it to the list of files to include in the JAR file for Smiley.

EXAMPLE 12.3   **Smiley.java**

```
/* A bean with a smiley face. Pressing its nose
 * generates an ActionEvent.  Smiley's age is a
 * bound property.
 */

package smiley;                                              // Note 1
import java.awt.*;
import java.awt.event.*;
import java.beans.*;
import java.io.Serializable;
import java.util.Vector;

public class Smiley extends Canvas implements Serializable {
    private int age = 10;
    private PropertyChangeSupport changes = new PropertyChangeSupport(this);
    private Vector listeners = new Vector();
    public Smiley() {
        setSize(new Dimension(100,100));
        addMouseListener(new MouseAdapter () {              // Note 2
          public void mousePressed(MouseEvent evt) {
            int x = evt.getX();
            int y = evt.getY();
            if( x >= 45 && x <= 55 && y >= 45 && y <= 55)
                if(listeners != null) fireAction();          // Note 3
          }
        } );
    }
    public void paint(Graphics g) {
        g.drawOval(20,20,60,60);
        g.fillOval(30,30,10,10);
        g.fillOval(60,30,10,10);
        g.fillRect(45,45,10,10);
        g.drawArc(40,50,20,20,170,200);
    }
    public void fireAction() {
        ActionEvent e = new ActionEvent(this, 0, null);     // Note 4
        for (int i = 0; i < listeners.size(); i++)
            ((ActionListener)listeners.elementAt(i)).actionPerformed(e); // Note 5
    }
```

continued

```
        public void setAge(int a) {
           age = a;
           Integer newage = new Integer(a);
           changes.firePropertyChange("age", oldage, newage);
        }
        public int getAge() {
           return age;
        }
        public void addActionListener(ActionListener 1) {          // Note 6
           listeners.addElement(1);
        }
        public void removeActionListener(ActionListener 1) {
           listeners.removeElement(1);
        }
        public void addPropertyChangeListener(PropertyChangeListener 1) {
           changes.addPropertyChangeListener(1);
        }
        public void removePropertyChangeListener(PropertyChangeListener 1) {
           changes.removePropertyChangeListener(1);
        }
     }
```

Note 1: We put Smiley in a package to keep all its files together. In the manifest file we use the forward slash, /, to name files, as in

```
smiley/Smiley.class
```

even on Windows systems. Executing the jar command from the command prompt in the directory containing the smiley package directory, we include the package name when naming files, as in

```
jar cfm smiley\Smiley.jar smiley\Smiley.mf
    smiley\Smiley.class smiley\Smiley$1.class
```

on Windows systems.

Note 2: This is an anonymous inner class. We define it right where it is used. Defining a named class MousePress would be fine too, but anonymous inner classes are often used to provide adapter classes for event handling as we do here.

Note 3: If the coordinates where the user pressed the mouse are within the rectangle defining Smiley's nose, and if there are any listeners waiting to be notified, we call the fireAction method to notify them.

Note 4: We create an ActionEvent to send to registered listeners. The first argument is the source of the event; we pass this representing Smiley. The second argument, of type **int**, is an ID for the type of event. We do not use the ID and arbitrarily pass 0. The AWT passes ActionEvent.ACTION_PERFORMED when it generates an ActionEvent, and we could alternatively have passed that value. The third argument, of type String, is the command; for example, a label for a button. We do not use this argument and pass **null** instead.

Note 5: We cast each listener in the listener's vector to an ActionListener (we added only action listeners to listeners so this cast is valid), and call its actionPerformed method which specifies how that listener wants to handle the nose press. In the `SmileApplet`, the BeanBox has hooked up the Juggler to respond to this action by stopping juggling.

Note 6: `public void addActionListener(ActionListener l)`

We implement the `addActionListener` method to save listeners in a vector. They will be notified when the user presses Smiley's nose.

☛ **A LITTLE EXTRA**   The `SmileApplet` uses four beans, each with its own JAR file composed of files needed to implement that bean. The `archive` attribute allows us to list JAR files in the applet tag of an HTML file. The browser or applet viewer will search these JAR files for the classes needed to load an applet. We use the `archive` attribute in addition to the `code` attribute which still specifies the applet's class file. For example the BeanBox generated the applet tag

```
<applet
    archive="./SmileApplet.jar,./support.jar
        ,./juggler.jar,./Smiley.jar,./buttons.jar
        ,./BoundTextBean.jar"
    code="SmileApplet" width=382 height=173>
```

for SmileApplet.

> **The Big Picture**
>
> In writing a bean we include `set` and `get` methods for each property. The `PropertyChangeSupport` object helps us manage property change listeners, but we still need to include `addPropertyChangeListener` and `removeProperty-tyChangeListener` methods if we have bound properties. We call the `firePropertyChange` method when the property value changes. If our bean generates an `ActionEvent` we call the `actionPerformed` method for all listeners, and include `addActionListener` and `removeActionListener` methods. We use the `jar` command to package our bean in a JAR file; the manifest indicates which files are beans.

TEST YOUR UNDERSTANDING

TRY IT YOURSELF   6. Modify TextBean by changing the name of the `setNumber` method to `numberSet` and `getNumber` to `numberGet`. Create the JAR file and add it to the BeanBox. When you drag the modified TextBean into the BeanBox does the number property appear in the Properties window?

TRY IT YOURSELF   7. Using the BeanBox, bind the number property of the BoundTextBean to the `priceInCents` of the JellyBean. In the BeanBox, change the number and check the Properties window of the JellyBean to see that its `priceInCents` changes to that same value.

8. An anonymous inner class can implement an interface as well as extend a class. Write the argument to the `addActionListener` method, called in the BoundTextBean of Example 12.2, as an anonymous inner class. Its start is

```
new ActionListener() {
```

9. Modify Example 12.3 not to implement `Serializable`. Follow the steps to create the `SmileApplet`. What happens?

12.4 ▪ Using BeanInfo

In the absence of a **BeanInfo class** for a bean, Java uses its reflection facility to inspect the bean's code and determine its characteristics. The bean creator, by adding a `BeanInfo` class, can configure the bean to present its properties in a more user friendly form than would result from letting Java figure them out by itself.

What Is BeanInfo?

The two BeanBox demo beans, ExplicitButton and OurButton, are very similar. ExplicitButton is just an OurButton with an `ExplicitButtonBeanInfo` class used to make the ExplicitButton more user friendly.[1]

The Events menu for the OurButton, shown in Figure 12.9, lists six event types, mouse, key, component, action, propertyChange, and focus, some of which are solely used internally by OurButton or its superclass, Component. By contrast, the Events menu for the ExplicitButton, shown in Figure 12.8, lists just button push (renamed from action to be more informative) and bound property change (renamed from propertyChange). The other event types, not meant for the end user, have been omitted.

The Properties window for an ExplicitButton (Figure 12.4) shows four properties; we leave it as an exercise to check that the similar window for an OurButton shows seven.

The BeanBox uses a `BeanInfo` class, if provided by the bean developer, to determine how to present that bean to the BeanBox user. If no `BeanInfo` class is available, the BeanBox uses the facilities of the `java.lang.reflect` package to analyze the code for the bean to determine its features to present to the BeanBox user. As users of beans, rather than developers of tools such as the BeanBox, we do not need to explore Java's reflection facilities in this text.

`BeanInfo` is an interface in the `java.beans` package containing eight methods. The `SimpleBeanInfo` class provides default implementations of the `BeanInfo` interface methods, so by extending `SimpleBeanInfo` we only have to override those methods of interest to us.

To illustrate the use of `BeanInfo`, we will make our Smiley bean more user friendly.

Friendly Bean

Friendly bean is just a Smiley bean with a `BeanInfo` class. Its code is

```
package smiley;
public class Friendly extends Smiley {
}
```

To make Friendly more user friendly, we specify that the only properties we want to display in the Properties window are background, foreground, and age. Figure 12.20 shows the simplification.

---

[1] We can read the BeanBox demo source files in the directory `c:\BDK\demo\sunw\demo`.

FIGURE 12.20 Comparing Properties windows

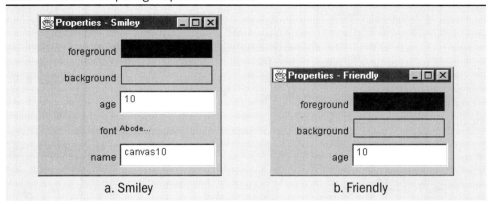

a. Smiley   b. Friendly

We will further specify that the only source events we want to display are nosePress and ageChange. Figure 12.21 shows the contrast.

FIGURE 12.21 Comparing source events

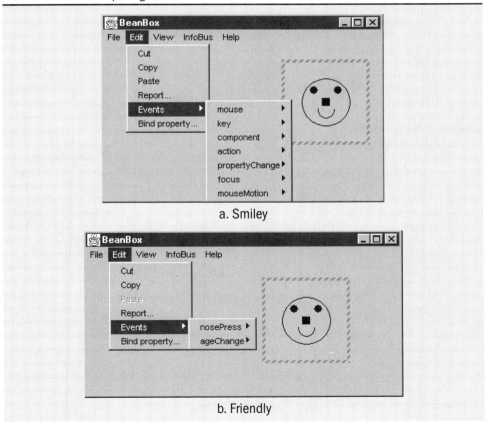

a. Smiley

b. Friendly

The third feature we make more user friendly is the list of target methods that the hookup class can call when handling a source event. When using the BeanBox to hook up a Friendly nosePress to a Smiley target method, Figure 12.22a shows that Smiley has many methods with no arguments (20 in all) inherited from its various superclasses. By contrast, when using the BeanBox to hook up a Smiley action event to a Friendly target method, Figure 12.22b shows that Friendly has only one method, fireAction, available. We have hidden the other 19 inherited methods, for which we have no use.

FIGURE 12.22 Comparing target methods

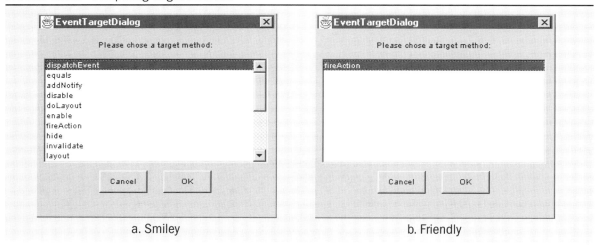

a. Smiley   b. Friendly

BeanInfo for Friendly

The FriendlyBeanInfo class makes Friendly more user friendly than Smiley. By creating PropertyDescriptor, EventDescriptor, and MethodDescriptor objects, we can get the BeanBox to create the friendlier display in Figures 12.20b, 12.21b, and 12.22b.

To specify the properties to display in the Properties window we override the getPropertyDescriptors method of SimpleBeanInfo. We create a PropertyDescriptor for each property we want to display, as, for example

```
PropertyDescriptor age =
    new PropertyDescriptor("age", Friendly.class);
```

where the first argument, age, is the name of the property, and the second argument is the bean class. Writing Friendly.class specifies the Class object for the Friendly class. Because age is a bound property we invoke

```
age.setBound(true);
```

so the BeanBox will list age as an item in the Edit, Bind Property menu.

To expose the background and foreground properties to the designer, we create a PropertyDescriptor for each. The return value for the getPropertyDescriptor method is a PropertyDescriptor array containing, in this case, the three property descriptors we created

```
PropertyDescriptor[] pd = {age,background,foreground};
```

The Properties window for Friendly displays these three properties in Figure 12.20b.

The getEventSetDescriptors method lets us simplify the Edit, Events menu in the BeanBox, describing the source events available to the designer. We create an EventSetDescriptor for each event we want the designer to see. For the ActionEvent generated by pressing Friendly's nose, we have

```
EventSetDescriptor nosePress =
    new EventSetDescriptor(Friendly.class, "nosePress",
        ActionListener.class, "actionPerformed");
```

where the arguments are:

Type	Meaning
Class	the bean class
String	the name for our event
Class	the listener type for that event
String	the listener method name to handle the event

In all, we expose two events for Friendly, the nosePress and the PropertyChangeEvent, created in response to age changes. Letting the getEventSetDescriptors method return the array

```
EventSetDesriptor[] esd = {ageChange, nosePress};
```

produces the shorter and more informative Event menu of Figure 12.21b.

Using BeanInfo can dramatically shorten the list of target methods displayed to the designer. We implement getMethodDescriptors to return just one method descriptor so we get the one-item list of Figure 12.22b rather than the 20-item list shown in Figure 12.22a for Smiley.

To create a MethodDescriptor, we need a Method object. Friendly.class is a class of type Class. Calling its getMethod method, as in

```
Method m = Friendly.class.getMethod("fireAction", null);
```

where the first argument names the method, and the second argument, of type Class[], gives the types of the arguments to the method, gives us the Method object we need. The fireAction method in Smiley, which Friendly inherits, generates and sends an ActionEvent to be handled by the actionPerformed method of listeners. We pass **null** for the second argument because fireAction has no arguments.

The ToolBox of Figure 12.1a shows some of the buttons with icons at their left. We use the getIcon method to return an image of the very friendly author to serve as an icon for Friendly.

EXAMPLE 12.4 **FriendlyBeanInfo.java**

```
/* Uses a BeanInfo class to make a more
 * user friendly version of Smiley.
 */

package smiley;
import java.beans.*;
import java.awt.*;
```

continued

```
import java.awt.event.*;
import java.lang.reflect.*;                                          // Note 1
public class FriendlyBeanInfo extends SimpleBeanInfo {
    public PropertyDescriptor[] getPropertyDescriptors() {
        try {
          PropertyDescriptor age =
              new PropertyDescriptor("age", Friendly.class);
          PropertyDescriptor background =
              new PropertyDescriptor("background", Friendly.class);    // Note 2
          PropertyDescriptor foreground =
              new PropertyDescriptor("foreground", Friendly.class);
          age.setBound(true);
          PropertyDescriptor[] pd = {age,background,foreground};
          return pd;
        } catch (IntrospectionException e) {                         // Note 3
           return null;
        }
    }
    public EventSetDescriptor[] getEventSetDescriptors() {
        try {
          EventSetDescriptor nosePress = new EventSetDescriptor(Friendly.class,
              "nosePress", ActionListener.class, "actionPerformed");
          EventSetDescriptor ageChange = new EventSetDescriptor(Friendly.class,
              "propertyChange", PropertyChangeListener.class,          // Note 4
              "propertyChange");
          EventSetDescriptor[] esd = {ageChange, nosePress};
          ageChange.setDisplayName("ageChange");                      // Note 5
          return esd;
        } catch (IntrospectionException e) {
           return null;
        }
    }
    public MethodDescriptor[] getMethodDescriptors() {
        try {
          Method m = Friendly.class.getMethod("fireAction", null);
          MethodDescriptor fireAction = new MethodDescriptor(m);      // Note 6
          MethodDescriptor[] md = {fireAction};
          return md;
        } catch(Exception e) {                                       // Note 7
           return null;
        }
    }
    public Image getIcon(int iconKind) {                             // Note 8
        return loadImage("gittleman.gif");
    }
}
```

Note 1: The `java.lang.reflect` package lets development tools, such as the BeanBox, find the fields and methods of a class. We use its `Method` class in this program.

Note 2: Creating a PropertyDescriptor for the background will cause the background property to appear in the Properties window. The designer will be able to change Friendly's background color. Unlike in the ExplicitButton, the background property is not bound.

Note 3: The `PropertyDescriptor` constructor may throw an `IntrospectionException`, meaning an error occurred when Java was gathering information about a program. Returning **null** tells the BeanBox not to use property descriptors, but rather to use reflection to find all the properties as it would if there were no `BeanInfo` class.

Note 4: For the nosePress EventDescriptor we were able to choose our own name, nosePress, for the event, which we passed as the second argument. For the ageChange EventDescriptor, we must use the `"propertyChange"` name to show the bean generates property change events so the BeanBox will add the Bind Property menu item. We use the `setDisplayName` method to make name propertyChange more user friendly (ageChange).

Note 5: The `setDisplayName` method displays its argument, ageChange, in the event list instead of the name in the second argument of the `EventDescriptor`, `propertyChange`.

Note 6: `MethodDescriptor fireAction = new MethodDescriptor(m);`

We only want to make one target method available to the designer, the `fireAction` method. The `MethodDescriptor` constructor takes a `Method` object as its argument.

Note 7: `} catch(Exception e) {`

The getMethod method may throw `NoSuchMethodException` or `SecurityException` exceptions. We handle them generically with the superclass `Exception`, again returning **null** to let the BeanBox use its default process for finding methods, as if there were no `BeanInfo` class.

Note 8: `public Image getIcon(int iconKind) {`

The `BeanInfo` class has four constants for different icon types. We ignore this argument, using the `loadImage` method from the `SimpleBeanInfo` class to always return the same image.

Packaging Friendly

Because Friendly is just a Smiley bean with BeanInfo, we put it in the `smiley` package. We package both Smiley and Friendly in the same JAR file so the manifest file, `Happy.mf`, is

```
Name: smiley/Smiley.class
Java-Bean: True

Name: smiley/Friendly.class
Java-Bean: True

Name: smiley/Smiley$1.class

Name: smiley/FriendlyBeanInfo.class

Name: smiley/gittleman.gif
```

We package all files used in Smiley and Friendly into a JAR file, `Happy.jar`, using the command

```
jar cfm smiley\Happy.jar smiley\Happy.mf
    smiley\Smiley.class smiley\Smiley$1.class
    smiley\Friendly.class smiley\FriendlyBeanInfo.class
    smiley\gittleman.gif
```

> **The Big Picture**
> ....
> We add a `BeanInfo` class for a bean to make it more friendly. The `getProperty Descriptors` method returns a property descriptor for each property we wish to display in the Properties window. The `getEventSetDescriptors` method returns an event set descriptor for each source event we wish to display in the menu. The `getMethodDescriptors` method returns a method descriptor for each target method we wish to display. The `getIcon` method lets us choose an icon to represent the bean. We name our `BeanInfo` class cnameBeanInfo, where cname is the name of the bean class. By overriding `SimpleBeanInfo` we only have to implement those methods of the `BeanInfo` interface in which we are interested.

TEST YOUR UNDERSTANDING

TRY IT YOURSELF   10. Compare the Properties window for the OurButton, which does not have a `BeanInfo` class, with that for ExplicitButton, which is an OurButton with a `BeanInfo` class.

TRY IT YOURSELF   11. Use the BeanBox to make an applet using one Friendly bean to start the Juggler juggling and another to stop the juggling.

TRY IT YOURSELF   12. Modify Example 12.4 to omit the `GetEventDescriptors` method. How does this effect how Friendly is presented in the BeanBox?

12.5 ▪ Programming with Beans

Java Beans are meant to be used with visual development tools, of which the BeanBox is a rudimentary example. Nevertheless beans are just Java programs, and we are free to use them in hand-coded programs developed without the use of any visual tools. In doing so, we use the event-handling techniques covered earlier in the text.

Using Beans in an Applet

Our example is similar to the `SmileApplet`, developed in Section 12.3 using the BeanBox. We write our program favoring the style of the BeanBox, creating a separate `hookup` class to handle each event. Figure 12.23 shows the `SmileyJug` applet. Pressing Smiley's nose alternately stops and starts the Juggler juggling. Entering a value for Smiley's age changes the animation rate of the Juggler.

To create beans we use the **new** operator, as in

```
private Smiley smile = new Smiley();
```

The Juggler is an applet, so after we construct it we need to call its `init` and `start` methods. Each `hookup` class handles a single connection from a source of the event to its target.

Bound properties generate a `PropertyChangeEvent` when they are changed. The `PropertyChangeListener` interface specifies the method

```
public void propertyChange(PropertyChangeEvent evt)
```

to handle property change events.

FIGURE 12.23 The SmileyJug applet

EXAMPLE 12.5 **SmileyJug.java**

```
/* Uses Juggler, Smiley, and BoundTextBean in a
 * Java program.  Entering Smiley's age changes the
 * Juggler's animation rate.  Clicking on Smiley's nose
 * stops and starts the Juggler.
 */

import sunw.demo.juggler.*;                                        // Note 1
import smiley.*;
import java.applet.*;
import java.awt.event.*;
import java.awt.*;
import java.beans.*;

public class SmileyJug extends Applet {
    private Juggler jug = new Juggler();
    private Smiley smile  = new Smiley();
    private BoundTextBean text = new BoundTextBean();
    private Label label = new Label("Smiley's Age");
    private boolean on = true;                    // Is Juggler juggling?
    public void init() {
        add(jug);
        add(smile);
        Panel p = new Panel();
        p.setLayout(new GridLayout(2,1));
        p.add(text);
        p.add(label);
        add(p);
        jug.init();
        jug.start();
        smile.addPropertyChangeListener
                    (new HookupAgeToAnimationRate());             // Note 2
        smile.addActionListener(new HookupNosePress());
        text.addPropertyChangeListener(new HookupNumberToAge());
    }
```

continued

```
class HookupNosePress implements ActionListener {
    public void actionPerformed(ActionEvent e) {
        if (on){                                          // Note 3
            jug.stopJuggling();
            on = false;
        }
        else {
            jug.startJuggling();
            on = true;
        }
    }
}
class HookupNumberToAge implements PropertyChangeListener {   // Note 4
    public void propertyChange(PropertyChangeEvent evt) {
        Object object = evt.getNewValue();
        int a = ((Integer)object).intValue();
        smile.setAge(a);
    }
}
class HookupAgeToAnimationRate
                    implements PropertyChangeListener {   // Note 5
    public void propertyChange(PropertyChangeEvent evt) {
        Object object = evt.getNewValue();
        int a = ((Integer)object).intValue();
        jug.setAnimationRate(a);
    }
}
}
```

Note 1: The Juggler is in the `sunw.demo.juggler` package. We also need to make sure this directory is on the classpath. In Windows we can add its parent directory `C:\BDK\demo` to the classpath using the command

```
set classpath=c:\BDK\demo;%classpath%
```

where `%classpath%` is the old classpath, which for the JDK1.1 must include the current directory, the path to user packages, and the path to the system packages. When using Java 2 (1.2) it works best to copy the `BDK\demo\sunw\demo\juggler` directory to a `sunw\demo\juggler` subdirectory of the directory containing this program.

Note 2: We could have let SmileyJug implement `ActionListener` and `Property ChangeListener` to handle these events, but we use hookup classes to more closely follow the technique of visual tools building applets from components. For simplicity, our hookup classes are inner classes. The BeanBox would make each a public class on a separate file.

Note 3: The variable on is **true** when the Juggler is juggling and **false** when not. This differs from the `SmileApplet` of Section 12.3 where Smiley stopped the Juggler and an ExplicitButton started it. The BeanBox connects an event, such as a nosePress, unconditionally to a single target method. Here we alternately choose the `stopJuggling` and `startJuggling` methods.

Note 4: When the user enters a desired value for Smiley's age in the BoundTextBean, that bean sets its number property which generates a `PropertyChangeEvent` that we handle here by calling the `setAge` method to change Smiley's age to the desired value.

Note 5: When Smiley's age changes it generates a `PropertyChangeEvent` which we handle here by changing the Juggler's animation rate.

Customizing Beans

Inside a tool such as the BeanBox we use property editors available in the Properties window to change properties of our beans as we build an applet with them. We can also write a Java program to change a bean's properties, saving the changed bean which we can use in its new configuration to build applets.

For example, suppose we would like an alternate version of Smiley bean with a red background. We can create a new Smiley in our program, set its background to red,

```
smile.setBackgound(Color.red);
```

and save the red Smiley.

Because Smiley implements `Serializable` we can save it using the `writeObject` method. We save it in the file `smiley/Smiley.ser`, where the `.ser` extension represents a serialized Java file.

EXAMPLE 12.6   **SaveSmiley.java**
....

```java
/* Saves a red Smiley in the
 * smiley/Smiley.ser file.
 */

import java.awt.*;
import java.io.*;
import smiley.*;

public class SaveSmiley {
  public static void main(String[] args) {
    Smiley smile = new Smiley();
    smile.setBackground(Color.red);
    try {
      FileOutputStream f = new FileOutputStream("smiley/Smiley.ser"); // Note 1
      ObjectOutputStream out = new ObjectOutputStream(f);
      out.writeObject(smile);
      out.flush();
    }catch(Exception e) {
        e.printStackTrace();
    }
  }
}
```

Note 1: We save `smile` in `smiley/Smiley.ser`, where the `smiley` directory is contained in the directory containing `SaveSmiley.class`. We could have used a program argument to specify the directory in which to save `smile`.

To use the customized version of Smiley in an applet we cannot just call its constructor. Calling new `Smiley()` would give us the original unmodified Smiley. The Beans class provides the `instantiate` method to load a class from a `.ser` file. In our RedSmiley applet, we use

```
smile=(Smiley)Beans.instantiate(null,"smiley.Smiley");
```

where the first argument specifies the class loader to use to load the class, and the second is the name of the bean. Passing **null** for the first argument lets Java use the standard system loader. The second argument, `smiley.Smiley`, tells Java to look first for a file `Smiley.ser` in the `smiley` directory (on the classpath) and if it cannot find it to look for `Smiley.class`, the original Smiley. Because `instantiate` returns an object, we cast the return value to have type Smiley.

Example 12.7 shows the changes to Example 12.6 needed to use the red Smiley. The complete code is on the disk included with this text. Running the RedSmiley applet shows a screen like that of Figure 12.23 except Smiley's background is red.

EXAMPLE 12.7

RedSmiley.java

```
/* Modifies Example 12.5 to use a serialized
 * red Smiley instead of the original.
 */

//   rest of code same as Example 12.5
public class RedSmiley extends Applet {
    public void init() {
        try {
            smile=(Smiley)Beans.instantiate(null,"smiley.Smiley");
        } catch (Exception e) {                                        // Note 1
            e.printStackTrace();
        }
          ...
    }
}
```

Note 1: The `instantiate` method may throw an `IOException` or a `ClassNotFound` `Exception`. We catch both with the superclass `Exception`.

The Big Picture

When using beans in our own Java program, we have to create the hookup classes that the BeanBox created. By using `set` methods, we can customize a bean in a program and write it, using object serialization, to a `.ser` file. This assumes that our bean class implements the `Serializable` interface. The `instantiate` method lets us load the customized bean and use it in our program.

TEST YOUR UNDERSTANDING

13. What additional capability does the `instantiate` method give, compared to the **new** operator for loading a bean into a program?

TRY IT YOURSELF   14. Modify Example 12.6 to use the `instantiate` method, instead of the **new** operator, to create the Juggler.

TRY IT YOURSELF   15. Modify Example 12.5 to omit the call `jug.start()`. What is the effect of this change?

Chapter 12 Summary

Java Beans are components we can use to build programs. Typically designers use a visual tool to build with beans. We illustrate this with the simple BeanBox provided by Sun to test beans. The ToolBox window contains a list of beans which the designer can drag into the BeanBox window. The Properties window lists properties of the selected bean, and provides editors for the designer to change these properties, customizing the bean for use in an applet.

Each bean shows three aspects of itself to interface with the BeanBox. In addition to its properties, a bean shows the events it can generate, which can be connected, using the BeanBox, to methods in a target bean. The event occurring in the source bean causes the chosen method to execute in the target bean. For example, we connected a button so pressing the button caused the Juggler to stop juggling. The BeanBox wrote the code to hook up the source event with the target action. The third feature a bean shows is the list of methods it can execute as the target of an event in a source bean.

Beans use standard syntax for properties, for example using a `getBackground` method to get its background color, and a `setBackground` method to set it. For a custom property such as age, the bean uses `getAge` and `setAge` methods. Properties can be bound or constrained. Using the BeanBox, the Bind Property menu allows the designer to bind a property of the source bean to a property of the same type in the target bean. When the value of that property changes in the source, it automatically changes in the target also. A constrained property allows the target to veto the change in the value of the source property.

A bean is a Java program. Our simple TextBean has a number property that we can get and set. We package a bean in a JAR file which includes all files needed by the bean so it may be downloaded or otherwise distributed as a component to use in building Java programs. We include a manifest file stating which of the files in the JAR file are beans.

A bean with a bound property calls the `firePropertyChange` method when that property is changed. A `PropertyChangeSupport` object handles the details of notifying property change listeners, which register using the `addPropertyChangeListener` method.

Our Smiley bean creates its own `ActionEvent` when the user presses the nose. This bean uses a vector to keep track of action listeners and notifies them when an action event occurs. It uses an anonymous inner class, defining the class, without a name, right where it is used, as a method argument. We often use anonymous inner classes in this way, defining them while registering them as event handlers.

By including a `BeanInfo` class with a bean, the bean developer can customize the list of properties, source events, and target methods that the bean designer sees when using a tool such as the BeanBox, making that bean easier to use.

Although visual tools are the intended way to develop programs using beans, we can use beans in hand-coded programs. Just as the BeanBox allows us to customize beans, we can write programs to customize beans and save the modified beans which can be reloaded in their changed state.

Component technology will have an increasingly important role in economically producing robust software.

Build Your Own Glossary

Find the uses, in this chapter, of the terms below. Enter a definition of each in the `glossary.txt` file on the disk that comes with this text.

anonymous inner class	component	manifest file
BDK	constrained property	Properties window
BeanBox	hookup class	ToolBox window
BeanInfo class	JAR file	
bound property	Java Bean	

Skill Builder Exercises

1. Match the activity on the left with the performer of that activity on the right.

    a. customize a bean   i.   the end user
    b. write the bean   ii. the bean developer
    c. build an applet from beans   iii. the bean designer
    d. run an applet built from beans

2. Match the file type on the left with the file function on the right

    a. `.ser` file   i.   Used to package an application
    b. manifest file   ii. Specifies the beans in a JAR file
    c. JAR file   iii. Stores persistent objects

3. Fill in the blanks in the following:

    Changing the value of a _____ property may cause the value of another property to change. A change to the value of a _____ property may be vetoed.

Critical Thinking Exercises

4. Which file would not be included in a JAR file for a bean?

    a. manifest file   d. Java source file
    b. class file   e. none of the above
    c. gif image file

5. The `SimpleBeanInfo` class

    a. implements the `BeanInfo` interface.
    b. extends the `BeanInfo` class.
    c. provides `BeanInfo` for `SimpleBean`.
    d. none of the above.
    e. all of the above.

6. An anonymous inner class

    a. can only be used as an event handler.
    b. can extend a class, but not implement an interface.

    c.   can only be used in a JavaBean.

    d.   none of the above.

    e.   all of the above.

7. The BeanBox is

    a.   the preferred tool for developing applets from beans.

    b.   a simple tool for testing beans, not intended for serious development.

    c.   the only way to use beans.

    d.   none of the above.

    e.   all of the above.

Debugging Exercise

8. The following program attempts to create a BankBean with a bound Rate property. Identify and correct any errors in it.

```java
import java.awt.*;
import java.awt.event.*;
import java.io.*;
import java.beans.*;

public class BankBean {
    private double rate = 5.0;
    private PropertyChangeSupport pChange =
                new PropertyChangeSupport(this);
    public void setRate(double r) {
        Double oldVal = new Double(rate);
        Double newVal = new Double(r);
        rate = r;
        pChange.firePropertyChange("rate",oldVal,newVal);
    }
    public double getRate() {
        return rate;
    }
}
```

Program Modification Exercises

9. Modify Example 12.2 to use an anonymous inner class to handle the `ActionEvent` generated.

10. Modify Example 12.3 to generate an `ActionEvent` when the user releases the mouse on Smiley's nose, rather than when the user presses it.

11. Modify Example 12.2 to make the background a bound property.

12. Modify Example 12.2 to catch the `NumberFormatException` thrown when the user enters an invalid value in the text field. Handle the exception by erasing the invalid value and leaving the focus in the text field so the user must enter another value.

13. Modify Example 12.5 to let SmileyJug handle the events it generates, rather than using hookup classes.

14. Modify Example 12.6 to save Smiley with a red background and an age of 50.

Program Design Exercises

15. Write an InsertSort bean which has a value property. When value is set, it should add the new value to an array of integers and display the array in sorted order in a bar chart, using rectangles in alternating colors. Assume the values range from 0 to 100. Package this bean and add it to the BeanBox. Add a BoundTextBean to the BeanBox. Configure the beans so when the user enters a value in the BoundTextBean it is displayed in sorted order in InsertSort. Make an applet from this configuration.

16. Write a Java program to make an applet, as described in Exercise 15, from a BoundTextBean and an InsertSort bean.

17. Write a `NewBoundTextBeanBeanInfo` class for the NewBoundTextBean which is like the BoundTextBean of Example 12.2, but which uses BeanInfo. Repackage NewBoundTextBean with its `BeanInfo` class and drag it into the BeanBox to test it.

18. Write a Boss bean that generates an `ActionEvent` when the user clicks its left eye and an `ItemEvent` when the user clicks its right eye. Construct an item event using

```
new ItemEvent
    (new Choice(),0,null,ItemEvent.ITEM_STATE_CHANGED)
```

Write a Worker bean that has methods `payRaised` and `payCut`. The `payRaised` method displays a drawing of a happy face whenever it is called while the `payLowered` method displays a sad face. Use the BeanBox to create an applet in which clicking the boss's left eye causes a pay raise for the worker, while clicking the right eye causes a pay cut.

19. Write a Java program to make an applet from, as described in Exercise 18, a Boss bean and a Worker bean.

Programming Projects

20. Create a Calculator bean from various bean components such as button beans and text field beans. You may use beans without any visible representation to hold intermediate values.

21. Create a Calendar bean which displays the days of the month which the user enters or selects.

22. Create a BarChart bean which displays data entered by the user in a bar chart.

23. Create a Database bean that uses JDBC to connect to the database specified by the user, using the driver specified, and execute an SQL query.

Compendium of Java Basics

Identifiers and Keywords

A Java identifier must start with a letter (including underscore, _, and dollar sign, $), followed by letters or digits or both. It can be of any length.

TABLE A.1 Java identifiers

Valid Identifiers
```
savings, textLabel, rest_stop_12, x,
    I3, _test, $soup
```

Not valid Identifiers
```
4you      //  Starts with a number
x<y       //  Includes an  illegal character, <
top-gun   //  Includes an illegal character, -
int       //  Reserved, see below
```

Unlike some programming languages (but like C and C++), Java is case-sensitive, meaning upper- and lowercase letters are different. Keywords are reserved for special uses and cannot be used as identifiers. Table 2 lists the Java keywords.

TABLE A.2 Java Keywords

abstract	do	implements	package	throw
boolean	double	import	private	throws
break	else	inner	protected	transient
byte	extends	instanceof	public	try
case	final	int	rest	var
cast	finally	interface	return	void
catch	float	long	short	volatile
char	for	native	static	while
class	future	new	super	
const	generic	null	switch	
continue	goto	operator	synchronized	
default	if	outer	this	

TEST YOUR UNDERSTANDING

1. Which of the following are valid identifiers? For each non-valid example, explain why it is not valid.

a.	Baby	e.	&car	i.	float
b.	_chip_eater	f.	GROUP	j.	intNumber
c.	any.time	g.	A103	k.	$$help
d.	#noteThis	h.	76trombones		

The Type Int

Java uses the keyword **int** for the integer data type, which in Java can range from –2,147,483,648 to 2,147,483,647. In Java the **int** type always uses 32 bits. Java supports these binary arithmetic operators:

+ addition

- subtraction

* multiplication

/ division

% remainder

Integer division truncates the result so 3/2 results in an integer value of 1.

The Type Double

Numbers of type **double**, which always use 64 bits, provide 16 decimal digits accurately. Using scientific notation, the exponents for **double** values can range from –324 to 308. A literal of type **double** may have a decimal point or an exponent or both, and it must have at least one digit. Some valid values of type **double** are:

22.7

4.123E-2   which is equivalent to .04123

36e2   which is equivalent to 3600.0

3.

0.54296

.1234

When we write a decimal literal such as 2.54, Java treats it as a value of type **double**. We can declare variables of type **double**, and we can use the arithmetic operators, +, −, *, and /, with **double** operands.

If we output values of type **double** using the `println` method, then Java will write the numbers in the most convenient form, using scientific notation for numbers greater than 10,000,000 or less than −10,000,000, and for numbers between −.001 and .001.

The Boolean Type

Type **boolean**, named for the British mathematician and logician, George Boole (1815–1864), provides two values, **true** and **false**, which we use to express the value of relational and logical expressions.

The Char Type

Java represents characters using single quotes, as in, for example, 'a', 'A', 'b', 'B' for letters, '0', '1' for numerals, '+', '-' for operators, and '?', ',' for punctuation. Internally, Java uses the Unicode™ character set which has 38,885 characters, including those needed for the world's major languages.[1] We will only need the ASCII (American Standard Code for Information Interchange) character set which has 128 characters. (See Appendix D for a table of the ASCII characters.) Each ASCII character has an equivalent Unicode character. Java converts from ASCII input to Unicode, and from Unicode to ASCII output; we will not need to use Unicode explicitly, because internationalizing Java is beyond the scope of this text.

We call the first thirty-two ASCII characters control characters; they are non-printing, but control functions such as formatting, including tab, newline, and return. For example, the ENTER key and the TAB key have no visible symbol, but they control the position of the next input. To represent these control characters, and other special characters, in our program we use the escape character, the backslash, '\'. When Java sees the backslash it escapes from its normal reading of printing characters and interprets the following character as a special character. Figure A.1 shows some of these special characters.

FIGURE A.1 Escape sequences for special characters

Special Character	Meaning
\n	newline, move to the start of the next line
\t	tab
\b	backspace
\r	return, move to the start of the current line
\"	double quote
\\	backslash

---

[1] See http://unicode.org for more information about Unicode.

Double quote and backslash are printing characters, but they have special functions in Java. Normally we enclose strings within double quotes, as in `"The result is "`. Occasionally, we want to use a string that itself uses double quotes, as in

```
"Do you like the movie \"Gone With The Wind\"? "
```

Using the backslash, in `\"`, tells Java the double quote is part of the string and not the terminating double quote.

The Long Type

The range of values for **int** variables is from –2,147,483,648 to 2,147,483,647. Java provides the type **long** which can represent integer values outside of this range. **Long** values, which use 64 bits, range from –9,223,372,036,854,775,808 to 9,223,372,036,854,775,807. By default, when we use a whole number such as 25 in a program, Java assumes it is an **int**. We can specify a **long** value by adding an l or L suffix, as in 25L. We can declare variables of type **long**, but would only use **long** variables instead of **int** when we need values the **int** type cannot handle. It takes more space to hold **long** values, and more time to process them.

The Byte, Short, and Float Types

Java has the types **byte** and **short** for small integers that are used for specialized purposes; we do not use these types in this text. Java provides a **float** type for decimal numbers which uses less precision than the type **double**. In contrast to the integer types, where the smaller type **int** is the default and the bigger type **long** is less often used, for floating point types, the larger type **double** is the default, and **float** is less often used. In the modern world, we often deal with decimal numbers requiring a wide range of values and high precision, whether in scientific calculations or financial transactions, and need the values type **double** provides.

   Float values are accurate to six digits and range from 1.4E-45 to 3.4028235E38. To represent a **float** literal add an F or an f, as in 3.14f. We can declare variables of type **float**, but must initialize them with **float** values. The declaration

```
float good = 4.25f;     // Valid
```

is fine, but the declaration

```
float bad = 4.25;       // Invalid
```

will cause an error, because the value 4.25 has type **double** by default. Java will not automatically convert from type **double** to type **float**, because, in general, a **double** value may be out of the range a **float** variable accepts. We **float** variables when we want to output values of seven digits rather than the sixteen digits of the type **double**.

TEST YOUR UNDERSTANDING

2. What will be the result of each division?

   a. `5.0 / 2.0`   b. `5 / 2`   c. `12 / 5`   d. `12.0 / 5.0`

3. Suppose the **double** variable, x, has the indicated value. Will `System.out.println(x)` display x in scientific notation? Show the result.

    a.   3456.789   d.   1234567890.987

    b.   .0000023456   e.   −234567.765432

    c.   .09876543

4.   Show the output from each of the following statements.

    a.
```
System.out.println
    ("I like \n\nto write Java programs.");
```

    b.
```
System.out.println
    ("Ali Baba said, \"Open, Sesame!\"");
```

    c.
```
System.out.println("12345\r678");
```

    d.
```
System.out.println("Find 3\\4 of 24");
```

Relational Operators and Expressions

Java provides relational and equality operators, listed in Figure A.2, which take two operands of a primitive type and produce a **boolean** result.

FIGURE A.2 Java relational and equality operators.

Operator Symbol	Meaning	Example	
<	less than	31 < 25	is false
<=	less than or equal to	464 <= 7213	is true
>	greater than	-98 > -12	is false
>=	greater than or equal to	9 >= 99	is false
==	equal to	9 == 12 + 12	is false
!=	not equal to	292 != 377	is true

Figure A.3 shows the Java symbols for the conditional operators.

FIGURE A.3 Conditional operators

Symbol	Meaning	Example
&&	conditional AND	(age > 20) && (age < 35)
||	conditional OR	(height > 78.5) || (weight > 300)

Note the operands of the conditional operators have type **boolean**. The expression age > 20 is either **true** or **false**, and so is age < 35.

Conditional AND

The conditional AND expression (age > 20) && (age < 35) will be **true** only when both of its operands are **true**, and **false** otherwise. Note when the first operand is **false**, as it is when age is 17, we know the conditional AND is **false** without even checking the value of the second operand. We say Java short-circuits the evaluation by not evaluating the second argument when it already knows the value of the expression.[2]

---

[2] This short-circuiting behavior is the reason why this operator is called the conditional AND, not simply AND. It evaluates the second argument on the condition that the first is **true**.

Conditional OR

The conditional OR expression (height > 78.5) || (weight > 300) is **true** if either one of its operands is **true**, or if both are **true**. As with the conditional AND operator, Java short-circuits the evaluation by not evaluating the second argument when it already knows the value of the expression.

Logical Complement

Java uses the symbol ! for the logical complement, or NOT, operator, which has only one operand. The logical complement negates the value of its operand. For example, if the **boolean** variable, on, has the value **true**, then !on is **false**, but if on is **false**, then !on is **true**.

TEST YOUR UNDERSTANDING

5. Write a relational expression in Java for each of the following:
   a. 234 less than 52
   b. 435 not equal to 87
   c. −12 equal to −12
   d. 76 greater than or equal to 54

6. What is wrong with the expression (3 < 4) < 5 in Java?

7. Explain the difference between x = 5 and x == 5.

8. Explain why the expression x > = 3 is not a correct Java expression to state x is greater than or equal to 3.

9. For each expression, find values for x and y that make it **true**.
   a. (x == 2) && (y > 4)
   b. (x <= 5) || (y >= 5)
   c. x > 10 || y != 5
   d. x > 10 && y < x + 4

10. For each expression in question 9, find values for x that allow Java to short-circuit the evaluation and not evaluate the right-hand argument. What is the value of the conditional expression?

11. For each expression, find a value for x that makes it **true**.
   a. !(x == 5)
   b. ! (x <= 10)
   c. !(x > 10 && x < 50)
   d. !(x == 5 || x > 8)

12. Omit any unnecessary parentheses from the following expressions.
   a. ((a > 1) || (c == 5))
   b. ((x < (y+5)) && (y > 2))
   c. !((x >2)||(y != 8))

Selection Statements

The if statement has the pattern

```
if (condition)
    if_true_statement
```

as in the example:

```
if (x > 2)
    y = x + 17;
```

The if-else statement has the form

```
if (condition)
    if_true_statement
else
    if_false_statement
```

For example,

```
if (x <= 20)
    x += 5;
else
    x += 2;
```

A switch statement chooses alternatives based upon the value of an **int** or **char** variable, and has the form

```
switch (test_expression)  {
    case expression1:
                statement1;
    case expression2:
                statement2;
        ...
    default:
                default_statement;
}
```

TEST YOUR UNDERSTANDING

13. What value will the variable x have after executing

```
x = 6;
if (k < 10)
    if (k < 5)
        x = 7;
    else
        x = 8;
```

if k has the value

a. 9   b. 3   c. 11   d. -2

14. What value will the variable x have after executing

```
x = 5;
switch(k) {
    case 2:
    case 3:             x = 6;
                        break;
    case 5:             x = 7;
                        break;
    case 9:             x = 8;
                        break;
    default:            x = 9;
}
```

if k has the value

| a. | 1 | b. | 3 | c. | 5 | d. | 6 | e. | 9 | f. | −5 | g. | 10 |

Repetition Statements

The `while` statement follows the pattern

```
while (condition)
    while_true_statement
```

where the condition evaluates to **true** or **false**, and the while_true_statement can be any Java statement including a code block. If the condition is **true**, then Java executes the while_true_statement and goes back to check the condition again. This process repeats until the condition is **false**.

We use a `for` statement when we know the number of repetitions. The `for` statement follows the pattern

```
for (initialize; test; update)
    for_statement
```

where the for_statement can be a simple statement or a block. The code

```
int sum = 0;
for (int i = 1; i <= 4; i++)
    sum += i;
```

uses a `for` statement to add the numbers from one to four.

The `while` statement lets us repeat a block of code; it checks the condition before executing the loop body. In some problems, when we know we will execute the body at least once, it is more natural to check the condition after executing the body. The do-`while` statement, having the form

```
do
    statement
while (condition) ;
```

lets us do that.

TEST YOUR UNDERSTANDING

15. Which of the following loops terminate? Assume x has the value 12 at the start of the loop.

    a. `while (x != 5)`   b. `while (x != 5)`   c. `while (x != 5)`
        `    x++;`   `    x--;`   `    x = 5;`

16. How many times will the body of each of following `while` loops be executed if x has the value 5 at the start of the loop?

    a. `while (x <= 10)`   b. `while (x == 2)`   c. `while (x > 1)`
        `    x +=3;`   `    x -= 7;`   `    x--;`

17. Write a `for` statement which will display the numbers from nine through three, in that order.

18. Write a `for` statement which will display the even numbers from four through twenty.

19. What value will the variable `sum` have after the execution of the following code?

```
int sum = 100;
for (int i = 20; i > 16 ; i--)
    sum -= i;
```

20. What value will the variable `sum` have after the execution of the following code?

```
int sum = 0;
for (int i = 1; i <= 20 ; i += 3)
    sum -= i;
```

21. What value will the variable `sum` have after the execution of the following code?

```
int sum = 100;
for (int i = 20; i > 16 ; i--)
    sum -= i;
```

22. Find the value of the variable `i` after the execution of the following code:

```
int i = 1;
int total = 0;
do {
    total += i;
    i++ ;
} while (total < 25);
```

The Java™ Class Hierarchy

We selected some of the Java classes, including all those used in this text, and some others. Subclasses are intended relative to their superclass. The complete documentation, including the fields and methods for each class, can be downloaded from http://java.sun.com/. This documentation is in HTML format and may be read using a browser, after unzipping it. (If this text includes the optional CDROM with the JDK, the documentation will also be included.)

```
class java.lang.Object
    class java.awt.AWTEventMulticaster
            (implements java.awt.event.ComponentListener,
            java.awt.event.ContainerListener, java.awt.event.FocusListener,
            java.awt.event.KeyListener, java.awt.event.MouseListener,
            java.awt.event.MouseMotionListener,
            java.awt.event.WindowListener, java.awt.event.ActionListener,
            java.awt.event.ItemListener, java.awt.event.AdjustmentListener,
            java.awt.event.TextListener)
    interface java.awt.event.ActionListener (extends java.util.EventListener)
    interface java.applet.AppletContext
    interface java.applet.AudioClip
    interface java.beans.BeanInfo
    class java.beans.Beans
    class java.lang.Boolean (implements java.io.Serializable)
    class java.awt.BorderLayout
            (implements java.awt.LayoutManager2, java.io.Serializable)
    class java.util.Calendar
            (implements java.io.Serializable, java.lang.Cloneable)
        class java.util.GregorianCalendar
    interface java.sql.CallableStatement (extends java.sql.PreparedStatement)
    class java.awt.CardLayout
            (implements java.awt.LayoutManager2, java.io.Serializable)
    class java.lang.Character (implements java.io.Serializable)
    class java.awt.CheckboxGroup (implements java.io.Serializable)
    class java.lang.Class (implements java.io.Serializable)
    interface java.lang.Cloneable
    class java.awt.Color (implements java.io.Serializable)
        class java.awt.SystemColor (implements java.io.Serializable)
    class java.awt.Component
            (implements java.awt.image.ImageObserver,
            java.awt.MenuContainer, java.io.Serializable)
        class java.awt.Button
        class java.awt.Canvas
        class java.awt.Checkbox (implements java.awt.ItemSelectable)
        class java.awt.Choice (implements java.awt.ItemSelectable)
```

```
        class java.awt.Container
            class java.awt.Panel
                class java.applet.Applet
            class java.awt.ScrollPane
            class java.awt.Window
                class java.awt.Dialog
                    class java.awt.FileDialog
                class java.awt.Frame (implements java.awt.MenuContainer)
        class java.awt.Label
        class java.awt.List (implements java.awt.ItemSelectable)
        class java.awt.Scrollbar (implements java.awt.Adjustable)
        class java.awt.TextComponent
            class java.awt.TextArea
            class java.awt.TextField
interface java.sql.Connection
interface java.beans.Customizer
interface java.io.DataInput
interface java.io.DataOutput
interface java.sql.DatabaseMetaData
class java.util.Date (implements java.io.Serializable, java.lang.Cloneable)
    class java.sql.Date
    class java.sql.Time
    class java.sql.Timestamp
class java.util.Dictionary
    class java.util.Hashtable
            (implements java.lang.Cloneable, java.io.Serializable)
        class java.util.Properties
class java.awt.Dimension (implements java.io.Serializable)
interface java.sql.Driver
class java.sql.DriverManager
class java.sql.DriverPropertyInfo
interface java.util.Enumeration
class java.awt.Event (implements java.io.Serializable)
interface java.util.EventListener
class java.util.EventObject (implements java.io.Serializable)
    class java.awt.AWTEvent
        class java.awt.event.ActionEvent
            class java.awt.event.FocusEvent
            class java.awt.event.InputEvent
                class java.awt.event.KeyEvent
                class java.awt.event.MouseEvent
            class java.awt.event.WindowEvent
        class java.awt.event.ItemEvent
        class java.awt.event.TextEvent
    class java.beans.PropertyChangeEvent
class java.beans.FeatureDescriptor
    class java.beans.BeanDescriptor
    class java.beans.EventSetDescriptor
    class java.beans.MethodDescriptor
    class java.beans.ParameterDescriptor
    class java.beans.PropertyDescriptor
        class java.beans.IndexedPropertyDescriptor
class java.io.File (implements java.io.Serializable)
```

continued

```
class java.awt.FlowLayout
        (implements java.awt.LayoutManager, java.io.Serializable)
class java.awt.event.FocusAdapter
        (implements java.awt.event.FocusListener)
interface java.awt.event.FocusListener (extends java.util.EventListener)
class java.awt.Font (implements java.io.Serializable)
class java.awt.FontMetrics (implements java.io.Serializable)
class java.text.Format
        (implements java.io.Serializable, java.lang.Cloneable)
    class java.text.DateFormat (implements java.lang.Cloneable)
        class java.text.SimpleDateFormat
    class java.text.MessageFormat
    class java.text.NumberFormat (implements java.lang.Cloneable)
        class java.text.ChoiceFormat
        class java.text.DecimalFormat
class java.awt.Graphics
class java.awt.GridBagConstraints
        (implements java.lang.Cloneable, java.io.Serializable)
class java.awt.GridBagLayout
        (implements java.awt.LayoutManager2, java.io.Serializable)
class java.awt.GridLayout
        (implements java.awt.LayoutManager, java.io.Serializable)
class java.awt.Image
interface java.awt.image.ImageObserver
class java.net.InetAddress (implements java.io.Serializable)
class java.io.InputStream
    class java.io.FileInputStream
    class java.io.FilterInputStream
        class java.io.BufferedInputStream
        class java.io.DataInputStream (implements java.io.DataInput)
        class java.io.LineNumberInputStream
        class java.io.PushbackInputStream
    class java.io.ObjectInputStream
            (implements java.io.ObjectInput, java.io.ObjectStreamConstants)
    class java.io.PipedInputStream
class java.awt.Insets
        (implements java.lang.Cloneable, java.io.Serializable)
class java.beans.Introspector
interface java.awt.event.ItemListener (extends java.util.EventListener)
interface java.awt.ItemSelectable
class java.awt.event.KeyAdapter (implements java.awt.event.KeyListener)
interface java.awt.event.KeyListener (extends java.util.EventListener)
interface java.awt.LayoutManager
interface java.awt.LayoutManager2 (extends java.awt.LayoutManager)

class java.util.Locale
        (implements java.lang.Cloneable, java.io.Serializable)
class java.lang.Math
class java.awt.MediaTracker (implements java.io.Serializable)
class java.awt.MenuComponent (implements java.io.Serializable)
    class java.awt.MenuBar (implements java.awt.MenuContainer)
    class java.awt.MenuItem
```

```
            class java.awt.CheckboxMenuItem
                    (implements java.awt.ItemSelectable)
            class java.awt.Menu (implements java.awt.MenuContainer)
                    class java.awt.PopupMenu
interface java.awt.MenuContainer
class java.awt.MenuShortcut (implements java.io.Serializable)
class java.awt.event.MouseAdapter
            (implements java.awt.event.MouseListener)
interface java.awt.event.MouseListener (extends java.util.EventListener)
class java.awt.event.MouseMotionAdapter
            (implements java.awt.event.MouseMotionListener)
interface java.awt.event.MouseMotionListener
            (extends java.util.EventListener)
class java.rmi.Naming
class java.lang.Number (implements java.io.Serializable)
    class java.math.BigDecimal
    class java.math.BigInteger
    class java.lang.Byte
    class java.lang.Double
    class java.lang.Float
    class java.lang.Integer
    class java.lang.Long
    class java.lang.Short
interface java.io.ObjectInput (extends java.io.DataInput)
interface java.io.ObjectOutput (extends java.io.DataOutput)
class java.io.ObjectStreamClass (implements java.io.Serializable)
class java.io.OutputStream
    class java.io.FileOutputStream
    class java.io.FilterOutputStream
        class java.io.BufferedOutputStream
        class java.io.DataOutputStream (implements java.io.DataOutput)
        class java.io.PrintStream
            class java.rmi.server.LogStream
    class java.io.ObjectOutputStream
            (implements java.io.ObjectOutput, java.io.ObjectStreamConstants)
    class java.io.PipedOutputStream
class java.awt.Point (implements java.io.Serializable)
class java.awt.Polygon (implements java.awt.Shape, java.io.Serializable)
interface java.sql.PreparedStatement (extends java.sql.Statement)
interface java.beans.PropertyChangeListener
            (extends java.util.EventListener)
class java.beans.PropertyChangeSupport
            (implements java.io.Serializable)
interface java.beans.PropertyEditor
class java.beans.PropertyEditorManager
class java.beans.PropertyEditorSupport
            (implements java.beans.PropertyEditor)
class java.util.Random (implements java.io.Serializable)
class java.io.RandomAccessFile
            (implements java.io.DataOutput, java.io.DataInput)
class java.io.Reader
    class java.io.BufferedReader
        class java.io.LineNumberReader
```

continued

```
            class java.io.CharArrayReader
            class java.io.FilterReader
                class java.io.PushbackReader
            class java.io.InputStreamReader
                class java.io.FileReader
            class java.io.PipedReader
            class java.io.StringReader
    class java.awt.Rectangle (implements java.awt.Shape, java.io.Serializable)
    interface java.rmi.Remote
    class java.rmi.server.RemoteObject
            (implements java.rmi.Remote, java.io.Serializable)
        class java.rmi.server.RemoteServer
            class java.rmi.server.UnicastRemoteObject
    interface java.sql.ResultSet
    interface java.sql.ResultSetMetaData
    interface java.lang.Runnable
    interface java.io.Serializable
    class java.net.ServerSocket
    interface java.awt.Shape
    class java.beans.SimpleBeanInfo (implements java.beans.BeanInfo)
    class java.net.Socket
    interface java.sql.Statement
    class java.io.StreamTokenizer
    class java.lang.String (implements java.io.Serializable)
    class java.lang.StringBuffer (implements java.io.Serializable)
    class java.util.StringTokenizer (implements java.util.Enumeration)
    class java.lang.System
    interface java.awt.event.TextListener (extends java.util.EventListener)
    class java.lang.Thread (implements java.lang.Runnable)
    class java.lang.ThreadGroup
    class java.lang.Throwable (implements java.io.Serializable)
        class java.lang.Error
            class java.lang.LinkageError
                class java.lang.IncompatibleClassChangeError
                    class java.lang.AbstractMethodError
                    class java.lang.IllegalAccessError
                    class java.lang.InstantiationError
                    class java.lang.NoSuchFieldError
                    class java.lang.NoSuchMethodError
                class java.lang.NoClassDefFoundError
        class java.lang.Exception
            class java.awt.AWTException
            class java.lang.ClassNotFoundException
            class java.lang.CloneNotSupportedException
                class java.rmi.server.ServerCloneException
            class java.io.IOException
                class java.io.EOFException
                class java.io.FileNotFoundException
                class java.io.InterruptedIOException
                class java.net.MalformedURLException
                class java.io.ObjectStreamException
                    class java.io.InvalidClassException
```

```
            class java.io.InvalidObjectException
            class java.io.NotSerializableException
        class java.net.ProtocolException
        class java.rmi.RemoteException
            class java.rmi.server.SkeletonNotFoundException
            class java.rmi.StubNotFoundException
            class java.rmi.UnexpectedException
            class java.rmi.UnknownHostException
        class java.net.SocketException
            class java.net.ConnectException
        class java.net.UnknownHostException
    class java.lang.IllegalAccessException
    class java.lang.InstantiationException
    class java.lang.InterruptedException
    class java.beans.IntrospectionException
    class java.lang.NoSuchFieldException
    class java.lang.NoSuchMethodException
    class java.beans.PropertyVetoException
    class java.lang.RuntimeException
        class java.lang.ArithmeticException
        class java.lang.ArrayStoreException
        class java.lang.ClassCastException
        class java.util.EmptyStackException
        class java.lang.IllegalArgumentException
            class java.lang.IllegalThreadStateException
            class java.security.InvalidParameterException
            class java.lang.NumberFormatException
        class java.lang.IndexOutOfBoundsException
            class java.lang.ArrayIndexOutOfBoundsException
            class java.lang.StringIndexOutOfBoundsException
        class java.lang.NegativeArraySizeException
        class java.util.NoSuchElementException
        class java.lang.NullPointerException
    class java.sql.SQLException
        class java.sql.SQLWarning
            class java.sql.DataTruncation
    class java.rmi.server.ServerNotActiveException
class java.util.TimeZone
        (implements java.io.Serializable, java.lang.Cloneable)
    class java.util.SimpleTimeZone
class java.awt.Toolkit
class java.sql.Types
class java.net.URL (implements java.io.Serializable)
class java.net.URLConnection
    class java.net.HttpURLConnection
class java.util.Vector
        (implements java.lang.Cloneable, java.io.Serializable)
    class java.util.Stack
interface java.beans.VetoableChangeListener
            (extends java.util.EventListener)
class java.beans.VetoableChangeSupport
        (implements java.io.Serializable)
```

continued

```
class java.lang.Void
class java.awt.event.WindowAdapter
         (implements java.awt.event.WindowListener)
interface java.awt.event.WindowListener
            (extends java.util.EventListener)
class java.io.Writer
   class java.io.BufferedWriter
   class java.io.CharArrayWriter
   class java.io.FilterWriter
   class java.io.OutputStreamWriter
      class java.io.FileWriter
   class java.io.PipedWriter
   class java.io.PrintWriter
   class java.io.StringWriter
```

The AtmScreen of Example 1.5

EXAMPLE 1.5 **AtmScreen.java**

```java
/* Illustrates event-driven object-oriented programming, and can be
 * extended,  using networking, databases, and Java Beans, to provide
 * remote banking.
 */

import iopack.Io;
import personData.*;
import java.applet.Applet;
import java.awt.*;
import java.awt.event.*;
import java.text.NumberFormat;

public class AtmScreen extends Applet
                       implements ActionListener, ItemListener {
  private Button start = new Button("Start");
  private Button finish = new Button("Finish");
  private TextField dataEntry = new TextField(10);
  private CheckboxGroup group = new CheckboxGroup();
  private Checkbox savings = new Checkbox("Savings",false,group);
  private Checkbox checking = new Checkbox("Checking",false,group);
  private Choice transaction = new Choice();
  private String message = "Welcome to Art's Bank";
  private MyCanvas canvas = new MyCanvas();
  private Label textLabel = new Label("Enter Name");
  private Bank bank = new Bank();
  private Person[] person = Database.personData();
  private Teller teller;
  private Font font = new Font("Serif",Font.BOLD,18);
  private FontMetrics metrics = getFontMetrics(font);

  public void init() {
    setLayout(new BorderLayout());
    setFont(font);
    Panel text = new Panel();
    text.setLayout(new GridLayout(2,1));
    text.add(dataEntry);
    text.add(textLabel);
    Panel north = new Panel();
    north.add(text);
    Panel checkboxes = new Panel();
```

continued

```java
        checkboxes.setLayout(new GridLayout(2,1));
        checkboxes.add(savings);
        checkboxes.add(checking);
        north.add(checkboxes);
        transaction.add("Deposit");
        transaction.add("Withdraw");
        transaction.add("Balance");
        north.add(transaction);
        add(north,"North");
        Panel south = new Panel();
        south.add(start);
        south.add(finish);
        add(south,"South");
        canvas.setBackground(Color.yellow);
        add(canvas,"Center");
        dataEntry.addActionListener(this);
        savings.addItemListener(this);
        checking.addItemListener(this);
        transaction.addItemListener(this);
        start.addActionListener(this);
        finish.addActionListener(this);
        clear();
    }
    public void actionPerformed(ActionEvent e) {
        Object source = e.getSource();
        if (source==dataEntry){
            String s = textLabel.getText();
            if (s.equals("Enter Name")){
                String name = dataEntry.getText();
                if (name.equals("John Venn"))
                    teller.acceptCard(person[0]);
                else if (name.equals("Mabel Venn"))
                    teller.acceptCard(person[1]);
                else {
                    message = "Enter John Venn or Mabel Venn";
                    dataEntry.setText("");
                    canvas.repaint();
                }
            }
            else if (s.equals("Enter PIN"))
                    teller.acceptPIN(dataEntry.getText());
            else if (s.equals("Enter Amount"))
                    teller.acceptAmount
                        (new Double(dataEntry.getText()).doubleValue());
        }
        else if (source == finish)  clear();
        else if (source == start){
                dataEntry.setEnabled(true);
                dataEntry.requestFocus();
                textLabel.setEnabled(true);
                start.setEnabled(false);
        }
    }
```

```java
public void itemStateChanged(ItemEvent e) {
  Object item = e.getItemSelectable();
  if (item == transaction)
    teller.acceptTransaction(transaction.getSelectedIndex());
  else if (item == savings)
    teller.acceptType(Bank.SAVINGS);
  else if (item == checking)
    teller.acceptType(Bank.CHECKING);
}
public void enterPIN() {
  dataEntry.setText("");
  textLabel.setText("Enter PIN");
  message = "Enter your PIN number";
  canvas.repaint();
}
public void selectTransaction() {
  dataEntry.setText("");
  dataEntry.setEnabled(false);
  textLabel.setEnabled(false);
  transaction.setEnabled(true);
  message = "Select your transaction";
  canvas.repaint();
}
public void selectType() {
  savings.setEnabled(true);
  checking.setEnabled(true);
  transaction.setEnabled(false);
  message = "Select your account type";
  canvas.repaint();
}
public void specifyAmount() {
  dataEntry.setEnabled(true);
  dataEntry.requestFocus();
  textLabel.setEnabled(true);
  textLabel.setText("Enter Amount");
  savings.setEnabled(false);
  checking.setEnabled(false);
  message = "Specify the amount";
  canvas.repaint();
}
public void display(String s) {
  dataEntry.setText("");
  textLabel.setEnabled(false);
  dataEntry.setEnabled(false);
  checking.setEnabled(false);
  savings.setEnabled(false);
  message = s;
  canvas.repaint();
}
public void clear() {
  start.setEnabled(true);
  savings.setEnabled(false);
  checking.setEnabled(false);
```

continued

```
        transaction.setEnabled(false);
        textLabel.setText("Enter Name");
        textLabel.setEnabled(false);
        dataEntry.setText("");
        dataEntry.setEnabled(false);
        message = "Welcome to Art's bank";
        canvas.repaint();
        teller = new Teller(bank,this);
      }
      class MyCanvas extends Canvas {
        public void paint(Graphics g) {
          int w = metrics.stringWidth(message);
          g.drawString(message, (getSize().width-w)/2, getSize().height/3);
        }
      }
    }

    class Teller {
      public static final int DEPOSIT = 0;
      public static final int WITHDRAW = 1;
      public static final int BALANCE = 2;
      private String id;
      private int transType;
      private int  acctType;
      private Person user;
      private Bank bank;
      private Account account;
      private AtmScreen screen;
      private NumberFormat nf = NumberFormat.getCurrencyInstance();

      public Teller(Bank b, AtmScreen s) {
        bank = b;
        screen = s;
      }
      public void acceptCard(Person p) {
        user = p;
        screen.enterPIN();
      }
      public void acceptPIN(String s) {
        id = s;
        screen.selectTransaction();
      }
      public void acceptTransaction(int trans) {
        transType = trans;
        screen.selectType();
      }
      public void acceptType(int type) {
        acctType = type;
        bank.find(id,acctType,this);
      }
      public void acceptAccount(Account a) {
        account = a;
```

```
    if (account != null)
      if (transType == BALANCE){
        screen.display("The balance is " + nf.format(account.getBalance()));
      }
      else {
        if (transType == DEPOSIT || transType == WITHDRAW){
          screen.specifyAmount();
        }
      }
    else
      screen.display("No such account -- session terminated");
  }
  public void acceptAmount(double amount) {
    switch(transType) {
      case DEPOSIT :
        account.deposit(amount);
        screen.display("Deposit of " + nf.format(amount));
        break;
      case WITHDRAW:
        double taken = account.withdraw(amount);
        if (taken >= 0)
          screen.display("Withdrawal of " + nf.format(taken));
        else
          screen.display("Insufficient funds");
        break;
    }
  }
}


class Database {
  public static Person[] personData() {
    Name n1 = new Name("John","Venn");
    Address a1 = new Address( "123 Main St.", "Tyler","WY", "45654");
    Person p1 = new Person("123123123",n1,a1);
    Name n2 = new Name("Mabel","Venn");
    Person p2 = new Person("456456456",n2,a1);
    Person[] p = {p1,p2};
    return p;
  }
  public static Account [] accountData() {
    Person[] p = personData();
    Account p1Savings = new Savings(1500.00,p[0],4.0);
    Account p1Checking = new Checking(p[0],2500.00,.50);
    Account p2Savings = new Savings(1000.00,p[1],3.5);
    Account[] a = {p1Savings,p1Checking,p2Savings};
    return a;
  }
}
class Bank {
  public static final int SAVINGS = 1;
  public static final int CHECKING = 2;
  private Account [] accounts = Database.accountData();
```

continued

```java
        public void find(String id, int acctType, Teller teller) {
          for (int i=0; i<accounts.length; i++) {
            Account acct = accounts[i];
            if (acct.getId().equals(id))
              switch(acctType) {
                case SAVINGS:
                  if (acct instanceof Savings){
                    teller.acceptAccount(acct);
                    return;
                  }
                case CHECKING:
                  if (acct instanceof Checking){
                    teller.acceptAccount(acct);
                    return;
                  }
              }
          }
          teller.acceptAccount(null);
        }
      }
      abstract class Account {
        private double balance;
        private Person holder;
        public Account(Person p)    {
          this(0,p);
        }
        public Account(double initialAmount, Person p) {
          balance = initialAmount;
          holder = p;
        }
        public String getId() {
          return holder.getId();
        }
        public void deposit(double amount) {
          balance += amount;
        }
        public double withdraw(double amount) {
          if (balance >= amount){
            balance -= amount;
            return amount;
          }
          else
            return -1.0;
        }
        public double getBalance() {
          return balance;
        }
        public void setBalance(double amount) {
          balance = amount;
        }
      }
      class Checking extends Account {
```

```
    private double minBalance;
    private double charge;
    public Checking(Person p,double minAmount, double charge) {
      super(p);
      minBalance = minAmount;
      this.charge = charge;
    }
    public double processCheck(double amount)  {
      if (getBalance() >= minBalance)
        return super.withdraw(amount);
      else
        return super.withdraw(amount + charge);
    }
    public double withdraw(double amount) {
      return processCheck(amount);
    }
}
class Savings extends Account  {
    private double interestRate;
    public Savings(double amount, Person p, double rate) {
      super(amount,p);
      interestRate = rate;
    }
    public void postInterest()  {
      double balance = getBalance();
      double interest = interestRate/100*balance;
      setBalance(balance + interest);
    }
}
```

The ASCII Character Set

The first 32 characters and the last are control characters. We show only the printing characters.

ASCII	char	ASCII	char	ASCII	char	ASCII	char
32	blank	56	8	80	P	104	h
33	!	57	9	81	Q	105	i
34	"	58	:	82	R	106	j
35	#	59	;	83	S	107	k
36	$	60	<	84	T	108	l
37	%	61	=	85	U	109	m
38	&	62	>	86	V	110	n
39	'	63	?	87	W	111	o
40	(	64	@	88	X	112	p
41	)	65	A	89	Y	113	q
42	*	66	B	90	Z	114	r
43	+	67	C	91	[	115	s
44	,	68	D	92	\	116	t
45	-	69	E	93	]	117	u
46	.	70	F	94	^	118	v
47	/	71	G	95	_	119	w
48	0	72	H	96	`	120	x
49	1	73	I	97	a	121	y
50	2	74	J	98	b	122	z
51	3	75	K	99	c	123	{
52	4	76	L	100	d	124		
53	5	77	M	101	e	125	}
54	6	78	N	102	f	126	~
55	7	79	O	103	g		

Note: The ASCII values for the non-printing characters we use in this text are 8 for "\b", 9 for "t", 10 for "n" and 13 for "\r".

Using Java in Various Environments

We show how to compile and run the programs in this text using:

Java Development Kit™	Sun Microsystems, Inc.
JBuilder™	Borland International, Inc. (now Inprise)
VisualCafe™	Symantec Corporation

Consult the documentation included with these or other environments for more complete information on their use.

JBuilder

We used the JBuilder™ University Edition version 1.01. Check

```
www.borland.com/education/jbuilder/jbue.html
```

for information about the University Edition which is available free to faculty for labs and distribution to students.

Creating the Project in JBuilder

After opening JBuilder, click on the File menu and then New Project to create a project which will contain all the files used in the given applet or application. The default directory is

```
c:\jbuilder\myprojects\untitled\untitled.jpr
```

where `jbuilder` is the directory in which JBuilder is installed and `untitled` is the directory created to hold the files for this project. Change `untitled` to a name representing this project. We demonstrate with Example 1.1, so we will call our project `ex1_1`. If the project files are in a package, use the package name as the directory name.

The left pane shows the files in the project. JBuilder includes a default HTML file, `ex1_1.html`. Our first program is an application, so remove this HTML file by highlighting it and clicking the minus (-) button just above the file list.

Adding a Program to the Project in JBuilder

To add the `Hello.java` program of Example 1.1 to the project:

1. Click on the plus (+) button above the left pane, which pops up a file dialog.

2. Browse in the file dialog to find `Hello.java`, and open it.

3. Use the File, Save menu to save a copy of this file in the `ex1_1` directory. This will allow modification of the file while keeping the original intact.

To compile `Hello.java`:

1. Click on File, Project Properties, Project tab.
2. Click on the Edit button of the Compiler Source Path field. Click on the Add Path button, select the directory, `ex1_1` in this example, containing the project files and click OK. Click OK in the Edit window and again in the outer window. These steps allow the compiler to find the source file.
3. Click on the Build, Make Project menu item.

Running the Project in JBuilder

To run the program:

1. The Console window does not stay open to read the output when we run a program from Windows. The output appears and the console window closes immediately. If the program does not use console input, we use the Execution Log window within JBuilder instead. For programs which input from the console we add a line at the end to pause the program before it finishes.

If the program does not use console input:

1. Click on File, Project Properties, Run/Debug tab.
2. At the bottom, select Send run output to Execution Log, and click on OK.
3. Click on View, Execution Log.
4. Click on Run, Debug. The output will appear in the Execution Log window.

If the program does use console input:

1. Add as the last line of the program

    ```
    Io.readString("Press any key to exit");¹
    ```
2. Click on File, Project Properties, Run/Debug tab.
3. At the bottom, select Send run output to Console Window and click on OK.
4. Click on Run, Debug. A console window will appear and remain until the user presses a key to exit after the program has finished.

Applets in JBuilder

We use Example 1.2 `HelloApplet.java` to illustrate running an applet in JBuilder.

1. Make a project, say `ex1_2.jpr`, as shown above for `ex1_1`. We have our own HTML file, so remove `ex1_2.html` from the project by clicking on it and clicking the minus (-) button.

---

¹ To make it convenient to use integrated development environments we added this line to each line of the programs that read from the console. This extra line does not appear in the text but is found in the programs on the disk that comes with this text.

2. Add `HelloApplet.java` and `HelloApplet.html` to the project as shown above for Hello.java.

3. Compile `HelloApplet.java` as shown above for `Hello.java`.

4. Click on Run, Run Applet in `"HelloApplet.html"`

Importing User-Defined Packages in JBuilder

The Project Properties defaults are set to allow the compiler to find the standard Java classes. To run programs such as Example 1.5 `AtmScreen.java` that use our own packages from the text, follow the instructions for compiling. Add the directories containing each of the packages to the classpath:

1. Click on File, ProjectProperties, Project tab.

2. Click on the Edit button of the Class Path field.

3. Click on the Add Path button, selecting the directory which contains the package directory (not the package directory itself) and click OK.

4. Click OK in the Edit window.

5. Click OK in the outer window.

6. Repeat these steps for each user-defined package used. For the `AtmScreen` example these are the `iopack` and `personData` packages. (If these directories are contained in the same parent directory then only add that parent once.)

Passing Program Arguments in JBuilder

To run a program such as Example 8.16, `ReadFileLines.java`, which uses program arguments, follow the steps for a console application, and to pass the arguments to the program:

1. Click on File, Project Properties, Run/Debug tab.

2. Add the arguments in the Command Line Parameters field and click OK.

Creating a New Program in JBuilder

Follow the steps for working with an existing program, but in the file dialog that pops up after clicking on the plus (+) button to add a file to the project:

1. Enter the file name of the file to be created in the File name field and click on Open.

VisualCafe

We used VisualCafe™ 3.0 Database Edition. See

`http://www.symantec.com/`

for more information about VisualCafe.

Creating a Project in VisualCafe

1. Click on File, New Project.
2. Select the Empty Project icon in the New Project window.
3. Click on File, Save as, enter a name for the project, and press Save.

Adding a Program to a Project in VisualCafe

To add `Hello.java` to the project:

1. Right-click the mouse in the Project window at the left.
2. Click on Insert Files.
3. Browse to select `Hello.java`
4. Press Add, then press OK.

Compiling in VisualCafe

To compile `Hello.java`

1. Click on Project, Build Application.

Running a Project in VisualCafe

To run the program:

1. The console window closes immediately after the output appears, not allowing time to read it. For programs with no console input, we use the message window instead. For programs which input from the console we add a line at the end to pause the program before it finishes.

If the program does not use console input:

1. Click on Project, Run in the Debugger.

If the program does use console input:

1. Add as the last line of the program

   ```
   Io.readString("Press any key to exit");
   ```
2. Add as the first line of the program

   ```
   import iopack.Io
   ```
3. Follow the instructions below for importing user-defined packages.
4. Click on Project, Execute.

Applets in VisualCafe

We use Example 1.2, `HelloApplet.java`, to illustrate running an applet.

1. Make a project, naming it and inserting the `HelloAppet.java` file.
2. Insert `HelloApplet.html` or let the system generate an HTML file.

3. Click on Project, Options.
4. Select Applet in the Project type field.
5. Press OK.
6. Compile by clicking on Project, Build Applet.
7. Run by clicking on Project, Execute.

Importing User-Defined Packages in VisualCafe

To run programs such as Example 1.5, `AtmScreen.java`, that use our own packages, follow the instructions for creating a project and adding a file to a project.

1. Click on Project, Options, Directories.
2. Select Input class files in the Show directories for field.
3, Select the New icon in the Directories area.
4. Press dir to browse, finding the directory containing the package directory (not the package directory itself). For the `AtmScreen` example add the directories containing the `iopack` and `personData` directories.
5. Press OK.
6. Compile and run as previously described.

Passing Program Arguments in VisualCafe

To run a program such as Example 8.16, `ReadFileLines.java`, which uses program arguments, follow the steps for a console application, and to pass the arguments to the program:

1. Click on Project, Options, Project tab.
2. In the Program arguments field, enter the arguments to pass to the program, separating them with a space.
3. Click OK.

Creating a New Program in VisualCafe

Follow the steps for working with an existing program, but instead of adding an existing file to the project,

1. Click on File, New File.
2. Enter the code for the file, save it and add it to the project.

Java Development Kit[2]

We used JDK™ 1.1.6 and JDK™ 1.2. See

```
http://java.sun.com
```

to download the Java Development Kit. The documentation requires a separate download.

---

[2] We have also included instructions for using the JDK™ in footnotes in the text.

Compiling Using the JDK

1. If using Windows, use the Start button to get a command (MS-DOS) prompt.
2. Enter the command

   ```
   javac Hello.java
   ```

   to compile the `Hello.java` file.

Running a Console Application Using the JDK

1. Enter the command

   ```
   java Hello
   ```

   to run the `Hello` application.

Applets Using the JDK

1. Compile using the command

   ```
   javac HelloApplet.java
   ```

2. Run using the command

   ```
   appletviewer HelloApplet.html
   ```

Importing User-Defined Packages Using the JDK

The default classpath allows the compiler to find the system classes. When using our own packages we need to augment the classpath. One method is to use the `-classpath` option. For example, using the JDK1.1 (in Windows) we compile with

```
javac -classpath .;c:\JAVA\lib\classes.zip;c:\gittleman AtmScreen.java
```

where JAVA is replaced by the directory in which the JDK is installed. This classpath has three directories, period (.), representing the current directory, `c:\JAVA\lib\classes.zip` containing the system classes, and `c:\gittleman` denoting the directory which contains the `iopack` and `personData` subdirectories containing the packages used in the `AtmScreen` example. Using Java 2 (formerly 1.2) we may omit the directory for the system classes because the classpath only refers to user packages and classes. On UNIX™ systems the classpath directories are separated by colons instead of semicolons, and would be different. When running the program add the same classpath option used with the compiler.

Another approach changes the CLASSPATH environment variable. In Windows, at the command prompt enter

```
set CLASSPATH=.;c:\JAVA\lib\classes.zip;c:\gittleman
```

if using the JDK 1.1 and

```
set CLASSPATH=.;c:\gittleman
```

if using Java 2 where JAVA is replaced by the directory in which the JDK is installed.

Unless you include this `set` command as part of the system startup you will need to execute it each time you open a command prompt window. To make the `set` command part of the system startup, on Windows 95 add it to the `autoexec.bat` file. On Windows™ NT, click on the Start button, then on Settings, then on Control Panel, then on the System icon, then on the Environment tab. Fill in CLASSPATH as the variable, fill in

`.;c:\JAVA\lib\classes.zip;c:\gittleman`

or

`.;c:\gittleman`

if using Java 2, as the value, and click the Set button.

Passing Program Arguments using the JDK

Add the program arguments to the end of the command, as in

`java ReadFileLines ReadFileLines.java 5`

which will display the first five lines of the file `ReadFileLines.java`.

Creating a New Program Using the JDK

Use a simple editor such as Notepad on Windows to create the `.java` and `.html` files.

Browsers

- Microsoft's Internet Explorer 4.0 and later support Java version 1.1 and will run the applets in this text.
- Netscape's™ Communicator 4.5 and later for Windows NT/95 support Java version 1.1.
- Sun's HotJava™ 1.1 supports Java version 1.1.

Operator Precedence Table

Highest	Postfix Operators	[ ] . ( ) ++ −−
	Unary Operators	++ (prefix) −− (prefix) + − ~ !
	Cast, Allocation	( ) new
	Multiplicative	* / %
	Additive	+ −
	Shift	<< >> >>>
	Relational	< > <= >= instanceof
	Equality	== !=
	Bitwise, Logical AND	&
	Bitwise, Logical XOR	^
	Bitwise, Logical OR	|
	Conditional AND	&&
	Conditional OR	||
	Conditional	?:
Lowest	Assignment	= += −= *= /= %= >>= <<= >>>= &= ^= |=

Binary and Hexadecimal Numbers

Our familiar decimal number system uses ten digits, 0, 1, 2, 3, 4, 5, 6, 7, 8, and 9, to represent numbers. We can only represent numbers up to nine with one digit. To write the number ten, we use two digits, 10, with the idea that the one in the ten's place represents ten. The number 387 represents 3x100 + 8x10 + 7 because the three is in the hundred's place, and the eight is in the ten's place. We call this the base ten system because each place represents a power of ten. The unit's place is $10^0 = 1$, the ten's place is $10^1 = 10$, the hundred's place is $10^2 = 100$, and so on.

In the binary number system we use two digits, 0 and 1 to represent integers. This is particularly suitable for computers where each hardware memory bit (binary digit) can represent a 0 or a 1. Using only two digits, the largest number we can represent in one digit is one. The first few binary numbers are

Binary	Decimal Equivalent
0	0
1	1
10	2 = 1x2 + 0
11	3 = 1x2 + 1
100	4 = 1x4 + 0x2 + 0x1
101	5 = 1x4 + 0x2 + 1x1
110	6 = 1x4 + 1x2 + 0x1
111	7 = 1x4 + 1x2 + 1x1
1000	8 = 1x8 + 0x4 + 0x2 + 0x1
1001	9 = 1x8 + 0x4 + 0x2 + 1x1
1010	10 = 1x8 + 0x4 + 1x2 + 0x1

Just as the decimal system is base 10, the binary system is in base 2, and has a unit's place, a two's place, a four's place, an eight's place, and so on as given by the sequence $2^0 = 1$, $2^1 = 2$, $2^2 = 4$, $2^3 = 8$, and so on.

While the computer hardware uses binary digits, it is hard for human readers to grasp say 16 bits of binary, as for example

0010110100110110

If we group the bits by fours, as in,

0010 1101 0011 0110

it's a bit easier to read but still cumbersome. Each group of four represents a number from 0 to 15 in the binary system. If we had a system with 16 digits, we could replace each group of four bits by a single digit. The hexadecimal system uses the digits 0,1,2,3,4,5,6,7,8,9,a,b,c,d,e, and f, where a=10, b=11, c=12, d=13, e=14, and f=15. We can use either lower- or uppercase letters A, B, C, D, E, and F. To write the above 16-bit number in hexadecimal replace each group of four by its corresponding hexadecimal digit, giving

```
binary        0010 1101 0011 0110
hexadecimal    2    d    3    6
```

We can specify hexadecimal constants in Java using 0x or 0X to prefix the number as, for example, 0x2d36.

The first 10 hexadecimal digits are the same as the decimal digits. The remaining six have the following binary and decimal equivalents.

Hexadecimal	Binary	Decimal
a or A	1010	10
b or B	1011	11
c or C	1100	12
d or D	1101	13
e or E	1110	14
f or F	1111	15

The main use of hexadecimal numbers is to give a shorter representation for binary numbers which, using only two digits, tend to get long. We could of course convert the binary number to base ten but that takes more work. If you are curious, 0010 1101 0011 0110 converts to

```
8192 + 2048 + 1024 + 256 + 32 + 16 + 4 + 2 = 11,574
```

in base 10.

☞ A LITTLE EXTRA   **Converting Between Base 10 and Binary**

As we have just seen, to convert a binary number to decimal, just add the powers of two corresponding to the positions of the 1s in the number. Converting 11001 we add 16 + 8 +1 = 25 because the 11001 has a one in the sixteen's place, the eight's place, and the one's place.

To convert a base 10 number to binary we use division to get the digits from right to left, starting with the one's digit. To convert 25, divide by two, and the remainder will be the one's digit. Divide the quotient, 12, by two, and the remainder will be the two's digit. Repeat this process until the quotient is zero. The example below shows how the base ten number 25 converts to the base two number 11001.

```
25 ÷ 2 =  12 r. 1   so the 1's digit is 1.
12 ÷ 2 =   6 r. 0   so the 2's digit is 0.
 6 ÷ 2 =   3 r. 0   so the 4's digit is 0.
 3 ÷ 2 =   1 r. 1   so the 8's digit is 1.
 1 ÷ 2 =   0 r. 1   so the 16's digit is 1.
```

TEST YOUR UNDERSTANDING

1. Express the value of each base 10 number using powers of 10. For example, 254 has the value 2x100 + 5x10 + 4.

   a. 38   b. 4179   c. 562   d. 94531   e. 306

2. Express the value of each base 2 number using powers of 2. For example, 101 has the value 1x4 + 0x2 + 1x1.

   a. 11   b. 1101   c. 101100   d. 11011   e. 11101010

3. Give the base 10 value for each of the binary numbers in Exercise 2.

4. Write each of the binary numbers in Exercise 2 in hexadecimal. (Hint: Group by fours, and add zeros in front if necessary. For example, 111010 would group as 0011 1010 and convert to 3a.)

A LITTLE EXTRA

5. Convert each of the following base 10 numbers to binary.

   a. 17   c. 38   e. 160   g. 4444
   b. 23   d. 86   f. 235

Bitwise and Shift Operators

We typically use base 10 integers in our program which are represented internally as binary numbers. Java requires that **int** values always use 32 bits. Thus Java represents the number 25 internally as

```
0000 0000 0000 0000 0000 0000 0001 1001
```

Bitwise Operators

Java has several bitwise operators that operate on each bit of an **int** value.

    &   bitwise AND
    | bitwise OR
    ^   bitwise XOR (exclusive or)
    ~   bitwise complement

Figure H.1 shows the effect of these operations on corresponding bits of each operand.

FIGURE H.1 The effect of bitwise operators

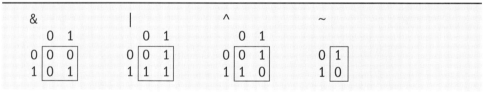

To illustrate these operators we use four bit values for simplicity, rather than work with the much longer 32-bit **int** values. To evaluate 1010 & 0011 we apply the & operator to the bits in corresponding positions in the left and right operands, according to the table in Figure H.1. Using the & operator on each pair of corresponding bits gives

    1 & 0 = 0 for the leftmost bits of 1010 and 0011,
    0 & 0 = 0 for the corresponding bits in the next position,
    1 & 1 = 1 for the next bits,
and 0 & 1 = 0 for the rightmost bits,

so the result is 0010.
Similarly using the tables in Figure H.1 we see

```
1010 | 0011 = 1011
1010 ^ 0011 = 1001
 ~ 1010      = 0101
```

We can apply the bitwise operators to variables. For example, if

```
int a = 10;
int b =  3;
```

then the 32-bit representations of a and b are

```
    0000 0000 0000 0000 0000 0000 0000 1010  for a
and 0000 0000 0000 0000 0000 0000 0000 0011  for b
```

so   a & b =   2 (0010 in binary)
      a | b = 11 (1011 in binary)
      a ^ b =   9 (1001 in binary)

Logical Operators

The operators &, |, and ^ can also operate on boolean values. The & and | operators are like the && and || operators, except the logical operators & and | always evaluate both operands, while the conditional operators && and || evaluate both operands only when necessary.

Shift Operators

The Java shift operators are:

   <<   Shift bits left, filling with zero bits on the right-hand side
   >>   Shift bits right, filling with the high bit on the left-hand side
   >>>   Shift bits right, filling with zero bits on the left-hand side

Using 8-bit numbers, rather than 32-bit, to illustrate

```
10101010 << 3   evaluates to 01010000
10101010 >> 3   evaluates to 11110101
10101010 >>> 3  evaluates to 00010101
```

where the left-hand operand is the number to shift, and the right-hand operand is the number of positions to shift. In evaluating 10101010 << 3, when we shift 10101010 three positions to the left, the rightmost three bits disappear and we replace the leftmost three bits with zeros. We can use the shift operator with **int** variables. For example given

```
int a = 0x00003A7D;
```

evaluating a << 7 will give 0x001D3E80 because 0x00003A7D in binary is

```
0000 0000 0000 0000 0011 1010 0111 1101
```

and shifting seven positions to the left gives

```
0000 0000 0001 1101 0011 1110 1000 0000
```

in binary, or 0x001D3E80 in hex.

TEST YOUR UNDERSTANDING

1. Find
   a.   1100 & 0011   b.   1111 & 0101   c.   0000 & 1010

2. Find
   a.   1100 | 0011   b.   1111 | 0101   c.   0000 | 1010

3. Find
   a.   1100 ^ 0011   b.   1111 ^ 0101   c.   0000 ^ 1010

4. Find
   a.   ~0010   b.   ~1011

5. Given
   ```
   int a = 27;
   int b = 53;
   ```
   find
   a.   a & b   b.   a | b   c.   a ^ b   d.   ~a

6. Evaluate, using 8-bit numbers
   a.   11001011 << 5   b.   11001011 >> 4   c.   11001011 >>> 4

Answers to Selected Exercises

Test Your Understanding Exercises

Chapter 1

1. C (and C++)

3. The Java Virtual Machine

9. Many answers are possible. Each represents responses to external events as well as internally generated chores. For example,

```
!!!    It's late.
       Go to sleep.
!!!    Alarm rings signaling a fire.
       Throw on some clothes.
       Slide down the fire pole.
       Drive fire engines to the fire.
       Execute put out fire procedure.
       Return to station.
!!!    It's time for breakfast.
       Execute make breakfast procedure.
!!!    Alarm rings signaling a medical emergency.
       Drive paramedic truck to the scene of the emergency.
       Execute paramedic procedures.
       Return to station.
```

11. A puppet show is more analogous to procedural programming. The puppeteer controls all the actions and speech of the puppets, the way that the procedural logic manipulates the data. In contrast, in a stage play each is responsible for his or her own lines, just as objects contain their own methods to meet their responsibilities, acting on their private data.

13. `public static void main(String[ ] args)`

    Only the parameter name, `args`, can be changed.

15. The `import` statement makes a class available with an abbreviated name. Without it we would have to use the fully qualified name in the program.

17. 16.0

Chapter 2

1. a. 27   b. h   c. 2   d. 10   e. three   f. THE THREE DID FEED THE DEER

3. a. negative   b. positive   c. zero

5. One possibility is:

The agent asks the customer to select a car.
The customer asks the agent to provide a mid-size convertible.
The agent asks the customer to specify a rental period.
The customer tells the agent that the rental is for one week.
The agent checks the computer system for a mid-size convertible.
The computer checks the inventory for a mid-size convertible.
The inventory shows a mid-size convertible is available.
The computer system checks the reservations for a conflict.
The reservations indicates no conflict.
The database tells the agent that such a car is available.
The agent requests a license and credit card from the customer.
The customer provides a license and a credit card.
The agent updates the system.
The system updates the inventory.
The agent provides the car to the customer.

Agent
   Accept request for car.
   Accept result of inquiry to system.

Customer
   Provide rental type.
   Provide license and credit card.

System
   Accept request for rental.
   Accept update.

Inventory
   Accept request for car availability.
   Accept update.

Reservation List
   Accept request for conflict check

7. We declare instance variables outside of any method. Example 2.9 uses the instance variable `balance`.

9. null

11. The `BankAccount` constructor with no arguments.

13. `TestBankAccount.java:12`: Variable balance in class `BankAccount` not accessible from class `TestBankAccount`.

15. Not allowed, since there is already a constructor with no arguments.

17. After the `serve burger` arrow add a `take burger` arrow from `aWaiter` to `aCustomer`. Similarly, add a `take soda` arrow after `serve soda`, and add a `take fries` arrow after `serve fries`.

19. `Address myAddress = new Address(street, city, state, zip)` where

    String street = 123 Main St.
    String city = Hometown
    String state = CA
    String zip = 12345

Chapter 3

1. a. BankAccount   b. Object   c. Object   d. Object   e. BankAccount

3. `CheckingAccount c = new CheckingAccount(1500,.35);`

5. Overloading. To override, the method must have exactly the same parameters as the method of the same name in the superclass. Here we added a boolean parameter not present in the superclass method.

7. c

9. Yes, because the class is still declared using the abstract keyword.

11. Implement the `RateChangeListener` interface. Implement the methods

    ```
    public void rateRaised(double amount);
    public void rateLowered(double amount);
    ```

13. `package stuff;`

15. Call `stuff.GoodStuff.doStuff();`

17. `Point` must be a public class. Without the `public` modifier it could only be used inside the package `java.awt` and the `Shape` class of Example 3.3 is not in the `java.awt` package.

19. We can only use `get4()` in the class `A` in which it is declared. We can use `get2()` anywhere in the package `visibility`.

21. The user asks the teller to accept an ATM card.
    The teller asks the user to enter a PIN.
    The user asks the teller to accept a PIN.
    The teller asks the user to select a transaction type.
    The user asks the teller to accept a withdraw.
    The teller asks the user to select an account type.
    The user asks the teller to accept a savings account type.
    The teller asks the bank to find the account of the chosen type for the user with the specified PIN.
    The bank gives the teller a reference to the account.
    The teller asks the user to specify an amount.
    The user asks the teller to accept an amount.
    The teller asks the account to withdraw the specified amount.
    The teller asks the user to select another transaction . . .

23. The user asks the teller to accept an ATM card.
The teller asks the user to enter a PIN.
The user asks the teller to accept a PIN.
The teller asks the user to select a transaction type.
The user asks the teller to accept a deposit.
The teller asks the user to select an account type.
The user asks the teller to accept a cancellation.

Chapter 4

1. ```
int[] intArray = {37,44,68,-12};
intArray[1] = 55;
```

3. ```
char[] charArray = {'s','y','t','c','v','w'};
```

5. ```
(int)(450*Math.random()) + 50;
```

7. a. `intArray`

b.

c.

   `doubleArray;`

9. ```
int[] b = new int[a.length];
System.arraycopy(a,0,b,0,a.length);
```

11. `display(myArray);`

13. `myArray`

  `anArray`

15. ```
int[][] results = {{55,66,87,76},{86,92,88,95}};
```

17. a. b   b. x   c. w   d. c

19. loop, stopping when item i >= item i-1 or index = 0 {
    interchange item i and item i-1;
    decrease i by 1;
}

21.
```
                52 38 6 97 3 41 67 44 15
Insert 38   38 52 6 97 3 41 67 44 15
Insert 6   38 6 52 97 3 41 67 44 15
                 6 38 52 97 3 41 67 44 15
Insert 97
Insert 3   6 38 52 3 97 41 67 44 15
                 6 38 3 53 97 41 67 44 15
```
```

```
                6   3  38  53  97  41  67  44  15
                3   6  38  53  97  41  67  44  15
Insert 41       3   6  38  53  41  97  67  44  15
                3   6  38  41  53  97  67  44  15
Insert 67       3   6  38  41  53  67  97  44  15
Insert 44       3   6  38  41  53  67  44  97  15
                3   6  38  41  53  44  67  97  15
                3   6  38  41  44  53  67  97  15
Insert 15       3   6  38  41  44  53  67  15  97
                3   6  38  41  44  53  15  67  97
                3   6  38  41  44  15  53  67  97
                3   6  38  41  15  44  53  67  97
                3   6  38  15  41  44  53  67  97
                3   6  15  38  41  44  53  67  97
```

23. `Vector v = new Vector(25,7);`

Chapter 5

1. Hypertext Transer Protocol (HTTP)

3. Hypertext Markup Language (HTML)

5. To link to another document.

7. The class `Applet` in the `java.applet` package.

9. `Graphics`, `Panel`, `Container`, and incidentally `Button`, `Checkbox`, and `TextField`.

11. The upper-left corner.

13. `g.drawLine(3,5,15,5);`

15. `g.drawRect(30,60,50,50);`

17. `g.drawArc(100,50,200,100,90,120);`

19. `Font f = new Font("Monospaced",Font.ITALIC,30);`

21. `g.drawString(s,175,50);`

23. `new Color(0,0,255);`

25. `int red = Color.black.getRed();`
    `int green = Color.black.getGreen();`
    `int blue = Color.black.getBlue();`

Chapter 6

1. From left to right and top to bottom. By default, components are centered when they do not fill the entire last row.

3. The canvas has a default size of 0x0 so we do not see it.

5. The extension is:

   The user presses the Clear button.
   The Clear button sends an action event as an argument to the canvas' `actionPerformed` method.

continued

The canvas' `actionPerformed` method saves the label of the button which the user pressed.

The canvas' `actionPerformed` method calls the `repaint` method, which asks Java to schedule a call the `update` method to redraw the canvas.

Java calls the `update` method of the canvas.

The canvas' `update` method clears the screen and calls the `paint` method.

The canvas' `paint` method uses the button's label to determine that the Clear button was pressed, and returns, taking no action.

7. The applet registers the canvas to listen for presses of the Draw button.

The user selects the color green.

The user selects the square shape.

The user presses the Draw button.

The Draw button sends an action event as an argument to the canvas' `actionPerformed` method.

The canvas' `actionPerformed` method calls the `repaint` method, which asks Java to schedule a call the `update` method to redraw the canvas.

The `update` method (inherited from `Component`) clears the screen and calls the `paint` method.

The `paint` method sets the color to the selected color, green, and draws the selected square shape.

9. Change the arrays `colorName` and `theColor` to include the four additional colors.

11. The flow layout always respects the preferred sizes of the components it is laying out. The grid layout never respects the preferred sizes of its components. It always makes the component fill the entire cell it is in.

13. The result is the same. The overall row weight is the maximum of the weights for each component in that row, so giving b1 a `weighty` of 1.0 and b2 a `weighty` of 0.0 is the same as the original in which these weights were reversed.

15. Buttons b2, b4, b5, and b6 each have a `weightx` of 1.0, and are in the same column because `gridx` is 1 for each, while b1 and b3 have a `weightx` of 0.0. Thus the middle column gets all the extra space.

Chapter 7

1. The applet registers as a mouse listener with the polygon.

The user presses the mouse inside the polygon.

The applet (the component in which the user pressed the mouse) passes a mouse event describing the mouse press to the applet's `mousePressed` method.

The applet's `mousePressed` method does nothing because the user pressed the mouse outside of the polygon.

The user releases the mouse inside the polygon.

The applet (the component in which the user pressed the mouse) passes a mouse event describing the mouse release to the applet's `mouseReleased` method.

The applet's `mouseReleased` method changes the foreground color to blue and asks Java to redraw it.

Java calls the applet's `update` method which clears the screen and calls the `paint` method which fills the polygon in blue.

5. For r : pressed, typed, released.
   For R : pressed, pressed, typed, released, released.

7. A compiler error results because the `WindowListener` interface is not fully implemented.

9. Closing the frame does not work, and no event terminates the program. The user must abort pressing (CTRL)-(C) while the console window is selected.

13. The polygon in which the mouse was last pressed will rotate around the point at which the mouse was last released which may be far away from the polygon. This is not what we need to solve the tangram puzzle.

Chapter 8

1. In Example 8.3 the output is `value=8`, the value set in the `catch` block when an exception occurs. Example 8.4 also prints the stack trace to show explicitly the error that has occurred.

3.
```
public class ArrayTest {
    public static void main(String[] args) {
        Object[] objectArray = new Object[3];
        objectArray[0] = new Integer(12);
        objectArray[1] = new Double(4.5);
//      objectArray[2] = 7;              Compiler error
    }
}
```

5. Using the prices.data file

   Milk 3 2.10
   Coffee 2.5 3.39
   Bread 3 1.89

   will cause the program to abort after the output of the total price for milk. The quantity 2.5 for coffee causes Java to throw a `NumberFormatException`. We could modify the program to handle the exception and continue executing good lines in the data file.

7. The program ignores the line containing the item `ice cream` because the number of tokens on the line is now four and it fails the test `strings.countTokens()==3`.

9.
```
import java.io.*;
public class FileCopy {
    public static void main(String [] args) {
        FileInputStream fis = null;
        FileOutputStream fos = null;
        try {
            File f = new File(args[0]);
            fis = new FileInputStream(f);
            fos = new FileOutputStream(args[1]);
            int length = (int)f.length();
            byte [] data = new byte[length];
            fis.read(data);
            fos.write(data);
```
continued

```
        }catch (IOException e) {
            e.printStackTrace();
        }finally {
            try {
                fis.close();
                fos.close();
            }catch(IOException e) {
                e.printStackTrace();
            }
        }
    }
}
```

11. The changed part is:

```
for (int i=0; i<10; i++)
    raf.writeDouble(i);
raf.seek(40);
double number = raf.readDouble();
System.out.println("The number starting at byte 40 is " + number);
raf.seek(8);
number = raf.readDouble();
System.out.println("The number starting at byte 8 is " + number);
```

13. Java throws a `FileNotFoundException` because the file `zxcvb.data` does not exist.

15. Without the `param` tags in the HTML file, `getParameter("size")` will return **null** and cause Java to throw a `NumberFormatException` which will abort the program because we have removed the exception handler.

17.
```
int[] data = {8,5,2,7,10,9,3,4};
mergeSort(data,1,8);
    mergeSort(data,1,4);
        mergeSort(data,1,2);
            mergeSort(data,1,1);
            mergeSort(data,2,2);
            merge(data,1,1,2,2); result {5,8}
        mergeSort(data,3,4);
            mergeSort(data,3,3);
            mergeSort(data,4,4);
            merge(data,3,3,4,4); result {2,7}
        merge(data,1,2,3,4);    result {2,5,7,8}
    mergeSort(data,5,8);
        mergeSort(data,5,6);
            mergeSort(data,5,5);
            mergeSort(data,6,6);
            merge(data,5,5,6,6); result {9,10}
        mergeSort(data,7,8);
            mergeSort(data,7,7);
            mergeSort(data,8,8);
            merge(data,7,7,8,8); result {3,4}
        merge(data,5,6,7,8);    result {3,4,9,10}
    merge(data,1,4,5,8);        result {2,3,4,5,7,8,9,10}
```

19. Try each list operation with a null list. Try the operations in various orders with a list of length one, and with a list of length greater than one.

21. The stack will be the same. The push operation is the inverse pop.

Chapter 9

1. main 1, Bonnie 1, Clyde 1, Clyde 2, Bonnie 2, Clyde 3, Bonnie 3, Clyde 4, Clyde 5, Bonnie 4, main 2, Bonnie 5, main 3, main 4, main 5

3. Main will print its name five times followed first by five Bonnies then by five Clydes. Each thread has more than enough time to print its name five times when it has the processor.

5. The output will vary depending upon the sleep times chosen and the order in which the threads are started.

7. Yes, we can edit another program while the animation is running and open another console window to run another Java program.

9. The image starts at a much smaller size and reaches a smaller final size because `imageWidth` and `imageHeight` each get the value -1 because they are called before the image has finished loading. By executing the `waitForID(0)` we get the correct values of `imageWidth` and `imageHeight`.

11. Omitting the call to `clearRect` causes parts of the image that were drawn in a previous frame to remain on the screen when the new image is drawn.

13. If we stop the applet, the whistle continues to play until we terminate the applet.

Chapter 10

1. Using the applet viewer shows a blank screen because the applet viewer does not display HTML documents but only runs applets embedded in those documents.

3. The client and the server both terminate.

5. The changes are:

```
URLConnection c = url.openConnection();
input = new BufferedReader
            (new InputStreamReader(c.getInputStream()));
```

7. The change is:

```
System.out.println("Content type: " + c.getHeaderField("Content-type"));
```

9. The output includes part of the request headers, which should be informational only to the client.

11. The client receives a message that it is connected, but no data because only headers were requested. We could use this to inspect the headers if we modified the program.

13. Clients will be active simultaneously, and will alternately receive data.

15. The port 3000 overrides the default port of 1099.

Chapter 11

1. The ID is unique, but the name may be duplicated.

3. ```
   INSERT INTO Salesperson VALUES ('12', 'Peter Patterson', '66 FifthSt.')
   INSERT INTO Salesperson VALUES ('98','Donna Dubarian', '77 Sixth St.')
   ```

5. ```
   DELETE FROM OrderItem
   WHERE OrderNumber = '5' AND ItemNumber = '444444'
   ```

7. ```
   SELECT OrderNumber FROM Orders, Salesperson
   WHERE Orders.SalesPersonID = Salesperson.SalesPersonID
   AND      SalespersonName = 'Peter Patterson'
   ```

9. Changes are:

   ```
   Class.forName(args[0]); (instead of new JdbcOdbcDriver())
   Connection con = DriverManger.getConnection(args[1], user, password);
   ```

11. ```
    SELECT DISTINCT SalespersonName, CustomerName
    FROM Salesperson, Customer, Orders
    WHERE  Orders.CustomerID = Customer.CustomerID
    AND        Orders.SalesPersonID = Salesperson.SalesPersonID
    ```

13. Change `rs.getString("CustomerName")` to `rs.getString(1)` in two places and change
    `rs.getString("OrderNumber")` to `rs.getString(1)`

15. ```
    SELECT DISTINCT CustomerName, CustomerAddress
    FROM Customer, Salesperson, Orders
    WHERE Customer.CustomerID = Orders.CustomerID
    AND        Salesperson.SalesPersonID = Orders.SalesPersonID
    AND        SalespersonName = 'Peter Patterson'
    ```

17. Use dbMetaData.getTables(null, null, "Order%", tables);

19. Add

    ```
    displayStrings("Type Info Column Names and Types", colNamesTypes(rs));
    after getTypeInfo().
    ```

21. java.sql.SQLException: [Microsoft][ODBC Microsoft Access 97 Driver]Invalid cursor state

25. After pressing the Display button, the new labels are truncated to have the same size as the generic table labels they are to replace.

27. The `fieldValue` text field remains 12 character wide, centered in its two-column field, rather than expanding to fill the two columns.

Chapter 12

1. Click on the background entry in the Properties window and choose orange in the color editor that pops up.

3. Bind using the Bind Properties item of the Edit menu. Higher prices make the juggler go slower.

5. The change is vetoed.
    ```

7. Bind using the Bind Properties item of the Edit menu.

9. The BeanBox cannot load `Smiley.jar` if `Smiley` does not implement `Serializable`.

11. Use the Events, nosePress item in the Edit menu to hookup one Friendly to the `startJuggling` method of the Juggler, and another to the `stopJuggling` method.

13. The `instantiate` method allows us to load a `.ser` file for a serialized bean whose state we changed by configuring in a tool such as the BeanBox. Using **new** loads only the original version of the bean.

15. The Juggler does not appear in the applet.

Appendix A

1. a, b, f, g, j, and k are valid. c,d,e use invalid characters. h starts with a digit, i is a keyword.

3. a. `3456.789`   d. `1.234567890987E9`
   b. `2.3456E-6`   e. `-234567.765432`
   c. `.09876543`

5. a. `234 < 52`   b. `435 != 87`   c. `-12 == -12`   d. `76 >= 54`

7. x=5 assigns 5 to x. x==5 tests whether x is equal to 5.

9. Many correct answers are possible, for example

   a. `x=2, y=7`   c. `x = 20, y = any integer`
   b. `x=0, y = any integer`   d. `x=11, y =0`

11. Many correct answers are possible, for example

    a. 3   b. 12   c. 3   d. 3

13. a. 8   b. 7   c. 6   d. 7

15. b and c

17. `for(int i=9; i>=3; i--)   System.out.println(i);`

19. 26

21. 55

Appendix G

1. a. $3 \times 10 + 8$ c. $5 \times 100 + 6 \times 10 + 2$ e. $3 \times 100 + 6$

3. a. 3   b. 13   c. 44   d. 27   e. 234

5. a.   `10001`   c.   `100110`   e.   `10100000`   g.   `1000101011100`
   b.   `10111`   d.   `1010110`   f.   `11101011`

Appendix H

1. a.   0000   b.   0101   c.   0000

3. a.   1111   b.   1010   c.   1010

5. a.   17   b.   63   c.   46   d. 4

Skill Builder and Critical Thinking Exercises

Chapter 1

1. 64
2. 2
3. if (i == 1) j += 2;
   ```
   else if (i == 3) j-= 5;
   else if (i==7 || i==10) j *= 17;
   else j = 0;
   ```
4. a. 3, 4   b. 2, 3
5. b
6. d
7. b

Chapter 2

1. a. iii   b. i   c. iv   d. ii
2. value, reference
3. hack
4. a
5. e
6. e
7. c

Chapter 3

1. class, interface, abstract, methods, overriding, class
2. a. no modifier   b. protected   c. public   d. private
3. 4, 25
4. b
5. a
6. d
7. d

Chapter 4

1. split(0,7) returns 4, nums = { 3, 23, 12, 11, 45, 88, 67, 77}
2. {{21, 21, 21, 21}, {21, 21, 21, 21} {21, 21, 21, 21}}
3. {{34}, {34, 34}, {34, 34, 34}, {34, 34, 34, 34}, {34, 34, 34, 34, 34}}
4. a. i   b. iv   c. ii   d. iii

5. c

6. c

7. b

Chapter 5

1. a. iv   b. v   c. i   d. iii   e. ii
2. The applet displays a centered rectangle whose dimensions are half those of the applet.
3. `<applet code=DrawFill.class width=500 height=320>`
4. c
5. a
6. c
7. d

Chapter 6

1. Action, Item, Item
2. a. ii   b. i   c. iii
3. implements, ActionListener, actionPerformed, implements, ItemListener, itemStateChanged
4. b
5. c
6. b
7. c

Chapter 7

1. KEY_PRESSED, KEY_PRESSED, KEY_TYPED, KEY_RELEASED, KEY_RELEASED

2.
```
import java.awt.*;
import java.awt.event.*;
import java.applet.Applet;
public class MouseClickRed extends Applet {
    int x = 25, y = 25;
    public void init()  {
        setForeground(Color.red);
        addMouseListener(new MouseHandler());
    }
    public void paint(Graphics g) {
        g.fillOval(x-25,y-25,50,50);
    }
    class MouseHandler extends MouseAdapter {
        public void mouseClicked(MouseEvent e) {
            x = e.getX();
            y = e.getY();
            repaint();
        }
    }
}
```

3.
```java
import java.awt.*;
import java.awt.event.*;
public class MouseClickRedAlone extends Frame {
    int x = 25, y = 25;
    public MouseClickRedAlone()  {
        setForeground(Color.red);
        addMouseListener(new MouseHandler());
        addWindowListener(new WindowClose());
    }
    public void paint(Graphics g) {
        g.fillOval(x-25,y-25,50,50);
    }
    class MouseHandler extends MouseAdapter {
        public void mouseClicked(MouseEvent e) {
            x = e.getX();
            y = e.getY();
            repaint();
        }
    }
    class WindowClose extends WindowAdapter {
        public void windowClosing(WindowEvent e) {
            System.exit(0);
        }
    }
    public static void main(String[] args) {
        MouseClickRedAlone m = new MouseClickRedAlone();
        m.setSize(300,200);
        m.show();
    }
}
```

4. a

5. c

6. d

7. b

Chapter 8

1. a. iii   b. i   c. ii   d. iii

2. BufferedReader, FileReader, "text.data"

3. PrintWriter, FileWriter, "text.out"

4. d

5. a

6. b

7. b

Chapter 9

1. a. iii   b. ii   c. i   d. ii
2. a. ii   b. iii   c. ii   d. i
3. a. ii   b. iv   c. i   d. iii
4. c
5. e
6. d
7. a

Chapter 10

1. `rmic, rmiregistry`
2. `InetAddress, ServerSocket, URL, URLConnection, Socket MalformedURLException`
3. a. iii   b. i   c. ii, iii   d. iii
4. e
5. b
6. a
7. b

Chapter 11

1. a. iv   b. ii   c. i   d. iii   e. v
2. Fred Flynn
   Darnell Davis
   Carla Kahn
3. The first two arguments are **null** because our database system does not use schemas. In the `getTables` method, the `"%"` argument matches any of the user tables in the database. The tables arguments limits the tables to those of types mentioned in the tables array which we set to `{TABLE}` to return only out data tables. In the `getColumns` method, the `Customer` argument indicates that we only want the columns of the Customer table. The fourth argument specifies all columns of the Customer table.
4. a. c
5. c
6. b
7 b

Chapter 12

1. a. iii   b. ii   c. iii   d. i
2. a. iii   b. ii   c. i
3. bound, constrained
4. d
5. a
6. d
7. b

Index